Power System Transients

THEORY AND APPLICATIONS

Power System Transients

THEORY AND APPLICATIONS

Akihiro Ametani • Naoto Nagaoka
Yoshihiro Baba • Teruo Ohno
Koichi Yamabuki

CRC Press
Taylor & Francis Group
Boca Raton London New York

CRC Press is an imprint of the
Taylor & Francis Group, an **informa** business

CRC Press
Taylor & Francis Group
6000 Broken Sound Parkway NW, Suite 300
Boca Raton, FL 33487-2742

First issued in paperback 2020

© 2017 by Taylor & Francis Group, LLC
CRC Press is an imprint of Taylor & Francis Group, an Informa business

No claim to original U.S. Government works

ISBN-13: 978-1-4987-8237-1 (hbk)
ISBN-13: 978-0-367-73667-5 (pbk)

Contents

Preface

When lightning strikes a building or transmission tower, a lightning current flows into its structures, which are made of electrically conducting materials, such as steel and copper. The lightning current produces a high voltage called "overvoltage" (or "abnormal voltage") that can damage or break electrical equipment installed in the building or the power transmission system. The breakdown of equipment may shut down the electrical room of the building, resulting in a complete blackout. If the breakdown occurs in a substation within a high-voltage power transmission system, a city that receives its electricity from the substation can also experience a blackout.

An overvoltage can also be generated by switching operations of a circuit breaker or a load switch, both of which are electrically identical to breakers found in the homes of customers.

A period in which a short-lived overvoltage appears due to lighting and switching operation is called a "transient," while an equilibrium condition of electricity being supplied is called a "steady state." In general, a transient dies out and reaches a steady state within approximately 10 μs (10^{-6} s) in the lighting transient case and within approximately 10 ms (10^{-3} s) in the switching transient case. Occasionally, a transient sustains itself for a few seconds if it involves the resonant oscillation of circuit parameters (mostly inductance and capacitance) or mechanical oscillation of the steel shaft of a generator (called "subsynchronous resonance").

A transient analysis is a crucial subject for electrical engineers and researchers, especially in the field of electric power engineering when determining the strength of equipment in order to maintain safety.

First, this book will illustrate a transient on a single-phase line from a physical viewpoint, and how it can be solved analytically by an electric circuit theory. The impedance and admittance formulas of an overhead line will also be described. Approximate formulas that can be computed using a pocket calculator will be explained to show that a transient can be analytically evaluated via hand calculation. Since a real power line contains three phases, a theory to deal with a multiphase line will be developed. Finally, the book describes how to tackle a real transient in a power system. A computer simulation tool is necessary for this—specifically the well-known simulation tool ElectroMagnetic Transients Program (EMTP), originally developed by the U.S. Department of Energy, Bonneville Power Administration—which is briefly explained in Chapter 1.

In Chapter 2, wave propagation characteristics and transients in an overhead transmission line are described. The distributed parameter circuit theory is applied to solve the transients analytically. The EMTP is then applied to calculate the transients in a power system composed of the overhead line and a substation. Various simulation examples are demonstrated, together with the comparison of field test results.

Chapter 3 examines the transients in a cable system. A cable system is, in general, more complicated than an overhead line system, because one phase of the cable is composed of two conductors called a metallic core and a metallic sheath. The former carries a current and the latter behaves as an electromagnetic shield against the core current. Another reason why a cable system is complicated is that most long cables are cross-bounded, that is, the metallic sheaths on phases "a," "b," and "c" in one cable section are connected to that of phases "b," "c," and "a" in the next section. Each section is called a "minor section" and the length of each normally ranges from some 100 m to 1 km. Three minor sections comprise one major section. The sheath impedances of three phases thus become nearly equal to each other. As a result, a transient on a cable system is quite different from that on an overhead line system.

As in Chapter 2, a basic characteristic of wave propagation on a cable is explained analytically based on the distributed parameter circuit theory, along with EMTP simulation examples.

Two of the most attractive subjects in recent years are the so-called clean energy (or sustainable energy) and smart grids; as a result, wind farms and "mega solar" energy plants have become well-known. In Chapter 4, transients in a wind farm are explained based on EMTP simulations. Since the output voltage of most wind generators is about 600 V, wind generators are connected to a low-voltage transmission (distribution) line. Because their individual generating capacity is small, a number of wind generators are connected together, thus the term "wind farm," in a substation where the voltage is stepped up for power transmission. In the case of an offshore wind farm, the generated power is sent to an onshore connection point through submarine cables. A transient analysis in wind farms, the "mega solar" or the smart grid requires a vastly different approach in comparison to that for an overhead line or a cable. A transient in an overhead line or a cable is directly associated with traveling waves whose traveling time is on the order of 10 μs up to 1 ms in most cases; the maximum overvoltage appears within a few milliseconds. In contrast, a transient in a wind farm involving power electronics circuits is affected by the dynamic behavior of power transistors/thyristors, which are a basic element of the power electronic circuit. In the case of photovoltaic (PV) generation, the output voltage and power generation vary by time according to the amount of sunshine the photo cells receive, which depends on time and weather. A power conditioner and a storage system such as a battery are thus essential to operate the PV system. In the last section in Chapter 4, voltage regulation on equipment in a DC railway is described when a Li-ion battery is adopted, since the Li-ion battery is expected to be a storage element of the PV and wind farm generation systems.

In Chapters 1 to 4, a transient analysis/simulation is based on a circuit theory derived by the TEM mode of wave propagation. When a transient involves non-TEM mode propagation, a circuit-theory-based approach cannot give an accurate solution. A typical example is the arcing horn flashover considering the mutual coupling between power lines and tower arms, a transient in a grounding electrode, and an induced voltage from a lightning channel.

For solving these kinds of transients, a numerical electromagnetic analysis (NEA) is the most commonly used approach. In Chapter 5, the basic theory of the NEA is described first. There are various methods of NEA, either in a frequency domain or in a time domain, for example. A brief summary of the methods is given, and application examples are demonstrated. Some of the examples compare field test results with EMTP simulation results.

Chapter 6 deals with problems related to electromagnetic compatibility (EMC) in a low-voltage control circuit in a power station and a substation. Electromagnetic disturbances experienced in Japanese utility companies within a span of 10 years are summarized and categorized depending on the cause, that is, a lightning surge or a switching surge, and the incoming route. The influence of the disturbances on system operation and the countermeasures are explained together with case studies. Lightning-based disturbances that affect both utility companies and private customers and also home appliances are explained based on the collected statistical data, measured results, and EMTP/FDTD simulation results. Finally, an analytical method for evaluating electromagnetic-induced voltages on a telecommunication line or a gas pipeline from a power line is described.

Chapter 7 describes "grounding" for electric power equipment and systems. Practical grounding methods in a gas pipeline, a transmission tower, GW, underground cable, etc. are explained in Section 7.2. In Section 7.3, modeling of grounding is explained for a steady-state and transient analysis. First, analytical and/or theoretical model of a grounding electrode is described. Second, modeling methods used in EMTP simulations are described. Also, the effect of simulation model in a finite-difference time-domain (FDTD) method is explained. Then, measurement of a grounding impedance, measured results, theoretical investigations, the effect of the electrode shape, etc. on the grounding impedance are described in Section 7.4.

At this time, there are a number of numerical simulation tools that are widely used throughout the world to analyze a transient in a power system. Among them, the most well-known and widely used tool is EMTP. The accuracy and reliability of the original EMTP has been confirmed by a number of test cases since 1968. However, there is no perfect simulation tool in this world. Any simulation tool has its own application limits and restrictions. As previously discussed, the EMTP, based as it is on a circuit theory under the assumption of TEM mode propagation, cannot give an accurate solution of a transient associated with non-TEM mode propagation. Such application limits and restrictions are discussed in Chapter 8 for both circuit-theory-based approaches and NEA methods.

Akihiro Ametani
Editor

For product information, please contact:

The MathWorks, Inc.
3 Apple Hill Drive
Natick, MA 01760-2098 USA
Tel: 508 647 7000
Fax: 508-647-7001
E-mail: info@mathworks.com
Web: www.mathworks.com

Authors

Akihiro Ametani earned his PhD from the University of Manchester (UMIST), Manchester, UK, in 1973. He was with the UMIST from 1971 to 1974, and with Bonneville Power Administration for summers from 1976 to 1981, and developed the EMTP (Electro-Magnetic Transients Program). Since 1985, he has been a professor at Doshisha University, Kyoto, Japan. In 1988, he was a visiting professor at the Catholic University of Leuven, Belgium. From April 1996 to March 1998, he was the director of the Science and Engineering Institute, Doshisha University, and dean of the Library and Computer/Information Center from April 1998 to March 2001. He was chairperson of the Doshisha Council until March 2014.

Since April 2014, Dr. Ametani has been an Emeritus Professor at Doshisha University, and is an invited professor at École Polytechnique Montreal, Montreal, Canada. He is an IEEE life fellow and CIGRE distinguished member.

Naoto Nagaoka earned BS, MS, and PhD degrees from Doshisha University, Kyoto, Japan, in 1980, 1982, and 1993, respectively. In 1985, he joined Doshisha University, where since 1999 he has been a professor. From April 2008 to March 2010, he was dean of the Student Admission Center, Doshisha University. From April 2010 to March 2012, he was director of the Liaison Office and the Center of Intellectual Properties, Doshisha University. Dr. Nagaoka is a member of the Institution of Engineering and Technology and the Institute of Electrical Engineers of Japan.

Yoshihiro Baba earned his PhD from the University of Tokyo, Tokyo, Japan, in 1999. Since 2012, he has been a professor at Doshisha University, Kyoto, Japan. He received the Technical Achievement Award from the IEEE EMC (Electromagnetic Compatibility) Society in 2014. He was the chairperson of Technical Program Committee of the 2015 Asia-Pacific International Conference on Lightning, Nagoya, Japan. He has been the convener of C4.37 Working Group of the International Council on Large Electric Systems since 2014. He has been an editor of the *IEEE Transactions on Power Delivery* since 2009, and an advisory editorial board member of the *International Journal of Electrical Power and Energy Systems* since 2016.

Teruo Ohno earned a BSc from the University of Tokyo, Tokyo in electrical engineering in 1996, an MSc from the Massachusetts Institute of Technology, Cambridge in electrical engineering in 2005, and a PhD from the Aalborg University, Aalborg, Denmark in energy technology in 2012.

Since 1996 he has been with the Tokyo Electric Power Company, Inc., where he is currently involved in power system studies, in particular, on cable systems, generation interconnections, and protection relays. He was a secretary of Cigré WG C4.502, which focused on technical performance issues related to the application of long HVAC (high-voltage alternating current) cables. He is a member of the IEEE (Institute of Electrical and Electronics Engineers) and the IEEJ (Institute of Electrical Engineers of Japan).

Koichi Yamabuki earned a PhD from Doshisha University, Kyoto, Japan, in 2000. Since 1999 he has been with National Institute of Technology, Wakayama College. He was a visiting researcher of the University of Bologna, Bologna, Italy from 2006 to 2007. His research area includes experimental and analytical investigation on surge phenomenon due to lightning strikes and grounding conditions on power and transportation facilities.

He is a committee member of IET (Innovation, Engineering, and Technology) Japan Network, a member of IEEE and IEEJ.

List of Symbols

The symbols used in this book are listed together with the proper units of measurement, according to the International System of Units (SI).

Angular frequency	Radians per second (rad/s)	ω
Conductance (admittance, susceptance)	Siemens (S)	$G\ (Y, B)$
Conductivity	Siemens per meter (S/m)	σ
Current density	Ampere per square meter (A/m²)	J
Decibel (dB)	Decibel is a dimensionless number expressing the ratio of two power levels, W_1 to W_2: dB = 10 log (W_1/W_2)	
	Further expressions of dB if both the voltages (U_1, U_2) or currents (I_1, I_2) are measured on the same impedance: dB = 20 log (U_1/U_2) dB = 20 log (I_1/I_2)	
Electric capacitance	Farad (F)	C
Electric current	Ampere (A)	I
Electric field strength	Volt per meter (V/m)	E
Electric resistance (impedance, reactance)	Ohm (Ω)	$R\ (Z, X)$
Frequency	Hertz (Hz)	f
Inductance: self, mutual	Henry (H)	L, M
Length	Meter (m)	d, D, R, x (distance)
		r (radius)
		ℓ (length)
		h (height)
		δ (skin depth)
		λ (wavelength)
Magnetic field strength	Ampere per meter (A/m)	H
Magnetic flux	Weber (Wb)	Φ
Magnetic flux density	Tesla (T)	B
Permeability	Henry per meter (H/m)	μ
Permittivity	Farad per meter (F/m)	ε
Potential difference, voltage, and electric potential	Volt	V, U
Power	Watt (W)	P
Resistivity	Ohm meter (Ω m)	ρ
Time, pulse rise time, and pulse width	Second (s)	t, τ
Velocity	Meter per second (m/s)	Y

List of Acronyms

The following list includes the acronyms that are frequently used in this book:

AIS	Air-insulated substation
ATP	Alternative transients program
CB	Circuit breaker
CM	Common mode
CT	Current transformer
DM	Differential mode
DS	Disconnector
EHV	Extra-high voltage (330–750 kV)
EMC	Electromagnetic compatibility
EMF	Electromotive force
EMI	Electromagnetic interference
EMTP	Electromagnetic transients program
ESD	Electrostatic discharge
GIS	Gas-insulated substation
GPR	Ground potential rise
HV	High voltage (1–330 kV)
IC	Integrated circuit
IEC	International Electrotechnical Commission
IKL	Isokeraunic level
LPS	Lightning protection system
LS	Lightning surge
SPD	Surge-protective device
SS	Switching surge
TE	Transverse electric
TEM	Transverse electromagnetic
TL	Transmission line
TLM	Transmission line model
TM	Transverse magnetic
UHV	Ultrahigh voltage ($\geq$800 kV for AC and DC transmission)
VT	Voltage transformer

International Standards

1. IEC 61000-4-5 2005. *Electromagnetic Compatibility (EMC)—Part 4–5: Testing and Measurement Technique—Surge Immunity Test*, 2nd edn.

2. IEC 60364-5-54 2011. *Low-Voltage Electrical Installations—Part 5–54: Selection and Erection of Electrical Equipment—Earthing Arrangements and Protective Conductors*, Edition 3.0.

3. IEC 61000-4-3 2008. *Electromagnetic Compatibility (EMC)—Part 4-3: Testing and Measurement Technique—Radiated, Radio Frequency Electromagnetic Field Immunity Test*, Edition 3.1.

4. IEC 60050-161. *International Electrotechnical Vocabulary—Chapter 161: Electromagnetic Compatibility (EMC)*, 1st edn. (1990), Amendment 1 (1997), Amendment 2 (1988).

5. IEC 60050-604 1987. *International Electrotechnical Vocabulary—Chapter 604: Generation, Transmission and Distribution of Electricity—Operation*, Edition 1.0.

1

Theory of Distributed-Parameter Circuits and Impedance/Admittance Formulas

1.1 Introduction

When investigating transient and high-frequency steady-state phenomena, conductors such as a transmission line, a machine winding, and a measuring wire demonstrate a distributed-parameter nature. Well-known, lumped-parameter circuits are an approximation of a distributed-parameter circuit for describing a low-frequency, steady-state phenomenon of the conductor. That is, a current in a conductor, even a short conductor, needs time to travel from its sending end to its remote end because of its finite propagation velocity ($300 \, \text{m/\mu s}$ in a free space). From this fact, it should be clear that a differential equation expressing the behavior of current and voltage along the conductor involves variables of distance x and time t or frequency f. Thus, it becomes a partial differential equation. However, a lumped-parameter circuit is expressed by an ordinary differential equation since no concept of the length or the traveling time exists. This is the most significant difference between the distributed-parameter circuit and the lumped-parameter circuit.

In this chapter, a theory of distributed-parameter circuits is explained starting from the approximate impedance and admittance formulas of an overhead conductor. The derivation of the approximate formulas is described from the viewpoint of the physical behavior of current and voltage on a conductor.

Then, a partial differential equation is derived to express the behavior of current and voltage in a single conductor by applying Kirchhoff's laws based on a lumped-parameter equivalence of the distributed-parameter line. The current and voltage solutions of the differential equation are derived by assuming (1) sinusoidal excitation and (2) a lossless conductor. From the solutions, the behavior of current and voltage is discussed. For this, the definition and the concept of a propagation constant (attenuation and propagation velocities) and a characteristic impedance are introduced.

As is well known, all alternating current (AC) power systems are basically three-phase circuits. This fact makes voltage, current, and impedance a 3-D matrix form. A symmetrical component transformation (i.e., Fortescue and

Clarke transformations) is well known to deal with three-phase voltages and currents. However, the transformation cannot diagonalize an $n \times n$ impedance/admittance matrix. In general, modal theory is necessary to deal with an untransposed transmission line. In this chapter, modal theory is explained. By adopting modal theory, an n-phase line is analyzed as n-independent single conductors so that the basic theory of a single conductor can be applied.

To analyze a transient in a distributed-parameter line, a traveling-wave theory is explained for both single- and multiconductor systems. A method to introduce velocity difference and attenuation in the multiconductor system is described together with field test results. Impedance and admittance formulas of unusual conductors, such as finite-length and vertical conductors, are also explained.

Application examples of the theory described in this chapter are given so the reader can understand the need for the theory.

Finally, the ElectroMagnetic Transients Program (EMTP), which is widely used all over the world, is briefly explained.

It should be noted that all of the theories and formulas in this chapter are based on transverse electromagnetic (TEM) wave propagation.

1.2 Impedance and Admittance Formulas

In general, the impedance and admittance of a conductor are composed of the conductor's internal impedance Z_i and the outer-media impedance Z_o. The same is applied to the admittance [1]:

$$Z = Z_t + Z_o \ (\Omega/\text{m}) \tag{1.1}$$

$$Y = j\omega C = j\omega 2\pi\varepsilon P^{-1}, \quad P = P_i + P_o \tag{1.2}$$

where
  Z_i is the conductor internal impedance
  Z_o the conductor outer-media (space/earth return) impedance
  $Y_i = j\omega 2\pi\varepsilon P_i^{-1}$ the conductor internal admittance
  $Y_o = j\omega 2\pi\varepsilon_0 P_o^{-1}$ the conductor outer-media (space/earth return) admittance
  P the potential coefficient matrix

It should be noted that the impedance and admittance in this equation become a matrix when a conductor system is composed of multiconductors. Remember that a single-phase cable is, in general, a multiconductor system because the cable consists of a core and a metallic sheath or a screen. In an overhead conductor, no conductor internal admittance Y_i exists, except a covered conductor.

1.2.1 Conductor Internal Impedance Z_i

1.2.1.1 Derivation of an Approximate Formula

Let us obtain the impedance of the cylindrical conductor illustrated in Figure 1.1 [1,2]. We know that the direct current (DC) resistance of the conductor is as given in the following equation:

$$R_{dc} = \frac{\rho_c}{S} \,(\Omega/m) \tag{1.3}$$

$$S = \pi\left(r_0^2 - r_i^2\right) \tag{1.4}$$

where
  S is the cross-sectional area (m²)
  r_i the inner radius of the conductor (m)
  r_0 the outer radius of the conductor (m)
  $\rho_c = \sigma_c$ the resistivity of the conductor (Ω m)
  σ_c the conductor conductivity (S/m)
  μ_c the permeability (H/m)

Also, it is well known that currents concentrate near the outer surface area of the conductor when the frequency of an applied (source) voltage (or current) to the conductor is high. This phenomenon is called the "skin effect." The depth d_c of the cross-sectional area where most of the currents flow is given approximately as the (complex) penetration depth or the so-called skin depth in the following form:

$$d_c = \sqrt{\frac{\rho_c}{j\omega\mu_c}} = \frac{1}{\sqrt{j\omega\mu_c\sigma_c}} \,(m) \tag{1.5}$$

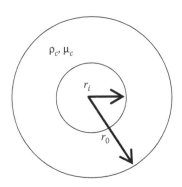

FIGURE 1.1
A cylindrical conductor.

Penetration depth is physically defined as the depth of an electromagnetic wave penetrating a conductor when the wave hits the conductor's surface. The physical concept of penetration depth is very useful for explaining the behavior of current and voltage on a conductor and also for deriving impedance and admittance formulas of various conductor shapes and geometrical configurations. However, keep in mind that the concept is based on TEM wave propagation and thus is not applicable to non-TEM propagation. Also, remember that it is just an approximation.

By considering the penetration depth, the internal impedance Z_i in a high-frequency region can be derived in the following manner.

Following the definition of conductor resistance in Equation 1.3, the internal impedance is given by the ratio of the resistivity ρ_c and the cross-sectional area S, which is expressed as

$$S = \pi \left\{ r_0^2 (r_0 - d_c)^2 \right\} = \pi \left(2 r_0 d_c - d_c^2 \right).$$

In a high-frequency region, d_c is far smaller than the conductor's outer radius r_0. Thus, the following approximation is satisfied:

$$S \cong 2\pi r_0 d_c \quad \text{for } r_0 \gg d_c$$

Substituting this equation and Equation 1.5 into Equation 1.3 gives

$$Z_{hf} \doteq \frac{\rho_c}{2\pi r_0 d_c} = \frac{\sqrt{j\omega\mu_c\rho_c}}{2\pi r_0} \, (\Omega/\text{m}) \quad \text{for a high frequency} \tag{1.6}$$

This formula for transmission lines, power engineering, and transients can be found in many textbooks.

Knowing the low-frequency and high-frequency formulas, the internal impedance formula at any frequency can be given in the following form by applying Rolle's averaging theorem [2]:

$$Z_c = R_{dc} \sqrt{\left[1 + \left(\frac{j\omega\mu_c S}{R_{dc} \cdot l^2} \right) \right]} \tag{1.7}$$

where
S is the cross-sectional area of the conductor (m²)
l the circumferential length of the conductor outer surface (m)

It is clear that this equation becomes identical to Equation 1.3 in a low-frequency region when assuming a small ω and to Equation 1.6 when assuming a large ω. It is noteworthy that Equation 1.7 is applicable to an arbitrary cross-sectional conductor, not necessarily to a circular or cylindrical conductor, because the equation is defined by the cross-sectional area S and the circumferential length of the conductor l but not by the inner and outer radii.

At low frequency, R_{dc} is much greater than Z_{hf} in Equation 1.7. By adopting the approximation:

$$\sqrt{1+x} \doteq 1 + \frac{x}{2} \quad \text{for } x \ll 1$$

Equation 1.7 is approximated in the following form:

$$Z_i \doteq \frac{(R_{dc} + j\omega\mu_c)}{8\pi} \tag{1.8}$$

This formula is commonly known as conductor impedance at a power frequency.

EXAMPLE 1.1

Calculate the internal impedance (R_c and L_c) of a conductor with $r_1 = 1.974$ mm, $r_2 = 8.74$ mm, $\rho_c = 3.78 \times 10^{-8}$ Ω m, and $\mu_c = \mu_0$ at frequencies $f = 50$ Hz and 100 kHz.

Solution

$$S = \pi\left(r_2^2 - r_1^2\right) = \pi(8.74 + 1.974)(8.74 - 1.974) \times 10^{-6}$$
$$= 10.714 \times 6.766 \times 10^{-6}\pi = 72.49\pi \times 10^{-6}$$
$$\ell = 2\pi r_2 = 17.48\pi \times 10^{-3}, \quad R_{dc} = \frac{\rho_c}{S} = 0.166 \times 10^{-3}\,\Omega/\text{m}$$

To calculate the square root of a complex number $a + jb$, it is better to rewrite the number in the following form so that we need only a real number calculation:

$$a + jb = A \cdot e^{j\varphi}, \quad A = \sqrt{a^2 + b^2}, \quad \varphi = \tan^{-1}\left(\frac{b}{a}\right)$$

Thus,

$$\sqrt{a + jb} = \sqrt{A} \cdot e^{j\varphi/2}$$

In the internal impedance case, $a = 1$, $b = \omega\mu_c S/R_{dc}\,\ell^2$.

At 50 Hz:

$$b = 100\pi \times 4\pi \times 10^{-7} \times 72.49 \times 10^{-6} \; \pi / 0.166 \times 10^{-3} \times 17.48 \times 10^{-6} \; \pi^2$$

$$= \frac{4\pi \times 72.49 \times 10^{-2}}{0.166 \times 17.48^2} = 0.180$$

$$A = \sqrt{1 + b^2} = 1.016, \quad \sqrt{A} = 1.00797, \quad \varphi = 10.2°$$

$$Z_c = 0.1673 \angle 5.1° = 0.1667 + j14.87 \times 10^{-3} \, \Omega/m$$

$$\therefore R_c = 0.1667 \, \Omega/km, \quad L_c = 0.0473 \, mH/km$$

If we use Equation 1.8, that is, $Z_c = R_{dc} + j\omega L_c \therefore R_c = 0.166 \, \Omega/km$, then

$$L_c = \frac{\mu_c}{8\pi} = 0.5 \times 10^{-7} \, H/km = 0.05 \, mH/km$$

Using the same method at 100 kHz,

$$b = 359.19, \quad Z_c = 3.1452 \angle 44.92° = 2.227 + j2.221 \, \Omega/km.$$

$$L_c = 3.53 \times 10^{-3} \, mH/km$$

In this case

$$b \gg 1 \quad \therefore Z_c \fallingdotseq R_{dc}\sqrt{jb} = \frac{(1+j)\sqrt{R_{dc}f\mu_c}}{2} (\Omega/m) = (1+j)2.225 \, \Omega/km$$

It is clear that further approximation at low and high frequencies gives a satisfactory accuracy.

These results correspond to those given in Table 1.1 for the 500 kV transmission line shown in Figure 1.25. Because a phase wire is composed of four bundles, the analytical results are four times the internal impedance given in the table. The analytical results agree well with those in the table, which are calculated using the accurate formula in Section 1.2.1 (Equation 1.9).

1.2.1.2 Accurate Formula by Schelkunoff

The accurate formula for the internal impedance of the cylindrical conductor in Figure 1.1 was derived by Schelkunoff in 1934 [3].

1. Inner surface impedance z_i is

$$z_i = j\omega \left(\frac{\mu_c}{2\pi} \right) \frac{\{I_0(x_1) \cdot K_1(x_2) + K_0(x_1) \cdot I_1(x_2)\}}{(x_1 D)}$$

2. Mutual impedance between the inner and outer surfaces z_m is

$$z_m = \frac{\rho_c}{(2\pi r_1 r_2 D)}$$

TABLE 1.1

Self-Impedance of Phase a (Figure 1.25: $h_p = 16.67$ m, $h_e = 23.33$ m)

	(a) Conductor Internal			(b) Earth Return			(c) Space	(d) Total		
f (Hz)	R_c (Ω/km)	L_c (mH/km)	$X_c = \omega L_C$ (Ω/km)	R_e	L_e (mH/km)	$X_e = \omega L_e$ (Ω/km)	$X_s = \omega L_S$ (Ω/km)	R (Ω/km)	L (mH/km)	$X = \omega L$ (Ω/km)
50	0.0416	1.14E − 02	3.57E − 03	0.048	0.739	0.232	0.333	0.0896	1.8110	0.5689
1 k	0.0653	8.56E − 03	5.38E − 02	0.883	0.455	2.86	6.65	0.9483	1.5216	9.5604
10 k	0.186	2.79E − 03	0.175	7.28	0.261	16.4	66.5	7.466	1.3218	83.051
100 k	0.566	8.85E − 04	0.556	46.9	0.117	73.7	665	47.47	1.1759	738.84
1 M	1.77	2.80E − 04	1.76	218	0.0419	263	6650	219.77	1.1002	6912.8

Note: $\ln(2h_p/rP_e) = 5.2292$, $L_s = 1.058$ mH/km, $C = 10.62$ nF/km.

3. Outer surface impedance z_0 is

$$z_0 = j\omega\left(\frac{\mu_c}{2\pi}\right)\frac{\{I_0(x_2)\cdot K_1(x_1)+K_0(x_2)\cdot I_1(x_1)\}}{(x_2 D)} \tag{1.9}$$

where

$$x_1 = \frac{r_i}{d_c}$$

$$x_2 = \frac{r_0}{d_c}$$

$$D = I_1(x_2)\cdot K_1(x_1) - I_1(x_1)\cdot K_1(x_2)$$

$I_n(x)$ and $K_n(x)$ are the modified Bessel functions of the first and second kind, respectively, with order n.

As is clear from Equation 1.9, three-component impedances exist for a cylindrical conductor. In the case of a circular solid conductor, $z_i = z_m = 0$ for $r_i = 0$, and z_0 becomes

$$z_0 = j\omega\left(\frac{\mu_c}{2\pi x_2}\right)\frac{I_0(x_2)}{I_1(x_2)} \tag{1.10}$$

The conductor internal impedance Z_i, which means z_0, is the outer surface impedance given in Equation 1.9—or in Equation 1.10 as far as an overhead line is concerned. However, in the case of a cable, Z_i is composed of a number of component impedances, as in Equation 1.9, and also an insulator imped-ance between metallic conductors because the cable is, in general, composed of a core conductor carrying current and a metallic sheath (shield or screen) for a current return path [4,5].

1.2.2 Outer-Media Impedance Z_0

1.2.2.1 Outer-Media Impedance

The outer media of an overhead conductor are the air and the earth since the conductor is isolated by the air from the earth, which is a conducting medium. Therefore, the outer-media impedance Z_0 of the overhead conduc-tor is composed of the following component:

$$Z_0 = Z_s + Z_e \quad \text{for an overhead line} \tag{1.11}$$

where
Z_s is the space impedance
Z_e the earth-return impedance

The outer-media impedance of an underground cable (insulated conductor) is the same as the earth-return impedance because the underground cable is surrounded by only the earth:

$$Z_0 = Z_e \quad \text{for an underground cable} \tag{1.12}$$

When considering the mutual impedance between an overhead conductor and an underground cable or a buried gas and/or water pipeline, the self-impedance of the overhead conductor is given by Equation 1.11, while that of the underground conductor is given by Equation 1.12. Mutual impedance will be explained in Section 1.2.2.3.

1.2.2.2 Overhead Conductor

1.2.2.2.1 Derivation of an Approximate Formula

Using the penetration depth h_e for the earth, the outer-media impedance of an overhead conductor is readily obtained based on image theory. Figure 1.2 illustrates a single overhead conductor and its image:

$$h_e = \sqrt{\frac{\rho_e}{j\omega\mu_e}} \tag{1.13}$$

where
  ρ_e is the earth resistivity
  μ_e the earth permeability

In most cases, $\mu_e = \mu_0$.

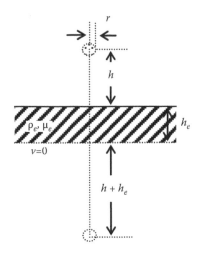

FIGURE 1.2
A single overhead conductor and its image.

Because the earth is not perfectly conducting, its surface is not a zero potential plane. Instead, the zero potential plane is located at a depth h_e from the earth's surface. Then, image theory gives the following inductance L_e [6]:

$$L_e = \left(\frac{\mu_0}{2\pi}\right) \ln\left\{\frac{2(h+h_e)}{r}\right\} \tag{1.14}$$

Thus, the outer-media impedance of a single overhead conductor is given by

$$Z_0 = Z_e = j\omega L_e = j\omega\left(\frac{\mu_0}{2\pi}\right) \cdot \ln\left\{\frac{2(h+h_e)}{r}\right\} \tag{1.15}$$

For the multiconductor illustrated in Figure 1.3, the outer-media impedance is obtained in the same manner as Equation 1.15 [6]:

$$Z_{0ij} = Z_{eij} = j\omega\left(\frac{\mu_0}{2\pi}\right) \cdot P_{ij}, \quad P_{ij} = \ln\left(\frac{S_{ij}}{d_{ij}}\right) \tag{1.16}$$

where

$$S_{ij}^2 = (h_i + h_j + 2h_e)^2 + y_{ij}^2, \quad d_{ij}^2 = (h_i - h_j)^2 + y_{ij}^2 \tag{1.17}$$

In this equation, h_i and h_j are the heights of the ith and jth conductors, respectively, and y_{ij} is the horizontal separation between the ith and jth conductors.

Remember that the penetration depth is not a real value but rather a complex value; thus, the zero potential plane at depth h_e is just a concept and does not exist in physical reality.

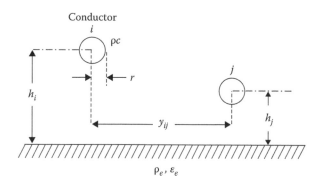

FIGURE 1.3
A multiconductor overhead line.

When the earth is perfectly conducting, that is, $\rho_e = 0$, then $h_e = 0$ in Equation 1.13.

Therefore, Equation 1.16 becomes

$$Z_{0ij} = j\omega\left(\frac{\mu_0}{2\pi}\right) \cdot P_{\theta ij}, \quad P_{\theta ij} = \ln\left(\frac{D_{ij}}{d_{ij}}\right) \tag{1.18}$$

where

$$D_{ij}^2 = (h_i + h_j)^2 + y_{ij}^2 \tag{1.19}$$

This impedance is well known as the space impedance of an overhead conductor, that is

$$Z_{0ij} = Z_{sij} = j\omega\left(\frac{\mu_0}{2\pi}\right)\ln\left(\frac{D_{ij}}{d_{ij}}\right) \tag{1.20}$$

For a single conductor:

$$D_{ij} = 2h_i, \quad d_{ij} = r$$

This is the reason space impedance is often confused with earth-return impedance. In fact, when earth-return impedance is derived from Maxwell's equation, space impedance appears as a part of earth-return impedance [7,8].

EXAMPLE 1.2

Calculate the earth-return impedance of a conductor with $r = 0.1667$ m, $h = 16.67$ m, and $\rho_e = 200\ \Omega$ m at $f = 50$ Hz.

Solution

Similar to Example 1.1, the impedance formula is rewritten in the following form so that only a real number calculation is necessary:

$$Z_e = 2\pi f \times 10^{-7}\left\{\varphi + j\ln\left(\frac{A}{d_2}\right)\right\}$$

where

$$A = \sqrt{a^2 + b^2}, \quad \varphi = \tan^{-1}\left(\frac{b}{a}\right), \quad a = D^2 + 2H_1H_e, \quad b = 2H_e(H_1 + H_e),$$

$$H_1 = h_1 + h_2, \quad D^2 = H_1^2 + y^2, \quad d^2 = (h_1 - h_2)^2 + y^2, \quad H_e = \sqrt{\frac{2\rho_e}{\omega\mu_0}}$$

For self-impedance, $d = r$, $D = H_1 = 2h$, $a = H_1(H_1 + 2H_e)$.

$D = H_1 = 33.34$, $H_e = 1006.6$, $a = 68.232 \times 10^3$, $b = 2.094 \times 10^6$,
$A = 2.095 \times 10^6$

$\varphi = 88.13° = 1.538$ rad, $\ln\left(\dfrac{A}{d^2}\right) = \ln\left(\dfrac{A}{r^2}\right) = 18.14$

$\therefore Z_e = 0.0483 + j0.570 = 0.0483 + j(0.237 + 0.333)$ Ω/km

The result agrees with that in Table 1.1, which is calculated by Carson's accurate formula using EMTP cable constants (see Section 1.8, Table 1.14b [5,9]).

Earth-return impedance at a low frequency can be easily evaluated by an approximate formula derived from Equation 1.16 under the assumption that $h_e \gg h_1, h_2$:

$$Z_e = f + if\left\{8.253 + 0.628 \cdot \ln\left(\frac{\rho_e}{f \cdot d^2}\right)\right\} \text{ (m}\Omega\text{/km)} = 0.05 + j0.568 \text{ (m}\Omega\text{/km)}$$

This approximate result agrees with that calculated by Deri's formula.

1.2.2.2.2 Accurate Formula by Pollaczek

Pollaczek derived the following earth-return impedance in 1926 [7]:

$$Z_{eij} = j\omega \frac{\mu_0}{2\pi}[P_{oij} + (Q - jP)], \quad Q - jP = 2\int_0^\infty F(x) \cdot dx \tag{1.21}$$

$$F(x) = \frac{\exp\{-(h_i + h_j)\}\cos(y_{ij} \cdot x)}{\left(x + \sqrt{x^2 + m_1^2 - m_0^2}\right)s} \tag{1.22}$$

where

$$m_0^2 = j\omega\mu_0 \cdot j\omega\varepsilon_0, \quad m_1^2 = j\omega\mu_e(\sigma_e + j\omega\varepsilon_e), \quad \varepsilon_e\text{: earth permittivity} \tag{1.23}$$

In Equation 1.21, $Q - jP$ is often called the correction term of earth-return impedance, or the earth-return impedance correction. It should be clear that P_{oij} gives the space impedance. m_1 is called the intrinsic propagation constant of the earth.

The infinite integral of Pollaczek's impedance is numerically very unstable and often results in numerical instability. However, the integral can be numerically calculated by commercial software such as MAPLE and MATLAB® if special care is taken (e.g., logarithmic integration) [10].

1.2.2.2.3 Carson's Earth-Return Impedance

There were no computers in the 1920s; thus, it was impossible to use Pollaczek's impedance [8]. Carson derived the same formula as Pollaczek, neglecting the earth permittivity (i.e., $\varepsilon_e = \varepsilon_0$ in Equation 1.23), and he further derived a series expansion of the infinite integral in Equation 1.21. The details of Carson's expansion formula are explained in many publications, for example, Reference 11.

1.2.2.2.4 Admittance

Almost always, the following well-known admittance is used in steady-state and transient analyses of overhead lines:

$$[Y] = j\omega[C] = j\omega[P_0]^{-1}, \quad P_{0ij} = \frac{1}{2\pi\varepsilon_0} \ln\frac{D_{ij}}{d_{ij}} \tag{1.24}$$

For a single conductor:

$$Y_0 = j\omega C, \quad C = \frac{2\pi\varepsilon_0}{\ln(2h/r)}$$

Wise derived an admittance formula considering an imperfectly conducting earth in 1948 [12]:

$$[Y] = [Y_e] = j\omega[P]^{-1} \tag{1.25}$$

$$P_{ij} = P_{0ij} + M + jN, \quad M + jN = 2\int_0^\infty (A + jB)dx \tag{1.26}$$

$$A + jB = \exp\{-(h_i + h_j)x\}\frac{1}{(a+bx)} \cdot \cos(y_{ij} \cdot x) \tag{1.27}$$

where

$$a^2 = x^2 + m_1^2 - m_0^2, \quad b = \frac{m_1}{m_0} \tag{1.28}$$

$$m_0^2 = j\omega\mu_0 \cdot j\omega\mu_0, \quad m_1^2 = j\omega\mu_e(\sigma_e + j\omega\varepsilon_e) \quad \text{see Equation 1.23}$$

Because of the complicated infinite integral in Equation 1.26, similar to Pollaczek's impedance, Wise's admittance is, in most cases, neglected. However, depending on earth resistivity and conductor height, the admittance for the imperfectly conducting earth should be considered, especially in a high-frequency region, say, above a few megahertz. When a transient involves a transition between a TEM wave and a TM/TE wave, Wise's admittance should be considered. Then, the attenuation constant differs significantly from that calculated with Equation 1.24.

The numerical integration of Equation 1.26 can be carried out in a similar manner to that of Pollaczek's impedance using MAPLE or MATLAB.

1.2.2.2.5 Impedance and Admittance Formulation of an Overhead Conductor System

Summarizing Sections 1.2, 1.2.1, 1.2.2.1, and 1.2.2.2.1 through 1.2.2.2.4, impedance and admittance of an overhead conductor system are given in the following form:

$$[Z] = [Z_i] + [Z_e]\,(\Omega/\mathrm{m}) \quad [Y] = [Y_0]\,(\mathrm{S/m}) \tag{1.29}$$

Z_{ijj}: Equation 1.7 or the last equation of Equation 1.9
$Z_{ijk} = 0$
Z_{eij}: Equation 1.15, Equation 1.21, or Carson's formula
Y_{ij}: Equation 1.24 or Equation 1.25

Remember that Equations 1.7 and 1.15 are approximate formulas for Z_i and Z_e, respectively. Also, Equation 1.24 is used almost always as an outer-media admittance.

1.2.2.3 Pollaczek's General Formula for Overhead, Underground, and Overhead/Underground Conductor Systems

Pollaczek derived a general formula that can deal with earth-return impedances of overhead conductors, underground cables, and multiconductor systems composed of overhead and underground conductors in the following form [7,13]:

$$Z_e = Z(i,j) = j\omega\left(\frac{\mu_0}{2\pi}\right)\left[K_0(m_i d) - K_0(m_i D) + \int_{-\infty}^{\infty} F_1(x)\cdot\exp(jyx)\cdot dx\right] \tag{1.30}$$

$$F_1(x) = \exp\frac{\left\{-\,|\,h_a\,|\,\sqrt{x^2 + m_i^2} - |\,h_b\,|\,\sqrt{x^2 + m_j^2}\right\}}{\left(\sqrt{x^2 + m_1^2} + \sqrt{x^2 + m_2^2}\right)} \tag{1.31}$$

where
$m_1 = j\omega\mu_0\,(\sigma_1 + j\omega\varepsilon_1) = j\omega\mu_0\sigma_1 - \omega^2\mu_0\varepsilon_1$
$m_2 = j\omega\mu_0\,(\sigma_2 + j\omega\varepsilon_2) = j\omega\mu_0\sigma_2 - \omega^2\mu_0\varepsilon_2$

σ is the conductivity

ε the permittivity

μ_0 the permeability in free space

y the horizontal separation between conductors a and b

h the conductor height/depth, $d^2 = (h_a - h_b)^2 + y^2$, $D^2 = (h_a + h_b)^2 + y^2$

i, j are subscripts corresponding to media 1 and 2 in Figure 1.4

Assuming medium 1 is air, $\sigma_1 = 0$ and $\omega^2\,\mu_0\,\varepsilon_1 = \omega^2\,\mu_0\varepsilon_0 \ll 1$ yield $m_1 = 0$. Thus,

$$F_2(x) = \exp\frac{\left\{-|\,h_a\,|\sqrt{x^2 + m_i^2} - |\,h_b\,|\sqrt{x^2 + m_j^2}\right\}}{\left(\sqrt{x^2 + m^2} + |\,x\,|\right)} \tag{1.32}$$

where $m^2 = j\omega\mu_0\,(\sigma_e + j\omega\varepsilon_e) \cong j\omega\mu_0\sigma_e = j\alpha$ for soil.

Equation 1.30 is rewritten depending on the positions of conductors a and b. For example,

1. Overhead lines h_a, $h_b \geq 0$; $i = j = 1$ (air)

$$K_0(m_i d) - K_0(m_i D) = \ln\left(\frac{2}{rm_1 d}\right) - \ln\left(\frac{2}{rm_1 D}\right) = \ln\left(\frac{D}{d}\right) \quad \text{for } m_1 = 0$$

$$\therefore Z_e = Z(1,1) = j\omega\left(\frac{\mu_0}{2\pi}\right)\left[\ln\left(\frac{D}{d}\right) + \int_{-\infty}^{\infty}\left[\frac{\exp\{jyx - (h_a + h_b)\,|\,x\,|\}}{\sqrt{x^2 + m^2} + |\,x\,|}\right]\cdot dx\right] \tag{1.33}$$

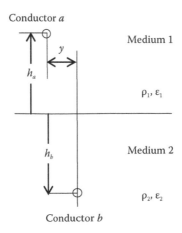

Conductor a

Medium 1

h_a

ρ_1, ε_1

h_b

Medium 2

ρ_2, ε_2

Conductor b

FIGURE 1.4

A conductor system.

2. Underground cable h_a, $h_b \leq 0$; $i = j = 2$ (soil)

$$Z_e = Z(2,2) = j\omega\left(\frac{\mu_0}{2\pi}\right)$$

$$\times \left[K_0(md) - K_0(mD) + \int_{-\infty}^{\infty} \left[\frac{\exp\left\{jyx - |h_a + h_b| \sqrt{x^2 + m^2}\right\}}{\left(\sqrt{x^2 + m^2} + |x|\right)} \right] \cdot dx \right] \quad (1.34)$$

3. Overhead/underground $h_a \geq 0$, $h_b \leq 0$; $i = j = 2$

$$K_0(m_i d) - K_0(m_i D) = 0 \quad \text{for } i \neq j$$

$$Z_e = Z(1,2) = j\omega\left(\frac{\mu_0}{2\pi}\right)\int_{-\infty}^{\infty} \left[\frac{\exp\left\{jyx - h_a |x| + h_b \sqrt{x^2 + m^2}\right\}}{\left(\sqrt{x^2 + m^2} + |x|\right)} \right] \cdot dx \quad (1.35)$$

It should be noted that Pollaczek and Carson assumed uniform current distribution along a conductor and neglect the earth permittivity. If those are considered, the denominator of the integral in Equation 1.33 is to be rewritten by $a_w = \sqrt{s^2 + j\omega\mu_0\{\sigma_e + j\omega\varepsilon_0(\varepsilon_r - 1)\}} + |x|$ as explained in References 14 and 15.

PROBLEMS

1.1 Calculate resistance R_c (Ω/km) and inductance L_c (mH/km) of the conductor in Figure 1.1 with radius $r_0 = 1$ cm, $r_1 = 0$, resistivity $\rho_c = 2 \times 10^{-8}$ Ω m, and permeability $\mu_c = \mu_0 = 4\pi \times 10^{-7}$ (H/m) at frequency $f = 50$ Hz and 100 kHz.

1.2 Calculate resistance R_c (Ω/km) and inductance L_c (mH/km) of a conductor with a cross-sectional area $S = 3.14 \times 10^{-4}$ (m²), circumferential length $\ell = 6.28$ cm, $\rho_c = 2 \times 10^{-8}$ Ω m, and $\mu_c = \mu_0$ at $f = 50$ Hz and 100 kHz.

1.3 Obtain a cylindrical conductor equivalent to a square conductor of 2×2 cm.

1.4 Calculate R_e (Ω/km) and L_e (mH/km) of the earth-return impedance for an overhead line with radius $r = 1$ cm, $h = 10$ m, $\rho_e = 100$ Ω m, and $\mu_e = \mu_0$ at $f = 50$ Hz and 100 kHz.

1.5 Discuss the difference between conductor internal impedance and earth-return impedance based on the results of Problems 1.1 and 1.4.

1.6 Derive a low-frequency approximate formula of the earth-return impedance from Equation 1.15 under the condition that $|h_e| \gg h_i, h_j$.

1.7 Derive a high-frequency approximate formula of Equation 1.15 under the condition that $|h_e| \ll h_i, h_j.$ by using the relation of $\ln(1+x) \doteqdot x$ for $x \ll 1$.

1.3 Basic Theory of Distributed-Parameter Circuit

1.3.1 Partial Differential Equations of Voltages and Currents

Considering the impedance and admittance explained in Section 1.2, the single distributed-parameter line in Figure 1.5a is represented by a lumped-parameter equivalence, as in Figure 1.5b.

Applying Kirchhoff's voltage law to the branch between nodes P and Q, the following relation is obtained:

$$v - (v + \Delta v) = R \cdot \Delta x \cdot i + L \cdot \Delta x \cdot \frac{di}{dt}$$

Rearranging this equation results in the following:

$$\frac{-\Delta v}{\Delta x} = R \cdot i + L \cdot \frac{di}{dt}$$

By taking the limit of Δx to zero, the following partial differential equation is obtained:

$$\frac{-\partial v}{\partial x} = R \cdot i + L \cdot \frac{\partial i}{\partial t} \tag{1.36}$$

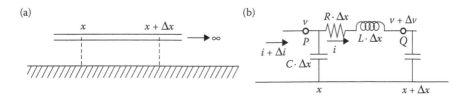

FIGURE 1.5
A single distributed-parameter line. (a) A distributed-parameter line. (b) A lumped-parameter equivalence.

Similarly, applying Kirchhoff's current law to node P, the following equation is obtained:

$$-\frac{\partial i}{\partial x} = G \cdot v + C \cdot \frac{\partial v}{\partial t} \tag{1.37}$$

A general solution of Equations 1.36 and 1.37 is derived in Section 3.2.1.

1.3.2 General Solutions of Voltages and Currents

1.3.2.1 Sinusoidal Excitation

Assuming v and i as sinusoidal steady-state solutions, Equations 1.36 and 1.37 can be differentiated with respect to time t. The derived partial differential equations are converted to ordinary differential equations, which makes it possible to obtain solutions to Equations 1.36 and 1.37. By expressing v and i in a polar coordinate, that is, in an exponential form, the derivation of the solution becomes straightforward.

By representing v and i in a phasor form, then

$$\dot{V} = \dot{V}_m \exp(j\omega t), \quad \dot{I} = \dot{I}_m \exp(j\omega t) \tag{1.38}$$

Either real or imaginary parts of these equations represent v and i. If imaginary parts are selected, then

$$v = \text{Im}[\dot{V}] = V_m \sin(\omega t + \theta_1)$$
$$I = \text{Im}[\dot{I}] = I_m \sin(\omega t + \theta_2) \tag{1.39}$$

where

$$\dot{V}_m = \dot{V}_m \exp(j\theta_1)$$
$$\dot{I}_m = \dot{I}_m \exp(j\theta_2)$$

Substituting Equations 1.36 and 1.37 and differentiating partially with respect to time t, the following ordinary differential equations are obtained:

$$-\frac{d\dot{V}}{dx} = R\dot{I} + j\omega L\dot{I} = (R + j\omega L)\dot{I} = \dot{Z}\dot{I}$$
$$-\frac{d\dot{I}}{dx} = G\dot{V} + j\omega C\dot{V} = (G + j\omega C)\dot{V} = \dot{Y}\dot{V} \tag{1.40}$$

where

$$R + j\omega L = \dot{Z}: \text{series impedance of a conductor}$$
$$G + j\omega C = \dot{Y}: \text{shunt admittance of a conductor} \tag{1.41}$$

Differentiating Equation 1.40 with respect to x gives

$$-\frac{d^2\dot{V}}{dx^2} = \dot{Z}\frac{d\dot{I}}{dx'}, \quad -\frac{d^2\dot{I}}{dx^2} = \dot{Y}\frac{d\dot{V}}{dx}$$

Substituting Equation 1.40 into this equation gives

$$-\frac{d^2\dot{V}}{dx^2} = \dot{Z}\dot{Y}\dot{V} = \dot{\Gamma}_v^2\dot{V}, \quad -\frac{d^2\dot{I}}{dx^2} = \dot{Y}\dot{Z}\dot{I} = \dot{\Gamma}_i^2\dot{I} \tag{1.42}$$

where

$$\dot{\Gamma}_v = (\dot{Z}\dot{Y})^{1/2}: \text{propagation constant with respect to voltage } [\text{m}^{-1}]$$
$$\dot{\Gamma}_i = (\dot{Y}\dot{Z})^{1/2}: \text{propagation constant with respect to current } [\text{m}^{-1}] \tag{1.43}$$

When Z and Y are matrices, the following relation is given in general:

$$[\Gamma_v] \neq [\Gamma_i] \quad \text{since } [Z][Y] \neq [Y][Z] \tag{1.44}$$

Only when Z and Y are perfectly symmetrical matrices (symmetrical matrices whose diagonal and nondiagonal entries are equal), is $[\Gamma_v] = [\Gamma_i]$ satisfied. In the case of a single-phase line, because Z and Y are scalars, then

$$\dot{\Gamma}_v^2 = \dot{\Gamma}_i^2 = \dot{\Gamma}^2 = \dot{Z}\dot{Y} = \dot{Y}\dot{Z} \quad \text{and} \quad \dot{\Gamma} = \sqrt{\dot{Z}\dot{Y}} \tag{1.45}$$

Substituting this equation into Equation 1.42 gives

$$\frac{d^2\dot{V}}{dx^2} = \dot{\Gamma}^2\dot{V}, \quad \frac{d^2\dot{I}}{dx^2} = \dot{\Gamma}^2\dot{I} \tag{1.46}$$

A general solution is obtained by solving one of these equations. Once it is solved for V or I, Equation 1.40 can be used to derive the other solution.

The general solution of Equation 1.46 with respect to voltage is given by

$$\dot{V} = \dot{A}\exp(-\dot{\Gamma}x) + \dot{B}\exp(\dot{\Gamma}x) \tag{1.47}$$

where $\dot{A}, \dot{B}$ are the integral constants determined by a boundary condition.

The first part of Equation 1.40 gives the general solution of current in the following differential form:

$$\dot{I} = -\dot{Z}^{-1}\dot{\Gamma}\frac{d\dot{V}}{dx} = \dot{Z}^{-1}\dot{\Gamma}\{\dot{A}\exp(-\dot{\Gamma}x) - \dot{B}\exp(\dot{\Gamma}x)\} \tag{1.48}$$

The coefficient of this equation is rewritten as

$$\frac{\dot{\Gamma}}{\dot{Z}} = \frac{\sqrt{\dot{Y}\dot{Z}}}{\dot{Z}} = \sqrt{\frac{\dot{Y}}{\dot{Z}}} = \frac{\dot{Y}}{\sqrt{\dot{Z}\dot{Y}}} = \frac{\dot{Y}}{\dot{\Gamma}} = \dot{Y}_0$$

where

$$\dot{Y}_0 = \sqrt{\frac{\dot{Y}}{\dot{Z}}} = \frac{1}{\dot{Z}} : \text{characteristic admittance } [S] \tag{1.49}$$

$$\dot{Z}_0 = \sqrt{\frac{\dot{Z}}{\dot{Y}}} : \text{characteristic impedance}(\Omega)$$

In general cases, when Z and Y are matrices, then

$$[Z_0] = [\Gamma_v]^{-1}[Z] = [\Gamma_v][Y]^{-1}$$
$$[Y_0] = [Z_0]^{-1} = [Z]^{-1}[\Gamma_v] = [Y][\Gamma_v]^{-1} \tag{1.50}$$

Substituting Equation 1.49 into Equation 1.48, the general solution of Equation 1.46 with respect to current is expressed as

$$\dot{I} = \dot{Y}_0\{\dot{A}\exp(-\dot{\Gamma}x) - \dot{B}\exp(-\dot{\Gamma}x)\} \tag{1.51}$$

Exponential functions in Equations 1.47 and 1.51 are convenient for dealing with a line of infinite length (infinite line), but hyperbolic functions are preferred for dealing with a line of finite length (finite line). To obtain an expression using hyperbolic functions, new constants C and D are defined as

$$\dot{A} = \frac{\dot{C} - \dot{D}}{2}, \quad \dot{B} = \frac{\dot{C} + \dot{D}}{2}$$

Substituting this into Equations 1.47 and 1.51 gives

$$\dot{V} = \dot{C}\left\{\frac{\exp(\dot{\Gamma}x) + \exp(-\dot{\Gamma}x)}{2}\right\} + \dot{D}\left\{\frac{\exp(\dot{\Gamma}x) - \exp(-\dot{\Gamma}x)}{2}\right\}$$

$$\dot{I} = -\dot{Y}_0\left[\dot{C}\left\{\frac{\exp(\dot{\Gamma}x) - \exp(-\dot{\Gamma}x)}{2}\right\} + \dot{D}\left\{\frac{\exp(\dot{\Gamma}x) + \exp(-\dot{\Gamma}x)}{2}\right\}\right]$$

From the definitions of hyperbolic functions:

$$\dot{V} = \dot{C}\cosh\dot{\Gamma}x + \dot{D}\sinh\dot{\Gamma}x, \quad \dot{I} = -\dot{Y}_0(\dot{C}\sinh\dot{\Gamma}x + D\cosh\dot{\Gamma}x) \tag{1.52}$$

Constants A, B, C, and D defined here are arbitrary constants determined by boundary conditions.

1.3.2.2 Lossless Line

When a distributed line satisfies $R = G = 0$, the line is called a "lossless line." In this case, Equations 1.36 and 1.37 can be written as

$$-\frac{\partial v}{\partial x} = L\frac{\partial i}{\partial x \partial t}, \quad -\frac{\partial i}{\partial x} = C\frac{\partial v}{\partial x \partial t} \tag{1.53}$$

Differentiating these equations with respect to x gives

$$-\frac{\partial^2 v}{\partial x^2} = L\frac{\partial(\partial i/\partial x)}{\partial t}, \quad \frac{\partial^2 i}{\partial x^2} = C\frac{\partial(\partial v/\partial x)}{\partial t}$$

Similar to the sinusoidal excitation case, the following equations are obtained for voltage and current:

$$-\frac{\partial^2 v}{\partial x^2} = L\frac{\partial(\partial i/\partial x)}{\partial t} = L\frac{\partial(-C\partial v/\partial t)}{\partial t} = -LC\frac{\partial^2 v}{\partial t^2}$$

$$\therefore \frac{\partial^2 v}{\partial x^2} = LC\frac{\partial^2 v}{\partial t^2} = \frac{1}{c_0^2}\frac{\partial^2 v}{\partial t^2} \quad \text{and} \quad \frac{\partial^2 i}{\partial x^2} = LC\frac{\partial^2 i}{\partial t^2} = \frac{1}{c_0^2}\frac{\partial^2 i}{\partial t^2} \tag{1.54}$$

where $c_0 = 1/\sqrt{(LC)}\,(\text{m}/\text{s})$.

From Equation 1.14, with $h_e = 0$, and Equation 1.24:

$$LC = \frac{(\mu_0/2\pi)\ln(2h/r)\cdot 2\pi\varepsilon_0}{\ln(2h/r) = \mu_0\varepsilon_0 = (1/c_0^2)}$$

Thus,

$$c_0 = \frac{1}{\sqrt{LC}} = \frac{1}{\sqrt{\mu_0 \varepsilon_0}} = 3 \times 10^8 \,(\text{m/s}): \text{light velocity in free space} \qquad (1.55)$$

The parts of Equation 1.54 are linear second-order hyperbolic partial differential equations called wave equations. The general solutions for the wave equations were given by D'Alembert in 1747 [16] as

$$\begin{aligned} v &= e_f(x - c_0 t) + e_b(x + c_0 t) \quad \text{as a function of distance} \\ i &= Y_0 \{ e_f(x - c_0 t) - e_b(x + c_0 t) \} \end{aligned} \qquad (1.56)$$

$$\begin{aligned} v &= E_f\left(t - \left(\frac{x}{c_0}\right) \right) + E_b\left(t + \left(\frac{x}{c_0}\right) \right) \quad \text{as a function of time} \\ i &= Y_0 \left\{ E_f\left(t - \left(\frac{x}{c_0}\right) \right) - E_b\left(t + \left(\frac{x}{c_0}\right) \right) \right\} \end{aligned} \qquad (1.57)$$

where

$$\begin{aligned} c_0 C &= \frac{1}{\sqrt{LC}} C = \sqrt{\frac{C}{L}} = Y_0: \text{surge admittance (S)} \\ Z_0 &= \frac{1}{Y_0} = \sqrt{\frac{L}{C}}: \text{surge impedance } (\Omega) \end{aligned} \qquad (1.58)$$

Surge impedance Z_0 and surge admittance Y_0 in these equations are extreme values of the characteristic impedance and admittance in Equation 1.49 for frequency $f \to \infty$.

This solution is known as a wave equation, which behaves as a wave traveling along the x-axis with velocity c_0. It should be clear that the values of functions e_f, e_b, E_f, and E_b do not vary if $x - c_0 t = \text{constant}$ and $x + c_0 t = \text{constant}$. Since e_f and E_f show a positive traveling velocity, they are called "forward traveling waves": $c_0 = x/t$ along the x-axis in a positive direction.

On the contrary, e_b and E_b are "backward traveling waves," which means the waves travel in the direction of $-x$, that is, the traveling velocity is negative:

$$c_0 = \frac{-x}{t}$$

Having defined the direction of the traveling waves, Equation 1.56 is rewritten simply as

$$v = e_f + e_b, \quad i = Y_0(e_f - e_b) = i_f - i_b \qquad (1.59)$$

where

e_f, e_b are the voltage traveling waves
i_f, i_b are the current traveling waves

This is a basic equation to analyze traveling-wave phenomena, and the traveling waves are determined by a boundary condition. More details are given in Section 1.6.

1.3.3 Voltages and Currents on a Semi-Infinite Line

Here, we consider the semi-infinite line shown in Figure 1.6. The AC constant voltage source is connected to the sending end ($x = 0$), and the line extends infinitely to the right-hand side ($x = +\infty$).

1.3.3.1 Solutions of Voltages and Currents

From the general solutions in Equations 1.47 and 1.51, the solutions of voltages and currents on the semi-infinite line in Figure 1.6 are obtained by using the following boundary conditions:

$$\dot{V} = \dot{E} \text{ at } x = 0 \quad \text{and} \quad \dot{V} = 0 \text{ at } x = \infty \qquad (1.60)$$

The boundary condition in the second part of Equation 1.60 is obtained from the physical constraint in which all physical quantities have to be zero at $x \to \infty$.

Substituting the condition into Equation 1.47 gives

$$0 = \dot{A}\exp(-\dot{\Gamma}\infty) + \dot{B}\exp(\dot{\Gamma}\infty)$$

Since $\exp(\dot{\Gamma}\infty) = \infty$, constant B has to be zero in order to satisfy this equation:

$$\dot{B} = 0$$

Thus,

$$0 = \dot{A}\exp(-\dot{\Gamma}\infty)$$

Substituting the first part of Equation 1.60 into this equation, constant A is obtained as

$$\dot{A} = \dot{E} \qquad (1.61)$$

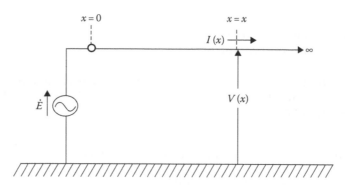

FIGURE 1.6
Semi-infinite line.

Substituting constants A and B into the general solutions, that is, Equations 1.47 and 1.51, voltages and currents on a semi-infinite line are given in the following form:

$$\dot{V} = \dot{E}\exp(-\dot{\Gamma}x), \quad \dot{I} = \dot{Y}_0\dot{E}\exp(-\dot{\Gamma}x) = \dot{I}_0\exp(-\dot{\Gamma}x) \tag{1.62}$$

where $\dot{I}_0 = \dot{Y}_0\dot{E}$.

1.3.3.2 Waveforms of Voltages and Currents

Since $\dot{\Gamma}$ is a complex value, it can be expressed as

$$\dot{\Gamma} = \alpha + j\beta \tag{1.63}$$

Substituting this into the voltage of Equation 1.62, then

$$\dot{V} = \dot{E}\exp\{-\alpha + j\beta x\} = \dot{E}\exp(-\alpha x)\exp(-j\beta x) \tag{1.64}$$

If the voltage source at $x = 0$ in Figure 1.6 is a sinusoidal source, then

$$\dot{E} = E_m\sin(\omega t) = \text{Im}\{E_m\exp(j\omega t)\} \tag{1.65}$$

The voltage on a semi-infinite line is expressed by the following equation:

$$v = \text{Im}(\dot{V}) = \text{Im}\{E_m\exp(j\omega t)\exp(-\alpha x)\exp(-j\beta x)\}$$
$$\therefore v = E_m\exp(-\alpha x)\sin(\omega t - \beta x) \tag{1.66}$$

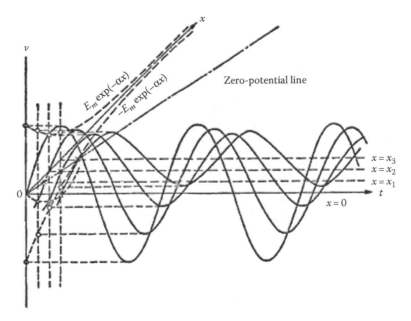

FIGURE 1.7
Three-dimensional waveforms of the voltage.

Figure 1.7 shows the voltage waveforms whose horizontal axes take different values of time when the observation point is shifted from $x = 0$ to x_1, x_2, ...

The figure illustrates that as the observation point shifts in the positive direction, the amplitude of the voltage decreases due to $\exp(-\alpha x)$, and the angle of the voltage lags due to $\exp(-j\beta x)$.

The horizontal axis is changed to the observation point, and voltage waveforms at different times can be seen in Figure 1.8.

Rewriting Equation 1.66, then;

$$v = -E_m \exp(-\alpha x)\sin\beta\left(x - \frac{\omega t}{\beta}\right) \tag{1.67}$$

Figure 1.8 illustrates how the voltage waveform travels in the positive direction of x as time passes, according to Equation 1.67.

1.3.3.3 Phase Velocity

Phase velocity is found from two points on a line whose phase angles are equal. For example, in Figure 1.8, x_1 (point P_1) and x_2 (point Q_1) determine the phase velocity.

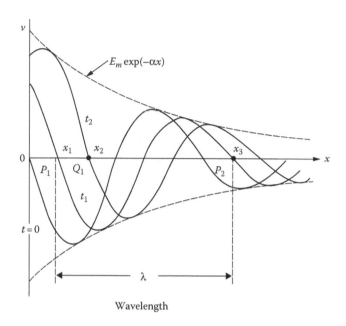

FIGURE 1.8
Voltage waveforms along the *x*-axis at different times.

From Equation 1.67, the following relationship is satisfied as the phase angles are equal:

$$x_1 - \frac{\omega t_1}{\beta} = x_2 - \frac{\omega t_2}{\beta}$$

The phase velocity c is found from Equation 1.67 as

$$c = \frac{x_2 - x_1}{t_2 - t_1} = \frac{\omega}{\beta} \tag{1.68}$$

This equation shows that the phase velocity is found from ω and β and is independent of location and time.

For a lossless line,

$$\dot{Z} = j\omega L, \quad \dot{Y} = j\omega C \tag{1.69}$$

$$L = \left(\frac{\mu_0}{2\pi}\right) \cdot \ln\left(\frac{2h}{r}\right), \quad c = \frac{2\pi\varepsilon_0}{\ln(2h/r)}$$

From Equation 1.45:

$$\dot{\Gamma} = \sqrt{\dot{Z}\dot{Y}} = j\omega\sqrt{LC} = j\beta \quad \text{and} \quad \beta = \omega\sqrt{LC} \tag{1.70}$$

As a result, for a lossless line, the phase velocity (Equation 1.55) is found from Equations 1.68 and 1.70.

The phase velocity in a lossless line is independent of ω.

1.3.3.4 Traveling Wave

When a wave travels at constant velocity, it is called a traveling wave. The general solutions of voltages and currents in Equations 1.56 and 1.57 are traveling waves. In more general cases, $\exp(-\dot{\Gamma}x)$ and $\exp(\dot{\Gamma}x)$ in the general solutions, that is, Equations 1.47 and 1.51, also express traveling waves.

The existence of traveling waves is confirmed by various physical phenomena around us. For example, when we drop a pebble in a pond, waves travel in all directions from the point where the pebble was dropped. These are known as traveling waves. If a leaf is floating in the pond, it does not travel along with the waves; it only moves up and down according to the height of the waves. Figure 1.9a shows the movement of the leaf and the water surface on the x and y axes. Here, x is the distance from the origin of the wave and y is the height. Figure 1.9b illustrates the movement of the leaf with time. Figure 1.9 also demonstrates that the movement of the leaf coincides with the shape of the wave.

This observation implies that the water in the pond does not travel along with the wave. What is traveling in the water is the energy from the dropping

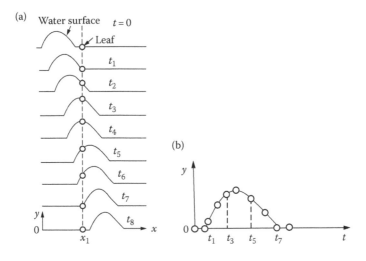

FIGURE 1.9
Movement of a leaf on a water surface. (a) Position of the leaf as a function of x, y, and t. (b) $y - t$ curve of leaf movement.

of the pebble, and the water (medium) in the pond only transmits the energy. In other words, traveling waves represent traveling energy—the medium itself does not travel.

Maxwell's wave equation can thus be considered an expression of traveling energy, which means that the characteristics of energy transmission can be analyzed as those of traveling waves. For example, the propagation velocity of traveling waves corresponds to the propagation velocity of energy.

1.3.3.5 Wavelength

Wavelength is found from two points on a line whose phase angles are 360° apart at a particular time. For example, x_1 (point P_1) and x_3 (point P_2) in Figure 1.8 determine the wavelength λ at $t = 0$:

$$\lambda = x_3 - x_1 \tag{1.71}$$

Since the phase angles of the two points are 360° apart, the following equation is satisfied from Equation 1.66:

$$(\omega t_1 - \beta x_1) - (\omega t_1 - \beta x_3) = 2\pi$$
$$\therefore \beta(x_3 - x_1) = 2\pi$$

The wavelength is found, from this equation and Equation 1.71, as

$$\lambda = \frac{2\pi}{\beta} \tag{1.72}$$

Equation 1.72 shows that the wavelength is a function of β and is independent of location and time.

For a lossless line, using Equation 1.70, then

$$\lambda = \frac{2\pi}{\omega\sqrt{LC}} = \frac{1}{f\sqrt{LC}} = \frac{c_0}{f} \tag{1.73}$$

1.3.4 Propagation Constants and Characteristic Impedance

1.3.4.1 Propagation Constants

The propagation constant Γ is expressed as follows:

$$\Gamma = \sqrt{\dot{Z}\dot{Y}} = \alpha + j\beta \tag{1.74}$$

where
  α is the attenuation constant (Np/m)
  β is the phase constant (rad/s)

Let us look at the meaning of the attenuation constant using the semi-infinite line case as an example. From Equation 1.62 and the boundary conditions:

$$\dot{V}_0 = \dot{V}(x=0) = \dot{E} \quad \text{at} \quad x=0$$
$$\dot{V}_x = \dot{V}(x=x) = \dot{E}\exp(-\Gamma x) \quad \text{at} \quad x=x$$

The attenuation after the propagation of x is

$$\frac{\dot{V}_x}{\dot{V}_0} = \exp(-\dot{\Gamma}_x) = \exp(-\alpha x)\exp(-j\beta x), \quad \frac{|\dot{V}_x|}{|\dot{V}_0|} = \exp(-\alpha x) \qquad (1.75)$$

From this equation:

$$\alpha x = a_T = -\text{In}\frac{|\dot{V}_x|}{|\dot{V}_0|}(N_p) \qquad (1.76)$$

The attenuation per unit length is

$$\alpha = \frac{\alpha_T}{x} = -\frac{1}{x}\text{In}\frac{|\dot{V}_x|}{|\dot{V}_0|}(N_p/m)$$

This equation shows that the attenuation constant gives the attenuation of voltage after it travels for a unit length.

Now, we will find propagation constants for a line with losses, that is, a line whose R and G are positive. From Equation 1.63:

$$\dot{\Gamma}^2 = \dot{Z}\dot{Y} = (R+j\omega L)(G+j\omega C) = \alpha^2 - \beta^2 + 2j\alpha\beta$$
$$\therefore \alpha^2 - \beta^2 = RG - \omega^2 LC, \quad 2\alpha\beta = \omega(LG+CR)$$

Also,

$$\alpha^2 + \beta^2 = \sqrt{(R^2+\omega^2 L^2)(G^2+\omega^2 C^2)}$$

From these equations, the following results are obtained:

$$2\alpha^2 = \sqrt{(R^2+\omega^2 L^2)(G^2+\omega^2 C^2)} + (RG-\omega^2 LC)$$
$$2\beta^2 = \sqrt{(R^2+\omega^2 L^2)(G^2+\omega^2 C^2)} - (RG-\omega^2 LC)$$

Since $\alpha\beta$ is positive, α and β have to have the same sign, that is, both positive:

$$\left.\begin{array}{l} \alpha = \sqrt{\dfrac{\left\{\sqrt{(R^2+\omega^2L^2)(G^2+\omega^2C^2)}+(RG-\omega^2LC)\right\}}{2}} \\[4mm] \beta = \sqrt{\dfrac{\left\{\sqrt{R^2+\omega^2L^2)(G^2+\omega^2C^2)}-(RG-\omega^2LC)\right\}}{2}} \end{array}\right\} \tag{1.77}$$

Here, we find the characteristics of α and β defined earlier. First, when $\omega = 0$, then

$$\alpha = \sqrt{RG}, \quad \beta = 0 \tag{1.78}$$

For $\omega \to \infty$, using the approximation $\sqrt{1+x} \approx 1+x/2$ for $x \ll 1$, then

$$\sqrt{R^2+\omega^2L^2} = \omega L\sqrt{1+\dfrac{R^2}{\omega^2L^2}} \approx \omega L\left(1+\dfrac{R^2}{2\omega^2L^2}\right)$$

$$\sqrt{G^2+\omega^2C^2} \approx \omega C\left(1+\dfrac{G^2}{2\omega^2C^2}\right)$$

Substituting this into Equation 1.77 gives

$$\alpha = \dfrac{\sqrt{C/LR}+\sqrt{L/CG}}{2}, \quad \beta = \omega\sqrt{LC}; \quad \omega \to \infty \tag{1.79}$$

Considering Equations 1.78 and 1.79, the frequency responses of α and β are found as shown in Figure 1.10.

Equation 1.79 shows that the propagation velocity at $\omega \to \infty$ is

$$\lim_{\omega \to \infty} c = \dfrac{1}{\sqrt{LC}} = c_0 \tag{1.80}$$

The propagation velocity c_0 in this equation is equal to the propagation velocity for a lossless line in Equation 1.55.

1.3.4.2 Characteristic Impedance

For a single-phase lossless overhead line in air, the characteristic impedance is found from Equations 1.49 and 1.69:

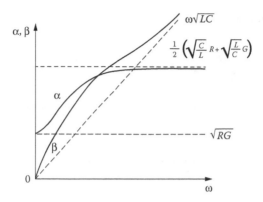

FIGURE 1.10
Frequency characteristics of α and β.

$$\dot{Z}_0 = \sqrt{\frac{\dot{Z}}{\dot{Y}}} = \sqrt{\frac{L}{C}} = \sqrt{\frac{\mu_0/2\pi}{2\pi\varepsilon_0}}\ln\frac{2h}{r} \approx 60\ln\frac{2h}{r}\,(\Omega) \tag{1.81}$$

This equation shows that the characteristic impedance becomes independent of frequency for a lossless line, and it is called surge impedance, as defined in Equation 1.58.

For a line with losses, the characteristic impedance is found as

$$\dot{Z}_0 = \sqrt{\frac{R+j\omega L}{G+j\omega C}} = \sqrt{\frac{(R+j\omega L)(G-j\omega C)}{G^2+\omega^2 C^2}} \tag{1.82}$$

which is defined as

$$\dot{Z}_0 = r + jx \tag{1.83}$$

The real part r and the imaginary part x of the characteristic impedance are found in the same way as we found α and β earlier:

$$\left. \begin{array}{l} r = \sqrt{\dfrac{\left\{\sqrt{(R^2+\omega^2 L^2)(G^2+\omega^2 C^2)}+(RG+\omega^2 LC)\right\}}{\left\{2(G^2+\omega^2 C^2)\right\}}} \\[4ex] x = \sqrt{\dfrac{\left\{\sqrt{(R^2+\omega^2 L^2)(G^2+\omega^2 C^2)}-(RG+\omega^2 LC)\right\}}{\left\{2(G^2+\omega^2 C^2)\right\}}} \end{array} \right\} \tag{1.84}$$

From this equation:

$$r = \sqrt{\frac{R}{G}}, \quad x = 0 \quad \text{that is } Z_0 = \sqrt{\frac{R}{G}}; \quad \omega = 0 \atop r = \sqrt{\frac{L}{C}}, \quad x = 0 \quad \text{that is } Z_0 = \sqrt{\frac{L}{C}}; \quad \omega \to \infty \right\}$$

(1.85)

Equation 1.85 shows that the characteristic impedance for $\omega \to \infty$ coincides with the surge impedance of a lossless line in Equation 1.81.

1.3.5 Voltages and Currents on a Finite Line

1.3.5.1 Short-Circuited Line

In this section, we consider a line with a finite length (finite line) whose remote end is short-circuited to ground, as illustrated in Figure 1.11.

To deal with a finite line, the general solution in the form of hyperbolic functions, as in Equation 1.52, is convenient. The boundary conditions in Figure 1.11 are

$$\dot{V} = \dot{E} \quad \text{at } x = 0 \atop \dot{V} = 0 \quad \text{at } x = l \right\}$$

(1.86)

Substituting these conditions into Equation 1.52, unknown constants C and D are determined as $\dot{E} = \dot{C}$:

$$0 = \dot{C} \cosh \dot{\Gamma} l + \dot{D} \sinh \dot{\Gamma} l$$

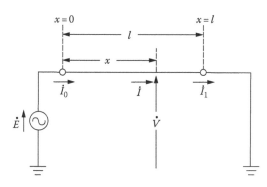

FIGURE 1.11
A short-circuited line.

Substituting C and D into Equation 1.52, the following solutions are obtained:

$$\dot{V} = \dot{E}\cosh\dot{\Gamma}x - \frac{\cosh\dot{\Gamma}l}{\sinh\dot{\Gamma}l}\dot{E}\sinh\dot{\Gamma}x$$

$$= \frac{\dot{E}(\sinh\dot{\Gamma}l\cosh\dot{\Gamma}x - \cosh\dot{\Gamma}l\sinh\dot{\Gamma}x)}{\sinh\dot{\Gamma}l} \qquad (1.87)$$

$$= \frac{\dot{E}\sinh\dot{\Gamma}(l-x)}{\sinh\dot{\Gamma}l}$$

Similarly,

$$\dot{I} = \frac{\dot{Y}_0\dot{E}\cosh\dot{\Gamma}(l-x)}{\sinh\dot{\Gamma}l} \qquad (1.88)$$

The current at the sending end ($x = 0$) is

$$\dot{I}_0 = \dot{I}(x = 0) = \frac{\dot{Y}_0\dot{E}\cosh\dot{\Gamma}l}{\sinh\dot{\Gamma}l} = \dot{Y}_0\dot{E}\coth\dot{\Gamma}l \qquad (1.89)$$

The solution for the current in Equation 1.88 is rewritten by using I_0:

$$\dot{I} = \frac{\dot{I}_0\cosh\Gamma(l-x)}{\cosh\dot{\Gamma}l} \qquad (1.90)$$

The current at the remote end ($x = l$) is

$$\dot{I}_l = \dot{I}(x = l) = \frac{\dot{Y}_0\dot{E}}{\sinh\dot{\Gamma}l} = \frac{\dot{I}_0}{\cosh\dot{\Gamma}l} \qquad (1.91)$$

The impedance of the finite line observed from the sending end is given as a function of line length l:

$$\dot{Z}(l) = \frac{\dot{E}}{\dot{I}_0} = \frac{1}{\dot{Y}_0\coth\dot{\Gamma}l} = \dot{Z}_0\tanh\dot{\Gamma}l \qquad (1.92)$$

Figure 1.12 shows an example of $|\dot{Z}(l)|$. For $l \to \infty$, since $\tanh(\infty) \to 1$:

$$\dot{Z}(l = \infty) = \dot{Z}_0$$

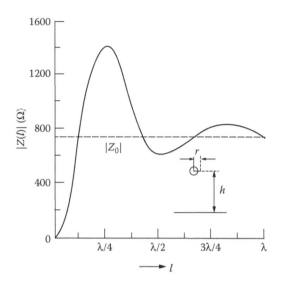

FIGURE 1.12
Input impedance $|Z(l)|$ of a short-circuited line $r = 1.3$ mm, $h = 30$ cm, and $f = 800$ Hz.

For a lossless line:

$$\dot{Z}_0 = \sqrt{\frac{L}{C}}, \quad \dot{\Gamma} = j\omega\sqrt{LC}$$

Using the relationships $\sinh jx = j \sin x$ and $\cosh jx = \cos x$, the solutions for voltage and current are expressed as

$$\dot{V} = \frac{\sin\{\omega\sqrt{LC}(l-x)\}}{\sin(\omega\sqrt{LC}l)}\dot{E}, \quad \dot{I} = -j\sqrt{\frac{C}{L}}\frac{\cos\{\omega\sqrt{LC}(l-x)\}}{\sin(\omega\sqrt{LC}l)}\dot{E} \qquad (1.93)$$

In Equation 1.93, voltage and current become infinite when the denominators are zero. This condition is referred to as the resonant condition. The denominators become zero when

$$\sin(\omega\sqrt{LC}l) = 0 \quad \therefore \omega\sqrt{LC}l = n\pi; \quad n: \text{positive integers} \qquad (1.94)$$

Therefore, natural resonant frequencies are found as

$$f_{Sn} = \frac{\omega}{2\pi} = \frac{n\pi}{2\pi\sqrt{LC}l} = \frac{n}{2\sqrt{LC}l} \qquad (1.95)$$

Infinite numbers of f_{Sn} exist for different n. The natural resonant frequency for $n = 1$ is called the fundamental resonant frequency.

Let us define τ as the propagation time for voltage and current on a line with length l, which is given by

$$\tau = \frac{l}{c_0} = \sqrt{LC}\,l \tag{1.96}$$

Using the propagation time τ, the natural resonant frequency and the fundamental resonant frequency are expressed as

$$f_{Sn} = \frac{n}{2\tau}, \quad f_{S1} = \frac{1}{2\tau} \tag{1.97}$$

The input impedance $Z(l)$ of the finite line seen from the sending end is also rewritten for a lossless line as follows:

$$\dot{Z}(l) = j\sqrt{\frac{L}{C}}\,\tan(\omega\sqrt{LC}\,l) \tag{1.98}$$

Figure 1.13 shows the relationship between $|\dot{Z}(l)|$ and $\theta = \sqrt{LC}\,l$ (or l) for a lossless line. The relationship coincides with Foster's reactance theorem. The line is in a resonant condition for $\theta = n\pi$; n: positive integers, and the line is in an antiresonant condition for $\theta = (2n - 1)\pi/2$.

1.3.5.2 Open-Circuited Line

In this section, we consider a finite line whose remote end is open, as shown in Figure 1.14.

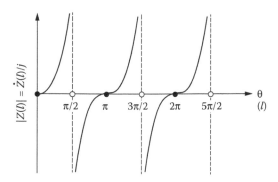

FIGURE 1.13
$|Z(l)| - \theta$ characteristic of a lossless short-circuited line.

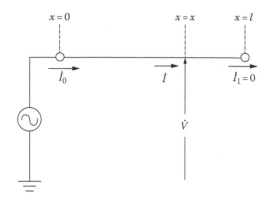

FIGURE 1.14
An open-circuited line.

For this line, the boundary conditions are defined as

$$\dot{V} = \dot{E} \quad \text{at } x = 0$$
$$\dot{I} = 0 \quad \text{at } x = l$$

(1.99)

In a similar manner to a short-circuited line, the solutions of voltage and current are obtained in the following forms:

$$\dot{V} = \frac{\dot{E}\cosh\dot{\Gamma}(l-x)}{\cosh\dot{\Gamma}l}$$
$$\dot{I} = \frac{\dot{Y_0}\dot{E}\sinh\dot{\Gamma}(l-x)}{\cosh\dot{\Gamma}l}$$

(1.100)

The input impedance of the finite line seen from the sending end is expressed as

$$\dot{Z}(l) = \frac{\dot{E}}{\dot{I_0}} = \dot{Z_0}\coth\dot{\Gamma}l$$

(1.101)

Figure 1.15 shows an example of the relationship between $|Z(l)|$ and l. For a lossless line, the solutions for voltage and current are expressed as

$$\dot{V} = \frac{\cos\{\omega\sqrt{LC}(l-x)\}}{\cos(\omega\sqrt{LC}l)}\dot{E}, \quad \dot{I} = j\sqrt{\frac{C}{L}}\frac{\sin\{\omega\sqrt{LC}(l-x)\}}{\cos(\omega\sqrt{LC}l)}\dot{E}$$

(1.102)

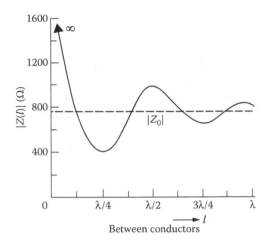

FIGURE 1.15
Input impedance of an open-circuited line. $r = 1.3$ mm, $y = 60$ cm between conductors, and $f = 800$ Hz.

The line is in a resonant condition when the denominator of Equation 1.102 is zero:

$$\omega\sqrt{LCl} = \frac{(2n-1)\pi}{2}; \quad n\text{: positive integers} \tag{1.103}$$

Therefore, the natural resonant frequency is found as

$$f_{On} = \frac{\omega}{2\pi} = \frac{(2n-1)(\pi/2)}{2\pi\sqrt{LCl}} = \frac{(2n-1)c_0}{4l} = \frac{2n-1}{4\tau} \tag{1.104}$$

The fundamental resonant frequency is

$$f_{O1} = \frac{1}{4\tau} \tag{1.105}$$

As $f_{s1} = 1/2\tau$ for a short-circuited line, then $f_{s1} = 2f_{O1}$.
The input impedance for a lossless line seen from the sending end is

$$\dot{Z}(l) = -j\sqrt{\frac{L}{C}}\cot(\omega\sqrt{LCl}) \tag{1.106}$$

Figure 1.16 shows the relationship between $|\dot{Z}(l)|$ and $\theta = \sqrt{LCl}$ for a lossless line. As for a short-circuited line, the relationship coincides with Foster's

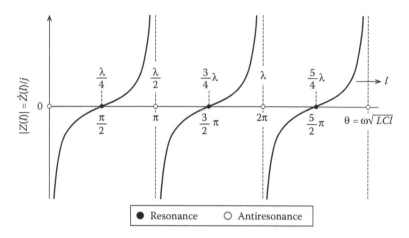

FIGURE 1.16
$|Z(l)|$ – θ characteristic of a lossless open-circuited line.

reactance theorem. The line is in a resonant condition for $\theta = (2n - 1)\pi/2$; n: positive integers, as in Equation 1.103.

PROBLEMS

1.8 Prove $\alpha(\omega = 0) < \alpha(\omega = \infty)$ in Figure 1.10.

1.9 Obtain the characteristic impedance $Z_0(\omega)$ for $\omega \to 0$ in R, L, C, and G lines.

1.10 Obtain the sending-end current I_s in R, L, C, and G lines for $\omega = 0$ in a short-circuited (finite length) line. In a real overhead line, $G \fallingdotseq 0$ in general. Then, what is the current I_s?

1.11 Calculate V_r for $E = 1000 \cdot \cos(\omega t)$ (V) with $f = 50$ (Hz), and $Z_0 = \sqrt{L/C} = 300\,(\Omega)$ on a lossless line with length $l = 300$ km by using (a) F-parameter and (b) π-equivalent circuit in the following case:

$$(1)\ Z_r = 1\,(\Omega),\ (2)\ Z_r = 300\,(\Omega),\ (3)\ Z_r = \infty.$$

1.4 Multiconductor System

1.4.1 Steady-State Solutions

Equations 1.40 through 1.42 hold true for the multiconductor system shown in Figure 1.17, provided that the coefficients Z, Y, R, L, G, and C are now

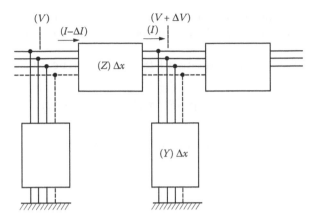

FIGURE 1.17
A multiconductor system.

matrices and the variables V and I are vectors of the order n in an n-conductor system.

The matrix P is defined as

$$P = ZY \tag{1.107}$$

where $P = [P]$: $n \times n$ matrix and in general $P \neq YZ$.

Since Z and Y are both symmetrical matrices, the transposed matrix of P is found as

$$P_t = (ZY)_t = Y_t Z_t = YZ \tag{1.108}$$

Here, the subscript t means the matrix is transposed, and $P_t = [P]_t$: $n \times n$ matrix.

From Equations 1.42 and 1.10,

$$\frac{d^2 V}{dx^2} = PV, \quad \frac{d^2 I}{dx^2} = P_t I \tag{1.109}$$

As in Equation 1.47, the general solution of Equation 1.109 is expressed as

$$V = \exp(-P^{1/2}x)V_f + \exp(P^{1/2}x)V_b \tag{1.110}$$

where V_f and V_b are arbitrary n-dimensional vectors.

The first term on the right-hand side of Equation 1.110 expresses wave propagation in the positive direction of x (forward traveling wave). The second term on the right-hand side corresponds to wave propagation in the negative direction of x (backward traveling wave). Equation 1.110 shows that voltage at any point on a line is the sum of the forward and backward traveling waves.

Since $I = -Z^{-1}\,dV/dx$ as in Equation 1.48, current can be given as

$$I = Z^{-1}P^{1/2}\{\exp(-P^{1/2}x)V_f - \exp(P^{1/2}x)V_b\} \tag{1.111}$$

For a semi-infinite line, since $V_b = 0$, Equations 1.110 and 1.116 are simplified to the following form:

$$V = \exp(-P^{1/2}x)V_f, \quad I = Z^{-1}P^{1/2}\exp(-P^{1/2}x)V_f = Z^{-1}P^{1/2}V \tag{1.112}$$

Equation 1.112 shows that the proportion of current to voltage at any point on a semi-infinite line, that is, the characteristic admittance matrix, is defined as follows:

$$Y_0 = Z^{-1}P^{1/2} \tag{1.113}$$

Since $Z_0 = Y_0^{-1}$, the characteristic impedance matrix is

$$Z_0 = P^{-1/2}Z \tag{1.114}$$

The general solution for current can also be found from the second part of Equation 1.109:

$$I = \exp\left(-P_t^{1/2}x\right)I_f + \exp\left(P_t^{1/2}x\right)I_b \tag{1.115}$$

Using the second part of Equation 1.40, the voltage in a semi-infinite line can also be found as follows, since $I_b = 0$:

$$V = Y^{-1}P_t^{1/2}I \tag{1.116}$$

From this equation, the characteristic impedance and admittance matrices are

$$Z_0 = Y^{-1}P_t^{1/2}, \quad Y_0 = P_t^{-1/2}Y \tag{1.117}$$

In general, the characteristic impedance and admittance matrices are expressed by Equations 1.113 and 1.109 using P instead of P_t.

Another way to express the characteristic impedance and admittance matrices is by integrating the second part of Equation 1.109:

$$I = -Y \int V dx \qquad (1.118)$$

For a semi-infinite line, substituting the first equation of Equation 1.112 into Equation 1.118 gives

$$I = -Y(-P^{-1/2}) \exp(-P^{1/2}x)V_f = YP^{-1/2}V \qquad (1.119)$$

Therefore, the characteristic impedance and admittance matrices are found as

$$Z_0 = P^{1/2}Y^{-1}, \quad Y_0 = YP^{-1/2} \qquad (1.120)$$

This equation produces the same matrices as in Equation 1.113. For example, for the characteristic admittance matrix:

$$Y_0 = ((YP^{-1/2})^{-1})^{-1} = (P^{1/2}Y^{-1})^{-1} = (P^{-1/2}PY^{-1})^{-1} = (P^{-1/2}(ZY)Y^{-1})^{-1}$$
$$= \left(P^{-1/2}Z\right)^{-1} = Z^{-1}P^{1/2} \qquad (1.121)$$

The characteristic impedance and admittance matrices are symmetrical matrices. For example, for the characteristic impedance matrix:

$$Z_{0t} = (P^{1/2}Y^{-1})_t = Y_t^{-1}P_t^{1/2} = Y^{-1}P_t^{1/2} = Z_0 \qquad (1.122)$$

Here, $Y = Y_t$ since Y is a symmetrical matrix.
Therefore,

$$Z_{0t} = Z_0, Y_{0t} = Y_0 \qquad (1.123)$$

P is not a symmetrical matrix in general, but Z_0 and Y_0 are always symmetrical matrices.

1.4.2 Modal Theory

Modal theory, which was established by L. M. Wedepohl in 1963 [17], provides an essential technique to solve for voltages and currents in a multiconductor system. Without modal theory, propagation constants and characteristic impedances of a multiconductor system cannot be precisely found, except for an ideally transposed line. One may assume an ideally transposed line or a perfectly conducting earth and find solutions of voltages and currents in a multiconductor system using symmetrical coordinate transformation [18,19].

However, it does not produce precise solutions of voltages and currents since an ideally transposed line and a perfectly conducting earth do not exist in an actual system. Before modal theory was established, propagation constants and characteristic impedances were found by expanding matrix functions to a series of polynomials.

This section discusses propagation constants and characteristic impedance and admittance matrices in the modal domain after reviewing modal theory.

1.4.2.1 Eigenvalue Theory

Let us define matrix P as a product of series impedance matrix Z and shunt admittance matrix Y for a multiconductor system:

$$[P] = [Z][Y] \tag{1.124}$$

where $[Z]$ and $[Y]$ are $n \times n$ off-diagonal matrices.

Applying eigenvalue theory, the off-diagonal matrix P can be diagonalized by the following matrix operation:

$$[A]^{-1}[P][A] = [Q] = [U](Q), \quad [A][Q][A]^{-1} = [P], \tag{1.125}$$

where
  $[Q]$ is the $n \times n$ eigenvalue matrix of $[P]$
  $[A]$ is the $n \times n$ eigenvector matrix of $[P]$
  (Q) is the eigenvalue vector
  $[U]$ is the identity (unit) matrix

The notations matrix [ ] and vector ( ) are hereafter omitted for simplification. Rewriting Equation 1.125 gives

$$PA = AQ,$$
$$\therefore PA - AQ = 0 \tag{1.126}$$

Since Q is the diagonal matrix, only the kth column of A is multiplied by the kth diagonal entry of Q when calculating AQ. Therefore, the following equation is satisfied for each k:

$$A_k Q_k = QkAk; \quad k = 1, 2, \ldots, n \tag{1.127}$$

The following equation is obtained for the kth column by substituting Equation 1.127 into Equation 1.126:

$$(P - Q_k U)A_k = 0 \tag{1.128}$$

This equation is a set of n equations with n unknowns. The determinant of $(P - Q_k U)$ must be zero in order to obtain the solution $A_k \neq 0$:

$$\det(P - Q_k U) = 0 \tag{1.129}$$

Equation 1.129 is an nth-order polynomial with unknown Q_k and is called a characteristic equation. Eigenvalues of P (i.e., Q_k) can be found as the solution of the characteristic equation.

Eigenvector A_k is found from Equation 1.126 for each eigenvalue of P. Since the determinant of $(P - Q_k U)$ is zero for the obtained Q_k, eigenvector A_k is not uniquely determined. Thus, one element of A_k can take an arbitrary value, and the other elements are determined according to it, satisfying the proportional relationship that eigenvectors A_k have to be linearly independent of each other. This is especially important when some eigenvalues of P are equal, that is, when the characteristic equation has repeated roots.

As discussed in Section 1.4.1, analysis of a multiconductor system requires a number of computations of functions. The application of eigenvalue theory makes it easy to calculate matrix functions. This is a major advantage of eigenvalue theory.

One way to calculate matrix functions without eigenvalue theory is to use series expansions. The following series expansions are often used to calculate matrix functions:

$$\sqrt{1+x} \approx 1 + \frac{x}{2}, \quad \sinh(x) \approx x + \frac{x^3}{6}, \quad \cosh(x) \approx 1 + \frac{x^2}{2}$$

$$\tanh(x) \approx 1 - \frac{x^3}{2}; \quad x \ll 1 \tag{1.130}$$

$$\exp(x) = 1 + x + \frac{x^2}{2!} + \cdots + \frac{x^n}{n!} + \cdots, \quad |x| < \infty$$

Using this equation, the exponential function of matrix P is found as

$$\exp([P]) \approx [U] + [P] + \frac{[P]^2}{2!}; \quad |P| \ll 1 \tag{1.131}$$

Using eigenvalue theory, a matrix function is given by

$$f([P]) = [A]f([Q])[A]^{-1} \tag{1.132}$$

where Q and A are the eigenvalue matrix and the eigenvector matrix of P, respectively.

For example, $[P]^{1/2}$ can be calculated simply with

$$[P]^{1/2} = [A][Q]^{1/2}[A]^{-1} \tag{1.133}$$

where

$$[Q]^{1/2} = \begin{bmatrix} Q_1^{1/2} & 0 & \cdots & 0 \\ 0 & Q_2^{1/2} & \cdots & 0 \\ \vdots & \vdots & & \vdots \\ 0 & 0 & \cdots & Q_n^{1/2} \end{bmatrix} = \begin{bmatrix} \sqrt{Q_1} & 0 & \cdots & 0 \\ 0 & \sqrt{Q_2} & \cdots & 0 \\ \vdots & \vdots & & \vdots \\ 0 & 0 & \cdots & \sqrt{Q_n} \end{bmatrix}$$

The exponential function exp $([P])$ can be calculated as

$$\exp([P]) = [A]\exp([Q])[A]^{-1} \tag{1.134}$$

where $\exp([Q]) = [U]\exp(Q)$; $\exp(Q) = (\exp Q_1, \exp Q_2,..., \exp Q_n)_t$.

Assuming eigenvalue matrix Q, eigenvector matrix A, and its inverse A^{-1} are found, then the propagation constant matrix can be calculated as in Equation 1.133:

$$\Gamma = P^{1/2} = AQ^{1/2}A^{-1} = A \cdot \gamma \cdot A^{-1} \tag{1.135}$$

where

Γ is the actual propagation constant matrix (off-diagonal)

$\gamma = \alpha + j\beta$ is the modal propagation constant matrix (diagonal)

Here, α is the modal attenuation constant and β is the modal phase constant. In Equation 1.135:

$$[\gamma] = [U](\gamma) = [U](Q)^{1/2} = [Q]^{1/2}$$

and

$$\gamma_k = Q_k^{1/2} = \sqrt{Q_k}; \quad k = 1, 2,..., n \tag{1.136}$$

The exponential function of the propagation constant matrix is found from Equation 1.134:

$$\exp(-\Gamma x) = A\exp(-\gamma x)A^{-1} \tag{1.137}$$

As a result, the voltage in a semi-infinite line given by Equation 1.112 can be calculated as

$$V = A \exp(-\gamma x) A^{-1} V_f \qquad (1.138)$$

Note that the computation of Equation 1.112 is made possible only by using eigenvalue theory, as in Equation 1.138.

In this section, we have discussed the method that directly applies eigenvalue theory. However, it is not efficient in terms of numerical computations as it requires the product of off-diagonal matrices. The method will be more complete with modal theory.

1.4.2.2 Modal Theory

Equation 1.138 is rewritten as:

$$A^{-1} V = \exp(-\gamma x) A^{-1} V_f \qquad (1.139)$$

Mode voltage (voltage in a modal domain) and modal forward traveling wave (forward traveling wave in a modal domain) are defined as follows:

$$v = A^{-1} V, \quad v_f = A^{-1} V_f \qquad (1.140)$$

where lowercase letters are modal components (components in a modal domain) and uppercase letters are actual or phasor components (components in an actual or phasor domain).

Using modal components, Equation 1.139 can be expressed as

$$v = \exp(-\gamma x) v_f \qquad (1.141)$$

In this equation, all components are expressed in a modal domain, including voltage vectors. Note that this equation in a modal domain takes the same form as Equation 1.112 in an actual domain. Similarly, relationships in an actual domain, for example, Ohm's law, are satisfied in a modal domain.

Using these relationships, the solutions in a modal domain are first derived, which can then be transformed to the solutions in an actual domain. For example, once the solution of Equation 1.141, that is, v, is found, its solution in the actual domain is found as

$$V = Av \qquad (1.142)$$

Applying modal theory, the solutions are derived as explained in Section 1.3. With modal theory, since the coefficient matrix in Equation 1.141 is a diagonal matrix, the equation is also written as

$$v_k = \exp(-\gamma_k x)v_{fk}; \quad k = 1, 2, \ldots, n \tag{1.143}$$

This equation shows that each mode is independent of the other modes; therefore, a multiconductor system can be treated as a single-conductor system in a modal domain. The solutions in a modal domain can be found by n operations, whereas solving Equation 1.138 in an actual domain requires time complexity of $o(n^2)$ since the coefficient matrix is an $n \times n$ matrix. Matrix A is called the voltage transformation matrix as it transforms the voltage in a modal domain to that in an actual domain.

1.4.2.3 Current Mode

In Section 1.4.2.2, we discussed voltage in a modal domain; in this section, we discuss current in a modal domain. We first need to find the eigenvalues of $P_t = YZ$ as the second equation of Equation 1.115 indicates. Since $P_t \neq P$ in general, we define Q' as the eigenvalue matrix of P_t and B as the eigenvector matrix of P_t:

$$P_t = BQ'B^{-1}, \quad Q' = B^{-1}P_tB \tag{1.144}$$

Since a matrix returns to its original form when it is transposed twice, then

$$\det(P - Q_kU) = \det(P_t - Q'_kU)_t = \det[(P_t)_t - (Q'_kU)_t] \tag{1.145}$$

Considering $(P_t)_t = P$ and $(Q'_kU)_t = Q'_kU$

$$\det(P - Q_kU) = \det(P - Q'_kU)$$
$$\therefore Q_k = Q'_k \tag{1.146}$$

This equation shows that the eigenvalues for voltage are equal to those for current. Since $\gamma = \sqrt{Q}$, propagation constants for voltage are also equal to those for current. These are important characteristics to consider when analyzing a multiconductor system, and they correspond to TEM mode propagation.

However, the current transformation matrix B is not equal to the voltage transformation matrix A. Transposing the first equation of Equation 1.144 gives

$$P = B_t^{-1}Q'B_t = B_t^{-1}QB_t \tag{1.147}$$

Also, from Equation 1.125:

$$P = AQA^{-1} = AD^{-1}DQA^{-1} = AD^{-1}QDA^{-1} \tag{1.148}$$

where D is an arbitrary diagonal matrix.

Comparing Equations 1.147 and 1.148 gives

$$B_t^{-1} = AD^{-1}, \quad B_t = DA^{-1} \tag{1.149}$$

This shows that the current transformation matrix can be found from the voltage transformation matrix. In general, D is assumed to be an identity matrix. Under this assumption:

$$B = (A^{-1})_t, \quad B^{-1} = A_t \tag{1.150}$$

1.4.2.4 Parameters in Modal Domain

By applying modal transformation, differential equations in a multiconductor are given as:

$$\left.\begin{aligned}
\frac{dV}{dx} &= \frac{d(Av)}{dx} = A\frac{dv}{dx} = -ZI = -ZBi \\
\frac{dI}{dx} &= \frac{d(Bi)}{dx} = B\frac{di}{dx} = -YV = -YAv
\end{aligned}\right\} \tag{1.151}$$

Modifying this set of equations gives

$$\frac{dv}{dx} = -zi, \quad \frac{di}{dx} = -yv \tag{1.152}$$

where

$$\left.\begin{aligned}
z &= A^{-1}ZB\text{: modal impedance} \\
y &= B^{-1}YA\text{: modal admittance}
\end{aligned}\right\} \tag{1.153}$$

or $Z = AzB^{-1}$, $Y = B \cdot yA^{-1}$.

Equation 1.152 in a modal domain takes the same form as that in a phase domain. In a modal domain, the impedance and admittance are defined by Equation 1.153.

From Equation 1.152:

$$\frac{d^2v}{dx^2} = zyv, \quad \frac{d^2i}{dx^2} = yzi \tag{1.154}$$

From previous discussions, we already know that

$$zy = yz = Q = \gamma^2, \quad \gamma = (zy)^{1/2}, \quad \gamma_k = \sqrt{z_k y_k} \tag{1.155}$$

In order for the product of two matrices to be a diagonal matrix, the two matrices have to be diagonal matrices. Since Q is a diagonal matrix, z and y are diagonal matrices [17].

For a semi-infinite line, the following equation is satisfied:

$$V = Z_0 I \tag{1.156}$$

Applying modal transformation to this equation gives

$$Av = Z_0 Bi \quad \therefore v = A^{-1} Z_0 Bi = z_0 i \tag{1.157}$$

The characteristic impedance and admittance in a modal domain are defined as follows:

$$\left. \begin{array}{l} z_0 = A^{-1} Z_0 B : \text{modal characteristic impedance} \\ y_0 = B^{-1} Y_0 A : \text{modal characteristic admittance} \end{array} \right\} \tag{1.158}$$

Rewriting these equations, the actual characteristic impedance and admittance (in phase domain) are given by

$$Z_0 = A z_0 B^{-1}, \quad Y_0 = B \cdot y_0 \cdot A^{-1} \tag{1.159}$$

From Equations 1.114 and 1.158:

$$z_0 = A^{-1} P^{-1/2} ZB$$

Using the relationships in Equations 1.135 and 1.153:

$$z_0 = A^{-1}(AQ^{-1/2}A^{-1})(AzB^{-1})B = Q^{1/2} z = \gamma^{-1} z = y^{-1}\gamma \tag{1.160}$$

This equation shows that z_0 is a diagonal matrix since γ and z are diagonal matrices. In the same way, it can be shown that y_0 is a diagonal matrix.

Equation 1.160 also shows that z_0 can be found from γ and z. Substituting Equation 1.155 into Equation 1.160 gives;

$$z_0 = \gamma^{-1} z = (zy)^{-1/2} z = y^{-1/2} z^{1/2} = (y^{-1/2} z)^{1/2} = (zy^{-1})^{1/2} \tag{1.161}$$

Therefore, the modal characteristic impedance and admittance are also found by

$$z_{0k} = \sqrt{\frac{z_k}{y_k}}, \quad y_{0k} = \frac{1}{z_{0k}} = \sqrt{\frac{y_k}{z_k}} \qquad (1.162)$$

1.4.3 Two-Port Circuit Theory and Boundary Conditions

The unknown coefficients V_f and V_b in the general solution expressed as Equation 1.110 are determined from boundary conditions. There are many approaches to obtain voltage and current solutions in a multiconductor system. The most well-known method is the four-terminal parameter (F-parameter) method of two-port circuit theory. The impedance parameter (Z-parameter) and the admittance parameter (Y-parameter) methods are also well known. It should be noted that the F-parameter method is not suitable for application in high-frequency regions, while the Z- and Y-parameter methods are not suitable in low-frequency regions because of the nature of hyperbolic functions.

1.4.3.1 Four-Terminal Parameter

The F-parameter of a two-port circuit, illustrated in Figure 1.18, is expressed in the following form:

$$\begin{pmatrix} V_s \\ I_s \end{pmatrix} = \begin{pmatrix} F_1 & F_2 \\ F_3 & F_4 \end{pmatrix} \begin{pmatrix} V_r \\ I_r \end{pmatrix} \qquad (1.163)$$

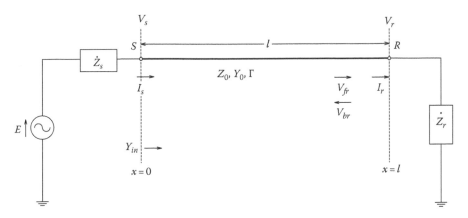

FIGURE 1.18
An impedance-terminated multiconductor system.

where

V_s and V_r are the voltage vectors at the sending and receiving ends in a multiconductor system

I_s and I_r are the current vectors at the sending and receiving ends

The coefficients F_1–F_4 in a multiconductor system are obtained in the same manner as in Equation 1.87, considering a matrix form from Equations 1.110 and 1.111:

$$F_1 = \cosh(\Gamma l), \quad F_2 = \sinh(\Gamma l) \cdot Z_0 \tag{1.164}$$

$$F_3 = Y_0 \sinh(\Gamma l), \quad F_4 = Y_0 \cosh(\Gamma l) \cdot Z_0$$

In this equation, Γ, Z_0, Y_0 are an $n \times n$ matrix for an n-conductor system.
It should be noted that the order of the products in this equation cannot be changed as has been done for a single conductor. That is

$$F_2 = Z_0 \cdot \sinh(\Gamma l), \quad F_4 = \cosh(\Gamma l) = F_1 \quad \text{only for a single conductor} \tag{1.165}$$

Equation 1.163 cannot be solved directly from a given boundary condition unless the coefficients in Equation 1.164 are calculated. By applying the modal transformation explained in Section 1.4.2, Equation 1.163 is rewritten as

$$A^{-1}V_s = v_s = A^{-1}F_1 A \cdot A^{-1}V_r + A^{-1}F_2 B \cdot B^{-1}I_r = f_1 v_r + f_2 i_r$$
$$B^{-1}I_s = i_s = f_3 v_r + f_4 i_r$$

Matrix form is;

$$\begin{pmatrix} v_s \\ i_s \end{pmatrix} = \begin{pmatrix} f_1 & f_2 \\ f_3 & f_4 \end{pmatrix} \begin{pmatrix} v_r \\ i_r \end{pmatrix} \tag{1.166}$$

In this equation, modal F-parameters are given by

$$f_2 = \sinh(\gamma l) \cdot z_0 = z_0 \sinh(\gamma l)$$
$$f_3 = y_0 \sinh(\gamma l) = \sinh(\gamma l) \cdot y_0$$
$$f_4 = y_0 \cosh(\gamma l) \cdot z_0 = y_0 z_0 \cosh(\gamma l) = \cosh(\gamma l) = f_1 \tag{1.167}$$

where z_0, y_0, and γ are defined in Section 1.4.2.4.

These modal parameters are easily obtained because every matrix, γ, z_0, and $y_0 = 1/z_0$, is a diagonal matrix. Then, the parameters in an actual phase domain are evaluated by

$$F_1 = Af_1A^{-1}, \quad F_2 = Af_2B^{-1}, \quad F_3 = Bf_3A^{-1}, \quad F_4 = Bf_4B^{-1} \tag{1.168}$$

It should be clear from these equations that F_1 is in the dimension of a voltage propagation constant, F_4 in the dimension of a current propagation constant, F_2 in the impedance dimension, and F_3 in the admittance dimension.

From Equations 1.167 and 1.168, the following relation is obtained:

$$\begin{aligned} F_2 &= A \cdot z_0 \sinh(\gamma l)B^{-1} = Az_0B^{-1}B\sinh(\gamma l)B^{-1} \\ &= Z_0\{A_t^{-1} \cdot \sinh(\gamma l)_t A_t\} = Z_{0t}\{A\sinh(\gamma l)A^{-1}\}_t \\ &= Z_{0t}\sinh(\Gamma l)_t = \{\sinh(\Gamma l) \cdot Z_0\}_t \end{aligned}$$

In comparison with Equation 1.164:

$$F_2 = F_{2t} \tag{1.169}$$

This relation means that F_2 is a symmetrical matrix. Similarly, F_3 is symmetrical and F_1 and F_4 have the following relations:

$$F_4 = F_{1t}, \quad F_1 \neq F_4, \quad f = f_4 \tag{1.170}$$

Note that F_1 and F_4 are not the same in a multiconductor system, but they are the same in the case of a single-conductor system.

1.4.3.2 Impedance/Admittance Parameters

The F-parameter formulation in Equation 1.163 is rewritten considering matrix algebra in the following forms:

$$V_s = \coth(\Gamma l)Z_0I_s - \mathrm{cosech}(\Gamma l)Z_0I_r \tag{1.171}$$

$$V_r = \mathrm{cosech}(\Gamma l)Z_0I_s - \coth(\Gamma l)Z_0I_r$$

Until now, current has been positive when it flows in the positive direction of x. It is more comprehensible to set the positive direction of the current to the direction of inflow (injection) to the finite line as shown in Figure 1.19.

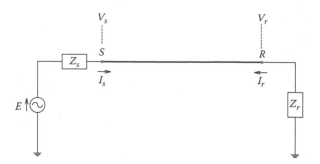

FIGURE 1.19
A multiconductor system for Z- and Y-parameters.

Since the positive direction of current has changed at the receiving end, I_r in Equation 1.171 has to be changed to $-I_r$:

$$\left.\begin{array}{l} V_s = \coth(\Gamma l)Z_0 I_s - \operatorname{cosech}(\Gamma l)Z_0 I_r \\ V_r = \operatorname{cosech}(\Gamma l)Z_0 I_s - \coth(\Gamma l)Z_0 I_r \end{array}\right\} \tag{1.172}$$

Matrix form is

$$\begin{bmatrix} V_s \\ V_r \end{bmatrix} = \begin{bmatrix} Z_{11} & Z_{12} \\ Z_{12} & Z_{11} \end{bmatrix} \begin{bmatrix} I_s \\ I_r \end{bmatrix} = \begin{bmatrix} \coth(\Gamma l)Z_0 & \operatorname{cosech}(\Gamma l)Z_0 \\ \operatorname{cosech}(\Gamma l)Z_0 & \coth(\Gamma l)Z_0 \end{bmatrix} \begin{bmatrix} I_s \\ I_r \end{bmatrix} \tag{1.173}$$

Here, Z_{ij} $(i, j = 1, 2)$ are Z-parameters.
Taking the inverse of the matrix gives

$$\begin{bmatrix} I_s \\ I_r \end{bmatrix} = \begin{bmatrix} Y_{11} & -Y_{12} \\ -Y_{12} & Y_{11} \end{bmatrix} \begin{bmatrix} V_s \\ V_r \end{bmatrix} = \begin{bmatrix} Y_0 \coth(\Gamma l) & -Y_0 \operatorname{cosech}(\Gamma l) \\ -Y_0 \operatorname{cosech}(\Gamma l) & Y_0 \coth(\Gamma l) \end{bmatrix} \begin{bmatrix} V_s \\ V_r \end{bmatrix} \tag{1.174}$$

Here, Y_{ij} $(i, j = 1, 2)$ are Y-parameters. Y-parameters are used more often than Z-parameters since a voltage source is typically given as a boundary condition.

Given the voltage source E in Figure 1.19, the voltage and current at the sending and receiving ends are found from Equation 1.174 and the boundary conditions:

$$\left.\begin{array}{l} V_s = \left\{ Y_s + Y_{11} - Y_{12}(Y_{11} + Y_r)^{-1}Y_{12} \right\}^{-1} Y_s E \\ V_r = (Y_{11} + Y_r)^{-1}Y_{12}V_s \\ I_s = Y_s(E - V_s) \\ I_r = -Y_r V_r \end{array}\right\} \tag{1.175}$$

where
$$Y_s = Z_s^{-1}$$
$$Y_r = Z_r^{-1}$$

The Z- and Y-parameter methods are stable for $\theta \to \infty$ since it is based on the convergence functions $\coth(\theta)$ and $\operatorname{cosech}(\theta)$. Thus, the method is suitable for transient analysis. However, it should not be used for analyzing low-frequency phenomena since $\operatorname{cosech}(\theta)$ becomes infinite for $\theta \to 0$, that is, $\omega \to 0$.

1.4.4 Modal Distribution of Multiphase Voltages and Currents

1.4.4.1 Transformation Matrix

When a three-phase transmission line is completely transposed or the impedance and admittance matrices are completely symmetrical, the following transformation matrices are widely used for both voltages and currents:

1. Fortescue's transformation [18]:

$$[A_F] = \begin{vmatrix} 1 & 1 & 1 \\ 1 & a^2 & a \\ 1 & a & a^2 \end{vmatrix}, \quad [A_F]^{-1} = \frac{1}{3} \cdot \begin{vmatrix} 1 & 1 & 1 \\ 1 & a & a^2 \\ 1 & a^2 & a \end{vmatrix},$$

$$a = \exp\left(\frac{j2\pi}{3}\right)$$

(1.176)

2. Clarke's transformation [19]:

$$[A_C] = \begin{bmatrix} 1 & 1 & 0 \\ 1 & -\dfrac{1}{2} & \sqrt{\dfrac{3}{2}} \\ 1 & -\dfrac{1}{2} & \sqrt{\dfrac{3}{3}} \end{bmatrix}, \quad [A_C]^{-1} = \frac{1}{3} \cdot \begin{vmatrix} 1 & 1 & 1 \\ 2 & -1 & -1 \\ 0 & \sqrt{3} & \sqrt{3} \end{vmatrix}$$

(1.177)

3. Karrenbauer's transformation [11]:

$$[A_k] = \begin{vmatrix} 1 & 1 & 1 \\ 1 & -2 & 1 \\ 1 & 1 & -2 \end{vmatrix}, \quad [A_k]^{-1} = \frac{1}{3} \cdot \begin{vmatrix} 1 & 1 & 1 \\ 1 & -1 & 0 \\ 1 & 0 & -1 \end{vmatrix}$$

(1.178)

4. Traveling-wave transformation [1]:

$$[A_T] = \begin{vmatrix} 1 & 1 & 1 \\ 1 & 0 & -2 \\ 1 & -1 & 1 \end{vmatrix}, \quad [A_T]^{-1} = \frac{1}{6} \cdot \begin{vmatrix} 2 & 2 & 2 \\ 3 & 0 & -3 \\ 1 & -2 & 1 \end{vmatrix} \quad (1.179)$$

Fortescue's transformation is well known in conjunction with symmetrical component theory [18]. Although it involves complex numbers, it has the advantage of generating only one nonzero modal voltage (positive-sequence voltage) if the source voltage is a three-phase symmetrical AC source. Clarke's transformation is also related to symmetrical component theory and is known as the $\alpha - \beta - 0$ transformation [19]. It involves only real numbers but generates positive- and negative-sequence voltages. Karrenbauer's transformation is adopted in the famous EMTP [11] and is easily extended to an n phase completely transposed line. This is also true for Fortescue's transformation. However, there are no transposed lines with a phase number greater than three. The traveling-wave transformation is often used when analyzing traveling waves on a three-phase line [1]. Its advantage is that the transformation can deal with not only a completely transposed line, but also an untransposed horizontal line.

1.4.4.2 Modal Distribution

Let us discuss modal current (voltage) distribution on a completely transposed line. It should be noted that modal voltage distribution is the same as that of current in the completely transposed line case because the impedance and admittance matrices are completely symmetrical. Assume that the phase currents are I_a, I_b, and I_c as illustrated in Figure 1.20. Using traveling-wave transformation, we obtain the following relation:

$$\begin{pmatrix} i_0 \\ i_1 \\ i_2 \end{pmatrix} = \frac{1}{6} \begin{bmatrix} 2 & 2 & 2 \\ 3 & 0 & -3 \\ 1 & -2 & 1 \end{bmatrix} \cdot \begin{pmatrix} i_a \\ i_b \\ i_c \end{pmatrix} = \frac{1}{6} \begin{bmatrix} 2I_a + 2I_b + 2I_c \\ 3I_a + 3I_b \\ I_a - 2I_b + I_c \end{bmatrix} \quad (1.180)$$

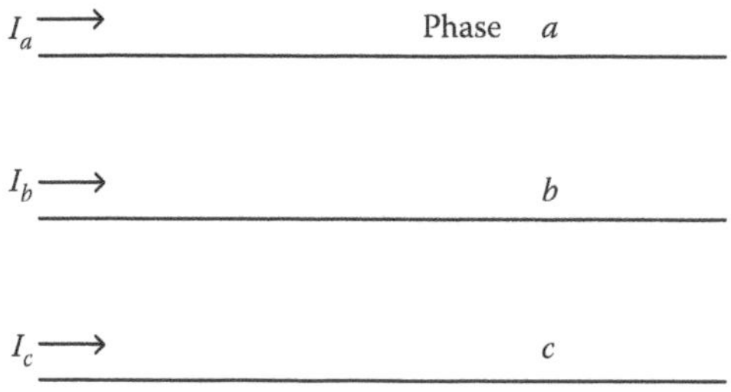

FIGURE 1.20
Actual phase current.

Assuming $I_a = I_b = I_c = I$ for simplicity, the following characteristics of the modal current are observed:

1. *Mode 0 (earth-return mode):* The current of $I/3$ flowing in the positive direction on each phase and the return current I have to flow back through the earth. Because the return current flows through the earth, the mode 0 component is called the earth-return component. A circuit corresponding to mode 0 can be drawn as shown in Figure 1.21a. The mode 0 component involves the earth-return path with an impedance that is far greater than the conductor internal imped- ance as explained in Section 1.5.1; the mode 0 propagation constant is much greater, that is, the mode 0 attenuation is much greater; and the mode 0 propagation velocity is smaller than that of the other modes.

2. *Mode 1 (first aerial mode):* The current of $I/2$ flows through phase a returning through phase c, with no current on phase b. Thus, the mode 1 circuit is composed of phases a and c as shown in Figure 1.21b. Because the mode involves no earth-return path in ideal cases, the mode is called the "aerial mode."

3. *Mode 2 (second aerial mode):* The current of $I/6$ flows through phases a and c, and the return current of $I/3$ flows back through phase b as illustrated in Figure 1.21c. The propagation characteristics of

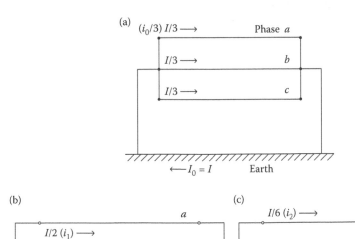

FIGURE 1.21
Modal current distribution for $I_a = I_b = I_c = I$. (a) Mode 0, (b) mode 1, and (c) mode 2.

mode 2 are identical to those of mode 1 in the case of a completely transposed line. If the line is untransposed, the characteristics are different.

The modal distribution can also be explained by applying the transformation matrix A rather than its inverse A^{-1}:

$$
\begin{pmatrix} i_a \\ i_b \\ i_c \end{pmatrix} = \begin{bmatrix} 1 & 1 & 1 \\ 1 & 0 & -2 \\ 1 & -1 & 1 \end{bmatrix} \cdot \begin{pmatrix} i_0 \\ i_1 \\ i_2 \end{pmatrix} = \begin{pmatrix} 1 \\ 1 \\ 1 \end{pmatrix} \cdot i_0 + \begin{pmatrix} 1 \\ 0 \\ 1 \end{pmatrix} \cdot i_1 \begin{pmatrix} 1 \\ -2 \\ 1 \end{pmatrix} \cdot i_2 \qquad (1.181)
$$

For example, the mode 0 current has a distribution equal to the actual current on each phase from Equation 1.181.

This same explanation can be given by applying the transformation matrices in Equations 1.176 through 1.178 in the case of a completely transposed line. When a line is untransposed, the transformation matrices are no longer useful — except Equation 1.179, which can be used as an approximation of a transformation matrix of an untransposed horizontal line. In the case of an untransposed line, the transformation matrix is frequency-dependent as explained in Section 1.5.1; thus, the modal voltage and current distributions vary as the frequency changes. Also, the current distribution differs from the voltage distribution.

PROBLEMS

1.12 Obtain a condition of reciprocity in Equation 1.163.

1.13 Obtain the eigenvalues and eigenvectors of the following matrices:

(a) $\begin{bmatrix} P_{11} & P_{12} & P_{13} \\ P_{21} & P_{22} & P_{21} \\ P_{13} & P_{12} & P_{11} \end{bmatrix}$ (b) $\begin{bmatrix} 35 & 8 & 5 \\ 8 & 32 & 8 \\ 5 & 8 & 35 \end{bmatrix}$ (c) $\begin{bmatrix} 35 & 5\sqrt{3} & 5 \\ 5\sqrt{3} & 35 & 5\sqrt{3} \\ 5 & 5\sqrt{3} & 35 \end{bmatrix}$

1.14 Explain why the modal propagation constants and the modal characteristic impedances for modes 1 and 2 (aerial modes) on a transposed three-phase line become identical.

1.15 Discuss how to obtain the inverse matrix of A when the transformation matrix A is singular.

Remember that a numerical calculation on a computer can give its inverse matrix.

1.5 Frequency-Dependent Effect

It is well known that current is distributed near a conductor's surface when its frequency is high. Under such a condition, the resistance (impedance) of the conductor becomes higher than that at a low frequency because the resistance is proportional to the cross section of the conductor. This is called frequency dependence of the conductor impedance. As a result, the propagation constant and the characteristic impedance are also frequency dependent.

1.5.1 Frequency Dependence of Impedance

Figure 1.22 illustrates a 500 kV horizontal transmission line, and Table 1.1 shows the frequency dependence of its impedance. It should be noted that Figure 1.22 is only for one of the towers in the 500 kV line with length about 83 km. The geometrical configuration of one tower differs from other ones because of the geographical features along the line. For example, separation distance y varies from 22 m to 25 m, although only 25 m is specified in Figure 1.22. It is observed that the resistance increases nearly proportional to $\sqrt{f}$, where f is the frequency. On the contrary, the inductance decreases as f increases. This phenomenon can be explained analytically based on the approximate impedance formulas in Equations 1.7 and 1.15:

1. For a low frequency: $f \ll f_c$ (f_c: critical frequency, which will be defined later):

$$Z = R + j\omega L, \quad R = R_{dc} = \frac{\rho}{S}, \quad L = \frac{\mu_0}{8\pi} \tag{1.182}$$

2. For a high frequency: $f \gg f_c$:

$$Z = (1+j)R, \quad R = \frac{\sqrt{\omega\mu_0\rho}}{2\sqrt{2\pi}r} \propto \sqrt{\omega}, \quad L = \frac{R}{\omega} \propto \frac{1}{\sqrt{\omega}} \tag{1.183}$$

where $\omega_c \mu_c S/R_{dc} \cdot l^2 = 1$.

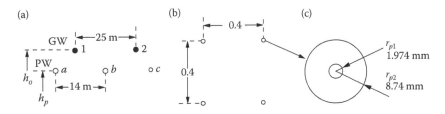

FIGURE 1.22
A 500 kV untransposed horizontal line. (a) Line configuration. (b) Phase wire (4 bundles). (c) Bundle conductor. $r_g = 6.18$ mm, $\rho_g = 5.36 \times 10^{-8}$ Ω m, $\rho_p = 3.78 \times 10^{-8}$ Ω m, $\rho_e = 200$ Ω m.

Thus,

$$\omega_c = 2\pi f_c = \frac{4\rho}{\mu_0 r^2} \tag{1.184}$$

For $\rho \doteqdot 2 \times 10^{-8}$ (Ω m), $\mu_0 = 4\pi \times 10^{-7}$ (H/m): $\omega_c = 0.2/\pi r^2$ (rad/s):

$$f_c = \frac{\omega_c}{2\pi} \doteqdot \frac{1}{100 r^2} \text{(Hz)}$$

For example, with $r = 0.5$ cm:

$$f_c = \frac{10^6}{25 \times 100} = 400\,\text{Hz} \tag{1.185}$$

Considering this equation, the frequency characteristics of R and L may be drawn as in Figure 1.23.

Similar to the conductor internal impedance explained earlier, the earth-return impedance in Equation 1.15 is frequency dependent as the penetration depth h_e is frequency dependent. Equation 1.15 is approximated considering $\ln(1 + x)$ for a small x by

$$Z_e \approx R_e + j\omega(L_e + L_0) = j\omega L_0 + (1 + j)R_e \tag{1.186}$$

$$R_e = \frac{\sqrt{(\omega \mu_0 \rho_e)}}{2\sqrt{2\pi h}} \propto \sqrt{\omega}, \quad L_e = \frac{R_e}{\omega} \propto \frac{1}{\sqrt{\omega}}$$

$$L_0 = \left(\frac{\mu_0}{2\pi}\right) \ln\left(\frac{2h}{r}\right) : \text{space inductance}$$

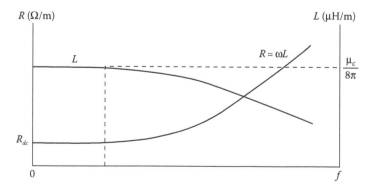

FIGURE 1.23
Frequency dependence of $Z = R + j\omega L$.

1.5.2 Frequency-Dependent Parameters

1.5.2.1 Frequency Dependence

Propagation constant Γ and characteristic impedance Z_0 of a conductor are frequency dependent as they are functions of the impedance of the conductor, as explained in Section 1.2. It should be noted that α and β in Section 1.3.4.1 (see Figure 1.10) are not frequency dependent (in the sense discussed in this section). The frequency dependence of the attenuation constant $\alpha(\omega)$ and phase constant $\beta(\omega)$ in Section 1.3.4.1 comes from the definition of impedance Z and admittance Y of a conductor:

$$Z = R + j\omega L, Y = j\omega C$$

In this section, we discuss frequency dependence, which comes from $R = R(\omega)$ and $L = L(\omega)$ as in Equation 1.183.

Figure 1.24 shows an example of the frequency dependence of attenuation constant α and propagation velocity c for the earth-return mode and the self-characteristic impedance Z_0 for a phase of a 500 kV overhead transmission line.

It is observed that α increases exponentially as frequency increases. Since a dominant factor of determining the attenuation constant is the conductor resistance, α is somehow proportional to $\sqrt{f}$ as explained in Section 1.5.1. The

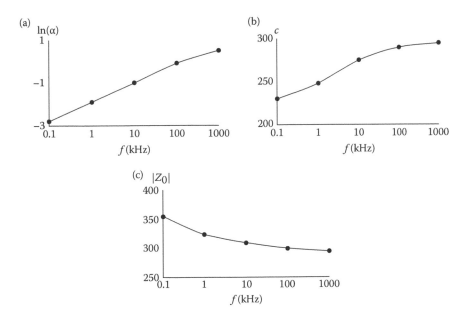

FIGURE 1.24
Frequency dependence of α, c, and Z_0 of a 500 kV line. (a) α (dB/km). (b) c (m/μs). (c) $|Z_0|$1 (Ω).

propagation velocity converges to the light velocity c_0 as frequency increases. On the contrary, the characteristic impedance (absolute value $|Z_0|$) decreases as frequency increases. This is readily explained from Equation 1.183 as

$$Z_0 = \sqrt{\frac{Z}{Y}} \propto \frac{\sqrt{\omega}}{\omega} = \frac{1}{\sqrt{f}} \tag{1.187}$$

In a multiconductor system, the transformation matrix A is also frequency dependent. Frequency dependence is significant in the cases of an untransposed vertical overhead line and an underground cable. In the former, more than 50% difference is observed between A_{ij} (i, jth element of matrix A) at 50 Hz and 1 MHz. In an untransposed horizontal overhead line, the frequency dependence is less noticeable.

Frequency dependence is very significant when an accurate transient simulation on a distributed-parameter line, such as an overhead line and an underground cable, is to be carried out from the viewpoint of insulation design and coordination in a power system. However, a simulation can be carried out neglecting frequency dependence if a safer-side result is required because simulation with frequency dependence, in general, results in a lower overvoltage than that neglecting frequency dependence.

1.5.2.2 Propagation Constant

The frequency-dependent effect is most noticeable in the propagation constant. Table 1.2 shows the frequency dependences of modal attenuations and velocities for untransposed and transposed lines. Figure 1.25 illustrates a vertical twin-circuit line. Figure 1.26 shows the frequency dependence of the modal propagation constant for the vertical twin-circuit line illustrated in Figure 1.25.

Large frequency dependence of the attenuation is clear from the table and figures. The propagation velocity of mode 0 shows significant frequency dependence, while the mode 1 velocity is not that frequency dependent. When a line is transposed, all of the aerial modes become identical. Thus, the number of different characteristic modes is reduced to two in the three-phase line case and to three in the twin-circuit line case. The different velocities of the aerial modes in an untransposed line cause a voltage spike on a transient voltage waveform, which is characteristic of an untransposed line.

1.5.2.3 Characteristic Impedance

The definition of the characteristic impedance of a single-phase line is

$$z_0 = \sqrt{\frac{Z}{Y}} = \sqrt{\frac{(R + j\omega L)}{j\omega C}} = \sqrt{\frac{|Z| \angle \varphi}{\omega C \angle 90°}} = |z_0| \angle \theta \tag{1.188}$$

TABLE 1.2

Frequency Responses of Modal Propagation Constant for the Horizontal Line in Figure 1.25

(a) Untransposed

f (kHz)	Attenuation (dB/km)			Velocity (m/µs)		
	α_0	α_1	α_2	c_0	c_1	c_2
0.1	1.389E–3	9.724E–5	6.314E–5	244.0	295.1	298.6
1	8.166E–3	3.041E–4	1.970E–4	255.3	295.9	299.3
10	7.941E–3	3.500E–3	6.798E–4	270.4	296.4	299.5
100	0.5723	5.446E–2	3.755E–3	285.7	297.6	299.6
1000	2.626	0.3932	3.000E–2	294.8	298.9	299.7

(b) Transposed

f (kHz)	Attenuation (dB/km)		Velocity (m/µs)	
	α_0	$\alpha_1 = \alpha_2$	c_0	$c_1 = c_2$
0.1	1.386E–3	8.191E–5	243.9	295.6
1	8.151E–3	2.558E–4	255.1	296.4
10	7.930E–2	2.204E–3	270.2	296.8
100	0.5723	3.124E–2	285.4	297.4
1000	2.628	0.2277	294.5	298.2

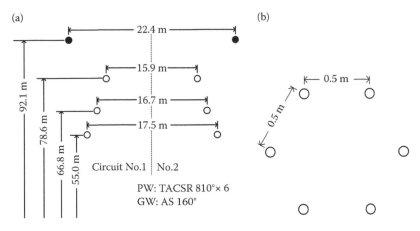

FIGURE 1.25

An untransposed vertical twin-circuit line. (a) Line configuration. (b) Phase wire.

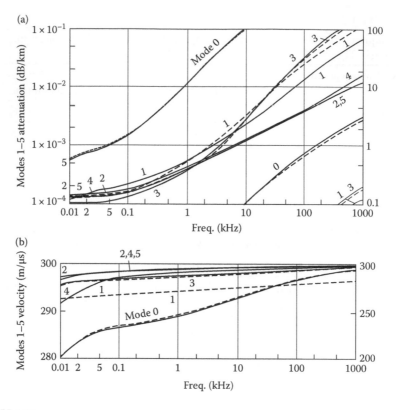

FIGURE 1.26
Frequency responses of modal propagation constants for the untransposed vertical twin-circuit line in Figure 1.25. (a) Attenuation. (b) Velocity.

where

$$|Z| = \sqrt{R^2 + \omega^2 L^2}$$

$$\varphi = \tan^{-1}\left(\frac{\omega L}{R}\right)$$

$$|z_0| = \sqrt{\frac{|Z|}{\omega C}}$$

$$\theta = \frac{(\varphi - 90°)}{2}$$

For R, $\omega L \geq 0$, $0 \leq \varphi \leq 90°$. Thus,

$$-45° \leq \theta \leq 0 \quad \text{or} \quad R_e |z_0| \geq 0, \quad I_m(z_0) \leq 0 \tag{1.189}$$

This fact in Equation 1.189 should be noted.

TABLE 1.3

Frequency Responses of Characteristic Impedances for the Horizontal Line in Figure 1.22

(a) Untransposed

Freq. (kHz)	Z_{aa} (Ω)	Z_{bb} (Ω)	Z_{ab} (Ω)	Z_{ac} (Ω)
0.1	$348 - j11$	$346 - j11$	$87.3 - j10$	$54.3 - j9.7$
1	$340 - j6.4$	$338 - j6.1$	$80.2 - j5.9$	$47.2 - j6.0$
10	$331 - j6.3$	$330 - j5.9$	$71.5 - j5.9$	$38.4 - j5.7$
100	$322 - j4.7$	$322 - j4.4$	$63.5 - j4.3$	$30.9 - j3.8$
1000	$317 - j2.2$	$317 - j2.1$	$59.1 - j2.0$	$27.3 - j1.6$
$\rho_e = 0\,\Omega\,m$	314.6	314.4	56.7	25.5

(b) Transposed

Freq. (kHz)	Z_s (Ω)	Z_m (Ω)
0.1	$347 - j11$	$76.5 - j9.9$
1	$339 - j6.3$	$69.4 - j5.9$
10	$330 - j6.1$	$60.6 - j5.8$
100	$321 - j4.6$	$52.8 - j4.1$
1000	$316 - j2.2$	$48.6 - j1.8$
$\rho_e = 0\,\Omega\,m$	314.5	46.3

$$
\text{(a)} \qquad\qquad\qquad\qquad \text{(b)}
$$

$$
[Z_0] = \begin{bmatrix} Z_{aa} & Z_{ab} & Z_{ac} \\ Z_{ab} & Z_{bb} & Z_{ab} \\ Z_{ab} & Z_{ab} & Z_{aa} \end{bmatrix} \qquad [Z_0] = \begin{bmatrix} Z_s & Z_m & Z_m \\ Z_m & Z_s & Z_m \\ Z_m & Z_m & Z_s \end{bmatrix}
$$

Table 1.3 shows the actual characteristic impedance of the horizontal line in Figure 1.27 for the vertical twin-circuit line illustrated in Figure 1.25. It is clear that the characteristic impedance is significantly frequency dependent. The variation reaches about 10% for self-impedance and about 50% for mutual impedance in the frequency range of 100 Hz to 1 MHz. The characteristic impedance decreases as frequency increases and tends to approach the value as in the case of the perfectly conducting earth and conductor.

In the case of a vertical line, it should be noted that the relation of magnitudes changes as frequency changes. For example, self-impedance is present in the following relations:

$$Z_{cc} > Z_{bb} > Z_{aa} \quad \text{for } f \leq 5\,\text{kHz}$$

$$Z_{bb} > Z_{cc} > Z_{aa} \quad \text{for } 20\,\text{kHz} \leq f \leq 1\,\text{MHz}$$

$$Z_{bb} > Z_{aa} > Z_{cc} \quad \text{for } 1\,\text{MHz} < f < \text{some MHz}$$

$$Z_{aa} > Z_{bb} > Z_{cc} \quad \text{for } f > \text{some MHz}$$

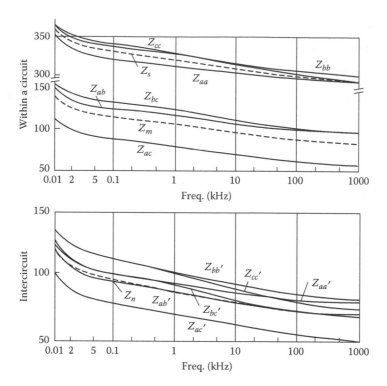

FIGURE 1.27
Frequency responses of characteristic impedance on an untransposed vertical twin-circuit line.

This phenomenon is due to the variation of return current distribution in ground wires (GW) and earth. In a very high-frequency region, if all of the currents are returning through the earth surface and the earth surface potential is becoming zero, then the image theory is applicable. Thus, the magnitude of self-impedance becomes proportional to the height of the conductor, that is, $Z_{aa} > Z_{bb} > Z_{cc}$.

The characteristic impedance of a transposed line becomes a mean value of that of an untransposed line.

1.5.2.4 Transformation Matrix

The importance of the frequency-dependent transformation matrix was not well recognized until recently, although it has been suggested for a long time [20,21].

1.5.2.4.1 Untransposed Horizontal Three-Phase Line

Table 1.4 shows the frequency-dependent transformation matrix of the untransposed horizontal line in Figure 1.22, neglecting GWs. The transformation matrix has the following form:

TABLE 1.4

Frequency Responses of a Transformation Matrix for the
Horizontal Line Shown in Figure 1.22

Freq. (kHz)	A_1	A_2
0.1	$0.996 - j0.002$	$-2.294 + j0.010$
1	$1.001 - j0.006$	$-2.267 - j0.026$
10	$1.015 - j0.012$	$-2.216 - j0.044$
100	$1.037 - j0.014$	$-2.136 - j0.050$
1000	$1.055 - j0.008$	$-2.074 - j0.026$

$$[A] = \begin{bmatrix} 1 & 1 & 1 \\ A_1 & 0 & A_2 \\ 1 & 1 & 1 \end{bmatrix} \tag{1.190}$$

It is observed that the frequency dependence of A_2 in Table 1.4 is less than 10% for the range of frequencies from 100 Hz to 1 MHz. The change is small compared with the parameters explained earlier;, thus, the frequency-dependent effect of the transformation matrix in the case of an untransposed horizontal line can be neglected. Then, the following approximation is convenient because it agrees with the traveling-wave transformation of Equation 1.179, explained in Sections 1.4.4.1 and 1.4.4.2:

$$A_1 \fallingdotseq 1, \quad A_2 \fallingdotseq -2 \tag{1.191}$$

In this case, the modal distribution is the same as that explained in Section 1.4.4.1. The current transformation matrix is given from Equation 1.179 by

$$[B] = [A]_i^{-1} = \frac{1}{6} \cdot \begin{bmatrix} 2 & 3 & 1 \\ 2 & 0 & -2 \\ 2 & -3 & 1 \end{bmatrix}, \quad [B]^{-1} = [A]_i = \begin{bmatrix} 1 & 1 & 1 \\ 1 & 0 & -1 \\ 1 & -2 & 1 \end{bmatrix} \tag{1.192}$$

It is observed from these equations that the modal current distribution is basically the same as the voltage distribution. Therefore, the modal circuit given in Figure 1.21 is also approximately applicable to the untransposed horizontal line.

Table 1.5 shows the transformation matrix of the untransposed horizontal line with GWs. No significant difference from Table 1.4 with the GWs is observed.

TABLE 1.5

Frequency Responses of a Transformation Matrix for a Horizontal Line with No GW

Frequency (kHz)	A_1		A_2	
	Real	**Imag**	**Real**	**Imag**
0.1	0.9934	0.0074	−2.2504	0.0120
1	1.0007	0.0126	−2.2280	0.0206
10	1.0312	0.0198	−2.1885	0.0326
100	1.0643	0.0167	−2.1339	0.0275
1000	1.0823	0.0079	−2.1043	0.0131

1.5.2.4.2 Untransposed Vertical Twin-Circuit Line

The current transformation matrix B is given in the following form in the case of the untransposed vertical twin-circuit line shown in Figure 1.25:

$$(i) = \begin{pmatrix} i_I \\ i_{II} \end{pmatrix} = \begin{bmatrix} [B_I] & [B_I] \\ [B_{II}] & -[B_{II}] \end{bmatrix} \cdot \begin{pmatrix} (I_I) \\ (I_{II}) \end{pmatrix} = [B]^{-1} \cdot (I) \tag{1.193}$$

where
(i) is the modal current
(I) is the actual current

and

$$(i_I) = \begin{pmatrix} i_0 \\ i_1 \\ i_2 \end{pmatrix}, \quad (i_{II}) = \begin{pmatrix} i_3 \\ i_4 \\ i_4 \end{pmatrix}, \quad (I_I) = \begin{pmatrix} I_a \\ I_b \\ I_c \end{pmatrix}, \quad (I_{II}) = \begin{pmatrix} I'_a \\ I'_b \\ I'_c \end{pmatrix}$$

It is clear from this equation that modal currents (I_I), group I, are currents when there is only one circuit, which is called the "internal mode." The currents (I_{II}) are generated due to the existence of the second circuit, which is called the "intercircuit mode." The internal mode has a plane of symmetry at the earth's surface, while the intercircuit mode has a plane of symmetry at the vertical center of the two circuits. Therefore, the polarities of currents on a phase of circuit *I* and the corresponding phase of circuit *II* are the same in the internal mode, and the polarities are opposite even though the amounts of the currents are the same in the intercircuit mode, as presented in Table 1.6. In this way, the transformation matrix given in Equation 1.193 is obtained.

In Table 1.6, the first kind of distribution corresponds to the so-called "zero-sequence mode" and is mode 0 in the same polarity case and mode 3 in the opposite polarity case. If the line is single circuit, there is no opposite polarity mode, and the first kind of distribution is the same as mode 0 distribution,

TABLE 1.6

Frequency Responses of Characteristic Impedances for a Vertical Line

		Kinds of Current Distribution Modes			
		First	Second	Third	Fourth
Internal (aerial) modes	Same polarity	ao+ +oa'	o+ +o	o+ +o	o+ +o
		bo+ +ob'	o+ +o	o− −o	o− −o
		co+ +oc'	o− −o	o+ +o	o− −o
	Mode no.	0	1	2	*
Intercircuit modes	Opposite polarity	o+ −o	o+ −o	o+ −o	o+ −o
		o+ −o	o+ −o	o− +o	o− +o
		o+ −o	o− +o	o+ −o	o− +o
	Mode no.	3	4	5	*

which was explained in Section 1.4.4.1. Mode 0 is often called the "first zero-sequence mode" (earth-return mode), and mode 3 is called the "second zero-sequence mode" (intercircuit zero-sequence mode).

The second, third, and fourth distributions correspond to an aerial mode. The second is the positive-sequence mode and the third is the negative-sequence mode. However, the pattern of current distribution varies as frequency changes; thus, the fourth distribution can become a negative-sequence mode at a certain frequency.

In the single-circuit case, no opposite polarity mode exists, and also the second and fourth distributions are the same. Thus, the number of current distribution patterns, that is, natural modes, is reduced to three.

Table 1.7 presents the frequency responses of the submatrices B_I and B_{II} of the transformation matrix given in Equation 1.193 only for the real part because the imaginary part is much smaller.

Rewriting Table 1.7 in the form of Equation 1.193 for $f = 50$ Hz, the following result is obtained:

$$\begin{pmatrix} \begin{pmatrix} i_0 \\ i_1 \\ i_2 \end{pmatrix} \\ \begin{pmatrix} i_3 \\ i_4 \\ i_5 \end{pmatrix} \end{pmatrix} = \begin{bmatrix} \begin{bmatrix} 0.751 & 0.878 & 1 \\ 1 & 0.158 & -0.570 \\ 0.272 & -1 & 0.361 \end{bmatrix} & \begin{bmatrix} 0.751 & 0.878 & 1 \\ 1 & 0.158 & -0.570 \\ 0.272 & -1 & 0.361 \end{bmatrix} \\ \begin{bmatrix} 0.466 & 0.751 & 1 \\ 1 & 0.589 & -0.782 \\ 0.850 & -1 & 0.289 \end{bmatrix} & \begin{bmatrix} -0.466 & -0.751 & -1 \\ -1 & -0.589 & 0.782 \\ -0.850 & -1 & 0.289 \end{bmatrix} \end{bmatrix} \cdot \begin{pmatrix} \begin{pmatrix} I_a \\ I_b \\ I_c \end{pmatrix} \\ \begin{pmatrix} I'_a \\ I'_b \\ I'_c \end{pmatrix} \end{pmatrix}$$

From this, it is clear that each mode has the following closed circuit:

Mode 0: all the phases to earth

Mode 3: first circuit to second circuit

TABLE 1.7

Frequency Responses of Current Transformation Matrix B^{-1} (Real Part)

(a) $[B_I] = B_I(j, k)$

	Mode 0		Mode 1		Mode 2	
Freq. (kHz)	(1,1)	(1,2)	(2,2)	(2,3)	(3,1)	(3,3)
0.05	0.751	0.878	0.158	−0.570	0.272	0.361
0.1	0.725	0.866	0.215	−0.558	0.325	0.339
1	0.691	0.848	0.364	−0.572	0.501	0.276
5	0.665	0.832	0.410	−0.546	0.542	0.264
10	0.653	0.825	0.427	−0.534	0.561	0.258
50	0.627	0.808	0.466	−0.506	0.601	0.246
100	0.618	0.802	0.488	−0.500	0.621	0.240
1000	0.600	0.790	0.511	−0.478	0.645	0.236

(b) $[B_{II}] = B_{II}(j,k)$

	Mode 3		Mode 4		Mode 5	
Freq. (kHz)	(1,1)	(1,2)	(2,2)	(2,3)	(3,1)	(3,3)
0.05	0.466	0.751	0.589	−0.782	0.850	0.289
0.1	0.454	0.659	0.580	−0.715	0.772	0.225
1	0.471	0.661	0.491	−0.671	0.674	0.256
5	0.462	0.666	0.513	−0.674	0.708	0.260
10	0.453	0.664	0.581	−0.699	0.782	0.236
50	0.425	0.646	0.712	−0.725	0.903	0.196
100	0.414	0.636	0.731	−0.717	0.924	0.189
1000	0.379	0.607	0.733	−0.658	0.900	0.197

Note: $B(1,3) = B(2,1) = −B(3,2) = 1$.

Mode 1: phases a and b to phase c

Mode 2: phases a and c to phase b

Mode 4: phases a, b, and c' to phases a', b', and c

Mode 5: phases a, c, and b' to phases a', c', and b

It is observed from Table 1.7 that the frequency dependence of the transformation matrix in the untransposed vertical twin-circuit case is significantly large. The largest frequency dependence is observed in $B_{II}(2,2) = B(5,2) = −B(5,5)$, and the variation reaches about 50% with reference to the smallest value, that is, the value at $f = 1$ kHz. Also, it should be noted that the value of the intercircuit mode (modes 3–5), that is, the value of the submatrix B_{II}, exhibits an oscillating nature. These may make transient calculations difficult.

1.5.2.5 Line Parameters in the Extreme Case

It has already been proved that line parameters at an infinite frequency are the same as those in the perfectly conducting earth and conductor case. It is quite useful to know the line parameters in such extreme cases because the parameters are a good approximation to the line parameters at a finite frequency with imperfectly conducting earth and conductor.

From what we have studied in Sections 1.1 through 1.4, 1.5.1, and 1.5.2.1 through 1.5.2.4, the following parameters have been known to exist at the infinite frequency or in the perfectly conducting media case.

1.5.2.5.1 Line Impedance and Admittance

$$[Z] = [Z_s] = j\omega\left(\frac{\mu_0}{2\pi} \cdot [P]\right)$$

$$[Y] = [Y_s] = j\omega 2\pi\varepsilon_0 \cdot [P]^{-1}$$

(1.194)

where

$$
\left.
\begin{aligned}
P_{ij} &= \ln\left(\frac{D_{ij}}{d_{ij}}\right) \quad \text{for } i \neq j \\
&= \ln\left(\frac{2h_i}{r_i}\right) \quad \text{for } i = j
\end{aligned}
\right\} \text{modified potential coefficient} \quad (1.195)
$$

and

$$D_{ji} = \sqrt{(h_i + h_j)^2 + y_{ij}^2}, \quad d_{ji} = \sqrt{(h_i + h_j)^2 + y_{ij}^2}$$

From this impedance and admittance, we can derive the following line parameters in the actual phase domain.

1.5.2.5.2 Actual Propagation Constant

$$[\Gamma] = ([Z][Y])^{1/2} = \left\{ j\omega\left(\frac{\mu_0}{2\pi}\right) \cdot [P] \cdot j\omega 2\pi\varepsilon_0 \cdot [P]^{-1} \right\}^{1/2} = j\omega\sqrt{\varepsilon_0\mu_0}[U]$$

$$= j\left(\frac{\omega}{c_0}\right)[U] = [\alpha] + j[\beta]$$

Therefore,

$$[\alpha] = [0], \quad [\beta] = \left(\frac{\omega}{c_0}\right) \cdot [U]$$

(1.196)

Thus, the propagation velocity is

$$[c] = c_0[U] \quad \text{or} \quad c_i = c_0 \quad \text{for phase } i$$

(1.197)

From these results, it is clear that the product $Z \cdot Y$ or the actual propagation constant matrix is diagonal and purely imaginary at infinite frequency or in the perfect conductor case. This results in the attenuation being zero and the propagation velocity being a light velocity in free space in any phase. Also, it is noteworthy that modal theory is not necessary as long as the propagation constant alone is concerned since it is already diagonal.

1.5.2.5.3 Actual Characteristic Impedance

Equation 1.159 gives

$$[Z_0] = [A] \cdot [z_0] \cdot [B]^{-1} = [A]([A]^{-1} \cdot [\Gamma]^{-1} \cdot [Z] \cdot [B]) \cdot [B]^{-1} = [\Gamma]^{-1} \cdot [Z]$$

$$= \left(\frac{c_0}{j\omega}\right) \cdot [U] \cdot j\omega\left(\frac{\mu_0}{2\pi}\right)[P] = \left(\frac{\mu_0 c_0}{2\pi}\right)[P] = 60[P] \qquad (1.198)$$

or

$$Z_{0ij} = 60\ln\left(\frac{D_{ij}}{d_{ij}}\right) \quad \text{for } i = j$$

$$= 60\ln\left(\frac{2h_j}{r_i}\right) \quad \text{for } i = j$$

It is clear that the actual characteristic impedance is constant independent of frequency.

1.5.2.5.4 Modal Parameters

Line impedance, admittance, and characteristic impedance matrices involve nonzero, off-diagonal elements or mutual coupling, although the propagation constant matrix is diagonal. If one needs to diagonalize these matrices, modal transformation is required.

In the case of a completely transposed three-phase line, any of the transformation matrices explained in Section 1.4.4.1 can be used. The current transformation matrix is the same as the voltage transformation matrix. Let us apply the traveling-wave transformation:

$$[z] = [A]^{-1} \cdot [Z] \cdot [A] = \begin{bmatrix} Z_s + 2Z_m & 0 & 0 \\ 0 & Z_s + Z_m & 0 \\ 0 & 0 & Z_s + Z_m \end{bmatrix}$$

or

$$z_0 = Z_s + 2Z_m, \, z_1 = z_2 = Z_s - Z_m$$

In the same manner:

$$y_0 = Y_s + 2Y_m, \quad y_1 = y_2 = Y_s - Y_m$$
$$z_{00} = Z_{0s} + 2Z_{0m}, \quad z_{01} = z_{02} = Z_{0s} - Z_{0m}$$
$$\gamma_0 = \gamma_1 = \gamma_2 = \frac{j\omega}{c_0} \quad \text{or} \quad c_0 = c_1 = c_2 \tag{1.199}$$

This same result can be obtained by applying the other transformation in Section 1.4.4.1.

1.5.2.5.5 Time-Domain Parameters

The parameters explained in Sections 1.1 through 1.5.2.5.4 are in the frequency domain. The parameters in the time domain are the same as those in the frequency domain in the perfect conductor case because they are frequency independent and, thus, time independent. In the case of the imperfectly conducting earth and conductor, only the parameters at infinite frequency are known analytically. These parameters should correspond to the parameters at $t = 0$ in the time domain from the initial value theorem of the Laplace transform, that is

$$\lim_{t \to +0} f(t) = \lim_{s \to \infty} \{sF(s)\} \tag{1.200}$$

Thus, we can obtain the time-domain parameters at $t = 0$ or in the perfect conductor case using the same methods as in Equations 1.198 and 1.199.

1.5.3 Time Response

1.5.3.1 Time-Dependent Responses

The time response of the frequency dependence explained in Section 1.5.2 is calculated by a numerical Fourier or Laplace inverse transform in the following form [1,20]:

1. Propagation constant:

$$e(t) = L^{-1}\left[\exp\frac{\{-\Gamma(s)x\}}{s}\right]: \text{step response of propagation constant} \tag{1.201}$$

where
$s = \alpha + j\omega$ is the Laplace operator
L^{-1} is the Laplace inverse transform

2. Characteristic impedance:

$$Z_0(t) = L^{-1}\left[\frac{Z_0(s)}{s}\right]: \text{surge impedance response} \tag{1.202}$$

3. Transformation matrix:

$$A_{ij}(t) = L^{-1}\left[\frac{A_{ij}(s)}{s}\right]: \text{time-dependent transformation matrix} \tag{1.203}$$

1.5.3.2 Propagation Constant: Step Response

The frequency dependence of the propagation constant appears as a wave deformation in the time domain. This is measured as a voltage waveform at distance x when a step (or impulse) function voltage is applied to the sending end of a semi-infinite line. The voltage waveform, which is distorted from the original waveform, is called "step (impulse) response of wave deformation" and is defined in Equation 1.201.

Figures 1.28 and 1.29 show modal step responses on the lines in Figures 1.22 and 1.25, respectively. It is clear from the figures that the wave front is distorted especially in mode 0, which has the largest attenuation and lowest velocity in the frequency domain. As time passes, the distorted waveform tends to reach 1 pu, the applied voltage. Figure 1.28 shows that wave deformation is greater when line length is greater. This is reasonable since an increase in line length results in greater distortion. Also, greater earth resistivity causes greater wave deformation because line impedance becomes greater. A GW reduces the wave deformation of mode 0 significantly. This is due to the fact that the earth-return current is reduced by the GW; thus, the line impedance is reduced.

The reason for the much smaller wave deformation in the aerial modes than in the earth-return modes is that the conductor internal impedance that contributes mainly to the aerial modes is far smaller than the earth-return impedance that mainly contributes to mode 0.

The line transportation does not significantly affect mode 0 wave deformation. It does, however, cause a noticeable effect on the aerial modes, as can be observed from Figure 1.29. The difference between transposed and untransposed lines is already clear from the frequency responses given in Table 1.2 and Figure 1.25. The significant difference in mode 1 propagation velocity in Figure 1.25b results in a difference in mode 1 wave deformation in Figure 1.29.

1.5.3.3 Characteristic Impedance

It should be noted that the definition of Equation 1.202 proposed by the author in 1973 is effective only for a semi-infinite line or for a time period of

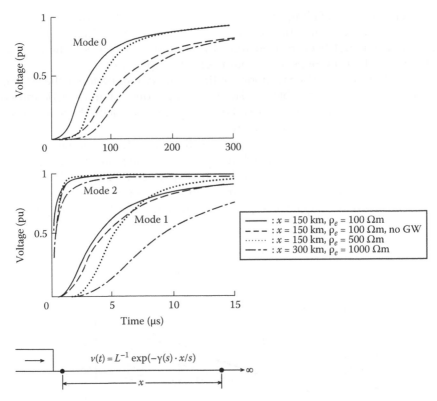

FIGURE 1.28
Modal step responses of wave deformation for a horizontal line.

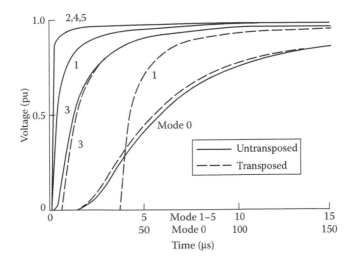

FIGURE 1.29
Modal step responses of wave deformation for a vertical line.

2τ, where τ is the traveling time of a line [20,21]. Also, the definition requires further study in conjunction with the wave equation in the time domain because it has not been proved that this definition expresses the physical behavior of the time-dependent characteristic impedance.

Table 1.8 shows the time response of the frequency-dependent characteristic impedance given in Table 1.3. The time-dependent characteristic impedance increases as time increases. This is quite reasonable because of the inverse relation of time and frequency.

Figure 1.30 shows the time response of the characteristic impedance of a vertical twin-circuit line illustrated in Figure 1.25. The relation of magnitudes corresponds to that in the frequency domain explained for Figure 1.27 considering the inverse relation of time and frequency. It is observed from comparing Figure 1.25 with the frequency response of Figure 1.27 that the time dependence

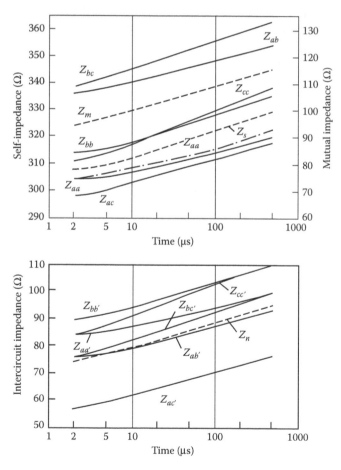

FIGURE 1.30
Time response of the characteristic impedance corresponding to Figure 1.27.

TABLE 1.8

Time Responses of the Characteristic Impedances in Table 1.3

Time (μs)	Z_{aa} (Ω)	Z_{bb} (Ω)	Z_{ab} (Ω)	Z_{ac} (Ω)
10	330.6	329.5	71.73	38.74
50	337.3	335.7	77.96	44.94
100	340.1	338.4	80.61	47.61
200	343.0	341.1	83.25	50.27
500	347.2	345.1	87.06	54.05

is greater than the frequency dependence of the characteristic impedance. For example, the variation of Z_{cc} is 8.5%, Z_{ac} 26.8%, and Z'_{ac} 31.3% for 2 μs $\leq t \leq$ 500 μs, while the variations of those in the frequency domain are 7.9%, 23.4%, and 27.8% for 500 kHz $\geq f \geq$ 2 kHz. Also, it should be noted that the time-dependent impedance is greater by about 5–15 Ω than the frequency-dependent one in general. The calculated result agrees with the measured result.

1.5.3.4 Transformation Matrix

The frequency dependence of the transformation matrix appears as time dependence in the time domain. The time-dependent transformation matrix is defined in Equation 1.203.

Table 1.9 shows the time response of the transformation matrix given in Equation 1.190 and Table 1.4 for the untransposed horizontal line in Figure 1.22 without GWs. A comparison of Table 1.9 and Table 1.4 shows that the time dependence is smaller than the frequency dependence as far as the results in the table are concerned. Also, it is clear that the values of A_1 and A_2 in Table 1.9 are not very different from the real values of A_1 and A_2 in Table 1.4. Time dependence is inversely related to frequency dependence, that is, A_1 and A_2 decrease as time increases, while they increase as frequency increases.

Figure 1.31 shows the frequency and time dependence of the voltage transformation matrix of an untransposed vertical single-circuit line. It is

TABLE 1.9

Time Response of the Transformation Matrix Given in Table 1.4

Time (μs)	A_1	A_2
10	1.0290	−2.1850
50	1.0108	−2.2151
100	1.0047	−2.2248
150	1.0018	−2.2296
200	0.9998	−2.2326

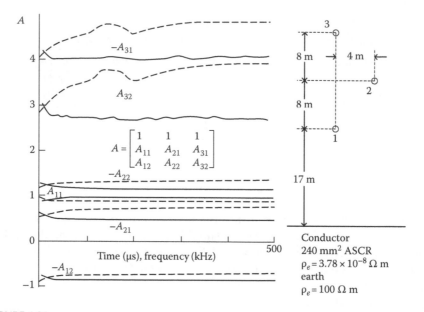

FIGURE 1.31
Time/frequency dependence of the transformation matrix of an untransposed vertical single-circuit line.

TABLE 1.10

Time Response of the Transformation Matrix Given in Table 1.7

(a) $[B_I] = B_I(j, k)$

	Mode 0		Mode 1		Mode 2	
Time (μs)	(1,1)	(1,2)	(2,2)	(2,3)	(3,1)	(3,3)
5	0.644	0.821	0.444	−0.524	0.568	0.253
10	0.653	0.825	0.425	−0.535	0.558	0.260
50	0.682	0.842	0.377	−0.562	0.514	0.272
100	0.694	0.849	0.353	−0.571	0.496	0.279
200	0.706	0.856	0.323	−0.578	0.469	0.290
500	0.725	0.864	0.285	−0.592	0.415	0.337

(b) $[B_{II}] = B_{II}(j, k)$

	Mode 3		Mode 4		Mode 5	
Time (μs)	(1,1)	(1,2)	(2,2)	(2,3)	(3,1)	(3,3)
5	0.443	0.657	0.468	−0.621	0.659	0.288
10	0.452	0.660	0.478	−0.642	0.670	0.278
50	0.467	0.658	0.500	−0.669	0.685	0.251
100	0.470	0.647	0.499	−0.666	0.672	0.241
200	0.472	0.631	0.503	−0.663	0.678	0.214
500	0.480	0.588	0.481	−0.662	0.653	0.180

Note: $B(1,3) = B(2,1) = -B(3,2) = 1$ normalized.

clear from the figure that the frequency dependence is greater than the time dependence of the transformation matrix. The maximum deviation from the average value is about 10% for the time dependence and about 30% for the frequency dependence. Also, the frequency and time dependencies are much greater in the vertical line case than in the horizontal line case.

Table 1.10 shows the time response of the transformation matrix given in Equation 1.203 and Table 1.7 for the untransposed vertical twin-circuit line in Figure 1.25.

The matrix deviation of each vector for $10 \ \mu s < t < 500 \ \mu s$ from the value at $t = 10 \ \mu s$ is

Mode 0: 11%, mode 1: 21%, mode 2: 30%

Mode 3: 11%, mode 4: 5.2%, mode 5: 35%

Assuming frequency is given as the inverse of time, this time range corresponds to the frequency range of $100 \ \text{kHz} > f > 2 \ \text{kHz}$. In this frequency range, the maximum deviation from the value at $f = 100 \ \text{kHz}$ is

Mode 0: 10%, mode 1: 21%, mode 2: 18%

Mode 3: 13%, mode 4: 33%, mode 5: 40%

From these results, it can be said that time dependence of the internal mode is greater and the intercircuit mode is smaller than the frequency dependence for an untransposed twin-circuit line. In general, the frequency dependence is greater than the time dependence of the transformation matrix.

PROBLEMS

1.16 Explain why it is not easy to obtain a transformation matrix in the cases of an infinite frequency and a perfectly conducting system.

1.17 Discuss the differences in modal components between three-phase transposed and untransposed horizontal lines.

1.6 Traveling Wave

1.6.1 Reflection and Refraction Coefficients

When an original traveling wave e_{1f} (equivalent to a voltage source) comes from the left to node P along line 1 in Figure 1.32, the wave partially refracts to line 2, and the remaining reflects to line 1, similar to light reflecting off a water surface [1,22].

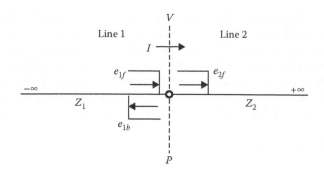

FIGURE 1.32
A conductor system composed of lines 1 and 2.

Let us define the refracted wave as e_{2f}, the reflected wave as e_{1b}, and the characteristic (surge) impedance of lines 1 and 2 as Z_1 and Z_2, respectively. Then, current I on line 1 is given from Equation 1.56 as

$$I = Y_1(e_{1f} - e_{1b}) = \frac{(e_{1f} - e_{1b})}{Z_1} \tag{1.204}$$

On line 2, which has no backward wave:

$$I = \frac{e_{2f}}{Z_2} \tag{1.205}$$

Voltage V at node P on line 1 is given from Equation 1.91 by

$$V = e_1 f + e_{1b} \tag{1.206}$$

On line 2:

$$V = e_{2f} \tag{1.207}$$

Substituting Equations 1.207 and 1.206 into Equation 1.205 gives

$$I = \frac{V}{Z_2} = \frac{(e_{1f} + e_{1b})}{Z_2}$$

Substituting this equation into Equation 1.204, e_{1b} is obtained as

$$e_{1b} = \theta \cdot e_{1f} \tag{1.208}$$

where

$$\theta = \frac{(Z_2 - Z_1)}{(Z_2 + Z_1)} : \text{reflection coefficient.} \tag{1.209}$$

Similarly, e_{2f} is given as

$$e_{2f} = \lambda \cdot e_{1f} \tag{1.210}$$

where

$$\lambda = \frac{2Z_2}{Z_2} + Z_1 = 1 + \theta : \text{refraction coefficient.} \tag{1.211}$$

It should be clear from Equations 1.208 and 1.210 that the reflected and refracted waves are determined from the original wave using reflection and refraction coefficients, which represent the boundary conditions at node P between lines 1 and 2 with surge impedances Z_1 and Z_2. The coefficients θ and λ give a ratio of the original wave (voltage) and the reflected and refracted voltages. For example:

1. Line 1 open-circuited ($Z_2 = \infty$): $\theta = 1$, $\lambda = 2$, $I = 0$, $V = 2e_{1f}$,
2. Line 1 short-circuited ($Z_2 = 0$): $\theta = -1$, $\lambda = 0$, $I = 2e_{1f}$, $V = 0$, and
3. Line 1 matched ($Z_2 = Z_1$): $\theta = 0$, $\lambda = 1$, $I = e_{1f}/Z_1$, $V = e_{1f}$.

These results show that the reflected voltage e_{1b} at node P is the same as the incoming (original) voltage e_{1f} and the current I becomes zero when line 1 is open-circuited. On the contrary, under the short-circuited condition, $e_{1b} = -e_{1f}$, and the current becomes maximum. Under the matching termination of line 1, there is no reflected voltage at node P.

1.6.2 Thevenin's Theorem

1.6.2.1 Equivalent Circuit of a Semi-Infinite Line

In Figure 1.33a, the following relation is obtained from Equations 1.205 and 1.207:

$$I = \frac{V}{Z_2} \tag{1.212}$$

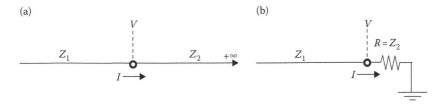

FIGURE 1.33
(a) A semi-infinite line. (b) An equivalent circuit.

This equation is the same as Ohm's law for a lumped-parameter circuit with resistance R. Thus, the semi-infinite line is equivalent to that in Figure 1.33b.

1.6.2.2 Voltage and Current Sources at the Sending End

A voltage source at the sending end of the line illustrated in Figure 1.34a is equivalent to that in Figure 1.34b because the traveling wave on the right in (b) is the same as that in (a). Then, (b) is rewritten based on Figure 1.34c, that is, the voltage source at the sending end is represented by a voltage source at the center of an infinite line.

Similarly, the current source in Figure 1.35a is represented by Figure 1.35b. Furthermore, by applying the result in Section 1.6.2.1, the voltage and current sources in Figures 1.34 and 1.35 are represented by Figure 1.36.

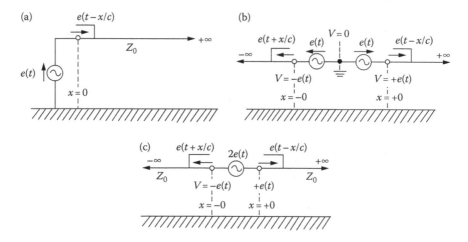

FIGURE 1.34
Equivalent circuit of a voltage source at the sending end. (a) A voltage source. (b) Equivalent circuit of (a). (c) Equivalent circuit of (b).

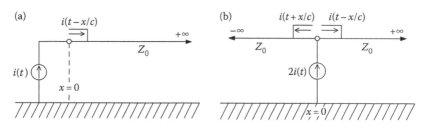

FIGURE 1.35
Equivalent circuit (b) of a current source ($R = Z_0$) (a) at the sending end.

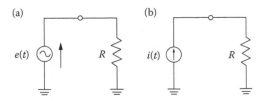

FIGURE 1.36
Lumped-parameter equivalent of a source at the sending end. (a) Voltage source. (b) Current source ($R = Z_0$).

1.6.2.3 Boundary Condition at the Receiving End

1. *Open-circuited line:* An open-circuited line Z_0 with an incoming wave $e(x - ct)$ from the left in Figure 1.37a is equivalent to an infinite line with the incoming wave from the left and another incoming wave $e(x + ct)$ from the right with the same amplitude and the same polarity as in Figure 1.37b.

2. *Short-circuited line:* A short-circuited line with an incoming wave $e(x - ct)$ as in Figure 1.38a is equivalent to an infinite line with $e(x - ct)$ and $-e(x + ct)$.

3. *Resistance-terminated line:* A resistance is equivalent to a semi-infinite line whose surge impedance is the same as the resistance as explained in Section 1.6.2.1 and in Figure 1.33. If the surge impedance of the semi-infinite line is taken to be the same as that of the

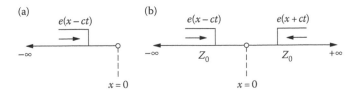

FIGURE 1.37
(a) An open-circuited line. (b) An equivalent circuit.

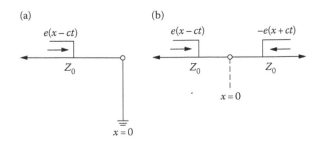

FIGURE 1.38
(a) Short-circuited line. (b) An equivalent circuit.

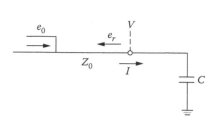

FIGURE 1.39
A capacitance-terminated line.

line to which the resistance is connected, then a backward traveling wave $e_b(x + ct) = e_b$ is to be placed on the semi-infinite line:

$$e_r(t) = \theta \cdot e(t), \quad \theta = \frac{(R - Z_0)}{(R + Z_0)} \tag{1.213}$$

4. *Capacitance-terminated line:* When a semi-infinite line Z_0 is terminated by a capacitance C as shown in Figure 1.39, node voltage V and current I are calculated in the following manner:

$$V = e_0 + e_r \quad \therefore e_r = V - e_0$$

$$I = \frac{(e_0 - e_r)}{Z0} = C \cdot \frac{dV}{dt}$$

Substituting e_r into I and multiplying with Z_0 gives

$$2e_0 = Z_0 C \cdot \frac{dV}{dt} + V$$

Solving this differential equation, the following is obtained:

$$V = K \cdot \exp\left(\frac{-t}{\tau}\right) + 2e_0, \quad \tau = Z_0 C$$

Considering the initial condition, $V = 0$ for $t = 0$, then

$$V = 2e_0\left\{1 - \exp\left(\frac{-t}{\tau}\right)\right\}, \quad e_r = e_0\left\{1 - 2\exp\left(\frac{-t}{\tau}\right)\right\} \tag{1.214}$$

In a similar manner, an inductance-terminated line either at the receiving end or at the sending end can be solved.

1.6.2.4 Thevenin's Theorem

Thevenin's Theorem is very useful when only voltage and current at a transition (boundary) point between distributed-parameter lines are to be

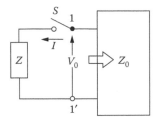

FIGURE 1.40
Thevenin's Theorem.

obtained. In Figure 1.40, the impedance seen from nodes 1 and 1′ to the right
is Z_0, and the voltage across the nodes is V_0.

When an impedance Z is connected to the nodes, a current I flowing into
the impedance is given by Thevenin's Theorem as:

$$I = \frac{V_0}{(Z_0 + Z)} \tag{1.215}$$

When an original traveling wave e comes from the left along a line Z_0 as
in Figure 1.41a, voltage V and current I at node P are calculated in an equivalent circuit (Figure 1.41b) where a voltage source $V_0(t)$ is given as $2e(t)$ by
Thevenin's Theorem.

There is no straightforward method to obtain a reflected traveling wave, e_r,
when Thevenin's Theorem is applied to calculate node voltage and current.
In such a case, the following relation is very useful to obtain the reflected
wave e_r from the node voltage V and the original incoming wave e:

$$e_r = V - e \tag{1.216}$$

By applying this relation, reflected waves in Figure 1.42 are easily evaluated:

$$e_{1b} = V - e_{1f}, \quad e_{2b} = V - e_{2f}, \quad e_{3b} = V - e_{3f}$$

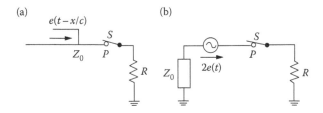

FIGURE 1.41
A resistance-terminated line with a voltage traveling wave. (a) Original circuit. (b) Equivalent
circuit.

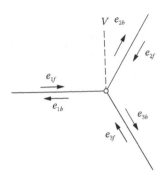

FIGURE 1.42
Reflected waves at a node with three lines.

1.6.3 Multiple Reflection

In a distributed-parameter circuit composed of three distributed lines as in Figure 1.43, node voltages V_1 and V_2 and currents I_1 and I_2 are evaluated analytically in the following manner.

The refraction coefficients λ at nodes 1 and 2 are given by

$$\lambda_{12} = \frac{2Z_2}{(Z_1 + Z_2)}, \quad \lambda_{21} = \frac{2Z_1}{(Z_1 + Z_2)}$$

$$\lambda_{23} = \frac{2Z_3}{(Z_2 + Z_3)}, \quad \lambda_{32} = \frac{2Z_2}{(Z_2 + Z_3)}$$

1. $0 \le t < \tau$

   For simplicity, assume that a forward traveling wave e_{1f} on line 1 arrives at node 1 at $t = 0$ Then, node voltage V_1 is calculated by

   $$V_1(t) = \lambda_{12} e_{1f}(t)$$

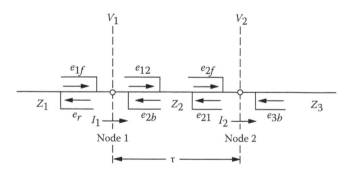

FIGURE 1.43
A three-line system.

The reflected wave e_r on line 1 is evaluated by Equation 1.216 as

$$e_r(t) = V_1(t) - e_{1f}(t)$$

The same is applied to the traveling waves on line 2:

$$e_{12}(t) = V_1(t) - e_{2b}(t)$$

For the moment, only an incoming wave from line 1 is assumed, and thus:

$$e_{2b}(t) = 0, \quad e_{12}(t) = V_1(t)$$

Current I_1 is evaluated by:

$$I_1(t) = \frac{\{e_{1f}(t) - e_r(t)\}}{Z_1} = \frac{e_{12}(t)}{Z_2}, \quad I_2(t) = 0$$

The refracted wave e_{12} travels to node 2 on line 2.

2. $\tau \leq t < 2\tau$

At $t = \tau$, e_{12} arrives at node 2 and becomes e_{2f} (incoming wave to node 2):

$$e_{2f}(t) = e_{12}(t - \tau)$$

e_{2f} produces a voltage V_2 at node 2, a reflected wave e_{21} on line 2, and a refracted wave e_{23}, which never returns to node 2 because line 3 is semi-infinite. Therefore, we can ignore e_{23} for now, giving

$$V_2(t) = \lambda_{23}e_{2f}(t), \quad e_{21}(t) = V_2(t) - e_{2f}(t)$$
$$I_2(t) = \frac{\{e_{2f}(t) - e_{21}(t)\}}{Z_2}$$

The reflected wave e_{21} travels to node 1.

3. $2\tau \leq t < 3\tau$

$$e_{2b}(t) = e_{21}(t - \tau)$$
$$V_1(t) = \lambda_{12}e_{1f}(t) + \lambda_{21}e_{2b}(t), \quad e_{21}(t) = V(t) - e_{2b}(t)$$

Repeating this procedure, node voltages V_1 and V_2 and currents I_1 and I_2 are calculated. The procedure is formulated in general as follows [1,20]:

a. Node equations for node voltages:

$$V_1(t) = \lambda_{12}e_{1f}(t) + \lambda_{21}e_{2b}(t)$$
$$V_2(t) = \lambda_{23}e_{2f}(t) + \lambda_{32}e_{3b}(t)$$

b. Node equations for traveling waves:

$$e_{12}(t) = V_1(t) - e_{2b}(t), \quad e_{21}(t) = V_2(t) - e_{2f}(t)$$

c. Continuity equations for traveling waves:

$$e_{2f}(t) = e_{12}(t - \tau), \quad e_{2b} = e_{21}(t - \tau)$$

d. Current equations:

$$I_1(t) = \frac{\{e_{12}(t) - e_{2b}(t)\}}{Z_2}, \quad I_2(t) = \frac{\{e_{2f}(t) - e_{21}(t)\}}{Z_2}$$

This procedure to calculate a traveling-wave phenomenon is called the "refraction coefficient method," which can be used to easily deal with multiphase lines, and it requires only a precalculation of the refraction coefficient [20]. The "Lattice diagram method" [22–24] is a well-known method, but it requires both the refraction and reflection coefficients; furthermore, it cannot be used for multiphase lines. There is a more sophisticated approach called the Schnyder–Bergeron (or simply Bergeron) method [25], which has been adopted in the well-known software EMTP [9,26] originally developed by the Bonneville Power Administration (BPA), U.S. Department of Energy. The method is very convenient for numerical calculation by a computer but not convenient for hand calculation that requires physical insight into the traveling-wave phenomenon.

EXAMPLE 1.3

Let us obtain voltages V_1 and V_2 and current I_1 for $0 \le t < 6\tau$ in Figure 1.44a.

Solution

$$\lambda_{12} = 2, \quad \lambda_{21} = 0, \quad \lambda_{23} = 2, \quad e_{1f}(t) = \frac{E}{2} = 50(\text{V})$$

(a)

(b)

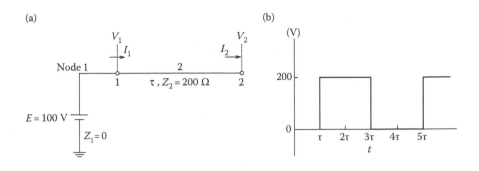

(c) (A)

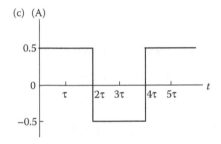

FIGURE 1.44
Voltage and current responses on an open-circuited line. (a) An open-circuited line. (b) $V_2(t)$. (c) $I_1(t)$.

1. $0 \leq t < \tau$

$$V_1(t) = e_{12}(t) = \lambda_{12} \cdot e_{2f} = E = 100(\text{V}), \quad I_1(t) = \frac{e_{12}(t)}{Z_2} = 0.5(\text{A})$$

2. $\tau \leq t < 2\tau$

$$e_{2f}(t) = e_{12}(t-\tau) = 100(\text{V}), \quad V_2(t) = \lambda_{23}e_{12}(t-\tau) = 200(\text{V})$$

$$e_{21}(t) = V_2(t) - e_{2f}(t) = 100(\text{V}), \quad I_2(t) = \frac{\{e_{2f}(t) - e_{21}(t)\}}{Z_2} = 0$$

3. $2\tau \leq t < 3\tau$

$$e_{2b}(t) = e_{21}(t-\tau) = 100(\text{V}), \quad V_1(t) = E + \lambda_{21}e_{2b}(t) = E = 100(\text{V})$$
$$e_{12}(t) = V_1(t) - e_{2b}(t) = 0(\text{V})$$
$$I_1(t) = \frac{\{e_{12}(t) - e_{2b}(t)\}}{Z_2} = -0.5(\text{A})$$

4. $3\tau \leq t < 4\tau$

$$e_{2f}(t) = e_{12}(t-\tau) = 0(\text{V}), \quad V_2(t) = 0(\text{V}), \quad I_2(t) = 0(\text{A}), \quad e_{21}(t) = 0(\text{V})$$

5. $4\tau \le t < 5\tau$

   $e_{2b}(t) = 0(V), \quad V_1(t) = 100(V), \quad e_{12}(t) = 100(V), \quad I_1(t) = 0.5(A)$

6. $5\tau \le t < 6\tau$

   $e_{2f}(t) = 100\,(V), \quad V_2(t) = 200\,(V), \quad I_2(t) = 0\,(A), \quad e_{21}(t) = 100\,(V)$

Based on these results, V_1, V_2, and I_1 are drawn as in Figure 1.44b and c.

1.6.4 Multiconductors

1.6.4.1 Reflection and Refraction Coefficients

The refraction and reflection coefficient matrices are given in the following form for the circuit in Figure 1.45:

$$\left.\begin{array}{ll} \text{Refraction coefficient} & [\lambda_{ij}] = 2[Z_j]([Z_i]+[Z_j]-1) \\ \text{Refraction coefficient} & [\theta_{ij}] = [\lambda_{ij}]-[U] \end{array}\right\} \tag{1.217}$$

where
  i, j are the ith and jth lines
  $[U]$ is the unit matrix

1.6.4.2 Lossless Two-Conductor Systems

Let us consider the lossless two-conductor system illustrated in Figure 1.46. The surge impedance matrices of the lines are given by

$$[Z_1] = \begin{bmatrix} R & 0 \\ 0 & \infty \end{bmatrix}, \quad [Z_2] = [Z_0] = \begin{bmatrix} Z_s & Z_m \\ Z_m & Z_s \end{bmatrix}, \quad [Z_3] = \begin{bmatrix} \infty & 0 \\ 0 & \infty \end{bmatrix} \tag{1.218}$$

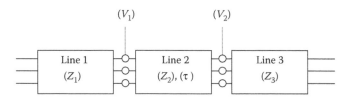

FIGURE 1.45
A multiconductor system.

The incoming traveling wave is given by

$$(E) = \begin{pmatrix} E \\ 0 \end{pmatrix} \tag{1.219}$$

The refraction coefficient matrix at each node is

$$[\lambda_{12}] = \left\{ \frac{2}{(R+Zs)} \right\} \begin{bmatrix} Z_s & 0 \\ Z_m & 0 \end{bmatrix}, \quad [\lambda_{23}] = \begin{bmatrix} 2 & 0 \\ 0 & 2 \end{bmatrix} \tag{1.220}$$

$$[\lambda_{21}] = \left\{ \frac{2}{(R+Zs)} \right\} \begin{bmatrix} R & 0 \\ -Z_m & (R+Z_s) \end{bmatrix}$$

Applying these refraction coefficients and the incoming wave, the node voltages in Figure 1.46 are calculated at every time step by using the refraction coefficient method discussed in Sections 1.6.3.1 and 1.6.3.2:

1. $0 \leq t < \tau$

$$(V_1) = [\lambda_{12}](E) = \frac{2}{R+Z_s} \begin{vmatrix} Z_s & 0 \\ Z_m & 0 \end{vmatrix} \begin{vmatrix} E \\ 0 \end{vmatrix} = \frac{2E}{R+Z_s} \begin{vmatrix} Z_s \\ Z_m \end{vmatrix} = (E_{12})$$

$$(I_1) = [Z_0]^{-1}(E_{12}) = \frac{1}{Z_s^2 - Z_m^2} \begin{vmatrix} Z_s & -Z_m \\ -Z_m & Z_s \end{vmatrix} \frac{2E}{R+Z_s} \begin{vmatrix} Z_s \\ Z_m \end{vmatrix}$$

$$= \frac{2E}{(R+Z_s)(Z_s^2 - Z_m^2)} \begin{vmatrix} Z_s^2 - Z_m^2 \\ -Z_s Z_m + Z_s Z_m \end{vmatrix} = \begin{vmatrix} \dfrac{2E}{R+Z_x} \\ 0 \end{vmatrix}$$

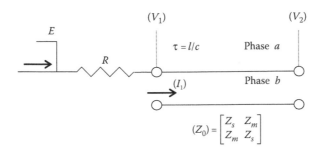

FIGURE 1.46
A lossless two-conductor system.

2. $\tau \leq t < 2\tau$

$$\{E_{2f(t)}\} = \{E_{12}(t - \tau)\}$$

$$(V_s) = [\lambda_{23}](E_{2f})$$

$$= \begin{vmatrix} 2 & 0 \\ 0 & 2 \end{vmatrix}\begin{vmatrix} E_{2fa} \\ E_{2fb} \end{vmatrix} = 2\begin{vmatrix} E_{2fa} \\ E_{2fb} \end{vmatrix} = 2(E_{2f}) = 2\{E_{12}(t - \tau)\} = \frac{4E(t - \tau)}{R + Z_s}\begin{vmatrix} Z_s \\ Z_m \end{vmatrix}$$

$$\{E_{21}(t)\} = \{V_2(t)\} - (E_{2f}(t)) = E_{2f(t)} = \{E_{12}(t - \tau)\} = \frac{2E(t - \tau)}{R + Z_s}\begin{vmatrix} Z_s \\ Z_m \end{vmatrix}$$

3. $2\tau \leq t < 3\tau$

$$\{E_{2b}(t)\} = \{E_{21}(t - \tau)\} = [E_{12}(t - 2\tau)]$$

$$(V_1) = [\lambda_{12}]\{E(t)\} + [\lambda_{21}]\{E_{2b}(t)\} = \frac{2}{R + Z_s}\left\{E(t) + \frac{2RE(t - 2\tau)}{R + Z_s}\right\}\begin{vmatrix} Z_s \\ Z_m \end{vmatrix}$$

$$\{E_{12}(t)\} = (V_1) - \{E_{2b}(t)\} = \frac{2}{R + Z_s}\left\{E(t) + \frac{R - Z_s}{R - Z_s}E(t - 2\tau)\right\}\begin{vmatrix} Z_s \\ Z_m \end{vmatrix}$$

$$(I_1) = [Z_0]^{-1}\{E_{12}(t)\} - \{E_{2b}(t)\} = \frac{2}{R + Z_s}\left\{E(t) - \frac{2Z_s}{R + Z_s}E(t - 2\tau)\right\}\begin{vmatrix} 1 \\ 0 \end{vmatrix}$$

Repeating this procedure, voltages (V_1) and (V_2) and current (I_1) are calculated.

EXAMPLE 1.4

Assuming that $R = 100\ \Omega$, $Z_s = 400\ \Omega$, $Z_m = 100\ \Omega$, and $E = 1$ pu, calculate (V_1), and (V_2) for $0 \leq t < 5\tau$:

Solution

1. $0 \leq t < \tau$

$$(E_{2f}) = \begin{vmatrix} 1.6 & 0 \\ 0.4 & 0 \end{vmatrix}\begin{vmatrix} 1 \\ 0 \end{vmatrix} = \begin{vmatrix} 1.6 \\ 0.4 \end{vmatrix} = (E_{12})$$

$$(I_1) = [Z_0]^{-1}(E_{12}) = \frac{1}{1500}\begin{vmatrix} 4 & -1 \\ -1 & 4 \end{vmatrix}\begin{vmatrix} 1.6 \\ 0.4 \end{vmatrix} = \begin{vmatrix} 4 \times 10^{-3} \\ 0 \end{vmatrix}$$

2. $\tau \leq t < 2\tau$

$$(E_{2f}) = \begin{vmatrix} 1.6 \\ 0.4 \end{vmatrix}, \quad (V_2) = \begin{vmatrix} 2 & 0 \\ 0 & 2 \end{vmatrix}\begin{vmatrix} 1.6 \\ 0.4 \end{vmatrix} = \begin{vmatrix} 3.2 \\ 0.8 \end{vmatrix}, \quad (E_{21}) = (V_2) - (E_{2f}) = \begin{vmatrix} 1.6 \\ 0.4 \end{vmatrix}$$

3. $2\tau \le t < 3\tau$

$$(E_{2b}) = \begin{vmatrix} 1.6 \\ 0.4 \end{vmatrix}, \quad (V_1) = \begin{vmatrix} 1.6 \\ 0.4 \end{vmatrix} \begin{vmatrix} 0 & 1 \\ 0 & 0 \end{vmatrix} + \begin{vmatrix} 0.4 \\ -0.4 \end{vmatrix} \begin{vmatrix} 0 & 1.6 \\ 2 & 0.4 \end{vmatrix} = \begin{vmatrix} 2.24 \\ 0.56 \end{vmatrix}$$

$$(E_{12}) = (V_1) - (E_{2b}) = \begin{vmatrix} 0.64 \\ 0.16 \end{vmatrix}, \quad (I_1) = \frac{1}{1500} \begin{vmatrix} 4 & -1 \\ -1 & 4 \end{vmatrix} \begin{vmatrix} 0.96 \\ 0.24 \end{vmatrix} = \begin{vmatrix} 2.4 \times 10^{-3} \\ 0 \end{vmatrix}$$

4. $3 \le t < 4\tau$

$$(E_{2f}) = \begin{vmatrix} 0.64 \\ 0.16 \end{vmatrix}, \quad (V_2) = \begin{vmatrix} 0.28 \\ 0.32 \end{vmatrix}, \quad (E_{21}) = \begin{vmatrix} 0.64 \\ 0.16 \end{vmatrix}$$

5. $4 \le t < 5\tau$

$$(E_{2b}) = \begin{vmatrix} 0.64 \\ 0.16 \end{vmatrix}, \quad (V_1) = \begin{vmatrix} 1.856 \\ 0.464 \end{vmatrix}$$

$$(E_{12}) = \begin{vmatrix} 1.126 \\ 0.304 \end{vmatrix}, \quad (I_1) = \begin{vmatrix} 1.44 \times 10^{-3} \\ 0 \end{vmatrix}$$

From these results, the drawing in Figure 1.47 is obtained.

1.6.4.3 Consideration of Modal Propagation Velocities

In a real transmission line, the line impedance becomes frequency dependent due to the skin effects of the conductor and the earth as explained in Section 1.5. The propagation velocity is also frequency dependent; furthermore, the modal velocities differ from one another. This velocity difference causes a significant effect on the voltage and current waveshapes along the line. The effect is included analytically in a calculation of voltage and current in the following manner:

1. *Traveling wave at the sending end:*

$$(V_1) = (E_{12}) = [\lambda_{12}](E) = \frac{2E}{R + Z_s} \begin{vmatrix} Z_s \\ Z_m \end{vmatrix} = \begin{vmatrix} E_{1a} \\ E_{1b} \end{vmatrix} \tag{1.221}$$

Transforming the traveling waves in the phase domain to the modal domain gives

$$\begin{vmatrix} e_{10} \\ e_{11} \end{vmatrix} = [A]^{-1} \begin{vmatrix} E_{1a} \\ E_{1b} \end{vmatrix} \tag{1.222}$$

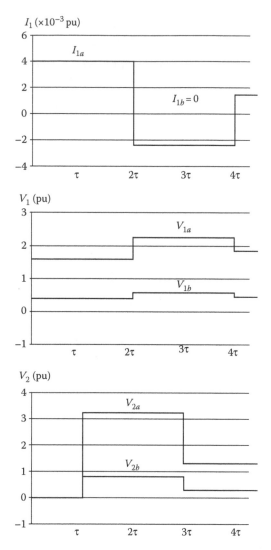

FIGURE 1.47
Analytical voltage waveforms on a two-conductor system.

2. *Propagation of modal traveling waves to the receiving end:* Modal traveling waves e_{10} and e_{11} propagate to the receiving end by propagation velocities c_0 and c_1, respectively. The traveling time of each is given by

$$\tau_0 = \frac{\ell}{c_0}, \quad \tau_1 = \frac{\ell}{c_1} \tag{1.223}$$

where $\tau_0 > \tau_1$ because $c_0 < c_1$.

Thus, the modal traveling wave at the receiving end is

$$e_{20} = e_{10}(t - \tau_0), \quad e_{21} = e_1(t - \tau_1) \tag{1.224}$$

Transforming this traveling wave back to the phase domain gives

$$\begin{vmatrix} E_{2a} \\ E_{2b} \end{vmatrix} = [A] \begin{vmatrix} e_{20} \\ e_{21} \end{vmatrix} \tag{1.225}$$

3. *Receiving-end voltage:* The receiving-end voltage (V_2) is obtained using traveling waves in the same manner as in Section 1.6.4.2:

$$(V_2) = [\lambda_{23}] \begin{vmatrix} E_{2a} \\ E_{2b} \end{vmatrix} \tag{1.226}$$

The differences in modal velocities obtained by repeating these steps are included in transient voltage and current calculations.

EXAMPLE 1.5

Assuming that $c_0 = 250$ m/μs, $c_1 = 300$ m/μs, and line length $\ell = 750$ m in Example 1.4, calculate (V_1), (V_2) for $0 \le t < 2\tau_1$.

Solution

$$\tau_0 = \frac{750}{250} = 3\,\mu s, \quad \tau_1 = \frac{750}{300} = 2.5\,\mu s$$

Traveling wave at $t = 0$ at the sending end is

$$(V_1) = (E_1) = \begin{vmatrix} 1.6 \\ 0.4 \end{vmatrix}.$$

The voltage transformation matrix A for a symmetrical two-conductor system is given by

$$[A] = \begin{vmatrix} 1 & 1 \\ 1 & -1 \end{vmatrix}, \quad [A]^{-1} = \frac{1}{2} \begin{vmatrix} 1 & 1 \\ 1 & -1 \end{vmatrix}.$$

Modal traveling waves e_{10} and e_{11} at the sending end are

$$\begin{pmatrix} e_{10} \\ e_{11} \end{pmatrix} = [A]^{-1}(E_1) = \frac{1}{2}\begin{vmatrix} 1 & 1 \\ 1 & -1 \end{vmatrix}\begin{vmatrix} 1.6 \\ 0.4 \end{vmatrix} = \begin{vmatrix} 1.0 \\ 0.6 \end{vmatrix}.$$

Modal traveling waves at the receiving end are given in the following forms:

$$e_{20}(t) = e_{10}(t - \tau_0) = 1.0u(t - \tau_0)$$
$$e_{21}(t) = 0.6u(t - \tau_1)$$

Transforming these modal components into an actual phase domain (phasor) component gives

$$\begin{vmatrix} E_{2a} \\ E_{2b} \end{vmatrix} = [A](e_2) = \begin{vmatrix} 1 & 1 \\ 1 & -1 \end{vmatrix}\begin{vmatrix} u(t - \tau_0) \\ 0.6u(t - \tau_1) \end{vmatrix} = \begin{vmatrix} u(t - \tau_0) + 0.6u(t - \tau_1) \\ u(t - \tau_0) - 0.6u(t - \tau_1) \end{vmatrix} = c.$$

Thus, the receiving-end voltage (V_r) is given by

$$\begin{vmatrix} V_{2a} \\ V_{2b} \end{vmatrix} = [\lambda_{23}](E_{2f}) = 2(E_{2f}) = \begin{vmatrix} 2u(t - \tau_0) + 1.2u(t - \tau_1) \\ 2u(t - \tau_0) - 1.2u(t - \tau_1) \end{vmatrix}.$$

The reflected waves at the receiving end are

$$\begin{vmatrix} E_{21a} \\ E_{21b} \end{vmatrix} = (V_2) - (E_{2f}) = (E_{2f}) = (E_{21}).$$

In the modal domain:

$$\begin{vmatrix} e_{210} \\ e_{211} \end{vmatrix} = [A]^{-1}(E_{21}) = (e_{21})\begin{vmatrix} u(t - \tau_0) \\ 0.6u(t - \tau_1) \end{vmatrix}.$$

Backward traveling waves at the sending end are given by

$$e_{1b0}(t) = e_{210}(t - \tau_0) = u(t - 2\tau_0)$$
$$e_{1b1}(t) = e_{211}(t - \tau_1) = 0.6u(t - 2\tau_1).$$

Transforming into the actual phase domain gives

$$\begin{vmatrix} c \\ E_{1bb} \end{vmatrix} = [A](e_{1b0}) = \begin{vmatrix} u(t - 2\tau_0) + 0.6u(t - 2\tau_1) \\ u(t - 2\tau_0) - 0.6u(t - 2\tau_1) \end{vmatrix}.$$

Thus, the sending-end voltages are

$$\begin{vmatrix} V_{1a} \\ V_{1b} \end{vmatrix} = (E_1) + \begin{bmatrix} \lambda_{21}(E_{1b}) = \begin{vmatrix} 1.6 \\ 0.4 \end{vmatrix} + \begin{vmatrix} 0.4 & 0 \\ -0.4 & 2 \end{vmatrix} \begin{vmatrix} u(t-2\tau_0)+0.6u(t-2\tau_1) \\ u(t-2\tau_0)-0.6u(t-2\tau_1) \end{vmatrix} \end{bmatrix}.$$

$$= \begin{vmatrix} 1.6 \\ 0.4 \end{vmatrix} + \begin{vmatrix} 0.4u(t-2\tau_0)+0.24u(t-2\tau_1) \\ 1.6u(t-2\tau_0)-1.11u(t-2\tau_1) \end{vmatrix}$$

These results are shown in Figure 1.48 by a real line in comparison with those in Figure 1.47, with constant velocity $c = 300$ m/μs ($\tau = 2.5$ μs) indicated with a dotted line. It can be observed from the figure that a negative voltage appears first at the receiving end on the induced phase (V_b) because the mode 1 traveling wave arrives at the receiving end at $t = \tau_1 = 2.5$ μs. Then, 0.5 μs later, the mode 0 wave arrives, and voltage V_b becomes positive and equal to the voltage neglecting the modal velocity

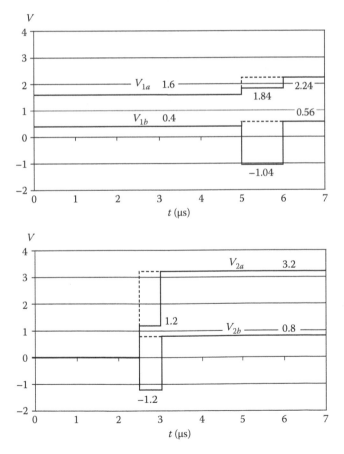

FIGURE 1.48
Analytical surge waveforms when considering modal velocities.

difference (dotted line). The negative voltage appears on the induced phase at the sending end when a refracted wave from the receiving end comes back to the sending end. The following phenomenon is clearly observed from these analytical calculations: phase b voltage becomes negative when modal voltages, which are positive, are transformed into the actual phase domain (see the transformation matrices).

This phenomenon was also observed in the field measurement of a 500 kV untransposed horizontal line [27], the line configuration of which is given in Figure 1.22. The measured result is shown in Figure 1.49.

The negative voltages on phases b and c in Figure 1.49 are explained by the analytical evaluation given earlier. The distorted waveform observed in the measured result is caused by the frequency-dependent attenuation and propagation velocities, as already explained in Figure 1.28, which is, in fact, a step response on the same line as that of Figure 1.49, that is, the 500 kV untransposed horizontal line in Figure 1.22.

Similar but more complicated behaviors are observed in an untransposed vertical twin-circuit line, as discussed in References 28 and 29.

1.6.4.4 Consideration of Losses in a Two-Conductor System

Traveling-wave deformation at a distance x from the sending end is defined in frequency domain by

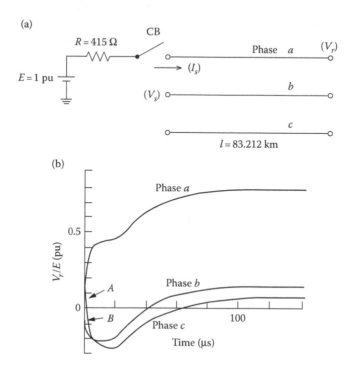

FIGURE 1.49
A field test circuit and test result from Reference 27. (a) Test circuit. (b) Test result V_r.

$$E_x(\omega) = \exp\{-\Gamma(\omega)x\}E_0(\omega)$$

or in Laplace domain with Laplace operator $s = j\omega + \alpha$ as

$$E_x(s) = \exp\{-\Gamma(s)x\}E_0(s) \tag{1.227}$$

where
$E_x(s)$ is the traveling wave at distance x
$E_0(s)$ is the original wave at $x = 0$

The inverse Laplace transform of this equation gives the following time response:

$$L^{-1}E_x(s) = L^{-1}\left[s\exp\left\{-\Gamma(s)\cdot\frac{x}{s}\right\}\cdot E_0(s)\right] = \left(\frac{d}{dt}\right)\int_0^t s(\tau)\cdot e_0(t-\tau)\cdot d\tau$$

or

$$e_x(t) = s(t) * e_0(t) \tag{1.228}$$

where

$$e_x(t) = L^{-1}E_x(s), \quad e_0(t) = L^{-1}E_0(s) \tag{1.229}$$

$s(t) = L^{-1}\exp\{-\Gamma(s)x/s\}$ is the step response of wave deformation
* is the real-time convolution

Figure 1.50 illustrates the step responses of wave deformation.
Figure 1.49 is a measured result of the step response on a three-phase line, when a step-function voltage $e_0(t) = 1$ is applied to phase a at the sending end $(x = 0)$ of the line.
Let us assume that the propagation constant $\Gamma(\omega)$ is given as a constant value at a frequency:

$$\Gamma = \alpha + j\beta = \text{const.}, \beta = \frac{\omega}{c}: \text{phase constant, } c\text{: velocity} \tag{1.230}$$

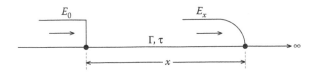

FIGURE 1.50
Step response of wave deformation.

Then, $s(t)$ is given by

$$s(t) = L^{-1}\left[\exp(-\alpha x)\cdot\exp\left(\frac{-s\tau}{s}\right)\right] = \exp(-\alpha x)\cdot u(t-\tau) \tag{1.231}$$

where

$$\tau = \frac{x}{c}, \quad u(t)\text{: unit step function} \tag{1.232}$$

Assuming $e_0(t)$ to be $e_0\cdot u(t)$, Equation 1.228 is rewritten as

$$e_x(t) = k\cdot e_0\cdot u(t-\tau) \tag{1.233}$$

where

$$k = \exp(-\alpha\cdot x)\text{: attenuation ratio}, \quad \alpha(\text{Np/m}) = \frac{\alpha}{8.686}(\text{dB/m}) \tag{1.234}$$

For a multiconductor system, this equation is applied to each modal wave.

EXAMPLE 1.6

For the problem in Example 1.5, calculate voltages considering attenuations $k_0 = 0.8$ and $k_1 = 0.98$.

Solution

From the solution of Example 1.5 at $t = 0$:

$$\begin{vmatrix} V_{1a} \\ V_{1b} \end{vmatrix} = \begin{vmatrix} 1.6 \\ 0.4 \end{vmatrix}, \quad \begin{vmatrix} e_{10} \\ e_{11} \end{vmatrix} = \begin{vmatrix} 1.0 \\ 0.6 \end{vmatrix}.$$

Modal traveling waves at the receiving end considering attenuation are given by

$$e_{20} = k_0 e_{10} U(t-\tau_0) = 0.8U(t-\tau_0)$$
$$e_{21} = k_1 e_{11} U(t-\tau_1) = 0.588U(t-\tau_1).$$

In an actual phase domain:

$$\begin{vmatrix} E_{2a} \\ E_{2b} \end{vmatrix} = \begin{vmatrix} 1 & 1 \\ 1 & -1 \end{vmatrix}\begin{vmatrix} 0.8u(t-\tau_0) \\ 0.558u(t-\tau_1) \end{vmatrix} = \begin{vmatrix} 0.8u(t-\tau_0)+0.558u(t-\tau_1) \\ 0.8u(t-\tau_0)-0.558u(t-\tau_1) \end{vmatrix}.$$

Thus, the receiving-end voltage is given as follows:

$$\begin{vmatrix} V_{2a} \\ V_{2b} \end{vmatrix} = 2 \begin{vmatrix} E_{2a} \\ E_{2b} \end{vmatrix} \begin{vmatrix} 1.6u(t-\tau_0)+1.176u(t-\tau_1) \\ 1.6u(t-\tau_0)-1.176u(t-\tau_1) \end{vmatrix}.$$

The reflected wave at the receiving end is

$$\begin{vmatrix} E_{21a} \\ E_{21b} \end{vmatrix} = \begin{vmatrix} E_{2a} \\ E_{2b} \end{vmatrix}.$$

Transforming into a modal domain gives

$$\begin{vmatrix} e_{210} \\ e_{211} \end{vmatrix} = \begin{vmatrix} e_{20} \\ e_{21} \end{vmatrix} = \begin{vmatrix} 0.8u(t-\tau_0) \\ 0.588u(t-\tau_1) \end{vmatrix}.$$

At the sending end, the traveling wave is attenuated as

$$e_{1b0}(t) = k_0 e_{210} U(t-\tau_0) = 0.64u(t-2\tau_0)$$
$$e_{1b1}(t) = 0.98 \times 0.588u(t-2\tau_1) = 0.576u(t-2\tau_1).$$

Transforming into a phasor domain gives

$$\begin{vmatrix} E_{1ba} \\ E_{1bb} \end{vmatrix} = \begin{vmatrix} 1 & 1 \\ 1 & -1 \end{vmatrix} \begin{vmatrix} e_{1b0}(t) \\ e_{1b1}(t) \end{vmatrix} = \begin{vmatrix} 0.64u(t-2\tau_0)+0.576(t-2\tau_1) \\ 0.64u(t-2\tau_0)-0.576(t-2\tau_1) \end{vmatrix}.$$

Thus, the sending-end voltage for $2\tau_1 \le t < 4\tau_1$ is obtained as

$$\begin{vmatrix} V_{1a} \\ V_{1b} \end{vmatrix} = \begin{vmatrix} 1.6 \\ 0.4 \end{vmatrix} + \begin{vmatrix} 0.4 & 0 \\ -0.4 & 2 \end{vmatrix} \begin{vmatrix} 0.64u(t-2\tau_0)+0.576(t-2\tau_1) \\ 0.64u(t-2\tau_0)-0.576(t-2\tau_\tau) \end{vmatrix}.$$
$$= \begin{vmatrix} 1.6 \\ 0.4 \end{vmatrix} + \begin{vmatrix} 0.25u(t-2\tau_0)+0.23(t-2\tau_1) \\ 1.024u(t-2\tau_0)-1.382(t-2\tau) \end{vmatrix}.$$

These results are illustrated in Figure 1.51.

1.6.4.5 Three-Conductor Systems

Let us consider the field test circuit in Figure 1.49a. The parameters of the line are given as

Characteristic impedance: $Z_{aa} = Z_{cc} \cong Z_{bb} = 331$, $Z_{ab} = Z_{bc} = 72$, $Z_{ac} = 34$ (Ω)

Modal velocity: $c_0 = 270.4$, $c_1 = 296.4$, $c_2 = 299.5$ (m/μs)

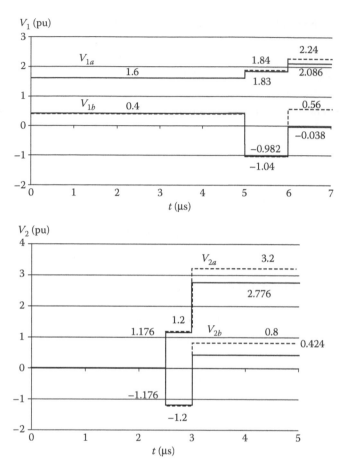

FIGURE 1.51
Analytical results of surge voltages on a two-phase line.

Modal attenuation: $\alpha_0 = 7.94 \times 10^{-2}$, $\alpha_1 = 3.5 \times 10^{-3}$, $\alpha_2 = 6.8 \times 10^{-4}$ (dB/km)

Attenuation ratio: $k_0 = 0.468$, $k_1 = 0.967$, $k_2 = 0.994$, (see Equation 1.234)

Line length: $\ell = 83.212$ km

Modal traveling time: $\tau = \ell/c$: $\tau_0 = 307.7$, $\tau_1 = 280.7$, $\tau_2 = 277.8$ (μs)

Assume $E = 1000$ (V)

When a circuit breaker (CB, a switch) is closed at $t = 0$, the phase a voltage at the sending end is calculated as the ratio of source resistance R and the phase a characteristic impedance Z_{aa} as

$$V_{sa} = \left\{ \frac{Z_{aa}}{(R + Z_{aa})} \right\} \cdot E = \left\{ \frac{331}{415 + 331} \right\} \times 1000 = 440 \text{ (V)}.$$

Then, the phase *a* current at the sending end is

$$I_{sa} = \frac{V_{sa}}{Z_{aa}} = \frac{440}{331} \, (\text{A}).$$

Phase *b* and *c* currents are zero because of the open-circuit condition:

$$I_{sb} = I_{sc} = 0.$$

Thus, the sending-end voltage (V_s) is calculated by using the characteristic impedance matrix from the three-phase currents:

$$(V_s) = [Z](I_s) \tag{1.235}$$

Thus, the following results are obtained:

$$V_{sb} = 97, \quad V_{sc} = 53 \, (\text{V})$$

Assume that the voltage transformation matrix is given by

$$[A] = \begin{bmatrix} 1 & 1 & 1 \\ 1 & 0 & -2 \\ 1 & -1 & 1 \end{bmatrix}, \quad [A]^{-1} = \left(\frac{1}{6}\right)\begin{bmatrix} 2 & 2 & 2 \\ 3 & 0 & -3 \\ 1 & -2 & 1 \end{bmatrix}.$$

Then, the modal traveling wave at the sending end is calculated as

$$e_{s0} = 197, \quad e_{s1} = 193.5, \quad e_{s2} = 49.8 \, (\text{V}).$$

Each modal wave arrives at the receiving end at different times (τ_i) and with different attenuations (k_i):

$$e_{r0} = k_0 \cdot e_{s0} \cdot u(t - \tau_0) = 92u(t - \tau_0),$$
$$e_{r1} = k_1 \cdot e_{s1} \cdot u(t - \tau_1) = 187u(t - \tau_1)$$
$$e_{r2} = k_2 \cdot e_{s2} \cdot u(t - \tau_2) = 49.5u(t - \tau_2)$$

The refraction coefficient at the receiving end is given by a unit matrix multiplied by factor 2 because the three phases at the receiving end are open-circuited. Thus,

$$v_{r0} = 2e_{r0} = 184u(t - \tau_0),$$
$$v_{r1} = 374u(t - \tau_1),$$
$$v_{r2} = 99u(t - \tau_2)$$

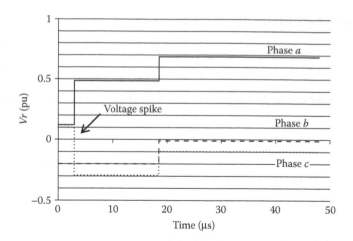

FIGURE 1.52

Analytical results of surge voltages on a three-phase line corresponding to Figure 1.49.

Transforming these modal voltages to phase voltages by $V = Av$ gives

$$V_{ra} = v_{r0} + v_{r1} + v_{r2} = 184u(t - \tau_0) + 374u(t - \tau_1) + 99u(t - \tau_2)$$
$$V_{rb} = v_{r0} - 2v_{r2} = 184u(t - \tau_0) - 198u(t - \tau_2)$$
$$V_{rc} = v_{r0} - v_{r1} + v_{r2} = 184u(t - \tau_0) - 374u(t - \tau_1) + 99u(t - \tau_2)$$

In drawing these results while considering the time difference τ_i, Figure 1.52 is obtained. It can be observed from the figure that a positive voltage spike appears on phase c for a time period of $0 \leq t \leq \tau_1 - \tau_2 = 2.9$ μs. This explains the voltage spike A in the measured results of Figure 1.49b. Similarly, the negative voltage B on phase b in Figure 1.49b is explained by the analytical calculation as in Figure 1.52.

1.6.4.6 Cascaded System Composed of Different Numbers of Conductors

It is often observed in practice that the number of conductors changes at a boundary, as shown in Figure 1.53, where phases a and b are short-circuited at node 1. In the case of a cross-bonded cable, three-phase metallic sheaths are rotated at every cross-bonding point. In such a case, it is required to reduce the order of an impedance matrix and/or to rotate the matrix elements.

In Figure 1.53, the following relations of voltages and currents are obtained:

$$(V) = \begin{pmatrix} V_a \\ V_b \\ V_c \end{pmatrix} = \begin{bmatrix} 1 & 0 \\ 1 & 0 \\ 0 & 1 \end{bmatrix} \begin{pmatrix} V_1 \\ V_2 \end{pmatrix} = [T_t](V') \tag{1.236}$$

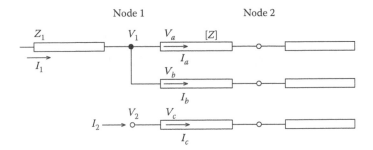

FIGURE 1.53
A single-conductor system connected to a three-conductor system.

$$(I') = \begin{pmatrix} I_1 \\ I_2 \end{pmatrix} = \begin{bmatrix} 1 & 1 & 0 \\ 0 & 0 & 1 \end{bmatrix} \begin{pmatrix} I_a \\ I_b \\ I_c \end{pmatrix} = [T](I) \tag{1.237}$$

By applying the relation $V = Z \cdot I$, I', this equation is rewritten as $(I') = [T]$ $(I) = [T][Z]^{-1}(V) = [T][Z]^{-1}[T_t](V)' = [Z']^{-1}(V')$:

$$\therefore [Z']^{-1} = [T][Z]^{-1}[T_t] \tag{1.238}$$

In this equation, Z is an original 3×3 matrix on the right of node 1, while Z' is reduced to a 2×2 matrix considering the short circuit of phases a and b. By using Z', the refraction coefficient at node 1, for example, can be calculated. Remember the fact that no inverse matrices exist for matrices T and T_t, so the correct sequence should be followed for calculating Equation 1.238.

PROBLEMS

1.18 Obtain voltage V, current I, and reflected voltage traveling wave e_r at node P when step-function voltage traveling wave e_0 arrives at node P at $t = 0$ in Figure 1.54 for $I(0) = 0$.

1.19 Obtain voltage V at node P when step-function voltage wave e_0 arrives at node P at $t = 0$ in Figure 1.55 for $V(0) = 0$.

1.20 When switch S is closed at $t = 0$ in Figure 1.56, obtain voltages V_2 and V_3 and current I_2 for conditions (a) and (b), and draw curves for V_2, V_3, and I_2 for $0\,0 \leq t \leq 15$ μs:

   a. $Z_2 = 60\ \Omega$, $c_2 = 150$ m/μs, $l_2 = 150$ m.

   b. $c_2 = c_1$, $Z_2 = Z_1$, $l_2 = 300$ m.

   c. Discuss the effect of a cable connected to an overhead line, assuming a surge impedance 300 Ω for the overhead line and 60 Ω for the cable.

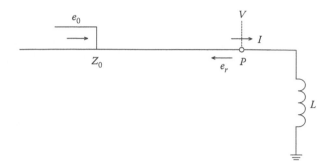

FIGURE 1.54
Inductance *L* terminated line.

1.21 Obtain receiving-end voltage (V_r) when a source voltage is applied to phase *b* in Figure 1.49a.

1.7 Nonuniform Conductors

There are a number of papers on nonuniform lines [30–49]. EMC-related transients or surges in a gas-insulated substation and on a tower involve nonuniform lines, such as short-line, nonparallel, and vertical conductors. Pollaczek's [7], Caron's [8], and Sunde's [50] impedance formulas for an overhead line are well known and have been widely used in the analysis of the transients mentioned earlier. However, it is not well known that these formulas were derived assuming an infinitely long and thin conductor, that is, a uniform and homogeneous line. Thus, impedance formulas are restricted to the uniform line where the concept of "per-unit-length impedance" is applicable.

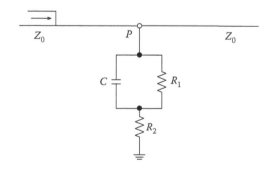

FIGURE 1.55
A tower model.

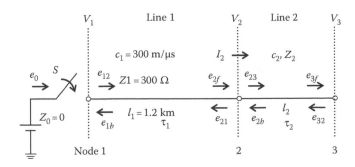

FIGURE 1.56
Two cascaded lines.

This section explains impedance and admittance formulas of nonuniform lines, such as finite-length horizontal and vertical conductors based on a plane wave assumption. The formulas are applied to analyze a transient on a nonuniform line by an existing circuit theory-based simulation tool such as the EMTP [9,11]. The impedance formula is derived based on Neumann's inductance formula by applying the idea of complex penetration depth explained earlier. The admittance is obtained from the impedance assuming the wave propagation velocity is the same as the light velocity in free space in the same manner as an existing admittance formula, which is almost always used in steady-state and transient analyses on an overhead line.

1.7.1 Characteristic of Nonuniform Conductors

1.7.1.1 Nonuniform Conductors

First, it is necessary to clarify a problem to be discussed in this section, that is, a nonuniform line or a nonhomogeneous line. Figure 1.57 shows a typical example of transient voltage responses measured on a vertical conductor with radius $r = 25$ mm and height $h = 25$ m [41,42].

Figure 1.57a is the current, (b) the voltage at the top of the conductor, and (c) the voltage at a height of 12 m. It should be clear from the figure that the voltage waveforms of (b) and (c) are distorted before a reflection from the bottom (earth surface) comes back. Also, the waveform in (c) is different from that in (b). The reason for these phenomena can be either the frequency-dependent effect of the conductor or the reflection of the traveling wave due to discontinuities of the characteristic (surge) impedance along a vertical conductor other than the earth's surface. Radiation is another cause of the distortion in such high-frequency regions. Also, the electromagnetic field is not perpendicular to the conductor surface. This chapter, however, is restricted to the TEM mode of propagation, and these phenomena can be translated or interpreted as the reflection/refraction of a traveling wave. It is hard to receive such noticeable frequency dependence when the distance between (b) and (c)

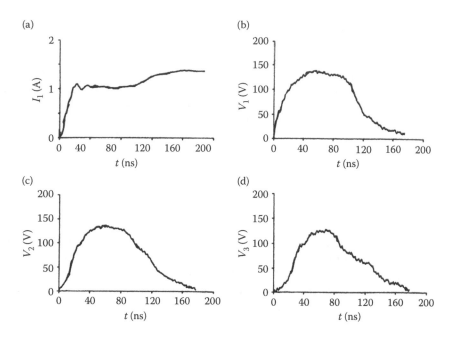

FIGURE 1.57
Measured transient responses at various positions of an artificial tower (h = 15 m, r = 25 mm).
(a) Current at the top. (b) Voltage at the top. (c) Voltage at the height of 12 m. (d) Voltage at the
height of 9 m.

is less than 10 m. Thus, it can be said that the voltage waveform at a vertical
conductor is distorted due to the nonuniformity of the vertical conductor at
every height (position). That is, the characteristic impedance (impedance and
admittance in general) of the vertical conductor is position dependent.

Figure 1.58 shows another example of a transient voltage at the top of
a 1100 kV transmission tower (height 140 m, average radius 6.3 m) [43].
Figure 1.58a is the measured result of an injected current and the voltage at
the top of the tower, and (b) is the step response of the voltage at the top of the
tower numerically evaluated from the measured current and voltage in (a).
The step response is heavily distorted before reflection from the bottom of
the tower, that is, time t is less than 0.933 μs. Although tower arms exist, their
effect on the average distortion is estimated to less than that in Figure 1.58
[42]. Most distortion is estimated based on the position-dependent surge
impedance of the tower.

Figure 1.59 shows a transient-induced voltage at the sending end of a horizontal nonparallel conductor when a steplike voltage is applied to the other
conductor through a resistance nearly equal to the conductor surge impedance. All of the other ends of the conductors are terminated by the surge
impedance. The radius, length, and height of the conductor are 5 mm, 4 m,
and 40 cm.

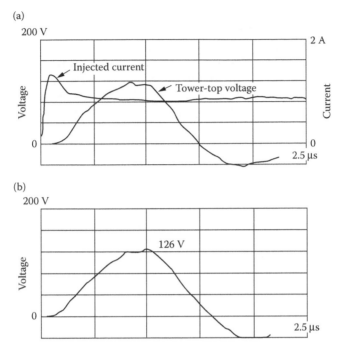

FIGURE 1.58
Measured transient response at the top of a 1100 kV transmission tower. (a) Injected current and voltage at the top of the tower. (b) Step response of the voltage at the top of the tower.

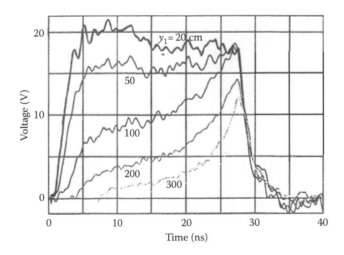

FIGURE 1.59
Measured transient induced voltages on a horizontal nonparallel conductor ($y_2 = 10$ cm, $y_1 = 20$–300 cm).

The separation y_2 at the receiving end between the conductors is 10 cm and that (y_1) at the sending end varies between 10 and 300 cm. A typical characteristic of a nonparallel conductor is observed in the figure due to a position (distance from the sending end)-dependent impedance [40]. The induced voltage gradually increases with time because the mutual impedance increases with time, and a positive reflection comes back to the sending end at every instance until a large negative reflection appears at about $t = 27$ ns from the receiving end.

1.7.1.2 Difference from Uniform Conductors

Carson's, Pollaczek's, and Sunde's formulas are well known and widely used for the impedances of overhead lines and underground cables isolated from the earth (soil). However, that the formulas are applicable only to a uniform or homogeneous line of which the impedance can be defined by per unit length does not seem to be understood. These formulas are based on an infinitely long line or an assumption that line length x is far greater than line height h and separation y between two lines and that h and y are far greater than radius r. The basic form of earth-return impedance is given by

$$Z_{ij} = \int_0^{x_1} \int_0^{x_2} [f(x_1, x_2)] \cdot d_{x1} \cdot d_{x2} \tag{1.239}$$

Assuming x_2 is infinite, the second integral is carried out, and the following expression is obtained:

$$Z_{ij} = \int_0^{x_1} [g(x_1)] \cdot d_{x1} \tag{1.240}$$

Again, assuming x_1 is infinite, Pollaczek's, Carson's, or Sunde's impedance is obtained:

$$Z_{ij} = \frac{j\omega\mu_0}{2\pi} \int_0^\infty \frac{\exp\{-(h_i + h_i)x\}}{\sqrt{x^2 + \gamma_0^2 + x}} \cdot \cos(y_{ij} \cdot x) \cdot dx \tag{1.241}$$

It should be noted that Equations 1.240 and 1.241 already contain the effect of mutual coupling due to the infinitely long line 2, or, in a single-line case, that is, in the self-impedance cases, the section of length d_{x1} has contained the mutual coupling effect due to the remaining part (infinitely long) of the line. Because Equation 1.241 has contained the mutual coupling effect of all the remaining parts of the line on the line section, we can define the impedance per unit length.

It should be clear from Equation 1.239 that the line impedance is a function of the line length x, and

$$\frac{Z_{finite} \text{ in Equation 1.239}}{x < Z_{infinite} \text{ in Equation 1.241 per unit length}} \tag{1.242}$$

Therefore, it is not possible to discuss wave propagation on a finite line by Pollaczek's, Carson's, or Sunde's formulas in Equation 1.241.

The characteristic impedance seen from an arbitrary position (distance x from origin) is always the same on an infinitely long line, as already explained, while it cannot be defined on a finite line. If we define the characteristic impedance $Z_0(\omega)$ as the ratio of voltage $V(\omega)$ and current $I(\omega)$ in a frequency domain at the sending end of the finite line within a time region of two travel times 2τ ($\tau = x/c$, $c = c(\omega)$: velocity), we find a difference between the characteristic impedances $Z_1(\omega)$ and $Z_2(\omega)$ of two lines with the same configuration but different lengths x_1 and x_2, that is, length dependence, because the series impedance of the two lines is different, as explained earlier. The situation is more noticeable in a vertical conductor because the finite length of the vertical line shows the height (distance)-dependent impedance/admittance, and the nonhomogeneity is easily understood from the physical viewpoint as explained in Section 1.7.1.1. It should be noted that "distance" (dependence) means the distance from the origin in a strict sense, not the distance from the earth's surface.

1.7.2 Impedance and Admittance Formulas

1.7.2.1 Finite-Length Horizontal Conductor

1.7.2.1.1 Impedance

The mutual impedance of a nonparallel conductor above the imperfectly conducting earth illustrated in Figure 1.60 is obtained in the following equation by applying the concept of the complex penetration depth to Neumann's inductance formula [39,44]:

$$Z_{ij} = j\omega\left(\frac{\mu_0}{2\pi}\right)P_{ij}(\Omega/m) \quad P_{ij} = (M_d - M_j)\cos\frac{\theta}{2} \tag{1.243}$$

$$M_d = \int_{x_{i1}}^{x_{i2}}\int_{x_{j1}}^{x_{j2}}\left(\frac{1}{S_d}\right)d_{s1}\cdot d_{s2}, \quad M_1 = \int_{x_{i1}}^{x_{i2}}\int_{x_{i1}}^{x_{i2}}\left(\frac{1}{S_i}\right)d_{s1}\cdot d'_{s2} \tag{1.244}$$

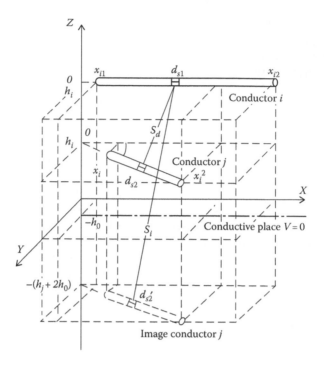

FIGURE 1.60
Nonparallel multiconductor system.

where

$$S_d = \sqrt{s_1^2 + s_2^2 - 2s_1 s_2 \cos\theta + H_1^2}.$$

$$S_i = \sqrt{s_1^2 + s_2'^2 - 2s_1 s_2' \cos\theta + H_2^2}$$

$$H_1 = |h_i - h_j|, \quad H_2 = h_i + h_j + 2h_e, \quad h_e = \sqrt{\dfrac{\rho_e}{j\omega\mu_0}}$$

Integrating Equation 1.244, the following solution is derived:

$$
M_d = x_{i2} \cdot \ln\!\left(\frac{A_{221}}{A_{211}}\right) + x_{j2} \cdot \ln\!\left(\frac{B_{221}}{B_{121}}\right) + x_{i1} \cdot \ln\!\left(\frac{A_{111}}{A_{121}}\right)
$$

$$
+\, x_{j1} \cdot \ln\!\left(\frac{B_{111}}{B_{211}}\right) + \frac{H_1(C_{121} + C_{211} - C_{221}C_{111})}{\sin\theta}
\tag{1.245}
$$

$$
M_i = x_{i2} \cdot \ln\!\left(\frac{A_{222}}{A_{212}}\right) + x_{j2} \cdot \ln\!\left(\frac{B_{222}}{B_{122}}\right) + x_{i1} \cdot \ln\!\left(\frac{A_{112}}{A_{122}}\right)
$$

$$
+\, x_{j1} \cdot \ln\!\left(\frac{B_{112}}{B_{212}}\right) + H_2(C_{122} + C_{212} - C_{222} - C_{112})\sin\theta
\tag{1.246}
$$

where

$$A_{kmn} = x_{jm} - x_{ik}\cos\theta + D_{kmn}, \quad B_{kmn} = x_{ik} - x_{jm}\cos\theta + D_{kmn}$$

$$C_{kmn} = \tan^{-1}\left\{\frac{\left(x_{ik}x_{jm}\sin^2\theta + H_n^2\cos\theta\right)}{(H_n\sin\theta\cdot D_{kmn})}\right\}$$

$$D_{kmn} = \sqrt{x_{ik}^2 + x_{jm}^2 - 2x_{ik}x_{jm}\cos\theta + H_n^2}, \quad H_3 = h_i + h_j$$

$$k, m = 1,2; \quad n = 1,2,3$$

In the case of perfectly conducting earth ($\rho_e = 0$), the substitution of $h_e = 0$ into M_i in Equation 1.246 gives the following expression:

$$P_{0ij} = \frac{(M_d - M_{i0})\cos\theta}{2} \tag{1.247}$$

$$M_{i0} = x_{i2}\cdot\ln\left(\frac{A_{223}}{A_{213}}\right) + x_{j2}\cdot\ln\left(\frac{B_{223}}{B_{123}}\right) + x_{i1}\cdot\ln\left(\frac{A_{113}}{A_{123}}\right)$$

$$+ x_{j1}\cdot\ln\left(\frac{B_{113}}{B_{213}}\right) + \frac{H_2(C_{123} + C_{213} - C_{223} - C_{113})}{\sin\theta}$$

Substituting $x_{i1} = 0$ and $x_{i2} = x_i$ and taking the limit of angle θ to zero with $\tan^{-1}z = \pi/2 - 1/z$ ($|z| > 1$) in Equation 1.245, the impedance formula of a parallel horizontal conductor is obtained in the following form:

$$M_d = x_i\ln\frac{x_i - x_{j1} + \sqrt{(x_i - x_{j1})^2 + d_{ij}^2}}{x_1 - x_{j2} + \sqrt{(x_i - x_{j1})^2 + d_{ij}^2}} - x_{i1}\ln\frac{x_i - x_{j1} + \sqrt{(x_i - x_{j1})^2 + d_{ij}^2}}{\sqrt{x_{j1}^2 + d_{ij}^2} - x_{j1}}$$

$$- x_{i2}\ln\frac{x_i - x_{j2} + \sqrt{(x_i - x_{j2})^2 + d_{ij}^2}}{\sqrt{x_{j2}^2 + d_{ij}^2} - x_{j2}} - \sqrt{(x_i - x_{j1})^2 + d_{ij}^2}$$

$$+ \sqrt{(x_i - x_{j2})^2 + d_{ij}^2} + \sqrt{x_{j1}^2 + d_{ij}^2} - \sqrt{x_{j2}^2 + d_{ij}^2} \tag{1.248}$$

$$M_i = x_i\ln\frac{x_i - x_{j1} + \sqrt{(x_i - x_{j1})^2 + S_{ij}^2}}{x_1 - x_{j2} + \sqrt{(x_i - x_{j1})^2 + S_{ij}^2}} - x_{j1}\ln\frac{x_i - x_{j1} + \sqrt{(x_i - x_{j1})^2 + S_{ij}^2}}{\sqrt{x_{j1}^2 + S_{ij}^2} - x_{j1}}$$

$$+ x_{j2}\ln\frac{x_i - x_{j2} + \sqrt{(x_i - x_{j2})^2 + S_{ij}^2}}{\sqrt{x_{j2}^2 + S_{ij}^2} - x_{j2}} - \sqrt{(x_i - x_{j1})^2 + S_{ij}^2}$$

$$+ \sqrt{(x_i - x_{j2})^2 + S_{ij}^2} + \sqrt{x_{j1}^2 + S_{ij}^2} - \sqrt{x_{j2}^2 + S_{ij}^2}$$

When x_{j1} is taken to be zero, this equation becomes identical to that given in Reference 39. Also, the substitution of $x_{j1} = 0$ and $x_{j2} = x_j$ into this equation gives the same formula as that derived in Reference 44.

In the case that $x_{i1} = x_{j1} = 0$ and $x_{i2} = x_{j2} = x$ in Equations 1.245 and 1.246, that is, the case of a parallel horizontal conductor with the same length, the formula is simplified as follows:

$$P_{ij} = \frac{(M_d - M_i)}{2} = x \ln \left[\frac{\left\{ 1 + \sqrt{1 + (d_{ij}/x)^2} \right\}}{\left\{ 1 + \sqrt{1 + (S_{ij}/x)^2} \right\}} \right]$$

$$+ x \ln \left(\frac{S_{ij}}{d_{ij}} \right) - \sqrt{x^2 + d_{ij}^2} + \sqrt{x^2 + S_{ij}^2} + d_{ij} - S_{ij} \qquad (1.249)$$

Taking a limit of distance x to infinity, the per unit length in Equation 1.249 is reduced to only the second term, that is:

$$Z_{ij} = j\omega \left(\frac{\mu_0}{2\pi} \right) \ln \left(\frac{S_{ij}}{d_{ij}} \right) (\Omega/m) \qquad (1.250)$$

This equation is identical to the impedance derived for an infinite horizontal conductor in Reference 6, which is an approximation of Carson's earth-return impedance, as already explained in Section 1.2.2.2.

From this observation, it should be clear that the impedance formulas in Equations 1.245 and 1.246 are the most generalized forms for a horizontal conductor, although they are approximate formulas based on the concept of penetration depth.

1.7.2.1.2 Admittance

The admittance of a finite-length horizontal conductor is evaluated by the potential coefficient of a perfectly conducting earth [1]:

$$[Y] = j\omega[C], \quad [C] = 2\pi\varepsilon_0 [P_0]^{-1} \qquad (1.251)$$

The element P_{0ij} of matrix $[P_0]$ is given in Equation 1.247.

1.7.2.2 Vertical Conductor

Let us consider the vertical multiconductor system illustrated in Figure 1.61. In the same manner as the finite-length horizontal conductor, the following impedance formula is obtained:

$$Z_{ij} = j\omega \left(\frac{\mu_0}{2\pi} \right) P_{ij} \qquad (1.252)$$

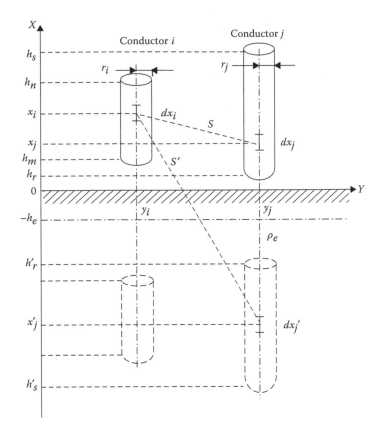

FIGURE 1.61
Vertical multiconductor system.

$$P_{ij} = (1-2X_{ij})[-h_1 \ln A_1 + h_2 \ln A_2] + h_3 \ln A_3 - h_4 \ln A_4 + a_1 - a_2 - a_3 + a_4$$
$$- (h_5 \ln A_5 - h_6 \ln A_6 - h_7 \ln A_7 + h_8 \ln A_8 - a_5 + a_6 + a_7 - a_8) \qquad (1.253)$$

where

$$X_{ii} = h_n - h_m, \quad X_{ij} = h_s - h_r, \quad X_{ij} = \frac{(X_{ii} + X_{jj})}{2}$$

$$h_1 = h_n - h_s, \quad h_2 = h_m - h_s, \quad h_3 = h_n - h_r$$

$$h_4 = h_m - h_r, \quad h_5 = h_n + h_s + 2h_e, \quad h_6 = h_m + h_s + 2h_e$$

$$h_7 = h_m + h_r + 2h_e, \quad h_8 = h_m + h_r + 2h_e$$

$$a_k = \sqrt{h_k^2 + d^2}, \quad A_k = a_k + h_k \quad (k = 1, 2, \dots, 8)$$

When conductors i and j are at the same vertical position, that is, $h_s = h_n$, $h_r = h_m$, Equation 1.253 is simplified to the following form:

$$P_{ij} = \ln\left[\frac{\left(\sqrt{d^2+X^2}+X\right)\left\{\sqrt{d^2+(H-2X)^2}+(H-2X)\right\}}{d\left\{\sqrt{d^2+(H-X)^2}+(H-X)\right\}}\right]+\left(\frac{H}{2X}\right)$$

$$\times \ln\left[\frac{\left\{\sqrt{d^2+(H-X)^2}+(H-X)\right\}^2}{\left(\sqrt{d^2+H^2}+H\right)\left\{\sqrt{d^2+(H-2X)^2}+(H-2X)\right\}}\right]$$

$$+\left(\frac{1}{2X}\right)\left[2d+\sqrt{d^2+H^2}+\sqrt{d^2+(H-2X)^2}\right.$$

$$\left. -2\sqrt{d^2+X^2}-2\sqrt{d^2+(H-X)^2}\right] \tag{1.254}$$

where
$H = 2(h+h_e)$
$h = h_n = h_s$
$h - X = h_m = h_r$

If the earth is assumed to be perfectly conducting, this equation can be further simplified as follows:

$$P_{0ij} = \ln\left[\frac{\left(\sqrt{d^2+X^2}+X\right)\left\{\sqrt{d^2+4(h-X)^2}+2(h-X)\right\}}{d\left\{\sqrt{d^2+(2h-X)^2}+(2h-X)\right\}}\right]$$

$$+\left(\frac{h}{X}\right)\ln\left[\frac{\left\{\sqrt{d^2+(2h-X)^2}+(2h-X)\right\}^2}{\left(\sqrt{d^2+4h^2}+2h\right)\left\{\sqrt{d^2+4(h-X)^2}+2(h-X)\right\}}\right]$$

$$+\left(\frac{1}{2X}\right)\left[2d+\sqrt{d^2+4h^2}+\sqrt{d^2+4(h-X)^2}\right.$$

$$\left. -\left[2\sqrt{d^2+X^2}-2\sqrt{d^2+(2h-X)^2}\right]\right. \tag{1.255}$$

When the bottom of a single conductor ($d = r$) is on the earth's surface, that is, $h - X = 0$, then

$$P_0 = \ln\left[\frac{\left\{\sqrt{r^2+h^2}+h\right\}^2}{\left\{\sqrt{r^2+4h^2}+2h\right\}r}\right]+\frac{\left\{\sqrt{r^2+4h^2}+3r/2-2\sqrt{r^2+h^2}\right\}}{h} \tag{1.256}$$

The admittance of the vertical conductor system is evaluated from Equation 1.251.

1.7.3 Line Parameters

1.7.3.1 Finite Horizontal Conductor

Figure 1.62 shows measured results (exp.) of the self-impedance and capacitance of a conductor with radius 1 cm and length 4 m above a copper plate,

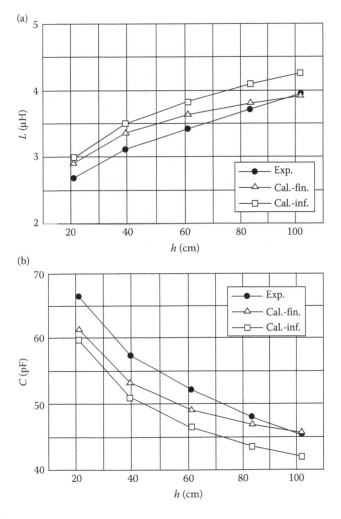

FIGURE 1.62
Self-inductance and capacitance of a finite horizontal conductor. (a) Inductance. (b) Capacitance. exp.: experimental result. cal.-fin.: calculated result of finite line impedance. cal.-inf.: calculated result of Carson's infinite line impedance.

when a step voltage with rise time 2 ns is applied, as a function of height h together with calculated results by the proposed formula (cal. fin.) of a finite-length conductor, Carson's formula (cal. fin.) of a finite-length conductor, and Carson's formula (cal. inf.) of an infinitely long conductor. In the figure, the calculated results by the proposed formula (average error 5.1%) show more accuracy than those by Carson's formula (average error 10.9%). It should be noted that the accuracy of the proposed formula increases as the conductor height increases. The degree of finite length is defined by the ratio of the conductor's length x and height h, x/h. As x/h decreases, the accuracy of the finite conductor formula increases. It should be noted that the inductance per unit length decreases as the length decreases. The reason for this is that the inductance of an infinitely long conductor includes the mutual inductance between the reference part of a conductor and the remaining part with infinite length as explained in Section 1.7.2.2.

Figure 1.63 shows the mutual inductance between two conductors with different lengths x_1 and x_2.

Because Carson's formula cannot deal with conductors of different lengths, three approaches to determine an effective length x are investigated: (a) x = shorter length x_2, (b) arithmetic mean distance $x = (x_1 + x_2)/2$, and (c) geometrical mean distance $x = \sqrt{x_1 \cdot x_2}$. It can be observed from Figure 1.63 that the proposed formula, in general, is accurate. For Carson's formula, approach (a), shorter length, seems to be best among the three. It should be noted that Carson's inductance evaluated even by approach (a) is greater than the measured result. The reason for this is as explained in Section 1.7.2.2. This observation is for a perfectly conducting earth, that is, for a high frequency. In power frequency regions, Carson's formula shows a rather poor accuracy. It has been pointed out in Reference 44 that the average error of Carson's formula was about 21%, while that of the proposed formula was 4%.

Table 1.11 shows measured and calculated results of the surge impedance of a horizontal conductor. It is clear that the proposed formula shows more accuracy than Carson's. The accuracy of the proposed formula increases as x/h decreases, corresponding to the characteristic of the inductance. A similar observation has been made in different measurements in Reference 44.

1.7.3.2 Vertical Conductor

Table 1.12 shows a comparison of measured and calculated surge impedances of a vertical single conductor with height h and radius r. Included in the table are results calculated by various formulas given in References 31,32,34,41. It is clear from the table that the accuracy of the proposed formula is higher than the other formulas in comparison with the measured results. Jordan's formula [31] also shows a high accuracy. It is worth noting that the proposed formula, Equation 1.256 for a single conductor, is identical to Jordan's formula. With $r \ll h$, Equation 1.256 is further simplified as follows:

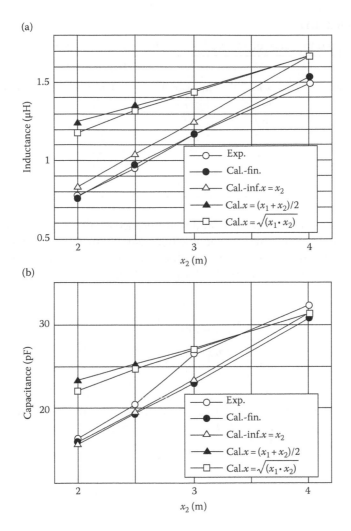

FIGURE 1.63
Mutual inductance between different-length horizontal conductors. (a) Inductance. (b) Capacitance.

$$P_0 \approx \ln\left(\frac{4h^2}{4hr}\right) + h - \left(\frac{2h}{h}\right) = \ln\left(\frac{h}{r}\right) - 1 = \ln\left(\frac{h}{er}\right) \qquad (1.257)$$

where $e = 2.71828$.

For the surge impedance case:

$$Z_s = 60\left\{\ln\left(\frac{h}{r}\right) - 1\right\} = \text{Jordan's formula} \qquad (1.258)$$

TABLE 1.11

Surge Impedance of an Overhead Conductor with Length $x = 4$ m

		Surge Impedance (Ω)		
h (cm)	*x/h*	Measured	Finite	Infinite
21.0	19.1	201	218	224
39.5	10.2	233	251	262
61.0	6.57	256	272	287
83.5	4.80	278	285	307
102	3.93	295	293	319
Error (%)	–	5.1	10.9	

Results calculated by the formula in Reference 42 also agree with the measured results. This is quite reasonable because the formula has been derived empirically from the measured results. It is interesting that the empirical formula agrees with Equation 1.258 by rewriting it as follows:

$$Z_h = 60\left\{\ln\left(\frac{2\sqrt{2h}}{(r-2)}\right)\right\} = 60\left\{\ln\left(\frac{h}{r}\right) - 0.96\right\}$$

$$= 60\left\{\ln\left(\frac{h}{r}\right) - 1\right\} + 2.4 \tag{1.259}$$

The fact that the empirical formula differs from the proposed formula (Equation 1.258) only with 2.4 Ω is proof of the high accuracy of the proposed formula compared with the measured results.

As observed in Table 1.12, the measured surge impedance Z_{mes} is roughly proportional to the parameter $\ln(h/er)$. This fact is further proof of the high accuracy of the proposed formula since the formula is directly proportional to $\ln(h/er)$.

1.7.3.3 Nonparallel Conductor

Figure 1.64 shows measured mutual inductance and capacitance of an overhead nonparallel conductor with radius $r_1 = r_2 = 1$ cm, length $x_1 = x_2 = 4$ m, and height $h_1 = h_2 = 0.4$ m as a function of separation y_2 at the receiving end with the parameter of separation y_1 at the sending end. Included are calculated results by the proposed formula in Section 1.7.2.1. The calculated results agree satisfactorily with the measured results.

Figure 1.65 shows the mutual impedance of a nonparallel conductor as a function of $y_2 - y_1$. In the figure, "fin" is the impedance evaluated by the formula proposed in this book and "inf" is that of the nonparallel conductor impedance derived from an infinite-length impedance [37,51]. The mutual impedance decreases as $y_2 - y_1$ and y_1 increase. The finite-length impedance is far smaller than that of the infinite length. This agrees with the tendency

TABLE 1.12

Measured and Calculated Surge Impedances of Vertical Conductors

Reference	Height h (m)	Radius r (mm)	Measured Z_{mes} (Ω)	$Z_{mes}/(\ln(h/er))$	Proposed	Approx.[a] = Jordan Reference 2	Wagner[b] Reference 3	Sargent[c] Reference 5	Hara[d] Reference 6
16	15.0	25.4	320.0	59.5	323.0	322.9	445.2	385.2	325.2
	15.0	2.5	459.0	59.6	462.0	462.0	584.4	524.4	464.4
	9.0	2.5	432.0	60.1	431.3	431.3	553.7	493.7	433.7
	6.0	2.5	424.0	62.5	407.0	407.0	529.4	469.4	409.4
	3.0	50.0	181.0	58.5	187.2	185.7	308.0	248.0	188.0
	3.0	25.0	235.0	62.3	228.0	227.2	349.6	289.6	229.6
	3.0	15.0	250.0	58.2	258.3	257.9	380.3	320.3	260.3
	3.0	2.5	373.0	61.2	365.5	365.4	487.8	427.8	367.8
	3.0	0.25	514.0	61.2	503.6	503.6	625.9	565.9	505.9
	2.0	2.5	345.0	60.7	341.2	341.1	463.5	403.5	343.5
	2.0	0.25	481.0	60.2	479.2	479.2	601.6	541.6	481.6
5	0.608	43.375	112.0	68.3	104.7	98.4	220.8	160.8	100.8
	0.608	9.45	180.0	56.9	191.2	189.8	312.2	252.2	192.2
	0.608	3.1125	250.0	58.5	256.9	256.5	378.9	318.9	258.9
	0.608	1.1750	310.0	59.1	315.1	314.9	437.3	377.3	317.3
Average of absolute error (%)				(60.4)	2.5	2.7	44.8	22.6	2.8

[a] $Z_j = 60 \ln(h/r) - 60 = 60 \ln(h/er)$.

[b] $Z_w = 60 \ln(2\sqrt{2}h/r) = Z_j + 122.4$.

[c] $Z_s = Z_w - 60 = Z_j + 62.4$.

[d] $Z_h = Z_w - 120 = Z_j + 2.4$.

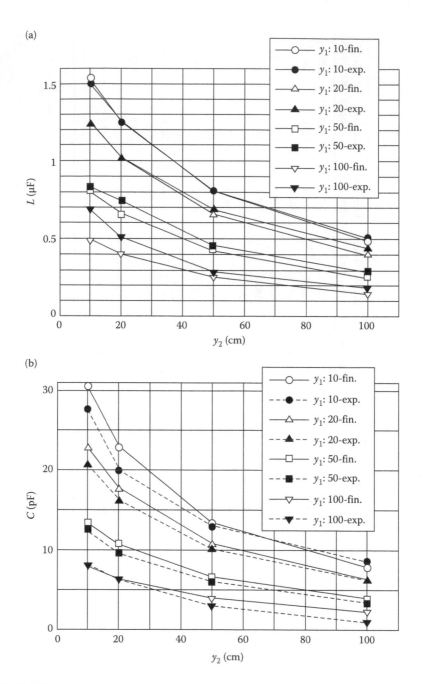

FIGURE 1.64
Mutual inductance and capacitance of a nonparallel conductor. (a) Inductance. (b) Capacitance.

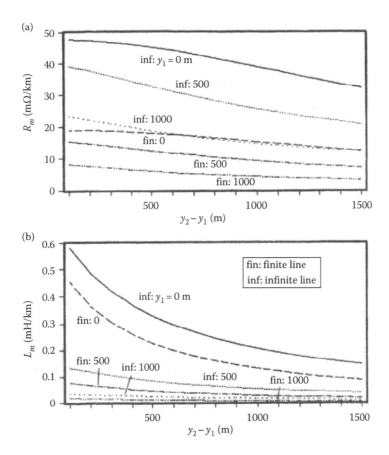

FIGURE 1.65
Mutual impedance of a nonparallel conductor. (a) Resistance. (b) Inductance.

observed between the impedances of finite and infinite conductors in Sections 1.7.3.1 and 1.7.3.2.

PROBLEMS

1.22 Calculate the surge impedances of the horizontal conductor with length $x = 4$ m given in Table 1.11 by using the following approximation, and confirm the results in the table:

$$Z_s = 60P_{ij}, \quad S_{ij} \cong 2h, \quad d_{ij} = r = 1 \, \text{cm}$$

Finite: Equation 1.249, infinite: Equation 1.250.

1.23 Calculate the surge impedance of a vertical conductor given in Table 1.12 by using the following approximation, and discuss the

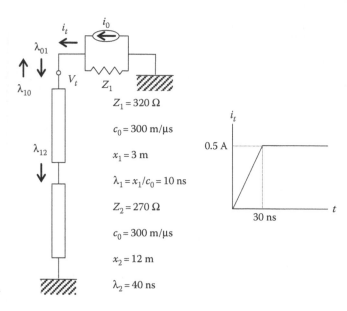

FIGURE 1.66
A tower model.

results in comparison with measured results, as well as Wagner's and Sargent's formulas, in the table:

$$Z_s = 60\left\{\ln\left(\frac{h}{r}\right) - 1\right\}.$$

1.24 Calculate voltage V_t at the top of the tower in Figure 1.66 for $0 \le t < 40$ ns, taking $\lambda_{01} = \lambda_{10} = 1$, and compare with the measured result shown in Figure 1.57b.

1.25 All the formulas in Section 1.7.2 are an approximation based on penetration depth. Discuss theoretical problems and engineering advantages of the formulas. If possible, develop a new formula that is theoretically better than those in Section 1.7.2.

1.8 Introduction to EMTP

1.8.1 Introduction

1.8.1.1 History of Transient Analysis

As is well known, many physical phenomena are expressed mathematically as a second-order partial differential equation. Electrical transients associated with a wave-propagation characteristic are mathematically represented by a hyperbolic partial differential equation. Therefore, solving electrical transients necessitates the solution of the differential equation with the given initial and boundary conditions.

The earliest solution of the partial differential equation was given by D'Alembert for the case of a vibrating string in 1750 [16]. At the same time, Bernoulli found a solution that was quite different from D'Alembert's solution. Bernoulli's solution is based on the eigenfunction and is comparable with the Fourier series.

Traveling-wave concepts and theories have been well developed since D'Alembert's solution. Allievi first applied the theory to the field of hydraulic engineering and established the general theory and idea of a graphical method, which was a direct application of the traveling-wave concept to engineering fields [52]. At a later stage, Bewley developed the traveling-wave theory and its application to various electrical transients [22]. The propagation of the traveling wave has been well analyzed using the modified Heaviside transform and the Sylvester theorem, which is the same as the eigenvalue theory of matrix algebra by Hayashi [53].

The graphical method developed by Allievi has been applied to the analysis of a water hammer by Schnyder [54], Bergeron [55], and Angus [56]. This was originally called the Schnyder–Bergeron method in the electrical engineering field. The name of the method has since been abbreviated, and it is nowadays called the Bergeron method, although Schnyder originated it. The detail and application of the graphical method are well described by Parmakian [57]. The graphical method corresponds to the method of characteristic to mathematically solve Maxwell's equation [58]. Similarly, the lattice diagram method based on the traveling-wave concept was developed by Bewley to solve electrical transients [22]. At a later stage, both the graphical method and the lattice diagram method were implemented on a digital computer for calculating electrical transients [25,26,59–66]. This technique is generally called the "traveling wave technique" or the "time domain method."

The numerical Fourier transform appeared in the electrical engineering field in the late 1950s [67], although the basis of the method was given by Bernoulli in 1750. Gibbs' phenomena and instability in a transform process, which are the inherent nature of the discrete Fourier transform, were greatly reduced with the development of the modified Fourier (Laplace) transform [68,69]. At a later stage, the modified Fourier/Laplace transform was applied to transient calculations by various authors [70–73]. Since the modified Fourier/Laplace transform provides high accuracy for obtaining a time solution, the analysis of a partial differential equation is rather easy in the frequency domain, and the implementation of the fast Fourier transform procedure into the modified Fourier/Laplace transform greatly improves computational efficiency [72]. The method has become one of the most accurate and efficient computer techniques for transient calculations.

1.8.1.2 Background of the EMTP

The EMTP has been widely used all over the world as a standard simulation tool for transient analysis not only in a power system, but also in an electronic circuit. The BPA of the U.S. Department of the Interior (later U.S. Department of Energy) started to develop a computer software for analyzing power system transients, especially switching overvoltage from the viewpoint of insulation design and coordination of transmission lines and substations in 1966, by inviting Dr. H. W. Dommel from Germany to be part of BPA's permanent staff. The EMTP development was a part of system analysis computerization, including a power/load flow analysis program and a stability analysis program in the BPA System Analysis Department led by Dr. W. Tinny. Before the EMTP, a transient network analyzer was used in the BPA.

The EMTP was based on the Schnyder–Bergeron method [54,55] of traveling-wave analysis in a hydraulic system, well known as a water hammer [52–57]. The Schnyder–Bergeron method was introduced to the field of electrical transients by Frey and Althammer [25]. The method was incorporated with a nodal analysis method by representing all of the circuit elements by a lumped resistance and the current source by Dommel [26]. This is the origin of the EMTP [9,11].

1.8.1.3 EMTP Development

Dommel developed a transient program at the Technical University of Munich based on Reference 25 for a distributed line and on the numerical integration of an ordinary differential equation for a lumped-parameter circuit. In 1966, he moved to the BPA to develop a generalized transient analysis program, and the original version of the EMTP called "transient program" was completed in 1968. In 1972, Scott-Meyer joined the BPA, and Dommel left the BPA in 1973. After that, the work on the EMTP was accelerated. Semlyen [74], Ametani [5,75], Brandwajn [76], and Dube [77] joined Scott-Meyer's team, and the EMTP Mode 31, which is basically the same as the present EMTP, was completed in 1981 [9]. Currently, four versions of the EMTP exist: (1) Alternative Transients Program (ATP) (royalty free) originated by Scott-Meyer and developed/maintained by the BPA, European EMTP User Group (EEUG) [78]; (2) EMTP RV (commercial) developed by Mahseredjian in Hydro-Quebec sponsored by EMTP Development Coordination Group (DCG)/EPRI [79]; (3) EMTDC/PSCAD (commercial) developed by Woodford and Goni at the Manitoba HVDC Research Center, which was founded by Wedepohl during his presidency of the Manitoba Hydro; and (4) the original EMTP (in public domain) developed by the BPA, the source code of which is available only from the Japanese EMTP Committee.

SPICE, similar to EMTP, is a well-known software, especially in the field of power electronics. The strength of SPICE is that the physical parameters of semiconductors are built into the software.

Table 1.13 summarizes the history of the EMTP from 1966 to 1991. Since the early 1990s, there have been many simulation tools related to or similar to the EMTP and many publications related to its development, which are not covered in this book.

1.8.2 Basic Theory of the EMTP

The basic procedure of an EMTP transient simulation involves the following steps:

1. Represent all the circuit elements including a distributed-parameter line by a current source and resistance.

TABLE 1.13

History of the EMTP

1961	Schnyder–Bergeron method by Frey and Althammer
1964	Dommel's PhD thesis in Tech. Univ. Munich
1966	Dommel started EMTP development in BPA
1968	Transients Program (TP = EMTP Mode 0), 4000 statements
1973	Dommel moved to Univ. British Columbia
	Scott-Meyer succeeded EMTP development
1976	Universal Transients Program File (UTPF) and Editor/Translator Program (E/T) completed by Scott-Meyer. UTPF and E/T made by EMTP could be used in any computers because those solved machine-dependent problems and prepared a platform for any researchers able to join EMTP development, not necessary to visit BPA
	Japanese EMTP Committee founded
1976	Semlyen, Ametani, Brandwajn, and Dube joined the team
1982	Marti joined EMTP development
1978	First EMTP workshop during IEEE PES meeting
1981	First EMTP tutorial course during IEEE PES meeting
	DCG proposed by BPA
1982	DCG 5-year project started, Chairman Vithayathil of BPA
	Members: BPA, Ontario Hydro, Hydro-Quebec, U.S. Bureau of Reclamation, WAPA
1983	EPRI joined DCG, and DCG/EPRI started. Copyright of EMTP to be given to EPRI/DCG
	EMTP Mode 39 completed and distributed
1984	EMTP development in BPA terminated Final version EMTP Mode 42
	Scott-Meyer started to develop ATP-EMTP independently on BPA and EPRI/DCG
1985	ATP development transferred to Leuven EMTP Center (LEC) in Belgium
	ATP ver. 2 completed. BPA joined LEC
1986	DCG/EPRI EMTP ver. 1 completed
1987	DCG/EPRI 5 year project to be terminated but DCG/EPRI project continued
	BPA resigned from DCG/EPRI
1991	ATP copyright transferred to Can/Am ATP User Group. BPA joined Can/Am User Group
	DCG/EPRI ver. 2 completed

2. Produce a nodal conductance matrix representing the given circuit.

3. Solve the nodal conductance matrix from given voltage sources (or current sources).

The procedure is the same for both steady states and transients. For transients, the procedure is repeated at every time step $t = n$ based on the known voltages and currents at time $t = t - \Delta t$ in the circuit. When the circuit configuration is changed due to switching operations, for example, the node conductance matrix is reproduced from a new circuit configuration.

1.8.2.1 Representation of a Circuit Element by a Current Source and Resistance

Figure 1.67 illustrates inductance L, capacitance C, and a distributed-parameter line Z_0 by a current source and resistance.

The representation is derived in the following manner. The voltage and the current of the inductance are related by

$$v = L \cdot \frac{di}{dt} \tag{1.260}$$

Integrating this equation from time $t = t - \Delta t$ to t gives

$$\int_{t-\Delta t}^{t} v(t) \cdot dt = L \int_{t-\Delta t}^{t} \left(\frac{di}{dt} \right) \cdot d_t = L[i(t)]_{t-\Delta t}^{t} = L\{i(t) - i(t - \Delta t)\}.$$

By applying the trapezoidal rule to the left-hand side of the equation, then

$$\int v(t) \cdot dt = \frac{\{v(t) + v(t - \Delta t)\}\Delta t}{2}.$$

From these two equations,

$$i(t) = \left(\frac{\Delta t}{2L} \right)\{v(t) + v(t - \Delta t)\} - i(t - \Delta t) = \frac{v(t)}{R_L} + J(t - \Delta t) \tag{1.261}$$

where
$J(t - \Delta t) = v \, (t - \Delta t)/R_L + i(t - \Delta t)$
$R_L = 2L/\Delta t, \Delta t$: time step

It is clear from this equation that current $i(t)$ at time t flowing through the inductance is evaluated by voltage $v(t)$ and current source $J(t - \Delta t)$, which was determined by the voltage and current at $t = t - \Delta t$. Thus, the inductance is represented by current source $J(t)$ and resistance R_L as illustrated in Figure 1.67a.

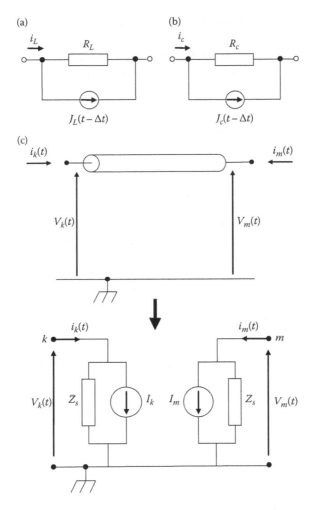

FIGURE 1.67
Representation of circuit elements by resistance and current source. (a) Inductance. (b) Capacitance. (c) Distributed line.

Similarly, Figure 1.67b for capacitance is derived from a differential equation expressing the relation between the voltage and the current of the capacitance.

Figure 1.67c is for a distributed-parameter line whose voltage and current are related in Equation 1.57:

$$v(x,t) + Z_0 \cdot i(x,t) = 2F_1\left(t - \left(\frac{x}{c}\right)\right), \quad v(x,t) - Z_0 \cdot i(x,t) = 2F_2\left(t + \left(\frac{x}{c}\right)\right) \quad (1.262)$$

where Z_0 is the characteristic impedance.

This equation is rewritten at nodes 1 and 2 as

$$v_1(t-\tau) + Z_0 i(t-\tau) = v_2(t) - Z_0 i_2(t)$$
$$v_1(t) - Z_0 i(t) = v_2(t-z) + Z_0 i_2(t-\tau)$$

(1.263)

where
$\tau = l/c$ is the traveling time from node 1 to node 2
l is the line length
c is the propagation velocity

It can be observed from Equation 1.263 that voltage $v_1(t)$ and current $i_1(t)$ at node 1, the sending end of the line, influence $v_2(t)$ and $i_2(t)$ at the receiving end for $t \geq \tau$, where τ is the traveling time from node 1 to node 2. Similarly, $v_2(t)$ and $i_2(t)$ influence $v_1(t)$ and $i_1(t)$ with time delay τ. In a lumped-parameter element, the time delay is Δt, as can be seen in Equation 1.261. In fact, the time delay Δt is not due to traveling-wave propagation, but it is a time step for time discretization to numerically solve a differential equation describing the relation between voltage and current of the lumped element. From Equation 1.263, the following relation is obtained:

$$i_1(t) = \frac{v_1(t)}{Z_0} + J_1(t-\tau), \quad i_2(t) = \frac{v_2(t)}{Z_0} + J_2(t-\tau)$$

(1.264)

where

$$J_1(t-\tau) = \frac{-v_2(t-\tau)}{Z_0 - i_2(t-\tau)}$$

$$J_2(t-\tau) = \frac{-v_1(t-\tau)}{Z_0 - i_1(t-\tau)}$$

These results give the representation of the distributed-parameter line in Figure 1.67c.

In Equation 1.263, Z_0 is the characteristic impedance, which is frequency dependent. When the frequency dependence of a distributed-parameter line as explained in Section 1.5 is to be considered, frequency-dependent line models such as Semlyen's and Marti's line models are prepared as a subroutine in the EMTP.

1.8.2.2 Composition of Nodal Conductance

In the EMTP, the nodal analysis method is adopted to calculate voltages and currents in a circuit. Figure 1.68 shows an example. By applying Kirchhoff's current law to nodes 1–3 in the circuit:

At node 1: $G(V - E_1 - V_3) + G_2(V_1 + E_2 - V_3) + G_3(V_1 - V_2) = 0$
At node 2: $G_3(V_1 - V_2) + G_5V_2 + G_6(V_2 - E_6) = 0$
At node 3: $G_4(V_3 - E_4) + G_2(V_3 - E_2 - V_1) + G_1(V_3 + E_1 - V_1) = 0$

where $G_i = 1/R_i$, $i = 1$ to 6.
  Rearranging this equation and writing in matrix form gives

$$\begin{bmatrix} G_1 + G_2 + G_3 & G_2 & -G_1-G_2 \\ -G_3 & G_3 + G_5 + G_6 & 0 \\ -G_1-G_2 & 0 & G_1 + G_2 + G_4 \end{bmatrix} \begin{pmatrix} V_1 \\ V_2 \\ V_3 \end{pmatrix} = \begin{pmatrix} J_1-J_2 \\ J_6 \\ -J_1 + J_2 + J_4 \end{pmatrix}$$

or

$$[G](V) = (J) \tag{1.265}$$

where
  $J_i = G_iE_i$, $i = 1, 2, 4, 6$
  $[G]$ is the node conductance matrix

It is clear from Equation 1.265 that once the node conductance matrix is composed, the solution of the voltages is obtained by taking the inverse of the matrix, as the current vector (J) is known. In the nodal analysis method, the composition of the nodal conductance is rather straightforward, as is well known in circuit theory. In general, nodal analysis gives a complex admittance matrix because of $j\omega L$ and $j\omega C$.
  However, in the EMTP, since all of the circuit elements are represented by a current source and resistance, it becomes a real matrix.

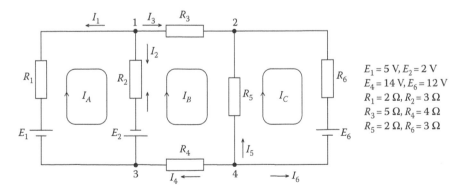

FIGURE 1.68
Nodal analysis.

1.8.3 Other Circuit Elements

Table 1.14 shows circuit elements and subroutines prepared in the EMTP. As observed from the table, most circuit elements are prepared for steady-state and transient simulations by the EMTP.

TACS and MODELS are computer languages by which a user can produce a computer code to use as an input data in the EMTP. These were, in a sense, pioneering software before MATLAB, MAPLE, and the like. If a user needs to develop a model circuit that is not available in the EMTP, it can be achieved by using TACS or MODELS.

The usage of circuit elements and supporting routines in the EMTP is explained in the EMTP rule books [9,78–80], and the theory is described in the theory book [11].

SOLUTIONS TO PROBLEMS

1.1 $S = \pi r^2 = \pi \times 10^{-4}$, $\ell = \pi \times 2 \times 10^{-2}$:

$$R_{dc} = \frac{\rho}{S} = \frac{2 \times 10^{-8}}{\pi \times 10^{-4}} = \left(\frac{2}{\pi}\right) \times 10^{-4} \doteqdot 0.637 \times 10^{-4} \, \Omega/\text{km}$$

At 50 Hz: $Z_c = R_{dc} + j\omega L_c$, $R_{dc} = 0.0637 \, \Omega/\text{km}$:

$$L_c = \frac{\mu_0}{8\pi} = \frac{4\pi \times 10^{-7}}{8\pi} = 0.5 \times 10^{-7} \, \text{H/m} = 5 \times 10^{-2} \, \text{mH/km}$$

At 100 kHz:

$$b = \frac{\omega \mu_c S}{R_{dc}} \cdot \ell^2 = \frac{2\pi \times 10^5 \times 4\pi \times 10^{-7} \times \pi \times 10^{-4}}{(2/\pi) \times 10^{-4} \cdot 4\pi^2 \times 10^{-4}} = (10\pi)^2 \gg 1$$

$$Z_c \doteqdot R_{dc} b \cdot \sqrt{j} = 10\pi \left(\frac{2}{\pi}\right) \times 10^{-4} \frac{(1+j)}{\sqrt{2}}$$

$$= \left(\frac{20}{\sqrt{2}}\right) \times 10^{-4} (1+j) \, \Omega/\text{m} = 1.414(1+j) \, \Omega/\text{km}$$

$$R_c = 1.414 \, \Omega/\text{km}, \quad L_c = \frac{1.414 \times 10^3}{2\pi \times 10^5} = 2.25 \times 10^{-3} \, \text{mH/km}$$

1.2 Same as Problem 1.1

1.3 $\ell = 8 \times 10^{-2} = 2\pi r_0 \therefore r_2 = (4/\pi) \times 10^{-2} = 1.273 \, \text{cm}$, $S = 4 \times 10^{-4} = \left(r_2^2 - r_1^2\right)$

TABLE 1.14

Circuit Elements and Subroutines Prepared in the EMTP

(a) Circuit Elements

Element	Model	Remark
Lumped R, *L*, *C*	Series, parallel	
Line/cable	Multiphase π circuit	Transposed, untransposed
	Distributed line with constant parameters	Overhead, underground
	Frequency-dependent line	Semlyen, Marti, Noda
Transformer	Mutually coupled *R–L* element	Single phase, three phase
	N winding, single phase	Saturation, hysteresis
	Three-phase shell type	
	Three-phase, three-leg core type	
Load, nonlinear	Staircase *R*(*t*) (type-97)	
	piecewise time-varying *R* (type-91, type-94)	
	Pseudo-nonlinear *R* (type-99)	
	Pseudo-nonlinear *L* (type-98)	
	Pseudo-nonlinear hysteretic *L* (type-96)	
Arrester	Exponential function $Z_n 0$	With gap, gapless
	Flashover-type multiphase *R*	
	TACS-controlled arc model	
Source	Step like (type-11)	Voltage source
	Piecewise linear (type-12, type-13)	Current source
	Sinusoidal (type-14)	
	Impulse (type-15)	
	TACS-controlled source	
Rotating machine	Synchronous generator (type-59)	Synchronous, induction, DC
	Universal machine	
Switch	Time-controlled switch	CB
	Flashover switch	
	Statistic/systematic switch	
	Measuring switch	
	TACS-controlled switch (type-12, type-13)	
Semiconductor	TACS-controlled switch (type-11)	Diode, thyristor
Control circuit	TACS	Transfer function
MODELS	Arithmetics, logics	

(b) Supporting Routines

Name	Function	Input Data
Line constants	Overhead line parameters	Frequency, configuration, physical parameters
Cable constants	Overhead/underground cable parameters	Frequency, configuration, physical parameters

(Continued)

TABLE 1.14 (*Continued*)

Circuit Elements and Subroutines Prepared in the EMTP

Name	Function	Input Data
Cable parameters	Overhead/underground cable parameters distributed Y, snaking	Frequency, configuration, physical parameters
XFORMER	Transformer parameters	Configuration, rating, %Z
BCTRAN	Transformer parameters	Configuration, rating, %Z
SATURATION	Saturation characteristics	Configuration, rating, %Z
HYSTERESIS	Hysteresis characteristics (type-96)	Configuration, rating, %Z
NETEQV	Equivalent circuit	Circuit configuration, Z, Y, frequency

$$\therefore r_1^2 = \left\{ \left(\frac{4}{\pi} \right)^2 - \frac{4}{\pi} \right\} \times 10^{-4}$$

$$r_1 = \left(\frac{4}{\pi} \right) \sqrt{\left\{ 1 - \frac{\pi}{4} \right\}} \times 10^{-2} = 0.5898 \text{ cm}$$

1.4 For $f = 50$ Hz, a low-frequency approximation is good enough (see Example 1.5):

$$Z_e = 50 + j50 \left\{ 8.253 + 0.628 \ln \left(\frac{100}{50} \cdot 10^{-4} \right) \right\} = 50 + j723.6 \text{ (m}\Omega/\text{km)}$$

For

$$f = 100\,\text{kHz}, D = H_1 = 20, H_e = 15.92$$

$$a = 1.0368 \times 10^3, b = 1.6506 \times 10^3$$

$$D^2 = 400, d^2 = r^2 = 1 \times 10^{-4}, A = 1.9492 \times 10^3, \varphi = 57.87° = 1.010\,\text{rad}$$

$$Z_e = 62.83 \times 10^{-3}(1.010 + j16.78) = 0.063 + j1.054 \text{ (}\Omega/\text{m)}$$

$$= 63 + j1054 \text{ }\Omega/\text{km}$$

1.5 The earth-return impedance is far greater than the conductor internal impedance; thus, the latter can be neglected. However, in a steady-state analysis such as fault and load flow calculations in a multiphase line, the positive-sequence (mode 1) component is important, and the conductor internal impedance is dominant for the positive-sequence component.

1.6 $h_e \gg h_i$, $S_{ij} = 2h_e$:

$$P_{eij} = \ln\left(\frac{2}{d_{ij}}\right) + \left(\frac{1}{2}\right)\ln\left(\frac{\rho_e}{j\omega\mu_0}\right) = \ln\left(\frac{2}{d_{ij}}\right) + \left(\frac{1}{2}\right)\left\{\ln\left(\frac{\rho_e}{\omega\mu_0}\right) + \ln e^{-j(\pi/2)}\right\}$$

$$= \frac{-j\pi}{4} + \ln 2 + \frac{\{\ln(\rho e/f \cdot d_{ij}^2) - \ln(8\pi^2 \times 10^{-7})\}}{2}$$

$$= \frac{-j\pi}{4} + 0.6931 - \frac{\{(-11.75) + \ln(\rho e/(f - d_{ij}^2))\}}{2}$$

$$\therefore Z_e = j(0.4\pi f)\left\{\frac{-j\pi}{4} + 6.5677 + \left(\frac{1}{2}\right)\ln\left(\frac{\rho_e}{f \cdot d_{ij}^2}\right)\right\}$$

$$\fallingdotseq f + jf\left\{8.253 + 0.628\ln\left(\frac{\rho_e}{f \cdot d_{ij}^2}\right)\right\}$$

1.7 $\ln \{2 (h + h_e)/r\} = \ln (2h/r) + \ln (1 + h_e/h) \fallingdotseq \ln (2h/r) + h_e/h = P_0 + h_e/h\}$

$$h_e = \sqrt{\frac{\rho_e}{j\omega\mu_0}} = (1 - j)\sqrt{\frac{\rho_e}{2\omega\mu_0}}$$

$$Z_e = j\omega\left(\frac{\mu_0}{2\pi}\right)\rho_e + (j+1)\sqrt{\frac{\rho_e/\omega\mu_0}{2/2\pi h}}$$

$$= Z_s + (1 + j)\sqrt{\frac{\rho_e\omega\mu_0}{2/2\pi h}}$$

1.8 $a^2 = R\sqrt{C/L}, b^2 = G\sqrt{L/C}; a^2 + b^2 \geq 2ab$

1.9 In Section 1.3.4.1, $\omega = 2\alpha\beta/(LG + CR)$. For $\omega \to 0(\omega = 0)$, $\alpha = \sqrt{RG}$. Thus, $c = \omega/\beta = 2\alpha(LG + CR) = 2\sqrt{RG}/(LG + CR)$.

1.10 $\Gamma = \sqrt{RG}$, $Y_0 = \sqrt{G/R}$ for $\omega = 0$. From Equation 1.124, $I_s = Y_0 \cdot E$ coth $(\Gamma\ell) = Y_0 E$ coth (θ).

For $G \to 0$, $\theta \to 0$; exp $(\theta) \to 1 + \theta$, coth $(\theta) \to 1/\theta$

$$I_s \fallingdotseq \frac{E\sqrt{G/R}}{(\sqrt{RG} \cdot \ell)} \fallingdotseq \frac{E}{(R \cdot \ell)} : \text{corresponding to Ohm's law}$$

1.11 $V_r = (A - BC/A)E - (B/AZ_r)V_r = E - (B/AZ_r)V_r; A \neq 0 \therefore V_r = Z_r E/(AZ_r + B)$

1. $A = \cosh\Gamma\ell = \cos(\omega\sqrt{LC\ell}) = \cos(\omega\ell/c_0) = \cos(\omega\tau) = \cos\theta,$

$$\tau = \frac{\ell}{c_0}, \quad c_0 = \text{light velocity}$$

$$B = Z_0 \sinh \Gamma\ell = jZ_0 \sin\theta, \; \theta \neq (2n-1)\,\pi/2, (n\!-\!1)\pi/2$$

$$\therefore V_r = \frac{Z_r E}{(Z_r \cos\theta + jZ_0 \sin\theta)}$$

2. $A = 1 + z\ell \cdot y\ell/2 = 1 - \omega^2 LC\ell^2/2 = 1 - \theta^2/2, \; B = Z\,\ell = j\omega L\ell = jZ_0\theta,$
    $Z = j\omega L, \; y = j\omega C$

$$V_r' = \frac{Z_r E}{\{(1-\theta^2/2)Z_r + jZ_0\theta\}}$$

a. $Z_r = 1\,\Omega$: $V_r = E/(\cos\theta + jZ_0 \sin\theta)$, $V_r' = E/\{(1-\theta^2/2) + j\theta Z_0$

b. $Z_1 = Z_0$: $V_r = E/(\cos\theta + j\sin\theta)Z_0 = E \cdot e^{-j\theta})/Z_0,$

$$V_r' = \frac{(E/Z_0)}{\{(1-\theta^2/2) + j\theta\}}$$

c. $Z_r = \infty$: $V_r = E/\cos\theta$, $V_r' = E/(1-\theta^2/2)$

   $f = 50\,\text{Hz}, \; \omega = 100\pi, \; \ell = 300\,\text{km}, \; \tau = 1\,\text{ms} = 10^{-3}, \; \theta = 0.1\pi = 0.3142,$
   $\cos\theta = 0.9511, \sin\theta = 0.3090, 1 - \theta^2/2 = 0.9506$

1.12 $BC - A \cdot D_t = U$

1.13 $Q_2 = P_{11} - P_{13}; Q_1, Q_3 = \dfrac{\left\{P_{11} + P_{22} + P_{13} \pm \sqrt{(P_{11} - P_{22} + P_{13})^2 + 8P_{12}P_{21}}\right\}}{2}$

a. $A_{1n} = 1(n = 1-3), A_{31} = A_{33} = 1, A_{32} = -1, A_{21} = \dfrac{-P_{21}}{P_{22} - Q_1}, A_{22} = 0,$

   $A_{32} = -1, A_{23} = -2$

b. $Q_1 = 48, \quad Q_2 = 30, \quad Q_3 = 24, \quad A_{n1} = 1(n = 1-3), \quad A_{12} = A_{13} = A_{33} = 1,$
   $A_{22} = 0, A_{32} = -1, A_{23} = -2$

c. $Q_1 = 50, \quad Q_2 = 30, \quad Q_3 = 25, \quad A_{1n} = 1, \quad A_{31} = A_{33} = 1, \quad A_{32} = -1,$
   $A_{21} = \dfrac{2}{\sqrt{3}}, A_{22} = 0, A_{23} = -\sqrt{3}$

1.14 Modal impedances and admittances for modes 2 and 3 are identical because all the off-diagonal elements of the series impedance and shunt admittance matrices are the same, and the diagonal elements are also the same in a transposed line.

1.15 Add a very small value to an element of matrix A. In a numerical calculation by a computer, $A_{ij} \neq A_{ji}$ if one observes the values of A_{ij} and A_{ji} for more than a certain number of digits.

1.16 At infinite frequency, an element of the impedance and admittance matrices becomes infinite. In a perfectly conducting system, $Z \cdot Y$ becomes diagonal.

1.17 Mode 0 (earth-return mode): no significant difference.
      Modes 1 and 2 (aerial modes) are identical in transposed lines, but they are different in untransposed lines.

1.18 $I = (2e_0/Z_0)\{1 - \exp(-t/\tau)\}$, $V = 2e_0 \cdot \exp(-t/\tau)$, $e_r = 2e_0$, $e_r = 2e_0\{2 \cdot \exp(-t/\tau) - 1\}$, $\tau = L/Z_0$

1.19 $V = (2e_0/a)\{R_1 + R_2 - (Z_0R_1/b)\exp(-t/\tau)\}$, $\tau = CR_1b/a$, $a = Z_0 + 2(R_1 + R_2)$
      $b = Z_0 + 2R_2$

1.20

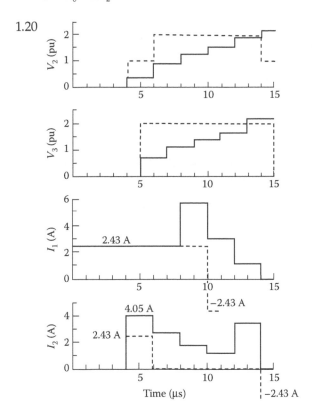

      (c) Because of lower surge impedance in a cable, the cable end voltage gradually increases with multiple reflections within the cable.

1.21 V_{rb}: similar waveform to V_{ra} in Figure 1.52
      $V_{ra} = V_{rc}$: similar waveform to the average of V_{rb} and V_{ra} in Figure 1.52

1.22 Results calculated by Equations 7.11 and 7.12 are given in Table. 1.11.

1.23 Calculated results are given in Table 1.12 as Approx. = Jordan.

1.24

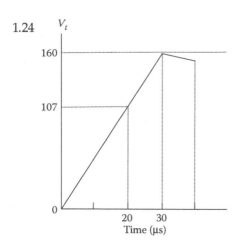

$$\lambda_{12} = \frac{2 \times 270}{320 + 270} = 0.195, \lambda_{01} = \lambda_{10} = 1$$

$V_t(t) = Z_i \cdot i_t(t)$, $e_{12}(t) = V_t(t)$ for $t < \tau_1 = 10\,\text{ns}$

for $\tau_1 \le t \le 3\tau_1$: $e_{2f}(t) = e_{12}(t - \tau_1)$, $V_1 = \lambda_{12} \cdot e_{2f} = 0.915 e_{12}(t - \tau_1)$

$e_{21}(t) = V_1(t) - e_{2f}(t) = -0.085 e_{12}(t - \tau_1)$

$2\tau_1 \le t < 4\tau_1$: $V_t = Z_i \cdot i_t(t) + \lambda_{10} \cdot e_{1b} = 320 i_t(t) - 0.085 e_{12}(t - 2\tau_1)$

$\quad = 320\{i_t(t) - 0.085 i_t(t - 2\tau_1)\}$

at $= 20\,\text{ns}$: $V_t = 320\left\{ 0.5 \times \frac{20}{30} - 0.085 \times 0 \right\} = 107\,\text{V}$

$\quad = 30\,\text{ns}$: $V_t = 320 \times 0.5 \left(1 - \frac{0.085}{3} \right) = 160 \times 0.9717 = 155\,\text{V}$

$\quad = 40\,\text{ns}$: $= 320 \times 0.5 \left(1 - 0.085 \times \frac{2}{3} \right) = 160 \times 0.9433 = 151\,\text{V}$

1.25 Formulas in Section 1.7.2 show a satisfactory accuracy in comparison with a number of measured results, and these are good enough from the viewpoint of engineering practice similar to the results in the *Electrician Handbook*. However, the concept of penetration depth is based on the theory of "electrostatics" within the TEM mode wave propagation. Thus, the formulas cannot be applied, in principle, to electromagnetic phenomena and non-TEM mode wave propagation.

References

1. Ametani, A. 1990. *Distributed—Parameter Circuit Theory*. Tokyo, Japan: Corona Pub. Co.

2. Ametani, A. and I. Fuse. 1992. Approximate method for calculating the imped-ances of multiconductors with cross sections of arbitrary shapes. *Electr. Eng. Jpn.* 111(2):117–123.
3. Schelkunoff, S. A. 1934. The electromagnetic theory of coaxial transmission line and cylindrical shields. *Bell Syst. Tech. J.* 13:523–579.
4. Wedepohl, L. M. and D. J. Wilcox. 1973. Transient analysis of underground power transmission systems. *Proc. IEE* 120(2):253–260.
5. Ametani, A. 1980. A general formulation of impedance and admittance of cables. *IEEE Trans. Power Appl. Syst.* 99(3):902–910.
6. Deri, A., G. Tevan, A. Semlyen, and A. Castanheira. 1981. The complex ground return plane: A simplified model for homogeneous and multi-layer earth return. *IEEE Trans. Power Appl. Syst.* 100(8):3686–3693.
7. Pollaczek, F. 1926. Über das Feld einer unendlich langen Wechsel-stromdurchflossenen Einfachleitung. *ENT* 3(9):339–359.
8. Carson, J. R. 1926. Wave propagation in overhead wires with ground return. *Bell Syst. Tech. J.* 5:539–554.
9. Scott-Meyer, W. 1982. *EMTP Rule Book*. Portland, OR: BPA.
10. Ametani, A. and K. Imanishi. 1979. Development of a exponential Fourier transform and its application to electrical transients. *Proc. IEE* 126(1):51–59.
11. Dommel, H. W. 1986. *EMTP Theory Book*. Portland, OR: Bonneville Power Administration (BPA).
12. Wise, W. H. 1948. Potential coefficients for ground return circuits. *Bell Syst. Tech. J.* 27:365–371.
13. Ametani, A. 2007. The history of transient analysis and the recent trend. *IEE J. Trans. EEE* (published by John Wiley & Sons, Inc.). 2:497–503.
14. Ametani, A., Y. Miyamoto, and J. Mahseredjian. 2014. Derivation of earth-return impedance of an overhead multi-conductor considering displacement currents. *IEE Japan Trans. PE*, 134:936–940.
15. Ametani, A., Y. Miyamoto, Y. Baba, and N. Nagaoka. 2014. Wave propagation on overhead multi-conductor in a high frequency region. *IEEE Trans. EMC* 56:1638–1648.
16. D'Alembert. 1747. Recherches sur la courbe que forme une corde tenduë mise en vibration.
17. Wedepohl, L. M. 1963. Application of matrix methods to the solution of travel-ling wave phenomena in polyphase systems. *Proc. IEE* 110(12):2200–2212.
18. Fortescue, C. L. 1918. Method of symmetrical coordinates applied to the solu-tion of polyphase network. *AIEE Trans.* Pt. II 37:1027–1140.
19. Clarke, E. 1943. *Circuit Analysis of A-C Power Systems, Vol. 1, Symmetrical and Related Components*. New York: Wiley.
20. Ametani, A. 1973. Refraction coefficient theory and surge phenomena in power systems. PhD dissertation. Victoria University of Manchester, Greater Manchester, U.K.
21. Ametani, A. 1973. Refraction coefficient method for switching surge calcula-tions on untransposed transmission line. *IEEE PES 1973 Summer Meeting*, C73-444-7, Vancouver, CA.
22. Bewley, L. V. 1951. *Traveling Waves on Transmission Systems*. New York: Wiley.
23. Ametani, A. 1980. Wave propagation characteristics on single core coaxial cables. *Sci. Eng. Rev. Doshisha Univ.* 20(4):255–273 (in Japanese).

24. Ametani, A. 1980. Wave propagation characteristics of cables. *IEEE Trans. Power Appl. Syst.* 99(2):499–505.

25. Frey, W. and P. Althammer. 1961. The calculation of electromagnetic transients on lines by means of a digital computer. *Brown Boveri Rev.* 48(5–6):344–355.

26. Dommel, H. W. 1969. Digital computer solution of electromagnetic transients in single- and multiphase networks. *IEEE Trans. Power Appl. Syst.* 88(4):388–398.

27. Ametani, A., T. Ono, Y. Honaga, and Y. Ouchi. 1974. Surge propagation on Japanese 500 kV untransposed transmission line. *Proc. IEE* 121(2):136–138.

28. Ametani, A. et al. 1981. Wave propagation characteristics on an untransposed vertical twin-circuit line. *IEE Japan* B-101(11):675–682.

29. Ametani, A., E. Ohsaki, and Y. Honaga. 1983. Surge characteristics on an untransposed vertical line. *IEE Japan* B-103(2):117–124.

30. Foster, R. M. 1931. Mutual impedance of grounded wires lying on or above the surface of the earth. *Bell Syst. Tech. J.* 10:408–419.

31. Jordan, C. A. 1934. Lightning computation for transmission line with overhead ground wires. *G. E. Rev.* 37(4):180–186.

32. Wagner, C. F. 1956. A new approach to calculation of lightning performance of transmission lines. *AIEE Trans.* 75:1233–1256.

33. Lundholm, R., R. B. Finn, and W. S. Price. 1957. Calculation of transmission line lightning voltages by field concepts. *AIEE Trans.* 76:1271–1283.

34. Sargent, M. A. and M. Darveniza. 1969. Tower surge impedance. *IEEE Trans. Power Appl. Syst.* 88:680–687.

35. Velzuez, R., P. H. Reynolds, and D. Mukhekar. 1983. Earth return mutual coupling effects in grounded grids. *IEEE Trans. Power Appl. Syst.* 102(6):1850–1857.

36. Okumura, K. and K. Kijima. 1985. A method for coupling surge impedance of transmission tower by electromagnetic field theory. *Trans. IEE Jpn.* 105-B(9):733–740 (in Japanese).

37. Ametani, A. and T. Inaba. 1987. Derivation of impedance and admittance of a nonparallel conductor system. *Trans. IEE Jpn.* B-107(12):587–594 (in Japanese).

38. Sarimento, H. G., D. Mukhedkar, and V. Ramachandren. 1988. An extension of the study of earth return mutual coupling effects in ground impedance field measurements. *IEEE Trans. Power Deliv.* 3(1):96–101.

39. Rogers, E. J. and J. F. White. 1989. Mutual coupling between finite length of parallel or angled horizontal earth return conductors. *IEEE Trans. Power Deliv.* 4(1):103–113.

40. Ametani, A. and M. Aoki. 1989. Line parameters and transients of a non-parallel conductor system. *IEEE Trans. Power Appl. Syst.* 4:1117–1126.

41. Hara, T. et al. 1988. Basic investigation of surge propagation characteristics on a vertical conductor. *Trans. IEE Jpn.* B-108:533–538 (in Japanese).

42. Hara, T. et al. 1990. Empirical formulas for surge impedance for a single and multiple vertical conductors. *Trans. IEE Jpn.* B-110:129–136 (in Japanese).

43. Yamada, T. et al. 1995. Experimental evaluation of a UHV tower model for lightning surge analysis. *IEEE Trans. Power Deliv.* 10:393–402.

44. Ametani, A. and A. Ishihara. 1993. Investigation of impedance and line parameters of a finite-length multiconductor system. *Trans. IEE Jpn.* B-113(8):905–913.

45. Ametani, A., Y. Kasai, J. Sawada, A. Mochizuki, and T. Yamada. 1994. Frequency-dependent impedance of vertical conductor and multiconductor tower model. *IEE Proc.-Gener. Transm. Distrib.* 141(4):339–345.

46. Oufi, E., A. S. Aifuhaid, and M. M. Saied. 1994. Transient analysis of lossless single-phase nonuniform transmission lines. *IEEE Trans. Power Deliv.* 9:1694–1700.
47. Correia de Barros, M. T. and M. E. Aleida. 1996. Computation of electromagnetic transients on nonuniform transmission lines. *IEEE Trans. Power Deliv.* 11:1082–1091.
48. Nguyen, H. V., H. W. Dommel, and J. R. Marti. 1997. Modeling of single-phase nonuniform transmission lines in electromagnetic transient simulations. *IEEE Trans. Power Deliv.* 12:916–921.
49. Ametani, A. 2002. Wave propagation on a nonuniform line and its impedance and admittance. *Sci. Eng. Rev. Doshisha Univ.* 43(3):135–147.
50. Sunde, E. D. 1968. *Earth Conduction Effects in Transmission Systems.* New York: Dover.
51. IEE Japan and IEICE Japan. Study Committee of Electromagnetic Induction. 1979. Recent movements and problems for electromagnetic interference. *IEE Japan and IEICE Japan* (in Japanese).
52. Allievi, L. 1902. *Teoria Generale del moto perturbato dell'acqua nei tubi in pressione.* Annali della Societa degli Ingegneri ed Architette Italiani (English translation, Halmos, E. E. 1925. *Theory of Water Hammer.* American Society of Mechanical Engineering).
53. Hayashi, S. 1948. *Operational Calculus and Transient Phenomena.* Kokumin Kogaku-sha, Tokyo, Japan.
54. Schnyder, O. 1929. Druckstosse in Pumpensteigleitungen. *Schweiz Bauztg* 94(22):271–286.
55. Bergeron, L. 1935. Etude das variations de regime dans les conduits d'eau: Solution graphique generale. *Rev. Generale de L'hydraulique* 1:12–69.
56. Angus, R. W. 1938. *Waterhammer in Pipes: Graphical Treatment.* Bulletin 152. University of Toronto, Toronto, CA.
57. Parmakian, J. 1963. *Waterhammer Analysis.* New York: Dover.
58. Sommmerferd. A. 1964. *Practical Differential Equations in Physics.* New York: Academic Press.
59. Baba, J. and T. Shibataki. 1960. Calculation of transient voltage and current in power systems by means of a digital computer. *J. IEE Jpn.* 80:1475–1481.
60. Barthold, L. O. and G. K. Carter. 1961. Digital traveling wave solutions, 1-Single -phase equivalents. *AIEE Trans.* Pt. 3. 80:812–820.
61. McElroy, A. J. and R. M. Porter. 1963. Digital computer calculations of transients in electrical networks. *IEEE Trans. Power Appl. Syst.* 82:88–96.
62. Arismunandar, A., W. S. Price, and A. J. McElroy. 1964. A digital computer iterative method for simulating switching surge responses of power transmission networks. *IEEE Trans. Power Appl. Syst.* 83:356–368.
63. Bickford, J. P. and P. S. Doepel. 1967. Calculation of switching transients with particular reference to line energisation. *Proc. IEE* 114:465–477.
64. Thoren, H. B. and K. L. Carlsson. 1970. A digital computer program for the calculation of switching and lighting surges on power systems. *IEEE Trans. Power Appl. Syst.* 89:212–218.
65. Snelson, J. K. 1972. Propagation of traveling waves on transmission lines— Frequency dependent parameters. *IEEE Trans. Power Appl. Syst.* 91:85–90.
66. Ametani, A. 1973. Modified traveling-wave techniques to solve electrical transients on lumped and distributed constant circuits: Refraction-coefficient method. *Proc. IEE* 120(2):497–504.

67. Lego, P. E. and T. W. Sze. 1958. A general approach for obtaining transient response by the use of a digital computer. *AIEE Trans.* Pt. 1. 77:1031–1036.
68. Day, S. J., N. Mullineux, and J. R. Reed. 1965. Developments in obtaining transient response using Fourier integrals. Pt. I: Gibbs phenomena and Fourier integrals. *Int. J. Elect. Eng. Educ.* 3:501–506.
69. Day, S. J., N. Mullineux, and J. R. Reed. 1966. Developments in obtaining transient response using Fourier integrals. Pt. II: Use of the modified Fourier transform. *Int. J. Elect. Eng. Educ.* 4:31–40.
70. Battisson, M. J. et al. 1967. Calculation of switching phenomena in power systems. *Proc. IEE* 114:478–486.
71. Wedepohl, L. M. and S. E. T. Mohamed. 1969. Multi-conductor transmission lines: Theory of natural modes and Fourier integral applied to transient analysis. *Proc. IEE* 116:1553–1563.
72. Ametani, A. 1972. The application of fast Fourier transform to electrical transient phenomena. *Int. J. Elect. Eng. Educ.* 10:277–287.
73. Nagaoka, N. and A. Ametani. 1988. A development of a generalized frequency-domain transient program—FTP. *IEEE Trans. Power Deliv.* 3(4):1986–2004.
74. Semlyen, A. and A. Dabuleanu. 1975. Fast and accurate switching transient calculations on transmission lines with ground return using recursive convolution. *IEEE Trans. Power Appl. Syst.* 94:561–571.
75. Ametani, A. 1976. A highly efficient method for calculating transmission line transients. *IEEE Trans. Power Appl. Syst.* 95:1545–1551.
76. Brandwajn, V. and H. W. Dommel. 1977. A new method for interfacing generator models with an electromagnetic transients program. *IEEE PES PICA Conf. Rec.* 10:260–265.
77. Dube, L. and H. W. Dommel. 1977. Simulation of control systems in an electromagnetic transients program with TACS. *IEEE PES PICA Conf. Rec.* 10:266–271.
78. European EMTP Users Group (EEUG) 2007. *ATP Rule Book.*
79. DCG/EPRI. 2000. *EMTP-RV Rule Book.*
80. Ametani, A. 1994. *Cable Parameters Rule Book.* Portland, OR: BPA.

2

Transients on Overhead Lines

2.1 Introduction

There are various kinds of transients in a power system. In general, over-voltages caused by transients are important in a power system because of its insulation against overvoltages. Table 2.1 summarizes overvoltages in a power system [1,2]. A temporary overvoltage is one caused by an abnormal system condition such as a line fault, and is evaluated by the steady-state solution of an abnormal system condition. Thus, temporary overvoltage is not considered a transient overvoltage. A typical example of temporary overvoltage caused by a line-to-ground fault is shown in Figure 2.1. Immediately after the initiation of the line-to-ground fault, a transient called "fault surge" occurs, but dies out in a few milliseconds. Then, a sustained dynamic overvoltage is observed. This is called the "temporary voltage."

A transient on a distributed parameter line is called a "surge" because the transient is caused by traveling waves. An overvoltage due to a surge is, in general, much greater than the temporary overvoltage, and thus, the insulation of a transmission system is mainly determined by the surge overvoltage [3].

In this chapter, the following surges on overhead lines are explained:

- Switching surges
- Fault initiation and fault-clearing surges
- Lightning surges

In addition, theoretical analyses of transients (using hand calculations) are explained. Finally, this chapter describes a frequency-domain (FD) method of transient simulations, a computer code of which can be readily developed by the reader.

TABLE 2.1

Classification of Overvoltages

	Overvoltage	Dominant Frequency/ Time Period	Remarks
Temporary overvoltage	1. Power frequency	50, 60 Hz/10 ms^{-1} s	Sustained overvoltage
	2. Harmonic	Some 100 Hz/1–100 ms	Resonance, nonlinear/ ferroresonance
	3. Low frequency	Lower than 50 Hz/some seconds	Subsynchronous resonance
Surge overvoltage	4. Lightning surge	0.1–10 MHz/0.1–some 10 μs	Direct lightning BFO induced lightning
	5. Switching surge	Some kHz/0.1–20 ms	Closing reclosing
	6. Fault surge	Some kHz/0.1–20 ms	Line-to-ground fault fault clearing

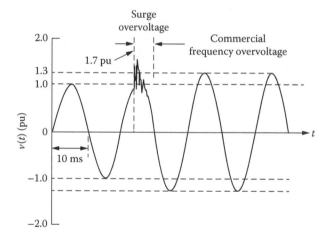

FIGURE 2.1
A temporary overvoltage due to a line-to-ground fault.

2.2 Switching Surge on Overhead Lines

2.2.1 Basic Mechanism of A Switching Surge

In a single distributed parameter line the remote end of which is open-circuited, as illustrated in Figure 2.2, the remote-end voltage, V_2, is given in the following equation as described in Section 1.6.1, when switch S is closed at $t = 0$:

$$V_2 = \lambda_{2f} \cdot E = 2E \quad \text{for } \tau \leq t < 3\tau$$

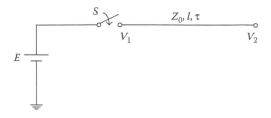

FIGURE 2.2
An open-circuited single conductor.

This voltage is called the "switching (surge) overvoltage," which reaches 2 pu (per unit voltage = V/E) in a single lossless line:

$$\text{Per unit voltage (pu)} = \frac{\text{Voltage}}{\text{Applied source (steady-state) voltage}} \quad (2.1)$$

In reality, the voltage is reduced to less than 2 pu because of the traveling-wave attenuation from the sending end to the remote end caused by a resistance in the conductor (see Section 1.6.4.4). However, a real transmission line has three phases, and a switching operation involves switching of three phase switches. Then, a switching surge on one phase induces a switching surge on the other phases. In addition, traveling waves are reflected at a boundary, such as a transformer or a series capacitor, and overlap the original waves. Thus, the maximum switching overvoltage can be higher than 3 pu, depending on the circuit breaker (CB) operation sequence, the phase angles of three-phase alternate current (AC) source voltages, line length, etc.

A switching surge voltage on a lossless line is given by a solid line in Figure 2.3 for a long time period. When attenuation constant α due to a line resistance is considered, the surge waveform is distorted by $\exp(-\alpha t)$ as time passes, and the dotted line in Figure 2.3 is obtained. After a certain time period, the oscillating surge voltage converges to the steady-state voltage of 1 pu ($= E$).

2.2.2 Basic Parameters Influencing Switching Surges

2.2.2.1 Source Circuit

2.2.2.1.1 Source Impedance Z_s

The voltage source E in Figure 2.2 has none of the internal impedance that an ideal voltage source with infinite capacity would have. In practice, no ideal source exists, and every source has its own internal impedance. Also, a transformer is connected to a generator to supply a higher voltage to a transmission line. The transformer has an inductance and is often represented in a transient calculation by its leakage inductance L [H], which is

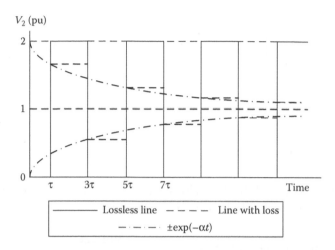

FIGURE 2.3
Switching surge waveforms on a single conductor.

evaluated by its capacity P [W], rated voltage V [V], and percent impedance (%Z) as

$$X = \omega L = (\%Z).\frac{V^2}{P}(\Omega) \tag{2.2}$$

where $\omega = 2\pi f$, f: source (power) frequency of 50 or 60 (Hz).
  For example, $P = 1000$ MW, $V = 275$ kV, %Z = 15, and $f = 50$ Hz:

$$X = 0.15 \times \frac{275^2}{1000} = 11.34\,\Omega, \quad L = \frac{X}{\omega} = 36.1\,\text{mH}$$

When there is a source inductance as illustrated in Figure 2.4, a switching surge waveform differs significantly from those shown in Figure 2.3. The sending and receiving end voltages in Figure 2.4 are obtained in a manner

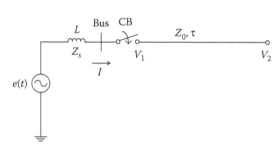

FIGURE 2.4
A single-phase line with an inductive source $e(t) = E\cos(\omega t)$.

similar to that in Section 1.6.2.3(4) or Problem 1.18, when the source voltage $e(t) = E$, the step function voltage, is applied to the sending end:

$$E = L\left(\frac{dI}{dt}\right) + Z_0 I \quad \therefore I(t) = \left(\frac{E}{Z_0}\right)\left\{1 - \exp\left(\frac{-t}{\tau}\right)\right\}$$

$$V_1(t) = Z_0 \cdot I(t), \quad V_2(t) = 2E\left\{1 - \exp\left(\frac{-t'}{\tau}\right)\right\} \quad \text{for } 0 \le t < 2\tau \qquad (2.3)$$

where $\tau = L/Z_0$, $t' = t - \tau$.
For example, if $\tau = 0.12$ ms for $L = 36$ mH and $Z_0 = 300\ \Omega$, then

$t = 0.12$ ms	$t = 0.5$ ms
$V_1 = 0.63\ E$	$V_1 = 0.98\ E$

Considering this result, V_1 and V_2 are drawn as in Figure 2.5, assuming that $2\tau = 0.5$ ms.

A case of no source inductance is included in Figure 2.3, which is shown by the dashed lines (– – –). Because of the inductance, the rise time of the wave front becomes longer, and the surge waveform is observed to be highly distorted.

When a number of generators, transformers, and/or transmission lines are connected to the bus in Figure 2.4, the source impedance $Z_s \doteqdot j\omega L$ becomes very small. Such a bus is called an "infinite bus," and the source circuit becomes equivalent to the ideal source in Figure 2.2. A switching overvoltage generated in the infinite bus case is the severest, as observed from Figure 2.3.

2.2.2.1.2 Sinusoidal AC Source

Equation 2.3 shows that a switching overvoltage increases as the source voltage E increases. A sinusoidal AC source voltage is often approximated by its amplitude E as a step function voltage because the time period of the

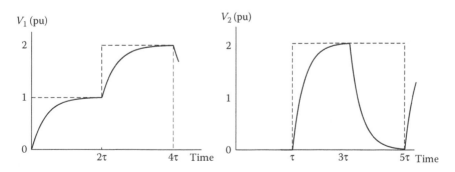

FIGURE 2.5
Effect of the source inductance on the switching surge.

switching surge is much smaller than the oscillating period T of the AC voltage.

$$e(t) = E \cdot \cos(\omega t) \qquad (2.4)$$

where $\omega = 2\pi f$, $f = 50$ or 60 (Hz):

$$T = \frac{1}{f}(s) = 16.7 \quad \text{or} \quad 20(\text{ms})$$

In such a time period, the AC voltage in Equation 2.4 can be regarded as a step voltage with amplitude E.

When calculations for a much longer time period are required, a source should be represented as a sinusoidal AC source. In the case of fault clearing and load rejection overvoltages, the observation time exceeds 1 s, and the mechanical and electrical characteristics (including generator control) of the generator should be considered. In such a case, the sinusoidal AC source is not appropriate, and Park's generator model is often used.

2.2.2.1.3 Impulse/Pulse Generator

An impulse generator (IG) and a pulse generator (PG) are often used to measure the transient response of a transmission line or cable, a grounding electrode, or a machine. The IG is composed of capacitors, and the recent PG is composed of a coaxial cable, which is a kind of capacitor. An impulse voltage and a pulse voltage are generated by charging the capacitors, which become a capacitive source. The simplest IG and PG model is an ideal source voltage with a given wave shape, that is, the wave front time T_f and tail T_t, or rise time T_f and pulse width T_w. When the IG is to be modeled accurately, all the elements of the IG, that is, the capacitances, resistances, and inductances, should be considered based on the circuit diagram. The PG is represented accurately as a charged cable in a transient calculation.

2.2.2.2 Switch

There are various types of switches. In a transmission system, a CB to interrupt current is the most common. For an ordinary switching surge calculation, the CB is modeled as an ideal switch controlled by time. When recovery and restriking transients are to be analyzed, the dynamic characteristics of a CB, especially for a vacuum CB or an interrupter, have to be considered. These characteristics are highly dependent on the material of the electrode and the CB's operation mechanism.

A line switch (LS) or a disconnector (DS) is used in a substation to interrupt voltage. Because the operating speed of an LS/DS is generally slow, it cannot be modeled as a time-controlled switch like a CB. Instead, an LS/DS is modeled as a voltage-controlled switch enabling it to handle a restriking voltage during a transient.

For analyzing a very fast front surge (switching surge in a gas-insulated bus) or a lightning surge, stray capacitances of a CB between poles and the ground have to be considered.

2.2.2.3 Transformer

As explained in Section 2.2.2.1.1, a transformer is represented by its leakage inductance for most switching surge analyses. When dealing with a fault surge, especially in a low voltage system with a transformer of high resistance grounding or an isolated neutral, the transformer winding, either the Y or Δ connection, has to be taken into account. Occasionally, the magnetizing impedance of the iron core has to be considered.

For a lightning surge analysis, stray capacitances between the primary and secondary windings, windings to the ground, and between phases, are to be considered especially in the case of a transferred surge.

2.2.2.4 Transmission Line

As described in Section 2.2.1, the attenuation of a traveling wave due to conductor resistance along a transmission line affects a switching surge waveform (see Figure 2.3). It is explained in Section 1.5.2 that the series impedance of a transmission line is notably frequency dependent, and so are the attenuation and propagation velocity of the traveling wave. As a result, the traveling wave is distorted as it travels along the line, as shown in Figures 1.28 and 1.49. Frequency dependence has a significant effect on the switching surge waveform. Therefore, the frequency-dependent effect should be included in an accurate calculation of a switching surge. This can be done by using the method explained in Section 1.5.3.1.

However, a source inductance also causes significant wave deformation, as in Figure 2.5. Then, considering only the resistance of a transmission line can give a reasonable result of a switching surge waveform. This is a safer option from the viewpoint of the insulation design and coordination of a transmission system and a substation, because the switching overvoltage is estimated to be more severe than that in an accurate calculation.

2.2.3 Switching Surges in Practice

2.2.3.1 Classification of Switching Surges

Switching surges are generated by a switching operation of a CB and are classified as follows [1]:

a. Closing surge—CB closing   (1) energization of a line
                                           (3) reenergization (reclosing surge)

b. Clearing surge—CB opening   (2) fault clearing

(1), (2), and (3) indicate the sequence of a CB operation.

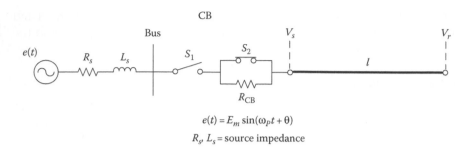

FIGURE 2.6
A model circuit of a switching surge analysis.

Closing surges are caused by closing switch contact S_1 of the CB, as illustrated in Figure 2.6, and the resultant overvoltage is called the "closing surge overvoltage." Reclosing surges are generated, in practice, after fault clearing, as explained earlier. That is, when a fault occurs in a power system, the faulty line has to be cleared from the source as soon as possible to avoid damage to the line and machineries in the system from a large fault current. Thus, the CB connected to the faulty line is opened. In most cases, a line fault is referred to as a "line-to-ground fault," originally due to lightning, and is sustained by an electrical arc the energy for which is supplied by a generator in the system. Therefore, when the energy from the generator is shut down by making the CB off, the arc is distinguished and the fault itself is cleared. At this stage, the fault-clearing surge is generated.

After the fault clearing, the system generally has to be returned to normal operation as soon as possible. As a consequence, the CB may be reclosed. When the time period from the fault clearing to the reclosing of the CB is not long enough for the trapped charge on the faulted line to be discharged, there will be a "residual charge voltage" on the line. This means that the reclosing surge requires the initial condition of the line to be closed. In other words, the difference between the reclosing and closing surges is the presence of an initial charge, or voltage.

It is a common practice to classify the fault-clearing surge not as a closing surge but as a fault surge since it occurs as a consequence of a fault.

It is clear from Section 2.1 that the maximum closing overvoltage is 2 pu on an ideal lossless single phase line. However, the reclosing overvoltage easily reaches 4 pu due to mutual coupling of the phases, CB closing time differences between the phases, and residual voltages in an actual multiphase line.

2.2.3.2 Basic Characteristics of a Closing Surge: Field Test Results

Figure 2.7b shows a field test result of a closing surge on the untransposed horizontal line in Figure 1.22 for the middle phase closing illustrated in Figure 2.7a [4,5]. The source voltage has a waveform of 1/4000 μs.

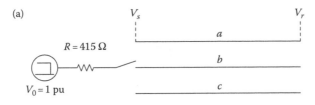

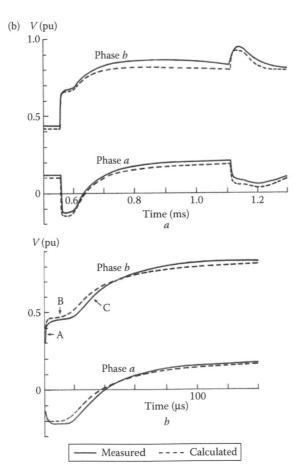

FIGURE 2.7
Switching surge waveforms on a 500 kV untransposed horizontal line. (a) Test circuit: $\ell = 83.212$ km (line configuration Figure 1.22). (b) Measured and calculated results. *(Continued)*

At the receiving end, phase *b* voltage rises very rapidly (A in Figure 2.7b) immediately after the arrival of the traveling wave ($t = 0$ in Figure 2.7b) and becomes nearly flat for about 20 µs (B). Then, it rises again gradually (C) as observed in Figure 2.7b. The induced phase (*a*, *c*) voltage becomes negative at the beginning and gradually increases to a positive value. This condition

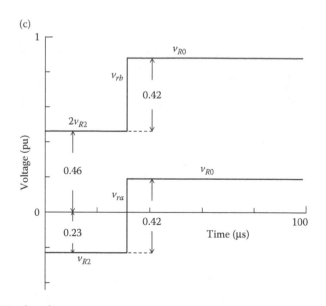

FIGURE 2.7 (Continued)
Switching surge waveforms on a 500 kV untransposed horizontal line. (c) Analytical waveforms corresponding to (b).

is a typical characteristic of a closing surge at the wave front and is due to
different propagation velocities of the earth return and aerial modes, as
explained in Section 1.6.4. Figure 2.7c shows the analytical results obtained
by the method in Section 1.6.4. The figure clearly explains the characteristics
of the closing surge observed in Figure 2.7b.

In Figure 2.7b, the dotted line is the result calculated by an FD method
based on the steady-state solution and the numerical Laplace transform [6].
The calculated result agrees quite well with the field test result.

A field test result for the phase "a" application in the test circuit of Figure
2.7a can be found in Chapter 1, Section 1.6.4. Additional explanations of this
field test are given in Appendices 2A.1 and 2A.2.

Figure 2.8b–d shows another field test result on the test circuit, illustrated
in Figure 2.8a, of an untransposed vertical double-circuit line [7,8]. The
applied source voltage has a waveform of 1/5000 μs. In Figure 2.8, the source
resistance R was 403 Ω in tests (b) and (c) and 150 Ω in test (d), in which all
six phases were short-circuited at the sending end.

It is observed in Figure 2.8b that twice the traveling time of the fastest wave
is about 681 μs, which results in a propagation velocity of 101.13 km/681 μs =
297 m/μs. Also, it is observed that the rising part of the wave front during the
time period of 680–700 μs has two events of voltage increase. The first rise is
due to the arrival of the fastest traveling wave at the sending end, which corresponds to modes 2, 4, and 5, as explained in Section 1.5.2. Then, the second-fast-
est wave corresponding to mode 1 arrives at the sending end, causing another

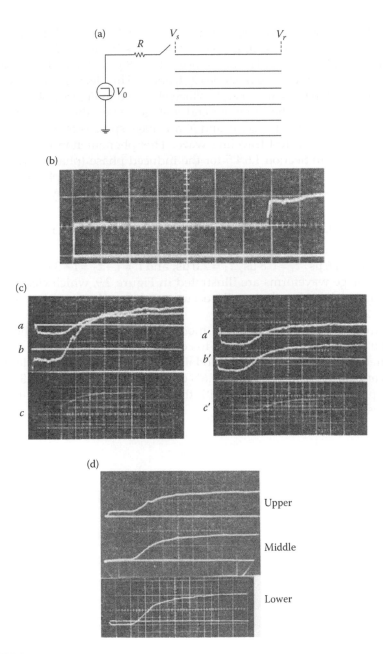

FIGURE 2.8
Field test results of closing surges on an untransposed vertical twin-circuit line in Figure 1.25 with a length of 101.13 km. (a) Test circuit. (b) Phase *a* sending end voltage for phase *a* application. (c) Receiving end voltages for phase *c* application. (d) Receiving end voltages for a source application to all the phases in the short circuit.

rise in the sending end voltage. The modal velocity for modes 2, 4, and 5 is estimated at about 299.2 m/μs, and that of mode 1 is estimated at about 298.4 m/μs at 1 kHz from Figure 1.26. The difference is the 1.8-μs delay in the arrival time of mode-1 wave from those of modes 2, 4, and 5. This delay is the cause of the "stair case" rise in the sending end voltage observed in Figure 2.8b.

Figure 2.8c shows the receiving end voltages when the source voltage is applied to phase c. On phases a' and b', a voltage spike is observed just after the arrival of the fastest traveling wave. This phenomenon is the same as that explained in Section 1.6.4.5 for the induced phase (phase c) voltage on an untransposed horizontal line. An analytical result of the phase a' voltage based on the parameters at $f = 2$ kHz of Figure 1.26 is given by

$$V_{ra'} = 0.285u(t' - \tau_0') - 0.172u(t' - \tau_1') - 0.115u(t' - \tau_3') + 0.130u(t') \quad (2.5)$$

where $\tau_0' = 64.6$ μs, $\tau_1' = 0.8$ μs, $\tau_3' = 2.0$ μs, and $t' = t - \tau_2 = t - 337.8$ μs.

The voltage waveforms are illustrated in Figure 2.9, which clearly shows the voltage spike. No voltage spike occurs on phase c' because the voltages of modes 2, 4, and 5 are canceled out.

Figure 2.8d indicates that aerial mode voltages exist even in the case of voltage application to all phases. This cannot be explained by conventional symmetrical component theory. Modal theory predicts aerial mode components at 28% on the upper phase, 7.6% on the middle phase, and 13% on the lower phase, which agrees well with the field test results of 20%, 4.3%, and 11%, respectively.

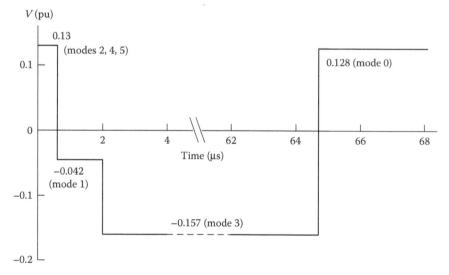

FIGURE 2.9
Analytical voltage waveform of phase a' corresponding to Figure 2.8c.

2.2.3.3 *Closing Surge on a Single Phase Line*

Figure 2.10 shows closing surges on a single phase line. The broken line in the figure is the case of neglecting the frequency-dependent effect (wave deformation) explained in Sections 1.5.2 and 1.5.3. It is observed that an oscillating surge voltage is sustained for a long time and the maximum overvoltage reaches –2.67 pu at $t = 10.2$ ms. In the case where the wave deformation given by the dotted line is included, the oscillation is damped as time increases, and the maximum voltage is reduced to about 2.0 pu. This clearly indicates the significance of the frequency-dependent effect on a switching surge analysis. If it is neglected, the insulation level against the switching surge will be overestimated, resulting in an economical inefficiency.

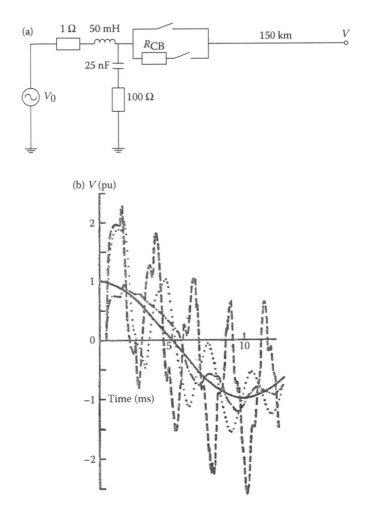

FIGURE 2.10
Single-phase closing surges. (a) Circuit diagram. (b) Calculated result.

In extra-high-voltage (EHV) and ultra-high-voltage (UHV) transmission systems, it is a common practice to control the closing surge overvoltage by means of an insertion resistor (closing resistor) or by synchronized switching of the CB. The effect of the closing resistor is shown by the dashed-dotted line (–.–.–) in Figure 2.10. It is quite clear how effective the closing resistor is in damping the overvoltage; the maximum overvoltage is reduced to 1.2 pu. The resistor is also used to damp a fault-clearing surge when necessary. Synchronized switching means that every phase is closed when the phase voltage is zero. Thus, no overvoltage appears in theory. Such CBs are widely used in Europe.

2.2.3.4 Closing Surges on a Multiphase Line

2.2.3.4.1 Wave Deformation

Figure 2.11 shows a closing surge due to a sequential closing of a three-phase CB on an untransposed horizontal line. The effect of wave deformation on the switching surge is also clear even in the multiphase line.

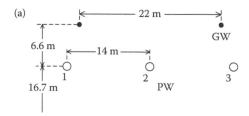

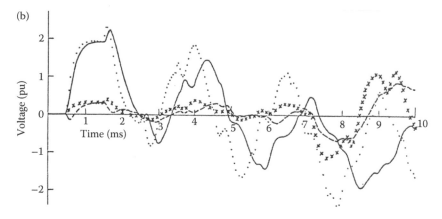

FIGURE 2.11
Switching surges at the receiving end for sequential closing. (a) Line configuration—PW: radius = 0.1785 m, resistivity = 3.78 × 10⁻⁸ Ω m; GW: radius = 8.8 mm, resistivity = 5.36 × 10⁻⁸ Ω m, line length = 150 km, and source inductance = 50 mH. (b) Calculated results: phase 1 closed at 90° (t = 0, v = 1 pu), phase 2 at 150°, phase 3 at 180°, no wave deformation; ——: phase 1, –.–.–: phase 3 wave deformation, earth resistivity = 100 Ω m; ……: phase 1, ×: phase 3.

2.2.3.4.2 *Closing Angle Distribution*

It is assumed in Figure 2.11 that each phase of the CB is closed sequentially at electrical angles of 90°, 150°, and 180°. In reality, it is not clear if a pole of any CB is closed as planned, because of the mechanical structure of the CB. Figure 2.12 shows a distribution of the CB closing angles. Figure 2.12a and b is an analytical distribution often used for a statistical analysis, and Figure 2.12c shows the measured distributions. Figure 2.12c1 is a measured result of time dispersion of the three phases of a CB. Figure 2.12c2 is another measured result of the time delay between the phases.

The calculated results of the statistical distribution of closing overvoltages for the four previously mentioned distributions of CB closing angles are shown in Figure 2.13. In the cases of uniform and Gaussian distributions,

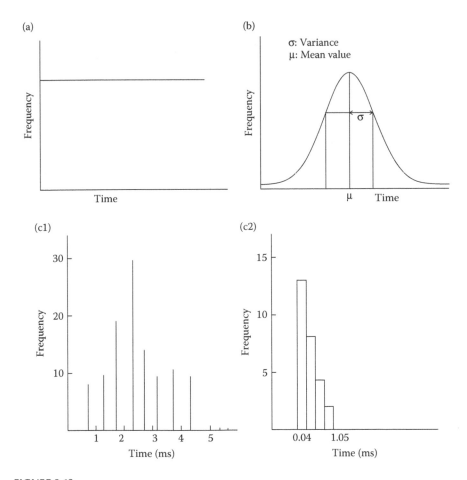

FIGURE 2.12
Various distribution characteristics of closing angles. (a) Uniform distribution. (b) Gaussian distribution. (c1, c2) Measured distribution in a field test.

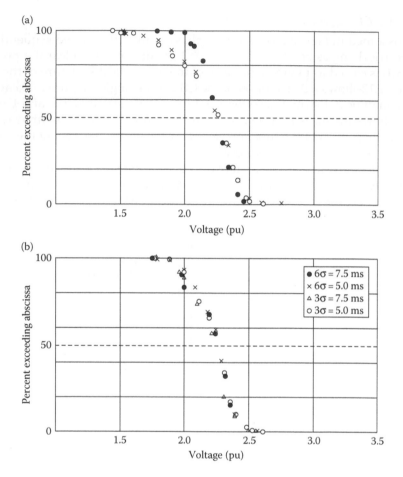

FIGURE 2.13
Statistical distribution curves of maximum switching overvoltages ($\ell = 150$ km, $L_s = 50$ mH in Figure 1.22). (a) Uniform distribution (Figure 2.12a), where × measured distribution (Figure 2.12c1), ● measured distribution (Figure 2.12c2). (b) Gaussian distribution (Figure 2.12b).

the maximum time delay of the closed first and third phases is assumed to be less than 7.5 ms. The total number of switching operations is taken as 100 for simplicity. For an insulation design, at least 1000 operations are required to obtain more detailed data of an overvoltage distribution. The calculated results indicate that the difference in the CB closing-angle distribution causes noticeable differences, especially in the maximum and minimum overvoltages. For example, the maximum overvoltage, that is, an overvoltage that has the lowest probability of occurrence, in Figure 2.12c1 is greater by about 0.3 pu than that in Figure 2.12a and c2. In the case of Gaussian distribution, a difference of 0.12 pu is observed in the maximum overvoltage for $6\sigma = 7.5$ ms and 5 ms, where σ is the standard deviation. This observation

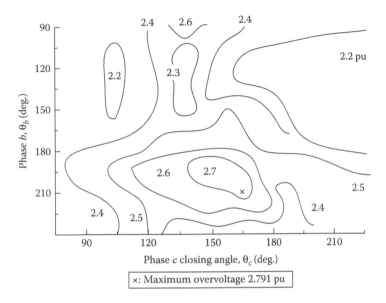

FIGURE 2.14
Distribution characteristics of maximum switching overvoltages ($\theta_a = 90°$, $\ell = 150$ km, and $L_s = 50$ mH).

makes it clear that the distribution of the CB closing angles affects the maximum overvoltage significantly, and thus one has to be careful when choosing the distribution.

Figure 2.14 shows a contour line expression of maximum overvoltages when the phase *a* closing angle θ_a is fixed at 90°. It can be observed from the figure that higher overvoltages distribute along the axes of $\theta_c \doteqdot 150°$ and $\theta_b \doteqdot 210°$. It can also be seen that higher overvoltages appear along the axis of $\theta_a \doteqdot 90°$ when θ_a is varied. Thus, it can be concluded that severe overvoltages appear along the axes of $\theta_a \doteqdot 90°$, $\theta_b = 210°$, and $\theta_c \doteqdot 150°$, as illustrated in Figure 2.15 in 3D space. This is quite reasonable because the source voltage on each phase peaks at the angle, and the voltage difference across the CB terminals of the source and line sides becomes the largest before closing the CB of the phase if the mutual coupling and the residual voltage are neglected. It is clear that the most severe overvoltage is generated when the CB is closed with the largest voltage difference across the CB. In reality, however, the closing angle that generates the most severe overvoltage is somewhat different to some degree due to the effects of the mutual coupling between the phases, the source inductance, and the residual voltage (if any).

2.2.3.4.3 Resistor Closing

It is a common practice to insert a closing resistor into a CB on EHV and UHV systems in Japan. Figure 2.16 shows the effect of resistor closing on the closing surge on the untransposed vertical double-circuit line in Figure 1.25

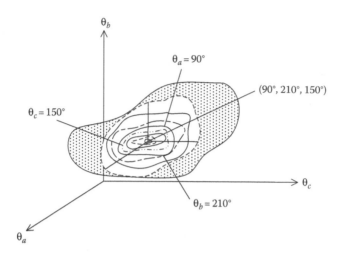

FIGURE 2.15
3D expression of maximum switching overvoltages.

with a line length of 101.13 km. It is quite clear that the overvoltage with no closing resistor is reduced to about half of that with a closing resistor.

Figure 2.17 shows the effect of closing resistors on maximum overvoltages for both closing and reclosing surges. It is observed that the surge over-voltage generated by resistor closing decreases as the value of the resistor increases. However, the overvoltage generated from short-circuiting the resistor increases as the value of the resistor increases.

Thus, an optimum value of the closing resistor exists. It is 300 Ω in the closing case in Figure 2.17a. Resistor closing is more effective in the reclosing case in Figure 2.17b. The maximum overvoltage of 3.5 pu is reduced to 1.3 pu in the reclosing case, while it is 2.2–1.2 pu in the closing case.

2.2.3.4.4 Closing Surge Suppression by an Arrester

The original purpose of an arrester is to protect power apparatuses from a lightning overvoltage. But it can also be applied to suppress a switching overvoltage on a UHV system thanks to the ZnO arrester, the development of which has greatly improved arrester performance. Figure 2.18a shows a typical characteristic of a ZnO arrester. Figure 2.19 shows an application example of the ZnO attester, which has the voltage–current characteristic shown in Figure 2.18b to control the closing surge overvoltage. The maximum overvoltage of 2.03 pu is suppressed to 1.56 pu by the arrester.

2.2.3.5 Effect of Various Parameters on a Closing Surge

As explained in Section 2.2.3.4.2, the maximum overvoltage is dependent on the closing angle of each phase, which in reality is probabilistic. Therefore, a deterministic analysis of the switching overvoltage is often found to be

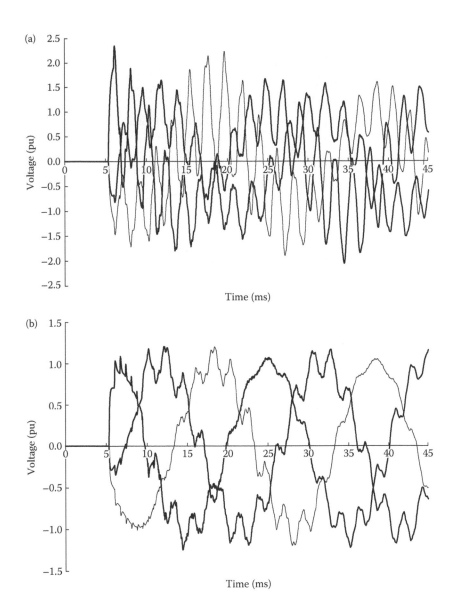

FIGURE 2.16
Three-phase closing surges on a 500 kV untransposed vertical line. (a) No resistor closing. (b) Resistor closing.

inadequate to investigate an insulation level. Nowadays, it is a common practice to analyze the switching overvoltage from a statistical viewpoint taking into account the probabilistic nature of the closing angles and insulation failure due to overvoltage. Thus, statistical analysis of the effect of various parameters on the closing overvoltage will be studied in this section.

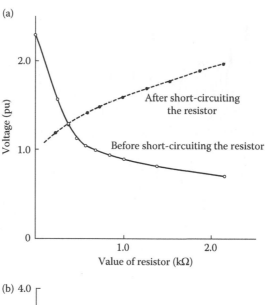

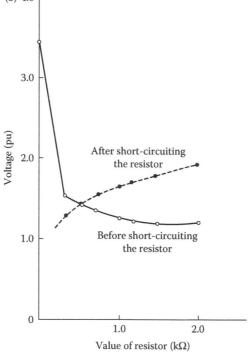

FIGURE 2.17
Effect of closing resistors on maximum overvoltages. (a) Closing surge. (b) Reclosing surge
residual voltages (–1, –1, 1 pu).

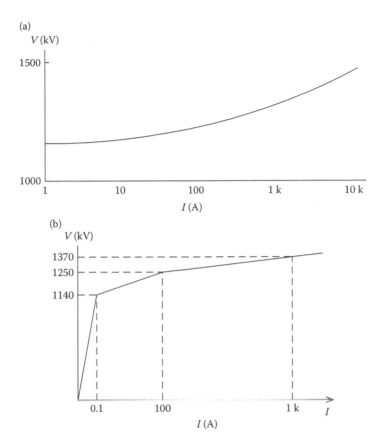

FIGURE 2.18
Voltage–current characteristic of a ZnO arrester. (a) Actual characteristic. (b) Linear approximation.

2.2.3.5.1 Effect of Earth Resistivity

Figure 2.20 shows statistical overvoltage distribution curves for various earth resistivities. It is clear that the highest overvoltages occur for the case of $\rho_e = 0$, that is, a perfectly conducting earth corresponding to no wave deformation. Also, it is observed that the overvoltage increases as the earth resistivity increases. The lowest overvoltage distribution is observed for $\rho_e = 10\ \Omega$ m. Thus, it should be clear that a resistivity that gives the lowest overvoltage distribution exists. This is explained by the fact that an increase in the attenuation constant due to the increase in earth resistivity reduces the current flowing into the soil, thereby decreasing the overall attenuation of the overvoltages. This results in a higher overvoltage. Conversely, a decrease in the attenuation constant due to the decrease in earth resistivity increases the current flowing into the soil, thereby increasing the overall attenuation of the overvoltages. This also results in a higher

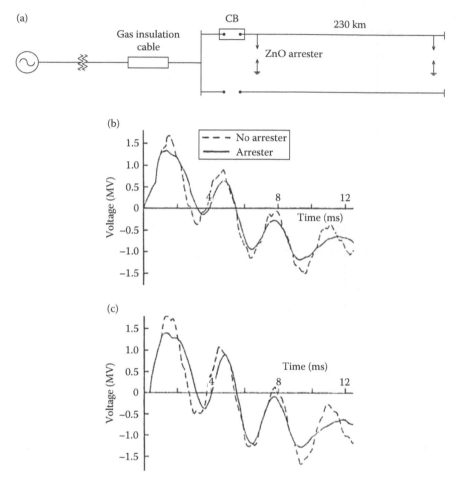

FIGURE 2.19
Suppression of the closing surge overvoltage by a ZnO arrester. (a) A 1100-kV line. (b) Sending end. (c) Receiving end.

overvoltage. This means that an earth resistivity that gives the lowest over-voltage distribution exists.

2.2.3.5.2 Effect of Line Length

Figure 2.21 shows a statistical distribution of overvoltages for various line lengths. In general, the overvoltage increases as line length increases. In practice, a long transmission line is compensated by a shunt reactor, and thus the overvoltage distribution differs from that shown in Figure 2.21.

2.2.3.5.3 Effect of Source Inductance

Figure 2.22 shows the effect of source inductance on a statistical distribution of overvoltages. No clear tendency is observed in the figure.

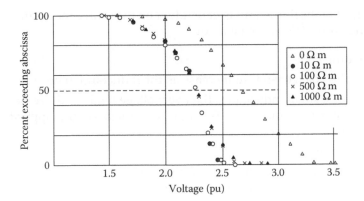

FIGURE 2.20
Effect of earth resistivity on statistical overvoltage distribution curves.

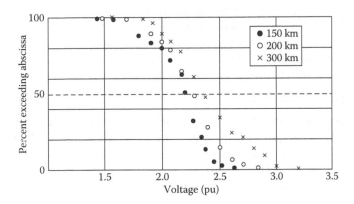

FIGURE 2.21
Effect of line length.

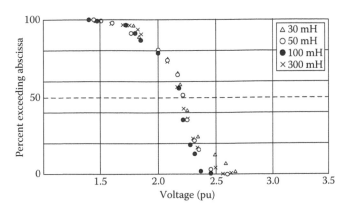

FIGURE 2.22
Effect of source inductance.

The effect of the line length is dependent on the source inductance, and the following observations can be made:

1. Sending end:
   a. The voltage tends to increase as the inductance increases.
   b. For a small inductance, the overvoltage tends to decrease as the line length increases. For a large inductance, the opposite tendency is observed.
2. Receiving end:
   a. For a long line, the overvoltage tends to increase as the inductance increases. No clear tendency is found for a short line.
   b. The overvoltage increases as the line length increases.

The reason the overvoltage tends to increase as the source inductance and line length increase is that the natural oscillating frequency of the line approaches the power frequency due to the increase in the inductance and line length. This results in a resonant oscillation of the system, and the overvoltage increases. This resonant overvoltage is a temporary overvoltage rather than a closing surge overvoltage.

2.2.3.5.4 Line Transposition

An untransposed line shows a noticeable difference in the voltage waveform from the transposed line case at the very beginning of the waveform. However, a statistical distribution of the overvoltages shows no significant difference, as observed from Figure 2.23. The difference is more noticeable at the sending end than at the receiving end. If surge waveforms with a specific sequence of closing angles are compared, a rather significant difference is often observed.

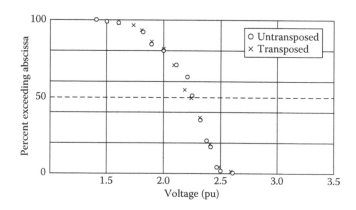

FIGURE 2.23
Effect of line transposition on statistical overvoltage distribution curves.

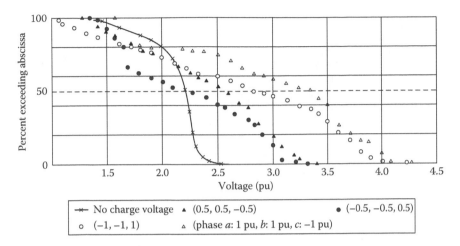

FIGURE 2.24
Effect of line charge voltage.

2.2.3.5.5 *Reclosing Overvoltage*

Figure 2.24 shows a statistical distribution of reclosing overvoltages. It is clear that the reclosing generates a higher overvoltage distribution. In general, the higher the residual charge voltage, the more severe the overvoltage distribution. The figure shows the polarity effect. This may vanish if the number of samples is large enough.

2.2.3.5.6 *Effect of Closing Resistor*

Figure 2.25 shows the effect of a closing resistor on a statistical distribution. It is clear how the closing resistor is effective in reducing the overvoltage in both the closing and reclosing cases. The damping effect is especially noticeable in the region of higher overvoltages. The 500-Ω resistor is effective in reducing the overvoltage by 1.7 pu.

2.3 Fault Surge

A fault surge is generated by a line fault. Fault surges are classified into two types: fault initiation surge (fault surge) and fault-clearing surge. The most probable fault is a phase single line-to-ground (SLG) fault. In this section, the SLG fault will be explained.

2.3.1 Fault Initiation Surge

The fault initiation surge (fault surge) is generated on a sound phase due to a single phase fault (faulty phase or line) [1]. The maximum overvoltage is

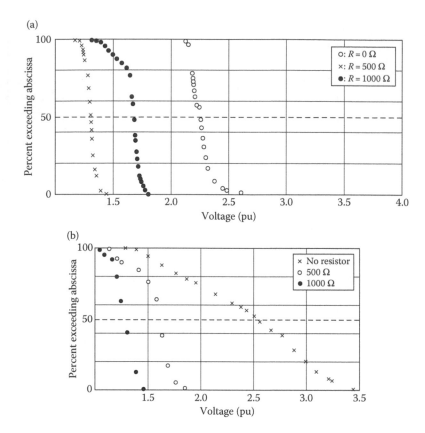

FIGURE 2.25
Effect of closing resistors on statistical overvoltage distribution curves. (a) Closing surges. (b) Reclosing surges (initial charge 0.5, 0.5, –0.5 pu).

generally much lower than that of a closing surge. Figure 2.26 is an example of a typical fault surge. The maximum voltage observed is –1.32 pu, which is far lower than the voltages discussed in Sections 2.1 and 2.2. Figure 2.1 is another example of a fault surge measured in an actual EHV transmission line. The maximum overvoltage in the measurement was 1.7 pu, as shown in the figure.

Figure 2.27 shows the calculated results of the fault surge on a 1100 kV transmission system. A single phase-to-ground fault occurs at the midpoint of the first circuit of line 1 (node L1-13) as illustrated in Figure 2.27a. Figure 2.27b shows the faulty circuit voltages and Figure 2.27c the sound circuit voltages. The maximum overvoltage is observed to be 1.67 pu (= 1500 kV) on phase *c* of the faulty circuit. These results are for the case where arresters are installed.

In general, the maximum fault overvoltage is expected to be less than 1.7 pu for a rather simple system. In a complicated system, such as that shown in

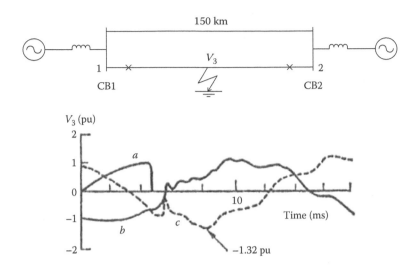

FIGURE 2.26
A fault surge.

Figure 2.27a, the maximum overvoltage may reach 1.8 pu. As the overvoltage is much lower than the closing overvoltages explained in Section 2.2, the fault surge is not an essential factor for determining the switching impulse withstanding level (SIWL) or the switching impulse insulation level (SIL) of a conventional transmission line below 700 kV. However, it becomes quite significant in a UHV system since the SIWL or SIL is expected to be lower than 1.7 pu, which is easily achieved by using a closing resistor as far as the closing surge is concerned. The fault surge, on the other hand, is hard to control, because it is impossible to predict where and when a fault might occur, or where the highest overvoltage occurs. Currently, the only way to control fault overvoltage is by installing so-called line arresters along the transmission line.

2.3.2 Characteristic of a Fault Initiation Surge

2.3.2.1 Effect of Line Transposition

Figure 2.28 compares the effect of a fault surge on untransposed and transposed double-circuit lines. The fault is initiated at the receiving end of one circuit as illustrated in the figure. The circuit configuration is equivalent to the model circuit illustrated in Figure 2.26 in which the fault occurs at the middle of the line.

The figure clearly shows that there is no significant difference between the untransposed and transposed lines. The observation is the same as that made for the closing surge case. As far as the surge overvoltage or line insulation is concerned, the effect of line transposition is not significant.

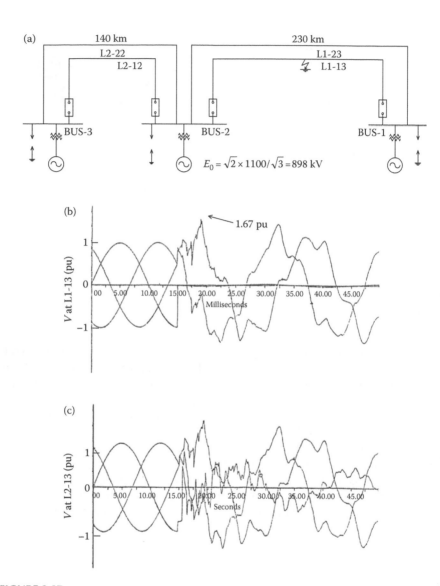

FIGURE 2.27
Calculated results of fault surge on a 1100-kV circuit. (a) A 1100-kV model circuit. (b) Faulty circuit. (c) Sound circuit.

2.3.2.2 Overvoltage Distribution

Figure 2.29 illustrates a maximum overvoltage due to a phase a-to-ground fault surge, when the fault position is changed from the sending end to the receiving end along an untransposed 500-kV line with a length of 200 km.

In the case of source inductances of 50 mH at both ends, Figure 2.29a, the highest maximum overvoltage of 1.56 pu occurs at the middle of the line,

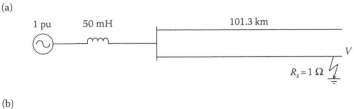

(a)

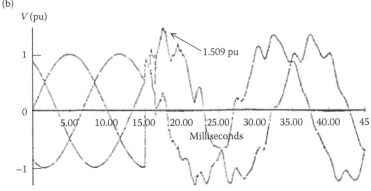

(b)

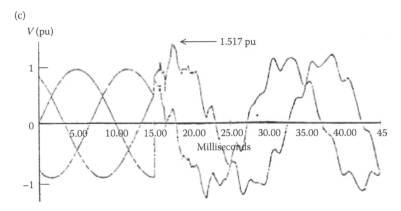

(c)

FIGURE 2.28
Effect of line transposition on a fault surge–fault circuit voltage. (a) A 500 kV model circuit. (b) Untransposed line. (c) Transposed line.

and the overvoltage decreases symmetrically toward the ends of the line. The reason for this condition is that positive reflected waves from both ends arrive at the middle of the line at the same instance and are superposed.

When an infinite source is connected to the right-hand side of the line, the highest overvoltage of 1.56 pu, the same as that in Figure 2.29a, occurs on the right-hand side, rather than the middle, of the line. This is due to the asymmetry of the circuit, that is, the electrical center being shifted to the right of the line.

Figure 2.30 shows a maximum overvoltage distribution along the line when the fault position is fixed. Figure 2.30a illustrates the case of the fault

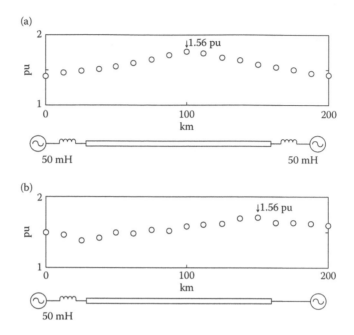

FIGURE 2.29
Overvoltages at the fault points. (a) 50-mH source. (b) 50 mH and an infinite bus.

at the middle, and Figure 2.30b at the 1/4 length from the sending end. In Figure 2.30b, the highest overvoltage of 1.53 pu occurs at the fault point and its symmetrical position to the center of the line, and the overvoltage distribution is flat within the two points. In Figure 2.30a, the highest overvoltage of 1.56 pu occurs at the center, and the overvoltage decreases rapidly toward both ends of the line. This observation demonstrates that the region of the line where the overvoltage is higher than a certain level is wider in the case of the fault occurring at a point apart from the line center, while the highest overvoltage occurs in the case of the middle point fault. This information may prove important in the design of future protection systems.

Table 2.2 shows the maximum overvoltage at various nodes of the system illustrated in Figure 2.27a. It is observed that the highest overvoltage occurs at the fault point on the sound phase of the faulty circuit. The overvoltage decreases as the distance from the fault point increases.

2.3.3 Fault-Clearing Surge

A fault-clearing surge is generated by clearing a fault by a CB. Generally, its overvoltage is smaller than a closing surge overvoltage and greater than that of a fault initiation surge overvoltage.

Figure 2.31 shows an example of a fault-clearing surge. The maximum overvoltage is observed to be 1.5 pu, which is greater than that of the 1.32 pu

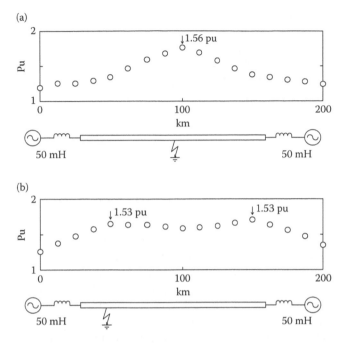

FIGURE 2.30
Overvoltage distribution along the line. (a) Fault at the middle. (b) Fault at the 1/4 point.

fault surge in Figure 2.26. The overvoltage increases to about 1.6 pu in the two phase-to-ground fault cases and to about 2.0 pu in the three phase-to-ground fault cases.

Figure 2.32 shows the effect of line transposition on fault-clearing overvoltage. It is observed that in both untransposed and transposed lines the maximum overvoltage exceeds 2 pu, which is greater by about 0.6 pu than

TABLE 2.2

Maximum Overvoltages at Various Points in the 1100-kV System in Figure 2.27a

	Phase		
Node	a	b	c
BUS-1	−1.06 (pu)	1.29	1.37
L1-13	–	1.55	1.67
L1-23	−1.07	1.49	1.64
BUS-2	−1.07	1.31	1.41
L2-12	−1.07	1.32	1.44
L2-22	−1.07	1.32	1.44
BUS-3	−1.14	1.29	1.39

$1\text{pu} = \sqrt{2} \times 1100/\sqrt{3} = 898 \text{ kV}$

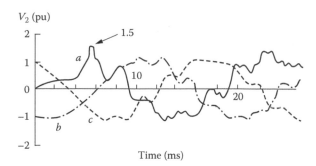

FIGURE 2.31
A fault-clearing overvoltage in the same circuit as in Figure 2.26.

the fault initiation overvoltages observed in Figure 2.28. The line transposition does not significantly affect the overvoltage, but affects the wave shape significantly as is clear from the figure. This phenomenon occurs because the attenuation of the aerial mode is greater and the velocity is lower in a transposed line than in an untransposed line, as explained in Section 1.5.2. As a result, the waveform is more distorted as time increases in the transposed line case. The difference, however, is not noticeable if an opening resistor of a CB is used.

Figure 2.33 shows the effect of the opening resistor of the CB on the fault-clearing overvoltage on a Japanese 1100-kV system. At $t = 0$, the system is in a steady state of three phase-to-ground fault at the middle of line 2 (140 km). At $t = 1$ ms, a resistor is inserted between the contacts of the CB, and the main contact is opened. Then, at $t = 30$ ms, the resistor is opened, and the three phase-to-ground fault is completely cleared.

The fault-clearing surge voltages at the middle of line 1 (230 km) on the faulty circuit are shown in Figure 2.33, from which it is clear that the maximum overvoltage reaches 1.71 pu in the case where the resistor is not opened (a). The overvoltage is decreased to 1.31 pu by the opening resistor in Figure 2.33b. Also, the voltage waveform tends to rapidly attain a sinusoidal steady state.

Table 2.3 shows the maximum overvoltages at various positions of the circuit illustrated in Figure 2.33. It is observed from the table that the highest voltage occurs at the middle of line 1 instead of the faulty line. This fact should be noted when a fault-clearing surge analysis is carried out. Otherwise, the highest overvoltage may be missed.

Figure 2.34 shows a relation between the fault-clearing overvoltage and an opening (insertion) resistor on the 1100-kV system discussed earlier. It is observed that an optimum resistor to control the overvoltage exists, similar to the closing overvoltage case. The degree of overvoltage reduction is lower in the fault-clearing surge case than in the closing surge case shown in Figure 2.17. This is due to the fact that the highest overvoltage occurs within the switched line in the closing surge case, while it may occur in the other

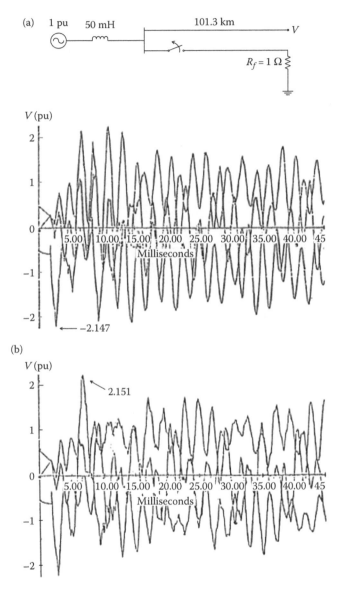

FIGURE 2.32
Effect of line transposition-sound circuit voltage. (a) Untransposed. (b) Transposed.

line connected to the faulty line in the fault-clearing surge case. In this circumstance, the effect of the insertion resistor would be minimal.

The overvoltage in Figure 2.34 is reduced to 1.42 pu, which is still greater than the 1.36 pu given in Table 2.3, the reason being that the overvoltage in Figure 2.34 takes into account all the voltages in the system studied.

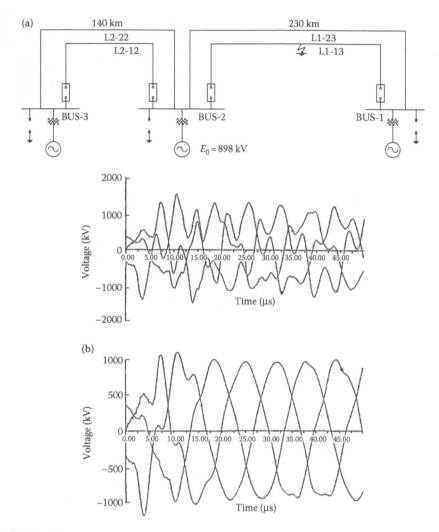

FIGURE 2.33
Effect of an opening resistor on a fault-clearing overvoltage. (a) No resistor. (b) Opening resistor.

A significant difference of the opening resistor from the closing resistor is its thermal requirement. Since a large fault current flows through the opening resistor, the resistance cannot be too small. Therefore, an optimum value of the opening resistor has to be determined not only by the degree of overvoltage reduction but also by the thermal requirement.

Figure 2.35 shows an example of a possible combination of opening and closing resistors. It is observed from the figure that a 400-Ω resistor is optimum for the closing surge (I-a, b) and the fault-clearing surge (II). In the reclosing case (I-a' and I-b), where the reclosing overvoltage is always higher than the fault-clearing overvoltage, the optimum resistor value may be

TABLE 2.3

Maximum Overvoltages at Various Points in the System

Node	Phase		
	a	*b*	*c*
(a) No Resistor			
BUS-1	−1.52 (pu)	1.42	1.53
L1-13	−1.70	1.48	1.71
L1-23	−1.70	1.48	1.71
BUS-2	−1.47	1.40	1.52
L2-12	−	−	−
L2-22	−1.54	1.55	−1.58
BUS-3	−1.44	1.43	−1.41
(b) $R_{CB} = 500\ \Omega$			
BUS-1	1.13 (pu)	1.11	−1.24
L1-13	1.19	1.12	−1.31
L1-23	1.19	1.12	−1.31
BUS-2	1.13	1.11	−1.23
L2-12	−	−	−
L2-22	1.19	1.11	−1.36
BUS-3	−1.12	1.10	−1.18

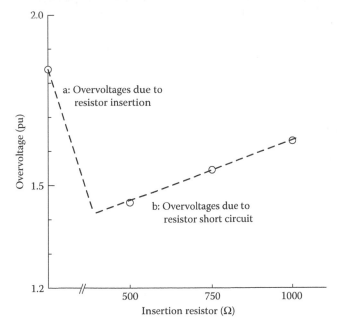

FIGURE 2.34
Relation between fault-clearing overvoltages and an insertion resistor.

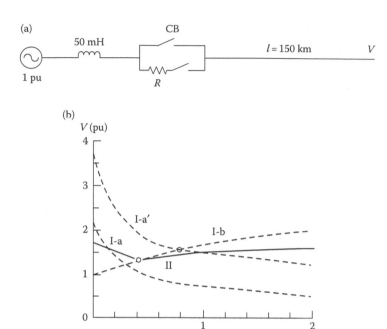

FIGURE 2.35
Reduction of surge overvoltages by an insertion resistor. *I*: closing surge. *a*: resistor insertion. *b*: resistor off. *a′*: resistor insertion, residual voltage. *II*: fault-clearing surge. (a) Model circuit. (b) Calculated results.

determined by the resistor corresponding to the lowest reclosing overvoltage, that is, about 800 Ω for 1.6 pu.

2.4 Lightning Surge

Of all surges, lightning to a tower or line causes the highest overvoltage. Numerous studies have been carried out [9–22], but there are still a number of unknown factors in the analysis of a lightning surge in power systems, for example: the tower footing impedance model, which is time- and current-dependent; the lightning source; and the lightning channel impedance [23]. Lightning surges are still a fruitful field of research.

2.4.1 Mechanism of Lightning Surge Generation

When lightning strikes a tower (or a GW) as illustrated in Figure 2.36, the lightning current flows into the tower and causes a sudden increase in tower voltage. When the voltage difference between the tower and a phase wire

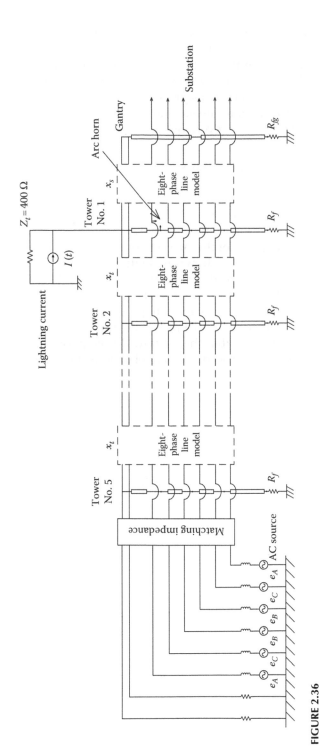

FIGURE 2.36

A representative model system for a lightning surge simulation.

(PW) reaches its electrical withstand voltage, a flashover from the tower to the PW occurs. This is called "back flashover" (BFO), because the tower voltage is higher than the PW voltage, unlike normal power system operation. Then, the lightning current flows into the PW and a traveling wave due to the current propagates to a substation. The traveling wave is partially reflected at and refracted into the substation. The traveling wave or its successive reflection generates a severe overvoltage at the substation.

2.4.2 Modeling of Circuit Elements

2.4.2.1 Lightning Current

Lightning is often modeled by a current source with a parallel resistance that represents the lightning channel impedance, as illustrated in Figure 2.36 [23]. Although the waveform of the current source is not conclusively known, it is defined by the waveform shown in Figure 2.37. In the figure, T_f is the time from the origin to the peak, called the "wave front length (time)," and T_t is the time from the origin to a half of the peak voltage, called the "wave tail length." In general, T_f is less than 10 μs, and T_t is less than 100 μs. A 1/40 μs ($T_f = 1$ μs, $T_f = 40$ μs) or a 2/70-μs wave is the usual standard. It should be clear from this condition that the sustained time period of the lightning surge is in the order of microseconds as explained in Table 2.1.

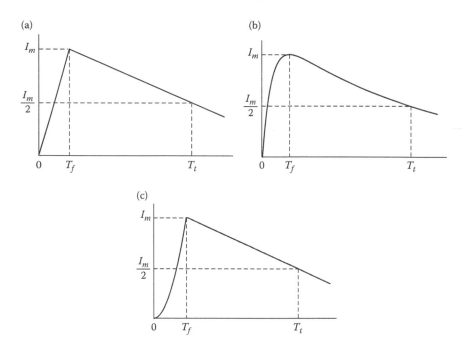

FIGURE 2.37
Waveforms of a lightning current. (a) Lump. (b) Double exponent. (c) CIGRE.

There are three different definitions of current waveforms as shown in Figure 2.37. The first definition is the "lump wave," which is expressed by two linear lines. The second is the "double exponential wave," expressed by

$$i(t) = I_m\{\exp(-at) - \exp(-bt)\} \tag{2.6}$$

The exponential wave has been widely used, especially when an insulation test of a power apparatus is carried out, because an IG inherently produces the exponential waveform. However, the exponential wave gives an excessive overvoltage compared to the lump wave and CIGRE if it is adopted in a lightning surge analysis. Thus, it is recommended that the double exponential wave not be used in the lightning surge analysis, or to take $T_f = 2\ \mu s$ instead of 1 μs in the case of a 1/40 μs waveform. The third waveform, called "Conseil International Des Grands Réseaux Electriques (CIGRE) wave," is based on the study of the CIGRE and is characterized by its negative di/dt at the wave front as in Equation 2.7. In contrast, other characterizations are $di/dt = 0$ for the lump wave and $di/dt > 0$ for the exponential wave.

$$
\left.
\begin{aligned}
i(t) &= I_m\left\{1 - \cos\left(\frac{\pi t}{2T_f}\right)\right\} && \text{for } t < T_f \\
&= I_m\frac{(2T_t - T_f - t)}{2(T_t - T_f)} && \text{for } T_f \leq t < 2T_t - T \\
&= 0 && \text{for } t \geq 2T_t - T_f
\end{aligned}
\right\} \tag{2.7}
$$

Measured examples of lightning current amplitudes are shown in Figure 2.38 [20–26]. It can be observed from the figure that the frequency of occurrence of a current greater than 50 kA is about 20%, and that of a current greater than 100 kA is about 5%. Therefore, assuming a lightning current of 100 kA is sufficient for analysis on an EHV transmission line [22,26]. In Japan, a current of 200 kA has been used for the insulation design of a 1100 kV transmission system [26]. Table 2.4 lists the recommended amplitudes of lightning currents for various voltage classes [26,27].

2.4.2.2 Tower and Gantry

A transmission tower is represented by four distributed parameter lines [28] as illustrated in Figure 2.39, where Z_{t1} is the tower top to the upper phase arm = upper to middle = middle to lower and Z_{t4} is the lowest to the tower bottom.

Table 2.4 gives a typical value of the surge impedance.

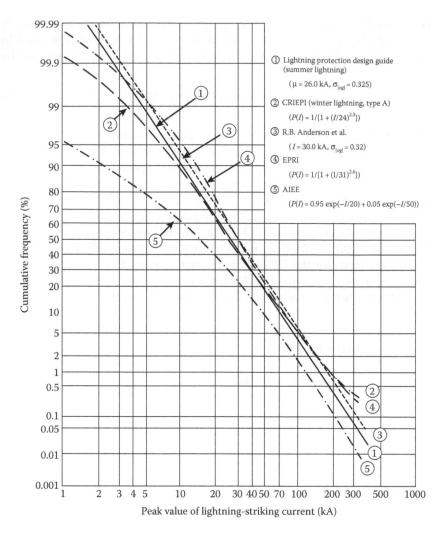

FIGURE 2.38
Measured results of lightning current amplitudes. (From *IEEE* Guide for Improving the Lightning Performance of Transmission Lines. 1997; Ametani, A. et al. 2002. *IEE Japan WG Report*. Technical Report No. 872; CRIPEI WG. 2003. Guide to transmission line protection against lightning. Report T72; Ametani, A. and T. Kawamura. 2005. *IEEE Trans. Power Deliv.* 20 (2):867–875; Anderson, R. B. and A. J. Eriksson. 1980. *Electra* 69:65.)

The propagation velocity c of a traveling wave along a tower is taken to be

$$c_0 = 300 \text{ m/µs} : \text{light velocity in free space} \qquad (2.8)$$

To represent the traveling-wave attenuation and distortion, an *RL* parallel circuit is added to each part, as illustrated in Figure 2.39. The values of *R* and *L* are defined in the following equation:

TABLE 2.4

Recommended Values for Lightning Parameters

System Voltage (kV)	Lightning Current (kA)	Tower Height/Geometry (m)				Surge Impedance (Ω)		Footing Resistance (Ω)
		h	x_1	x_2	x_3	Z_{t1}	Z_{t4}	R_f
1100	200	107	12.5	18.5	18.5	130	90	10
500	150	79.5	7.5	14.5	14.5	220	150	10
275	100	52.0	9.0	7.6	7.6	220	150	10
154 (110)	60	45.8	6.2	4.3	4.3	220	150	10
77 (66)	30, 40	28.0	3.5	4.0	3.5	220	150	10–20

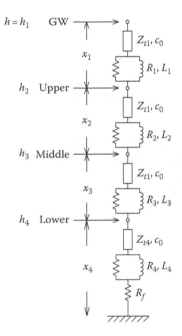

FIGURE 2.39
A tower model.

$$R_i = \Delta R_i \cdot x_i, \quad L_i = 2\tau R_i$$

$$\Delta R_1 = \Delta R_2 = \Delta R_3 = 2Zt_1 \cdot \ln \frac{(1/\alpha_1)}{(h - x_4)} \tag{2.9}$$

$$\Delta R_4 = 2Z_{t4} \cdot \ln \frac{(1/\alpha_4)}{h}$$

where
$\tau = h/c_0$ is the traveling time along the tower
$\alpha_1 = \alpha_4 = 0.89$ is the attenuation along the tower
h is the tower height

The *RL* parallel circuits in Figure 2.39 can be neglected in most lightning surge analyses, as explained later.

A substation gantry is represented by a single distributed line with no loss.

2.4.2.3 Tower Footing Impedance

In Japan, modeling a tower footing impedance as a simple linear resistance R_f is recommended, although a current-dependent nonlinear resistance is recommended by the IEEE and the CIGRE [16,18,20]. The inductive and capacitive characteristics of the footing impedance, as shown in Figure 2.40, are well known [10]. The recommended value of resistance for each voltage class is given in Table 2.4.

2.4.2.4 Arc Horn

The maximum voltage on a line and a substation due to lightning is highly dependent on the flashover voltage of an arc horn, which is installed between a tower arm and a PW along an insulator, as illustrated in Figure 2.36. The purpose of the arc horn is to control the lightning overvoltage on the line and the substation, and also to protect the insulator against mechanical damage due to electrical breakdown (= flashover). In general, an arc horn gap length (the clearance between the tower arm and the PW) that is shorter than the length of the insulator is considered so that the arc horn gap flashovers before the voltage between the tower arm and the PW, that

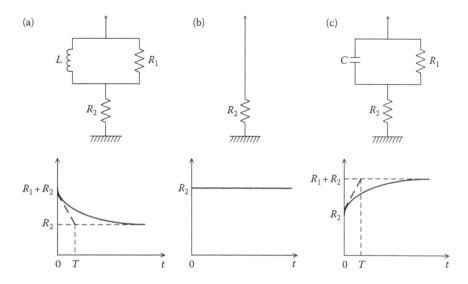

FIGURE 2.40

Footing impedance models and step responses. (a) Inductive. (b) Resistive. (c) Capacitive.

is, voltage across the insulator, reaches its breakdown voltage (= insulator withstand voltage). Thus, in theory, the insulator never breaks down and is protected from mechanical damage. Also, the voltages on the line and in the substation are dependent on the flashover voltage of the arc horn. Therefore, it should be noted that the so-called lightning overvoltage observed on a PW or in a substation is not the voltage of the lightning itself but the voltage dependent on or controlled by the arc horn. If the gap length of the arc horn is too small, too many lightning surges appear in the substation, resulting in excessive insulation of the substation apparatuses. On the other hand, if the gap is too large but smaller than the clearance, then the insulation of PWs from the tower (the clearance), becomes excessive. Therefore, determining an optimum gap length for an arc horn is quite difficult. It is entirely dependent on the overall construction cost of the transmission system and the philosophy behind the overall insulation design and coordination of the system.

An arc horn flashover is represented either by a piecewise linear inductance model with time-controlled switches as illustrated in Figure 2.41a, or by a nonlinear inductance as shown in Figure 2.41b, based on a leader progression model [29,30]. The parameters L_i ($i = 1$–3) and $t_i - t_{i-1}$, assuming the initial time $t_0 = 0$ in Figure 2.41a, are determined from a measured result of the V–I characteristics of an arc horn flashover. Then, the first simulation with no arc horn flashover is carried out, as in Figure 2.36, and the first flashover phase and the initial time t_0 are determined from the simulation results of the voltage waveforms across all the arc horns. By considering these parameters, the second simulation with the first flashover phase is carried out to determine the second flashover phase. By repeating this procedure until no flashover occurs, the lightning surge simulation by a piecewise linear model will be complete. Thus, a number of precalculations are necessary in the case of multiphase flashovers, while the nonlinear inductance model needs no precalculations and is easily applied to multiphase flashovers. Details of the leader progression model are explained in Reference 29 and the nonlinear inductance model is explained in Reference 30.

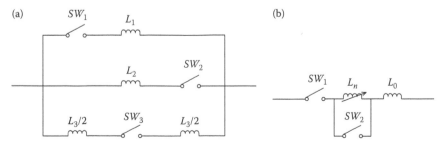

FIGURE 2.41
Arc horn flashover models. (a) A linear inductance model. (b) A nonlinear model.

2.4.2.5 Transmission Line

Most transmission lines in Japan are of double-circuit vertical configuration with two GWs and are thus composed of eight conductors. The use of a frequency-dependent line model is recommended in a numerical simulation, but a distributed line model with fixed propagation velocity, attenuation, and surge impedance, that is, the fixed-parameter distributed line model explained in Reference 23, is often used.

2.4.2.6 Substation

1. *Gas-insulated bus and cable:* A cable and a gas-insulated bus are represented either as three single-phase distributed lines with their coaxial mode surge impedance and propagation velocity or as a three phase distributed line system. As a gas-insulated substation involves a significant number of gas-insulated buses/lines, the pipes are, in most cases, eliminated by assuming zero voltage.
2. *CB, DS, transformer, and bushing:* A CB and a DS are represented by lumped capacitances between the poles and to the soil. A transformer is also represented by a capacitance to the soil unless a transferred surge to the secondary circuit must be calculated. A bushing is represented by a capacitance or, occasionally, a distributed line.
3. *Grounding mesh:* A grounding mesh is in general not considered in a lightning surge simulation and is regarded as a zero potential surface. When dealing with an incoming surge to a low-voltage control circuit, the transient voltage of the grounding mesh should not be assumed to be zero, and its representation becomes an important but difficult task.

2.4.3 Simulation Result of A Lightning Surge

2.4.3.1 Model Circuit

Figure 2.36 shows a representative model circuit for lightning surge analysis [23]. Lightning strikes the top of tower No. 1 in the substation vicinity. The lightning stroke is represented by a current source with a peak value of 200 kA and a waveform of 2/70 μs, which is the Japanese standard for a 1100-kV line, in the form of

$$i(t) = I_0 K_0 [\exp(-at) - \exp(bt)] \tag{2.10}$$

where $I_0 = 200$ kA, $K_0 = 1.0224$, $a = 1.024 \times 10^4$ s^{-1}, and $b = 2.8188 \times 10^6$ s^{-1}.

Five towers are included in the model. The span distance of the transmission line between adjacent towers is 450 m, and that from tower No. 1 to the substation is 100 m. The end of the transmission line is terminated with the surge impedance matrix or, approximately, with matching resistances: $R_p = 350$ Ω for a PW and $R_g = 560$ Ω for a GW.

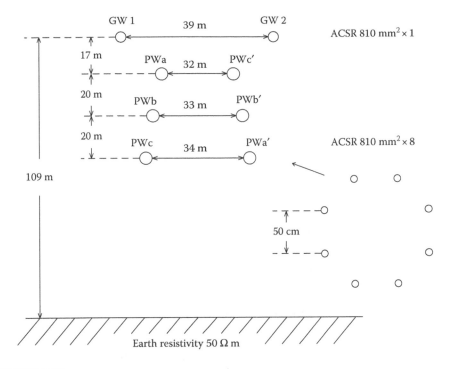

FIGURE 2.42
A 1100 kV twin-circuit line.

The transmission line is of double-circuit vertical configuration with two GWs, as shown in Figure 2.42. The total number of conductors is eight. The tower model is explained in Section 2.4.2 (see also Figure 2.39). It is divided at the cross-arm positions into four, and each section is modeled by a lossless distributed parameter line neglecting the RL parallel circuit. Data for the elements are given in Table 2.5. The tower cross-arms are neglected. The tower footing resistance is taken as 10 Ω. Figure 2.43 is a model of a UHV substation for one phase. C_b and C_s in the figure are capacitances that represent bushings and shunt reactors, respectively. A gas-insulated bus is represented by a lossless distributed parameter line with a surge impedance of 70 Ω and a velocity of 270 m/μs.

TABLE 2.5

Parameters of a 1100-kV Tower and a Structure

Tower	Structure
$Z_{t1} = 210\ \Omega$	$Z_{t3} = 125\ \Omega$
$Z_{t2} = 170\ \Omega$	$Z_{t4} = 125\ \Omega$
$c = 300$ m/μs	$c = 300$ m/μs

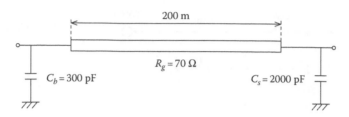

FIGURE 2.43
Single-phase expression of a substation model.

Figure 2.41a shows a lumped circuit model of an arc horn flashover proposed by Shindo and Suzuki [29]. Inductance and resistance values and closing times of switches are determined from a given voltage waveform across the arc horn gap based on a theory of a discharge mechanism.

2.4.3.2 Lightning Surge Overvoltage

Figure 2.44 shows a typical result of a lightning surge. It should be clear from the figure that the overvoltage generated by a lightning surge is far greater than the insulation level estimated from a switching surge overvoltage. For example, the insulation level against the switching surge on the Japanese 1100-kV line is considered to be less than 1.7 pu. The lightning overvoltage on the PW observed in Figure 2.44a is 7.04 MV. For the nominal operating voltage of the line is $\sqrt{2} \times 1100/\sqrt{3} = 898$ kV, the previous overvoltage is about 7.4 pu. If the lightning current is assumed to be 100 kA rather than 200 kA, the overvoltage is still 3.7 pu, which is much greater than the switching surge insulation level of 1.7 pu.

It can be observed from Figure 2.44b that the overvoltage at the substation entrance is about 4.6 MV, which exceeds the insulation of the substation apparatuses, such as a transformer. Thus, the apparatuses are protected

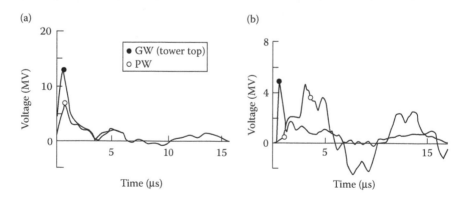

FIGURE 2.44
Lightning surge on the circuit of Figure 2.36. (a) Tower No. 1. (b) Substation entrance.

by lightning arresters from the excessive overvoltage. If no flashover is assumed, the PW overvoltage at the substation entrance is reduced to about 0.4 MV = 400 kV, which is low enough for substation insulations. However, the voltage difference between the tower arm (or GWs) and the PW reaches about 7–8 MV. The tower size might be too large to economically insulate the PW from the tower and GWs against the overvoltage. Therefore, analyzing lightning surges is very important when finding an optimum and economically feasible insulation design for a power system.

2.4.3.3 Effect of Various Parameters

2.4.3.3.1 Frequency Dependence of Line Parameters

It is well-known that the frequency dependence of a transmission line due to an imperfectly conducting earth causes a significant effect on surges traveling through a long transmission line. The frequency-dependent effect on a lightning surge can generally be neglected because the line length is short. This effect is investigated in this section.

The results obtained from using a frequency-dependent (distributed) line model and a frequency-independent line model are shown in Figure 2.45 for the case of no flashover of an arc horn. In the latter model, line parameters are calculated at the dominant transient frequency given by

$$f_t = \frac{1}{4\tau_0} = 750 \text{ kHz}$$

where $\tau_0 = l/c$, l is the distance from tower No. 1 to the substation = 100 m, and c is the velocity of light in free space.

Table 2.6 shows the maximum voltages calculated by the frequency-dependent Semlyen model and the frequency-independent distributed parameter line model of the EMTP. It is clear from Figure 2.45 and Table 2.6 that the results neglecting the frequency-dependent effect show a minor difference from the results including the effect. Thus, it can be concluded that the frequency-dependent model does not have a significant effect on a lightning surge.

The effect of various earth resistivities was also investigated, and it appears that there is no significant difference between the calculated results with the 50–1000 Ω m earth resistivity.

From this observation, it can be concluded that the frequency-dependent effect of a transmission line due to the imperfectly conducting earth is negligible in a lightning surge calculation. Thus, the lightning surge can be calculated with a reasonable accuracy using the frequency-independent distributed parameter line model. This approach is much more efficient in the computation of lightning surges, and it also becomes quite easy to explain a simulation result from a physical viewpoint.

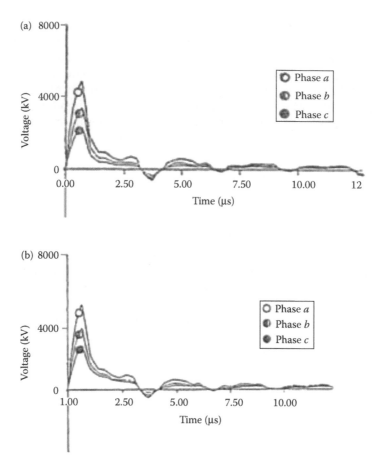

FIGURE 2.45
Effect of frequency dependence on lightning surge for the no-flashover case. (a) Frequency-dependent model. (b) Frequency-independent model.

2.4.3.3.2 *Tower Impedance and Footing Impedance*

It is well known that tower surge impedance and footing impedance affect lightning surges significantly. Tower surge impedance is a function of the height and radius of the tower as explained in Section 1.7.2.2, but in reality, it ranges from 80 to 250 Ω, as shown in Table 2.4, for towers of various voltage classes.

The tower-footing impedance is always represented as a resistance as in Table 2.4 recommended by guides on insulation design and coordination of transmission lines and substations [26]. However, it is not pure resistance, but shows an inductive or a capacitive nature (see Figure 2.40) as investigated by many authors [10,15,23,31].

Figure 2.46 shows the effect of tower footing impedance on the voltage at the top of the tower in comparison with a measured result on a 500 kV

TABLE 2.6

Maximum Voltages for the No-BFO Case

Line Model	Frequency-Dependent Line Model		Constant Frequency Line Model	
	(kV)	(µs)	(kV)	(µs)
Node 1				
Tower top	13,217	0.68	13,026	0.67
Upper arm	12,260	0.67	12,128	0.67
Middle arm	10,348	0.61	10,258	0.61
Lower arm	8,110	0.55	8,058	0.55
Upper horn	7,180	0.66	7,301	0.67
Middle horn	6,796	0.60	6,958	0.61
Lower horn	5,692	0.55	5,901	0.55
Upper phase	5,081	0.67	4,830	0.66
Middle phase	3,682	0.67	3,392	0.67
Lower phase	2,652	0.67	2,313	0.67
Node 2 (substation)				
Upper phase	433	2.42	433	2.42
(First peak)	241	0.70	241	0.70
Middle phase	332	15.5	335	15.4
(First peak)	95	0.70	94	0.70
Lower phase	261	15.5	253	15.4
(First peak)	48	0.70	45	0.70

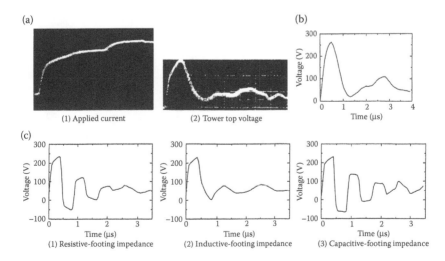

FIGURE 2.46

Influence of a tower model on a tower top voltage. (a) Measured results. (b) Frequency-dependent tower model with a resistive-footing impedance. (c) Distributed line tower model with various footing impedances.

transmission tower [23,31]. The inductive footing impedance shows a reasonable agreement with the measured result, but the resistive/capacitive impedance shows far more oscillatory wave shapes. In fact, many measured results of the grounding electrode impedance show the inductive characteristic [32]. It should be noted that the effect of the tower surge impedance, that is, the effect of the attenuation and the frequency dependence of the tower, is not significant if the footing impedance is inductive, as observed in Figure 2.46, although a number of papers discuss the modeling of tower surge impedance.

Figures 2.47 and 2.48 show the effect of tower footing resistance as a function of tower impedance on the tower-top voltage. When the footing impedance is modeled as a resistance, the effect of the tower surge impedance is

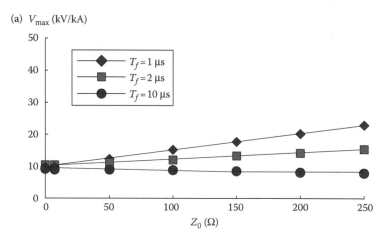

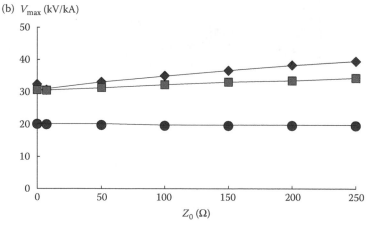

FIGURE 2.47
Effect of tower-footing resistance R_f on the tower top voltage for a 66-kV line. (a) $R_f = 10\ \Omega$. (b) $R_f = 50\ \Omega$.

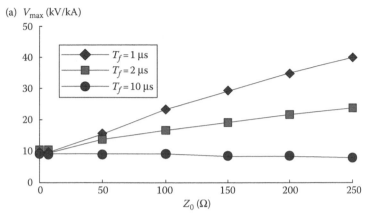

(a) V_{max} (kV/kA)

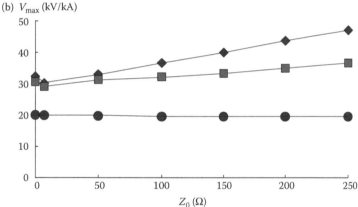

(b) V_{max} (kV/kA)

FIGURE 2.48
Effect of tower-footing resistance R_f on the tower top voltage for a 275-kV line. (a) $R_f = 10\ \Omega$. (b) $R_f = 50\ \Omega$.

clear. It should be noted that the effect of the surge impedance is less noticeable when the wave front duration T_f of the lightning current is large, and also when the footing resistance is high. Furthermore, the effect of the tower surge impedance on a surge voltage at a substation becomes less than that on the tower voltage, as observed in Figure 2.49.

2.4.3.3.3 AC Source Voltage

An AC source voltage is often neglected in a lightning surge simulation. However, it has been found that the AC source voltage affects a flashover phase of an arc horn, especially in the case of a rather small lightning current. Figure 2.50 is a measured result of arc horn flashover phases as a function of the AC source voltage on a 77 kV transmission line in Japan in the summertime [33]. The measurements were carried out in two 77-kV substations

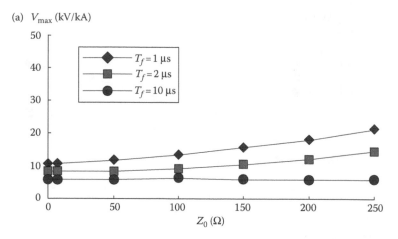

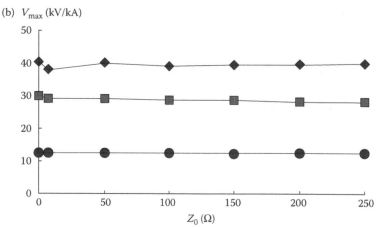

FIGURE 2.49
Effect of tower surge impedance Z_0 on the substation overvoltage for a 66-kV line. (a) $R_f = 10\ \Omega$.
(b) $R_f = 50\ \Omega$.

by surge recorders installed at the substations. From the recorded voltages
and currents, Figure 2.50 was obtained. The figure clearly shows that the arc
horn flashover phase is notably dependent on the AC source voltage, that is,
a flashover occurs at a phase where the AC voltage is in the opposite polarity of a lightning current. Table 2.7 shows a simulation result of arc horn
peak voltages (arc horn not operating) on a 77-kV line and a 500-kV line [34].
The simulation was carried out in a circuit similar to that in Figure 2.36, but
another five towers were added instead of a gantry and substation.

The parameters were the same as those in Table 2.4 for a 77-kV system,
except for a lightning current of 40 kA based on the field measurement [33].
The lower phase arc horn voltage was relatively smaller than those of the
other phases on the 500-kV (EHV) line compared with those on the 77-kV

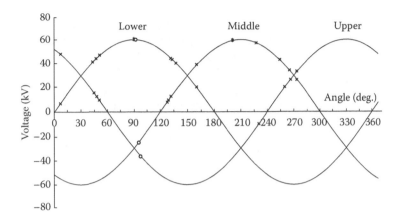

FIGURE 2.50

Measured results of arc horn flashover phases on a 77 kV transmission line. * Single-phase FO, × two-phase FO, and O three-phase FO.

line. Thus, an arc horn flashover phase on an EHV line is independent of the AC source voltage, and the lower phase flashover is less probable than those of the other phases. On the contrary, flashover probability is the same on each phase, and a flashover is dependent on the AC source voltage on a low voltage line.

Figure 2.51 shows the simulation results of arc horn flashover phases by a simple distributed line "tower model," that is, neglecting the RL circuit in Figure 2.39 with the parameters in Table 2.4, and by the recommended model illustrated in Figure 2.39. When this figure is compared with the field test result shown in Figure 2.50, it is clear that the recommended model cannot duplicate the field test result, while the simple distributed line model shows good agreement with the field test result. The reason for the poor accuracy of the recommended model [28] is that the model was developed originally for a 500-kV line on which the lower phase flashover was less probable, as explained in Section 2.4.2 [23]. Thus, the recommended tower model tends to result in a lower flashover probability of the lower phase arc horn. An $R–L$

TABLE 2.7

Maximum Arc Horn Voltages and the Time of Appearance

Transmission Voltage	Maximum Voltage (kV)/ Time of Appearance (μs)	
	77 kV	**500 kV**
Upper	873.0/1.012	4732/1.025
Middle	820.2/1.024	4334/1.073
Lower	720.0/1.035	3423/1.122

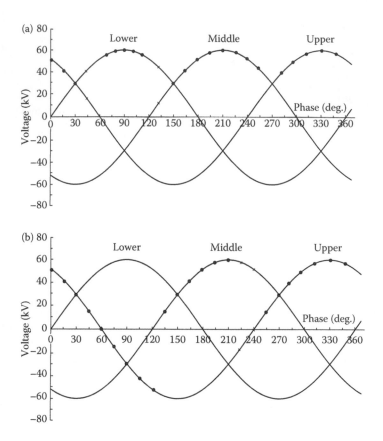

FIGURE 2.51
Simulation results of arc horn flashover phases corresponding to Figure 2.50. • Single-phase FO, × two-phase FO. (a) A simple distributed line model. (b) Recommended tower model.

parallel circuit between the two distributed lines in Figure 2.39 represents the traveling wave attenuation and distortion along a tower. The R and L values were determined originally based on a field measurement (α in Equation 2.9), and thus, they are correct only for the tower on which the measurement was carried out. Sometimes the R–L circuit generates unreal high frequency oscillations. This indicates a necessity of further investigation of the R–L circuit if the model is to be adopted.

2.5 Theoretical Analysis of Transients: Hand Calculations

In this section, examples of hand calculations of transients with a pocket calculator are explained by adopting (1) the traveling-wave theory described in Section 1.6 and (2) the Laplace transform by using a lumped parameter circuit equivalent to the distributed line [2]. These two approaches are the most powerful to analyze a transient theoretically by hand and they also correspond to the following representative simulation methods:

1. Time-domain method: EMTP [35]
2. FD method: frequency domain transient analysis program (FTP) [36]

2.5.1 Switching Surge on an Overhead Line

2.5.1.1 Traveling Wave Theory

EXAMPLE 2.1

Obtain switching surge voltages at the sending end and the open-circuited receiving end of the untransposed horizontal line illustrated in Figure 2.52.

Solution

The surge impedance matrix $[Z_s]$ of the source circuit and $[Z_r]$ at the right of node r are

$$[Z_s] = \begin{bmatrix} R & 0 & 0 \\ 0 & \infty & 0 \\ 0 & 0 & \infty \end{bmatrix}, \quad [Z_r] = \begin{bmatrix} \infty & 0 & 0 \\ 0 & \infty & 0 \\ 0 & 0 & \infty \end{bmatrix}$$

Thus, the refraction coefficient matrices are

$$[\lambda_{s1}] = 2[Z_0]([Z_s]+[Z_0])^{-1} = \begin{bmatrix} 4/5 & 0 & 0 \\ 1/5 & 0 & 0 \\ 3/25 & 0 & 0 \end{bmatrix}$$

$$[\lambda_{1s}] = 2[U]-[\lambda_{s1}] = \begin{bmatrix} 6/5 & 0 & 0 \\ -1/5 & 2 & 0 \\ -3/25 & 0 & 2 \end{bmatrix}, \quad [\lambda_{1r}] = \begin{bmatrix} 2 & 0 & 0 \\ 0 & 2 & 0 \\ 0 & 0 & 2 \end{bmatrix}$$

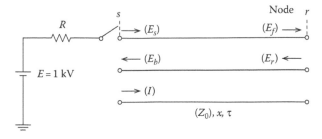

FIGURE 2.52
An untransposed horizontal line: $x = 100$ km.

The propagation time of the line is

$$\tau_0 = \frac{x}{c_0} = \frac{10^5}{250} = 370.4 \ \mu s, \quad \tau_1 = 339.0 \ \mu s, \quad \text{and} \quad \tau_2 = 333.3 \ \mu s$$

The time delay is

$$\tau_{12} = \tau_1 - \tau_2 = 5.7 \ \mu s, \quad \tau_{02} = \tau_0 - \tau_2 = 37.1 \ \mu s$$

Because of the symmetry of the line surge impedance $[Z_0]$, the voltage transformation matrices are

$$[A] = \begin{bmatrix} 1 & 1 & 1 \\ 1 & 0 & -2 \\ 1 & -1 & 1 \end{bmatrix}, \quad [A]^{-1} = \left(\frac{1}{6}\right) \begin{bmatrix} 2 & 2 & 2 \\ 3 & 0 & -3 \\ 1 & -2 & 1 \end{bmatrix}$$

1. $= 0 : (Vs) = [\lambda_{s1}] \left(\dfrac{E}{2}\right) = \begin{vmatrix} 400 \\ 100 \\ 60 \end{vmatrix} = (E_s)$: phasor-traveling waves

$$\text{Modal-traveling waves } (e_s) = [A]^{-1}(E_s) = \left(\frac{1}{6}\right) \begin{vmatrix} 1120 \\ 1020 \\ 260 \end{vmatrix}$$

2. The modal traveling waves arrive at the receiving end at $t = \tau_n$, and the modal waves at the receiving end are given with traveling time τ_n and attenuation k_n by $e_{fn} = k_n \cdot e_{sn} \cdot (t - \tau_n)$, where $u(t - \tau)$ is the unit step function with time delay τ, and $k_0 = 0.48$, $k_1 = 0.90$, and $k_2 = 0.96$ for modes 0–2:

$$(e_f) = \left(\frac{1}{6}\right) \begin{vmatrix} 0.48 \times 1120 u(t - \tau_0) \\ 0.90 \times 1020 u(t - \tau_1) \\ 0.96 \times 260 u(t - \tau_2) \end{vmatrix} = \begin{vmatrix} 8.96 u(t - \tau_0) \\ 153 u(t - \tau_1) \\ 41.6 u(t - \tau_2) \end{vmatrix} = \begin{array}{c|c} 89.6 & \tau_0 \\ 153 & \tau_1 \\ 41.6 & \tau_2 \end{array}$$

The receiving end voltage

$$(V_r) = [A] \cdot (V_r) = \begin{vmatrix} \dfrac{179.2}{\tau_0} + \dfrac{306}{\tau_1} + \dfrac{83.2}{\tau_2} \\[2mm] \dfrac{179.2}{\tau_0} \qquad\quad - \dfrac{166.4}{\tau_2} \\[2mm] \dfrac{179.2}{\tau_0} - \dfrac{306}{\tau_1} + \dfrac{83.2}{\tau_2} \end{vmatrix}$$

The reflected waves at the receiving end are $(E_r) = 2(E_f)$ $- (E_f) = (E_f)$. Thus, $(e_r) = (e_f)$.

3. $t = 2\tau_n$:

$$e_{bn} = k_n \cdot e_{rn} \cdot u(t - \tau_n), \quad \text{that is,} (e_b) = \begin{vmatrix} 0.48 \times 89.6 u(t - 2\tau_0) \\ 0.9 \times 153 u(t - 2\tau_1) \\ 0.96 \times 41.6 u(t - 2\tau_2) \end{vmatrix} = \begin{vmatrix} 43.0 \\ 2\tau_0 \\ 137.7 \\ 2\tau_1 \\ 39.9 \\ 2\tau_2 \end{vmatrix}$$

Thus, $(E_b) = [A](e_b) = \begin{vmatrix} \dfrac{43.0}{2\tau_0} + \dfrac{137.7}{2\tau_0} + \dfrac{39.9}{2\tau_0} \\ \dfrac{43.0}{2\tau_0} \quad - \dfrac{-79.8}{2\tau_0} \\ \dfrac{43.0}{2\tau_0} + \dfrac{137.7}{2\tau_0} + \dfrac{39.9}{2\tau_0} \end{vmatrix}$

The sending end voltage is given in the following form:

$$(V_s) = [\lambda_{s1}]\left(\frac{E}{2}\right) + [\lambda_{s1}](E_b) = \begin{vmatrix} 400 + \dfrac{51.6}{2\tau_0} + \dfrac{165.2}{2\tau_1} + \dfrac{47.9}{2\tau_2} \\ 100 + \dfrac{77.4}{2\tau_0} - \dfrac{27.5}{2\tau_1} - \dfrac{167.6}{2\tau_2} \\ 60 + \dfrac{80.8}{2\tau_0} - \dfrac{291.9}{2\tau_1} + \dfrac{75.0}{2\tau_2} \end{vmatrix}$$

These results are drawn in Figure 2.53.

EXAMPLE 2.2

In Example 2.1, consider the transposition of the line with $c_1 = c_2 = 298$ m/µs.

Solution

Considering the transposition, the surge impedance is given by

$$Z_{0s} = \frac{Z_{0aa} + Z_{0bb} + Z_{0cc}}{3} = 300 \ \Omega, \quad Z_{0m} = \frac{Z_{0ab} + Z_{0ac} + Z_{0bc}}{3} = 65 \ \Omega$$

: $x = 100$ km, $c_0 = 270$, $c_1 = 295$, and $c_2 = 300$ (m/µs)

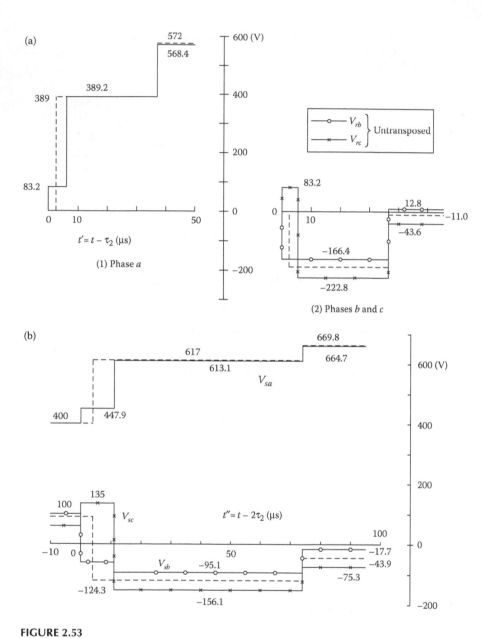

FIGURE 2.53
Analytical surge waveforms on a horizontal line. ⎯⎯ untransposed, ---- transposed
$(V_b = V_c)$. (a) Vr, (b) Vs.

$$(Z_0) = \begin{bmatrix} 300 & 75 & 45 \\ 75 & 300 & 75 \\ 45 & 75 & 300 \end{bmatrix} (\Omega)$$

$$[\lambda_{s1}] = \begin{vmatrix} \dfrac{4}{5} & 0 & 0 \\ \dfrac{13}{15} & 0 & 0 \\ \dfrac{13}{15} & 0 & 0 \end{vmatrix}, \qquad [\lambda_{1s}] = 2[U] - [\lambda_{s1}]$$

The propagation time is $\tau_0 = 370.4$ μs, $\tau_1 = \tau_2 = 335.6$ μs, and $t_{01} = \tau_0 - \tau_1 = 34.8$ μs.

The transformation matrix is the same as that in Example 2.1 (see Section 1.4.4):

1. $t = 0 : (V_s) = \begin{vmatrix} 400 \\ 86.7 \\ 86.7 \end{vmatrix} = (E_s), \qquad (e_s) = \begin{vmatrix} 191 \\ 157 \\ 52.2 \end{vmatrix}$

2. $t = \tau_n : (e_f) = \begin{vmatrix} \dfrac{91.7}{\tau_0} \\ \dfrac{146}{\tau_1} \\ \dfrac{48.8}{\tau_1} \end{vmatrix}, \qquad (V_r) = 2(e_f), (V_r) = \begin{vmatrix} \dfrac{183.4}{\tau_0} + \dfrac{389}{\tau_1} \\ \dfrac{183.4}{\tau_0} - \dfrac{194.5}{\tau_1} \\ \dfrac{183.4}{\tau_0} - \dfrac{194.5}{\tau_1} \end{vmatrix}$

3. $t = 2\tau_n : (e_b) = \begin{vmatrix} \dfrac{44.0}{2\tau_0} \\ \dfrac{136}{2\tau_1} \\ \dfrac{45.1}{2\tau_1} \end{vmatrix}, \qquad (E_b) = \begin{vmatrix} \dfrac{44}{2\tau_0} + \dfrac{181}{2\tau_1} \\ \dfrac{44}{2\tau_0} - \dfrac{90.5}{2\tau_1} \\ \dfrac{44}{2\tau_0} - \dfrac{90.5}{2\tau_1} \end{vmatrix},$

$$(V_s) = \begin{vmatrix} 400 + \dfrac{52.8}{2\tau_0} + \dfrac{217}{2\tau_1} \\ 86.7 + \dfrac{80.4}{2\tau_0} - \dfrac{212}{2\tau_1} \\ 86.7 + \dfrac{80.4}{2\tau_0} - \dfrac{212}{2\tau_1} \end{vmatrix}$$

The results are shown in Figure 2.53 by a dotted line considering the difference in time delays between the untransposed and transposed lines.

EXAMPLE 2.3

The calculation of the receiving end voltage (V_r) in an untransposed vertical twin-circuit line is illustrated in Figure 2.42 for $\tau_1 \le t < 3\tau_1$ under the condition that the source voltage $E = 1$ pu is applied to phase c of the first circuit at $t = 0$ with $c_0 = 251.2$, $c_1 = 298.4$, $c_3 = 297.4$, $c_2 = c_4 = c_5 = 299.2$ (m/μs), $x = 101.13$ km, $R = 403\ \Omega$, and the attenuation is zero. The surge impedance $[Z_0]$ and the voltage transformation matrices $[A]$ are

$$[Z_0] = \begin{bmatrix} [Z_1] & [Z_2] \\ [Z_2] & [Z_1] \end{bmatrix}, \quad [Z_1] = \begin{bmatrix} 311 & 117 & 78 \\ 117 & 325 & 122 \\ 78 & 122 & 325 \end{bmatrix}, \quad [Z_2] = \begin{bmatrix} 92 & 83 & 67 \\ 83 & 100 & 89 \\ 67 & 89 & 98 \end{bmatrix}$$

$$[A] = \begin{bmatrix} [A_1] & [A_1] \\ [A_2] & -[A_2] \end{bmatrix}, \quad [A_1] = \begin{bmatrix} 0.7 & 1.0 & 0.5 \\ 0.85 & 0.36 & 1.0 \\ 0.1 & -0.57 & 0.28 \end{bmatrix}$$

$$[A_2] = \begin{bmatrix} 0.47 & 1.0 & 0.67 \\ 0.66 & 0.49 & 1.0 \\ 1.0 & -0.167 & 0.26 \end{bmatrix}$$

$$[A]^{-1} = \begin{bmatrix} [A_3] & [A_4] \\ [A_3] & [A_5] \end{bmatrix}, \quad [A_3] = \begin{bmatrix} 0.762 & -0.642 & 0.931 \\ 0.354 & -0.0897 & -0.312 \\ -0.775 & 1.078 & -0.679 \end{bmatrix}$$

$$[A_4] = \begin{bmatrix} 1.267 & -1.126 & 1.067 \\ 0.522 & -0.301 & -0.187 \\ -1.227 & 1.511 & -0.726 \end{bmatrix}, \quad [A_5] = \begin{bmatrix} -0.112 & -0.0992 & -0.094 \\ -0.910 & 0.646 & -0.139 \\ 0.384 & -0.762 & 0.0166 \end{bmatrix}$$

Solution

When the phase c pole is closed at $t = 0$, voltage $V_{sc} = Z_{0cc}E/(R + Z_{0cc}) = 0.466$ pu.

Thus, current $I_{sc}/Z_{0cc} = 1/728$.

The sending end voltage (V_s) at $t = 0$ is obtained by using surge impedance $[Z_0]$ as

$$(V_s) = [Z_0](I_s)$$

where the currents on the phases, except that on phase c, are zero for the open-circuited case:

$$V_{sa} = 0.107, \quad V_{sb} = 0.168, \quad V'_{sa} = 0.092, \quad V'_{sa} = 0.122, \quad V'_{sc} = 0.135\ (\text{pu})$$
$$(E_s) = (V_s) \quad \text{for } t = 0$$

The modal traveling wave is as follows:

$$(e_s) = [A]^{-1}(E_s)$$
$$= [0.512,\ -0.130,\ -0.232,\ 0.379,\ -0.140,\ -0.261]_t$$

where t is the transposed matrix.

The modal propagation time is $\tau_0 = x/c_0 = 402.6$ (μs), $\tau_1 = x/c_1 = 338.9$, $\tau_2 = \tau_4 = \tau_5 = 338.0$, and $\tau_3 = 340.0$.

Neglecting the attenuation of the line, the modal traveling waves at the receiving end are the same as those at the sending end.

$$(e_r(t - t_i)) = (e_s(t)) \quad i = 0 \text{ to } 5 \quad \text{for modal components}$$

As the receiving end is open-circuited, the refraction coefficient matrix becomes a diagonal matrix of which all diagonal elements are equal to two. Thus, the modal voltages V_r are

$$(V_r) = 2(e_r)$$

or

$$V_{r0} = 1.024 \cdot u(t - \tau_0) = \frac{1.024}{\tau_0}, \quad V_{r1} = \frac{-0.260}{\tau_1}, \quad V_{r2} = \frac{-0.464}{\tau_2}$$

$$V_{r3} = \frac{-0.758}{\tau_2}, \quad V_{r4} = \frac{-0.280}{\tau_2}, \quad V_{r5} = \frac{-0.522}{\tau_2}$$

Transform these modal voltages into actual phasor voltages by using the transformation matrix [A]:

$$V_{ra} = 0.7V_{r0} + 1.0V_{r1} + 0.5V_{r2} + 0.7V_{r3} + 1.0V_{r4} + 0.5V_{r5}$$

$$= \frac{0.717}{\tau_0} - \frac{0.260}{\tau_1} - \frac{0.232}{\tau_2} + \frac{0.531}{\tau_3} - \frac{0.280}{\tau_2} - \frac{0.261}{\tau_2}$$

$$= \frac{0.717}{\tau_0} - \frac{0.260}{\tau_1} - \frac{0.773}{\tau_2} + \frac{0.531}{\tau_3}$$

In the same manner:

$$V_{rb} = \frac{0.870}{\tau_0} - \frac{0.0936}{\tau_1} - \frac{1.087}{\tau_2} + \frac{0.644}{\tau_3}$$

$$V_{rc} = \frac{0.102}{\tau_0} + \frac{0.148}{\tau_1} - \frac{0.116}{\tau_2} + \frac{0.758}{\tau_3}$$

$$V'_{ra} = \frac{0.481}{\tau_0} - \frac{0.060}{\tau_1} + \frac{0.319}{\tau_2} - \frac{0.356}{\tau_3}$$

$$V'_{rb} = \frac{0.676}{\tau_0} - \frac{0.127}{\tau_1} + \frac{0.195}{\tau_2} - \frac{0.500}{\tau_3}$$

$$V'_{rc} = \frac{1.024}{\tau_0} + \frac{0.174}{\tau_1} - \frac{0.173}{\tau_2} - \frac{0.758}{\tau_3}$$

Figure 2.54 illustrates these analytical surge waveforms.

2.5.1.2 Lumped Parameter Equivalent with Laplace Transform

It is well-known that a distributed parameter line is approximated by a lumped parameter circuit such as a *PI* equivalent and an *L* equivalent. For

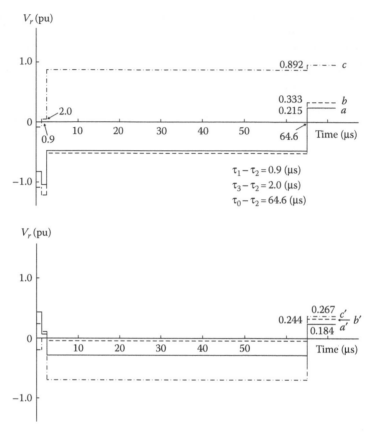

FIGURE 2.54
Analytical surge waveforms at the receiving-end on a vertical twin-circuit line.

example, the open-circuited line in Figure 2.2 is approximated by Figure 2.55 with the L equivalent. Let us analyze the switching surges in this circuit.

2.5.1.2.1 Single-Phase Line

In an L equivalent of the single-phase line illustrated in Figure 2.55, current I, when switch S is closed at $t = 0$, is defined with Laplace operator s as

$$E(s) = \left(\frac{sL_0 + R + 1}{sC} \right) \cdot I(s)$$

$$\therefore I(s) = \left(\frac{E(s)}{sL_0 + R + 1/sC} \right) \tag{2.11}$$

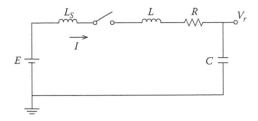

FIGURE 2.55
An L equivalent of an open-circuited line.

where

$E(s) = E/s$, $L_0 = L_s + L$

L_s is the source inductance (in most cases, transformer inductance)

L, R, and C are the inductance, resistance, and capacitance of the line with length x

Then, the sending and receiving end voltages, $V_s(s)$ and $V_r(s)$, are given by

$$V_s(s) = E(s) - sL_s \cdot I(s), \quad V_r(s) = \frac{I(s)}{sC} \tag{2.12}$$

Solving these equations and transforming them into the time domain by using the inverse Laplace transform, the following solutions are obtained:

$$\left. \begin{aligned} i(t) &= E \cdot \exp(-\alpha t) \cdot \sin\frac{(\omega_2 t)}{\omega_2 L_0} \\ v_s(t) &= E\left\{1 - \left(\frac{L_s}{L_0 \sin\varphi}\right)\exp(-\alpha t) \cdot \sin(\omega_2 t - \varphi)\right\} \\ v_r(t) &= E\left\{1 - \exp(-\alpha t)\frac{\sin(\omega_2 t - \varphi)}{\sin\varphi}\right\} \end{aligned} \right\} \tag{2.13}$$

where $\varphi = \tan^{-1}(\omega_2/\alpha)$, $\omega_2 = \sqrt{\omega_1^2 - \alpha^2}$, $\alpha = R/2L_0$, and $\omega_1 = \sqrt{L_0 C}$.

2.5.1.2.2 Single-Phase Line with a Residual (Line Charge) Voltage: Reclosing Surge

When there is a residual voltage V_0 (or charge Q_0) on the open-circuited line, a closing surge overvoltage becomes much higher than that with no residual voltage and is called a "reclosing surge" as explained in Section 2.2.3.5.5; see Figure 2.24. The derivation of the reclosing surge voltage is similar to the case of the closing surge, except that an initial value of the line voltage exists, and the following results for $v_r(t)$ are obtained:

$$v_r(t) = (E - V_0)\left\{1 - \exp(-\alpha t) \cdot \frac{\sin(\omega_2 t + \varphi)}{\sin \varphi}\right\} + V_0 \qquad (2.14)$$

In most real transmission lines, the following condition is satisfied:

$$\frac{R}{2} \ll Z_0 = \sqrt{\frac{L}{C}} : \text{line surge impedance (see Section 1.3.4.2)}$$

In such a case, Equations 2.13 and 2.14 are simplified by considering ω_1 to be far greater than α and nearly equal to ω_2. For example, Equation 2.14 is rewritten as

$$v_r(t) = E + (V_0 - E)\exp(-\alpha t) \cdot \cos(\omega_1 t) \qquad (2.15)$$

It is easily observed from this equation that

$$\left.\begin{array}{ll} v_r(t) \fallingdotseq E - 2E\cos(\omega_1 t); & v_{r\,max} = 3E \quad \text{for } v_0 = -E \\ v_r(t) \fallingdotseq E - E\cos(\omega_1 t); & v_{r\,max} = 2E \quad \text{for } v_0 = 0 \end{array}\right\} \qquad (2.16)$$

This result is a proof of the reason why the reclosing surge overvoltage is much higher than the closing surge overvoltage.

2.5.1.2.3 Sinusoidal AC Voltage Source

In the previous theoretical analysis, the source voltage was assumed to be a step function (or DC) voltage. This assumption is accurate enough as long as the observation time of a switching surge is less than 1 ms. If the observation time exceeds 5 ms, a sinusoidal AC voltage source should be taken into account. This makes the Laplace transform quite complicated.

Assume the following AC voltage source:

$$e_0(t) = E \times \sin(\omega_0 t + \theta) \qquad (2.17)$$

where
$\omega_0 = 2\pi f_0$, f_0 is the power frequency
θ is the closing angle of a CB

Considering that the overall solution is given as a superposition of the source voltage and the transient voltage at the instance of CB closing, the following result is obtained:

$$\left.\begin{array}{l} v_r(t) \fallingdotseq E\sin(\omega_0 t + \theta) + (V_0 - E\sin\theta) \cdot \exp(-\alpha t) \cdot \cos\{\omega_1(t - \tau)\} \\ v_{r\,max} \fallingdotseq E\sin(\omega_0 t + \theta) + (V_0 - E\sin\theta) \end{array}\right\} \qquad (2.18)$$

where
$\tau \doteq \ell/c_0$ is the traveling time of the line
c_0 is the light velocity
ℓ is the line length

2.5.1.2.4 Three-Phase Line

In the case of a three-phase line, Equation 2.11 becomes a matrix, and we need to apply modal theory, described in Section 1.4.

Assuming that the line is transposed and all the phases are simultaneously closed at $t = 0$, the positive sequence (aerial mode) voltage at the receiving end is given by

$$v_{r1} = E_a - E\sin\theta \cdot \exp(-\alpha_1 t)\cdot\cos\{\omega_1(t - \tau_1)\} \tag{2.19}$$

where
$\alpha_1 = R_1/2L_{01}$, $\omega_1 = 1/\sqrt{L_{01}C_1}$, $\tau_1 = \ell/c_1$, and $L_{01} = L_s + L_1$
R_1, L_1, and C_1 are the positive sequence components of the line resistance, inductance, and capacitance matrix
τ_1 is the propagation time of the positive sequence traveling wave
c_1 is the positive sequence propagation velocity

$$E_a = E\sin(\omega_0 t + \theta), \quad E_b = a^2 E_a, \quad E_c = aE_a \quad \text{where } a = \exp\left(\frac{j2\pi}{3}\right) \tag{2.20}$$

After transforming Equation 2.19 into a phase domain by $(V) = [A](v)$, the following voltages for the three phases are obtained:

$$\left.\begin{aligned}
V_{ra} &= E_a - E\sin\theta \cdot \exp(-\alpha_1 t)\cdot\cos\{\omega(t - \tau_1)\} \\
V_{rb} &= E_b - E\sin\left(\frac{\theta - 2\pi}{3}\right)\cdot\exp(-\alpha_1 t)\cdot\cos\{\omega(t - \tau_1)\} \\
V_{rc} &= E_c - E\sin\left(\frac{\theta + 2\pi}{3}\right)\cdot\exp(-\alpha_1 t)\cdot\cos\{\omega(t - \tau_1)\}
\end{aligned}\right\} \tag{2.21}$$

An example of switching surges on a three-phase line calculated by this equation is shown in Figure 2.56.

2.5.2 Fault Surge

A theoretical derivation of a fault surge voltage is, in principle, the same as that of a closing surge voltage, provided that the steady state voltage at $t = 0$ is superposed to the transient voltage similar to a reclosing surge.

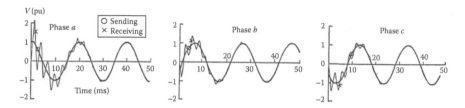

FIGURE 2.56
Switching surges due to simultaneous CB closing on a three-phase line.

Let us consider the multiphase circuit illustrated in Figure 2.57a. Assume that phase a is short-circuited to the ground (SLG) at node P at $t = 0$. Then the original circuit in (a) is represented by Figure 2.57b for a transient component. By applying an L-equivalent lumped parameter circuit to the distributed line circuit, Figure 2.57c is obtained. In the circuit, $[L]$, $[R]$, and $[C]$ are the inductance, resistance, and capacitance matrices of the original distributed line with length x. $[Z_0]$ is its surge impedance matrix. The circuit is similar to that of Figure 2.55 except that the source voltage $-E_a$ is applied to node P. Thus, it is quite possible to obtain a transient voltage (V_p) at node P in a similar manner to the switching surge on a three-phase line, as discussed in Section 2.5.1.

For example, neglecting the source inductance L_s and the surge impedance $[Z_0]$ at the right of node P with the source voltage $E_a = E \cos(\omega_0 t)$, the following phase b (sound phase) voltage is derived:

$$v_b(t) = E\left[C_1(\cos\varphi - 1)\frac{\{\cos(\omega_0 t) - \cos(\omega_1 t)\}}{(2C_0 + C_1)} \right.$$
$$\left. + \omega_0 \sin\varphi \cdot \frac{\{\cos(\omega_0 t)/\omega_0 - \cos(\omega_1 t)/\omega_1\}}{\{1 - (\omega_0/\omega_1)\}^2} \right] \qquad (2.22)$$

where
$\omega_1 = 1/\tau$, $\varphi = \tan^{-1}(\omega_1/\alpha) \doteq 2\pi/3$, α is the attenuation constant
τ is the propagation time of a traveling wave for the line length x_1
C_0, C_1 are zero- and positive-sequence capacitances for the line length x_1

Considering that $(\omega_0/\omega_1)^2$ is much smaller than 1 and $\varphi = 2\pi/3$, the following approximate solutions of the sound phases b and c voltages are obtained:

$$\left. \begin{array}{l} v_b(t) = E_b(t) - kE_a(t) - \{E_b(t = 0) - kE_a(t = 0)\}\cos(\omega_1 t) \\ v_c(t) = E_c(t) - kE_a(t) - \{E_c(t = 0) - kE_a(t = 0)\}\cos(\omega_1 t) \end{array} \right\} \qquad (2.23)$$

where $K = L_{ab}/L_{aa}$ is the ratio of the mutual and self-inductances.

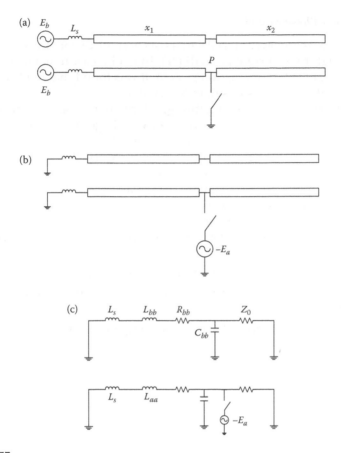

FIGURE 2.57
A circuit for a fault surge analysis. (a) Original circuit. (b) An equivalent circuit. (c) An *L* equivalent of (b).

In Equations 2.22 and 2.23, the damping factor $\exp(-\alpha t)$ is neglected. If this is included, the oscillating term in Equation 2.23 will die out for $t \to \infty$. Then, the equation becomes the steady-state voltage during the phase *a*-to-ground fault. For example, if $k = 0.4$, the maximum phase *b* voltage is given as

$$v_{b\,\text{max}} \doteq -1.48 \text{ pu} \quad \text{at } t = 1.0 \text{ ms } \omega_1 t = \frac{2\pi}{3}$$

2.5.3 Lightning Surge

2.5.3.1 Tower-Top Voltage

The transient voltage at a tower top, at which lightning strikes, is easily calculated by applying traveling wave theory. An example has been explained in Section 1.7 (see Problem 1.19).

2.5.3.2 Two-Phase Model

Lightning strikes a tower or a GW in most cases. Occasionally it strikes a PW when the lightning current is small [37]. In field measurements of lightning strikes on a 1100 kV transmission system, it was found that lightning strikes a PW when the current is less than 35 kA [25].

When a tower is struck by lightning, a large lightning current flows into the tower, the tower voltage exceeds the PW voltage, and BFO occurs. Then, a part of the lightning current flows through the PW and travels toward a substation. This traveling wave produces more overvoltage in the substation. Therefore, the analysis of the BFO and the resultant overvoltages are critical from the viewpoint of the insulation design and coordination of the substation and the transmission line. For this, it is imperative to consider both the PW and a GW including the tower. Thus, the analysis involves the two-phase circuit composed of a PW and a GW as illustrated in Figure 2.58, where

I_0, R_0 are the lightning current and channel impedance

I_t, V_t, Z_t, x_t are the tower current, voltage, surge impedance, and height (length)

V_f, R_f are the tower foot voltage and footing impedance

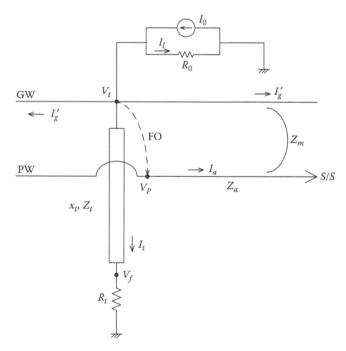

FIGURE 2.58
A two-phase model circuit for a BFO analysis.

I_g, Z_g are the GW current and surge impedance

I_a, V_a, Z_a are the PW current, voltage, and surge impedance

Z_m is the mutual impedance between GW and PW

For an original eight-conductor (two GWs and six PWs) system, the following relation exists:

$$
\begin{vmatrix} (V_g) \\ (V_1) \\ (V_2) \end{vmatrix} = \begin{bmatrix} [Z_g] & [Z_{g1}] & [Z_{g2}] \\ [Z_{g1}]_t & [Z_1] & [Z_{12}] \\ [Z_{g2}]_t & [Z_{12}]_t & [Z_2] \end{bmatrix} \begin{vmatrix} (I_g) \\ (I_1) \\ (I_2) \end{vmatrix} \tag{2.24}
$$

where g stands for GWs, 1 stands for circuit-1 PWs, and 2 stands for circuit-2 PWs.

Assuming that a BFO occurs on phase a of circuit-1, the following relation is derived from Equation 2.24:

$$
\begin{vmatrix} (V_{g1}) \\ (V_{g2}) \\ (V_a) \end{vmatrix} = \begin{bmatrix} [Z_{11}] & [Z_{12}] & [Z_{13}] \\ [Z_{12}] & [Z_{11}] & [Z_{23}] \\ [Z_{13}] & [Z_{23}] & [Z_{33}] \end{bmatrix} \begin{vmatrix} (I_{g1}) \\ (I_{g2}) \\ (I_a) \end{vmatrix} \tag{2.25}
$$

This equation is reduced to a 2×2 matrix for $V_{g1} = V_{g2} = V_g$ and $I_{g1} = I_{g2} = I_g$

$$
\begin{pmatrix} V_g \\ V_a \end{pmatrix} = \begin{bmatrix} Z_g & Z_m \\ Z_m & Z_a \end{bmatrix} \begin{pmatrix} 2I_g \\ I_a \end{pmatrix} = \begin{bmatrix} Z_g & Z_m \\ Z_m & Z_a \end{bmatrix} \begin{pmatrix} I'_g \\ I_a \end{pmatrix} \tag{2.26}
$$

where $Z_g = (Z_{11} + Z_{12})/2$, $Z_m = (Z_{13} + Z_{23})/2$, $Z_a = Z_{33}$, and $I_g = 2I_g$.

This equation is for a two-phase circuit model used to analyze a lightning surge.

2.5.3.3 No BFO

The GW voltage V_g and the GW current I_g are easily obtained from the following relation, keeping in mind that Z_g is an equivalent impedance of two GWs as in Equation 2.26:

$$
I_0 = I_t + I_\ell + 2I'_g, \quad V_g = Z_g I'_g = R_0 I_\ell = Z_t I_t \tag{2.27}
$$

Solving these equations,

$$
I'_g = \frac{R_0 \cdot Z_t \cdot I_0}{R_0 \cdot Z_g + 2R_0 Z_t + Z_t Z_g}, \quad V_g = Z_{in} I_0 = Z_g I'_g \quad \text{for } t \leq 2\tau \tag{2.28}
$$

where $1/Z_{in} = 1/R_0 + 1/Z_t + 2/Z_g$ is the impedance seen from the current source I_0 and $\tau = x_t/c$ is the traveling time along the tower.

If no channel impedance exists, that is, $R_0 = \infty$, then

$$I_g' = \frac{Z_t I_0}{Z_g + 2Z_t}, \quad V_g = Z_g I_g' \quad \text{for } t \leq 2\tau \tag{2.29}$$

At $t = \tau$, a traveling wave generated at $t = 0$ arrives at the tower bottom and is reflected back to the tower top:

$$e_{ft} = Vg \quad \text{at } t = 0 \tag{2.30}$$

The refraction coefficient λ_b is given by

$$\lambda_b = \frac{2R_f}{(R_f + Z_t)} \tag{2.31}$$

Thus, the following voltage appears at the tower bottom, that is, at the tower footing resistance:

$$V_f = \lambda_b \cdot e_{ft} = \frac{2V_g R_f}{(R_f + Z_t)} \quad \text{at } t = \tau \tag{2.32}$$

Then, the following reflected wave e_{bt} at the bottom travels back to the tower top arriving at $t = 2\tau$:

$$e_{bt} = V_f - e_{ft} \tag{2.33}$$

The refraction coefficient seen at the tower top is

$$\lambda_t = \frac{2Z_{in}'}{(Z_t + Z_{in}')} \tag{2.34}$$

where $1/Z_{in}' = 1/R_0 + 2/Z_g$.

Thus, the tower-top voltage $V_g(t)$ is changed to the following value:

$$V_g(t) = V_g(0) + \lambda_{bt} \times e_{bt} \quad \text{for } 2\tau \leq t < 4\tau \tag{2.35}$$

From Equations 2.28, 2.35, and 2.34, an analytical wave is drawn as shown in Figure 2.59. The waveform explains a numerical simulation result of the tower-top voltage when lightning strikes the tower top. This data has been published in many papers [14–22].

2.5.3.4 Case of a BFO

When a BFO occurs on phase a, the GW in Figure 2.58 is short circuited to phase a. Then, the total impedance Z_{BF} seen from the current source is given by

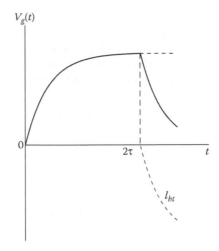

FIGURE 2.59
Analytical waveform of the tower top voltage.

$$\frac{1}{Z_{BF}} = \frac{1}{Z_{in}} + \frac{2}{Z_a} \tag{2.36}$$

where Z_{in} is the total impedance in the case of no flashover (see Equations 2.28 and 2.29).

The tower-top voltage is obtained by

$$V_g = Z_{BF} \times I_0 \tag{2.37}$$

2.5.3.5 Consideration of Substation

When lightning hits a tower or a GW, traveling waves generated by the lightning currents propagate to a substation along the ground and the PWs. When the waves arrive at the substation, they produce lightning surge overvoltages on the substation equipment.

Let us analyze lightning surges at a substation. Assume that lightning strikes the first tower next to the substation. The refraction coefficient matrix $[\lambda_s]$ from the line to the substation is given by

$$[\lambda] = 2[Z_s]([Z_s]+[Z_0]^{-1}) = 2\begin{bmatrix} R_g & 0 \\ 0 & R_s \end{bmatrix}\begin{bmatrix} R_g + Z_g & Z_m \\ Z_m & R_s + Z_a \end{bmatrix}^{-1}$$

$$= \left(\frac{2}{\Delta}\right)\begin{bmatrix} R_g(R_s + Z_a) & -R_g Z_m \\ -R_s Z_m & 2R_s(R_g + Z_g) \end{bmatrix}$$

$$\Delta = (R_g + Z_g)(R_s + Z_a) - Z_m^2 \tag{2.38}$$

where

[Z_s] is the substation impedance
[Z_0] is the line impedance defined in Equation 2.26
R_g is the surge impedance of the substation gantry
R_s is the phase a surge impedance of the substation

Then, the substation voltage (V_s) is calculated by

$$\begin{pmatrix} V_{sg} \\ V_{sa} \end{pmatrix} = [\lambda_s] \begin{pmatrix} E_{sg} \\ E_{sa} \end{pmatrix}$$

(2.39)

E_{sg} and E_{sa} are the traveling waves propagating from the tower and are given by

$$E_{sg}(t) = E_{tg} u(t - \tau), \quad E_{sa}(t) = E_{ta} u(t - \tau)$$

(2.40)

where

E_{tg} is the traveling wave on the GW at the tower
E_{ta} is the traveling wave on the PW at the tower
$u(t - \tau)$ is the unit step function with time delay τ

EXAMPLE 2.4

Calculate the substation entrance voltage under the following conditions:
$I_0 = 100$ kA step function, $Z_g = 332$ Ω, $Z_m = 128$ Ω, $Z_a = 349$ Ω, $Z_t = 210$ Ω, $R_g = 125$ Ω, $R_s = 70$ Ω: gas-insulated bus, and $R_0 = 400$ Ω

Solution

1. At the tower
   a. No BFO: Equations 2.28 and 2.26:

$$Z_{in} \left(\frac{1}{R_0} + \frac{1}{Z_t} + \frac{2}{Z_g} \right)^{-1} = 75.26 \ \Omega, \quad V_g = Z_{in} \cdot I_0 = 7.526 \ \text{MV},$$

$$I'_g = \frac{V_g}{Z_g} = 22.67 \ \text{kA}, \quad V_a = Z_m I'_g = 2.902 \ \text{MV}$$

   b. BFO:

$$Z_{BF} = \left(\frac{1}{Z_{in}} + \frac{2}{Z_a} \right)^{-1} = 52.58 \ \Omega, \quad V_g = V_a = Z_{BF} I_0 = 5.258 \ \text{MV},$$

$$I'_g = 15.84 \ \text{kA}$$

2. At the substation: Equations 2.38 through 2.40:

$$E_{tg} = V_g, \quad E_{ta} = V_a, \quad [\lambda_s] = \begin{bmatrix} 0.6 & -0.18 \\ -0.10 & 0.37 \end{bmatrix}$$

a. No BFO:

$$V_{sg} = 4.0 \text{ MV}, \quad V_{sa} = 0.32 \text{ MV}$$

b. BFO:

$$V_{sg} = 2.2 \text{ MV}, \quad V_{sa} = 1.42 \text{ MV}$$

It is observed from the refraction coefficient $[\lambda_s]$ at the substation that 37% of the incoming traveling wave E_{ta} on the PW enters the substation and determines overvoltages in the substation equipment. The remaining 63% reflects back to the transmission line. Of the traveling wave E_{tg} on the GW, -10% is induced to the PW at the substation entrance, which decreases the PW voltage. This effect has not been well realized but is very significant. If there is no negative induced voltage, the insulation of the substation equipment becomes much severe. This kind of an arrester is sometimes referred to as "nature's gift."

This analysis is based on a step function current. In reality, the lightning current has a much slower rise time at the wave front, and thus the lightning overvoltage becomes much lower. Such an analysis can be carried out considering the wave front, but a hand calculation of this would be quite tedious.

These analytical results clearly show that the PW voltage in the BFO case is much higher than that in the case of no flashover. This is the reason the insulation design/coordination of a substation is based on the result in the flashover case.

The GW (= tower) voltage is certainly much higher in the no-flashover case. In fact, because of this higher voltage, a BFO occurs in reality. Remember that we are assuming that there is no flashover in this analytical study for the purpose of clarity. It is noteworthy that a direct strike to a PW occasionally occurs. The magnitude of a lightning current is far smaller than that of a lightning strike to a tower or a GW. In a field test on a 1100-kV line in Japan, the lightning current was found to be less than 30 kA [25]. Assuming a direct strike with $I_0 = 30$ kA to phase a, the following results are obtained:

$$Z'_{in} = \left(\frac{2}{Z_a} + \frac{1}{R_0} \right)^{-1} = 121.5 \ \Omega, \ V_a = Z'_{in} \cdot I_0 = 3.64 \text{ MV}, \ V_g = Z_m \left(\frac{I_0}{2} \right) = 1.92 \text{ MV}$$

$$V_{sg} = 0.497 \text{ MV}, \quad V_{sa} = 1.155 \text{ MV}$$

The results have indicated that the direct strike to a PW with 30 kA produces an overvoltage at a substation comparable to that in the BFO case with $I_0 = 100$ kA. This fact should be carefully investigated, because this has not been considered in the standard insulation design and coordination of a substation.

2.6 Frequency-Domain (FD) Method of Transient Simulations

2.6.1 Introduction

There exist powerful simulation tools such as the EMTP [35]. These tools, however, involve a number of complex assumptions and application limits that are not easily understood by the user, and often lead to incorrect results. Quite often, a simulation result is not correct due to the user's misunderstanding of the application limits related to the assumptions of the tools. The best way to avoid this type of incorrect simulation is to develop a custom simulation tool. For this purpose, the FD method of transient simulations is recommended, because the method is entirely based on the theory explained in Section 2.5, and requires only numerical transformation of a frequency response into a time response using the inverse Fourier/Laplace transform [2,6,36–42]. The theory of a distributed parameter circuit, transient analysis in a lumped parameter circuit, and the Fourier/Laplace transform are included in undergraduate course curricula in the electrical engineering department of most universities throughout the world. This section explains how to develop a computer code of the FD transient simulations.

2.6.2 Numerical Fourier/Laplace Transform

A numerical calculation code of the Fourier/Laplace transform is prepared in commercial software such as MATLAB, MAPLE, or even Excel. Therefore, it is easy to carry out an inverse transform provided that all frequency responses are given by the user. Similarly, if the user can prepare the time response of a transient voltage, for example, as digital data of a measured result, then the user can easily obtain its frequency response using the software. However, it is better to first understand the basic theory of the Fourier/Laplace transform.

2.6.2.1 Finite Fourier Transform

Let us consider the following Fourier transform:

$$f(t) = \left(\frac{1}{2\pi}\right) \int_{-\infty}^{\infty} F(\omega) \cdot \exp(j\omega t) \cdot d\omega \tag{2.41}$$

A finite transform for $[-\Omega, \Omega]$ is defined as

$$f_1(t) = \left(\frac{1}{2\pi}\right)\int_{-\Omega}^{\Omega} F(\omega)\cdot\exp(j\omega t)\cdot d\omega \qquad (2.42)$$

where
$F(\omega)$ is the frequency response for $-\Omega \le \omega \le \Omega$.
$f_1(t)$ is the time response at time t evaluated by this equation, which is not accurate. An accurate solution $f(t)$ is obtained by the original infinite integral in Equation 2.41.

Assuming the following frequency function $G(\omega)$:

$$G(\omega) = 1 : |\omega| \le \Omega$$

$$0 : |\omega| > \Omega \qquad (2.43)$$

Equation 2.42 can be rewritten as

$$f_1(t) = \left(\frac{1}{2\pi}\right)\int_{-\infty}^{\infty} G(\omega)\cdot F(\omega)\cdot\exp(j\omega t)\cdot d\omega \qquad (2.44)$$

The time response $g(t)$ of $G(\omega)$ is given by

$$g(t) = \left(\frac{1}{2\pi}\right)\int_{-\Omega}^{\Omega} 1\cdot\exp(j\omega t)d\omega = \sin\left(\frac{\Omega t}{\pi t}\right) \qquad (2.45)$$

Expressing $f_1(t)$ by using time convolution (Duhamel's integral) of $f(t)$ and $g(t)$ under the condition that $f(t) = 0$ for <0,

$$f_1(t) = \int_{-\infty}^{t}\left[\sin\frac{(\Omega t)}{(\pi t)}\right]'\cdot f(t-\tau)\cdot dt \qquad (2.46)$$

or replacing $t - \tau$ by

$$f_1(t) = \int_{0}^{\infty}\left[\sin\frac{\{\Omega(u-t)\}}{\{\pi(u-t)\}}\right]\cdot f'(u)\cdot du \qquad (2.47)$$

it is possible to estimate the error of the approximate time solution $f_1(t)$ defined by Equation 2.42 in comparison with the accurate one $f(t)$ in Equation 2.41, which cannot be evaluated by numerical integration. When $f(u)$ changes suddenly, a noticeable oscillation, called "Gibbs oscillation," appears in $f_1(t)$. This is the error caused by the finite Fourier transform. A countermeasure to this is to take the average of the time region $[t - a, t + a]$ in the following form:

$$f_2(t) = \left(\frac{1}{2a}\right) \int_{t-a}^{t+a} f_1(\tau) \cdot d\tau \tag{2.48}$$

where $a = \pi/\Omega$.

Substituting Equation 2.42 into this equation and rearranging it, the following formula is obtained [38,39]:

$$f_2(t) = \left(\frac{1}{2\pi}\right) \int_{-\Omega}^{\Omega} F(\omega) \cdot \exp(j\omega t) \cdot \sigma(\omega) \cdot d\omega \tag{2.49}$$

$\sigma(\omega)$ in this equation is called a "sigma factor (weighting function)" and is expressed by

$$\sigma(\omega) = \frac{\sin(\omega a)}{\omega a} = \frac{\sin(\omega\pi/\Omega)}{(\omega\pi/\Omega)} \tag{2.50}$$

By taking Equation 2.49 rather than Equation 2.42 as a finite Fourier transform, the Gibbs oscillation due to the finite interval in a numerical calculation of the Fourier transform is reduced.

2.6.2.2 Shift of Integral Path: Laplace Transform

In the Fourier transform, integration is carried out along the imaginary axis $j\omega$, as is clear from Equations 2.41 and 2.42. Thus, the integration hits a singular point along the $j\omega$ axis. To avoid this, the integral path can be shifted to $j(\omega - j\alpha) = \alpha + j\omega$ rather than $j\omega$:

$$f_3(t) = \left(\frac{1}{2\pi}\right) \int_{-\Omega}^{\Omega} F(\omega - j\alpha) \cdot \exp\{j(\omega - j\alpha)t\} \cdot d\omega$$

$$= \left\{\frac{\exp(\alpha t)}{2\pi}\right\} \int_{-\Omega}^{\Omega} F(\omega - j\alpha) \cdot \exp(j\omega t) \cdot d\omega \tag{2.51}$$

This formulation is similar to the Laplace transform and, thus, can be called the "finite Laplace transform." The following value has been known to be empirically optimal as the constant α in Equation 2.51 [40]:

$$\alpha = \frac{2\pi}{T} \tag{2.52}$$

where T is the observation time.

2.6.2.3 Numerical Laplace Transform: Discrete Laplace Transform

On the basis of the explanations in Sections 2.6.2.1 and 2.6.2.2, the following form of the Laplace transform is obtained [6,40]:

$$f_4(t) = \left\{ \frac{\exp(\alpha t)}{\pi} \right\} \int_0^\Omega F(\omega - j\alpha) \cdot \exp(j\omega t) \cdot \alpha(\omega) \cdot d(\omega) \tag{2.53}$$

For numerical calculations, the following equation is used:

$$f_4(t) = f(k \cdot t_0) = \mathrm{Real}\left[\left\{ \frac{\exp(\alpha t)}{\pi} \right\} \cdot \omega_0 \right.$$
$$\left. \times \sum_{n=0}^{N-1} \left\{ F(n\omega_0 - j\alpha) \cdot \exp(jn\omega_0 t) \cdot \frac{\sin(n\pi/N)}{(n\pi/N)} \right\} \right] \tag{2.54}$$

where $\omega_0 = \Omega/N$, N is the total number of frequency (= time) samples $t = k \cdot t_0$, $t_0 = T/N$, T is the observation time, and $k = 1, 2, \ldots, N$.

The discretization of $F(\omega)$ by ω_0 in the numerical Laplace transform causes an error. The details of the numerical discretization error are discussed in References 38 and 39.

2.6.2.4 Odd-Number Sampling: Accuracy Improvement

In principle, the numerical Fourier transform is a kind of numerical integration. Therefore, the accuracy of the numerical Fourier transform is greatly dependent on its integration method. In this section, various methods of integration, including odd-number sampling developed by Wedepohl that gives quite a high accuracy but is not well-known [6,40,41], are investigated, and a method with the highest accuracy is introduced into the discrete Laplace transform (DLT).

The accuracy of various integration methods is investigated for the case of the conventional Fourier transform [41]. The conventional discrete Fourier transform (DFT) is given in the following form:

$$f(t) = \left(\frac{1}{\pi}\right) \int_0^\Omega F(\omega)\exp(j\omega t)\, d\omega = \left(\frac{1}{\pi}\right) \int_0^\Omega G(\omega,t)\, d\omega \qquad (2.55)$$

The numerical evaluation of this equation is carried out by the following methods:

1. Method (a):

$$f(t) = \left(\frac{1}{\pi}\right) \sum_{n=1}^N G(n\Delta\omega, t)\Delta\omega \qquad (2.56)$$

This is shown in Figure 2.60a. If the sampling of $G(\omega, t)$ shown in Figure 2.60b is applied, Equation 2.55 can be evaluated by

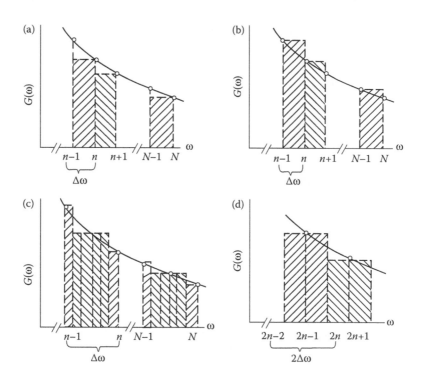

FIGURE 2.60
Various methods of integration. (a) Tropezoidal-1, (b) Tropezoidal-2, (c) Simpson's method, and (d) Odd-number sampling.

2. Method (b):

$$f(t) = \left(\frac{1}{\pi}\right) \sum_{n=1}^{N} G\{(n-1)\Delta\omega, t\}\Delta\omega \qquad (2.57)$$

Using Simpson's method of integration shown in Figure 2.60c, Equation 2.55 can be evaluated by

3. Method (c):

$$f(t) = \left(\frac{1}{\pi}\right) \sum_{n=1}^{N} G\left[(n\Delta\omega,t) + 4G\left\{\frac{(2n-1)\Delta\omega}{2,t}\right\} + G\{(n-1)\Delta\omega,t\}\right]\frac{\Delta\omega}{6} \qquad (2.58)$$

The following method of odd-number sampling was developed by Wedepohl [6,40,41]:

4. Method (d):

$$f(t) = \left(\frac{1}{\pi}\right) \sum_{n=1}^{N} G\{(2n-1)\Delta\omega,t\}2\Delta\omega \qquad (2.59)$$

This is shown in Figure 2.60d. This method can cover a frequency range twice as wide as that in methods (a) through (c) with the same number of frequency samples. If the maximum frequency is the same as in methods (a) through (c), then in method (d) the number of samples is halved.

The results of a unit step function calculated by these integration methods are shown in Figure 2.61 [41]. In the calculations, the weighting function is included. Table 2.8 shows a comparison of accuracy between the various methods. From the results, it is obvious that the odd-number sampling method (d) is the most accurate and efficient.

5. Application of the odd-number sampling: modified Laplace transform (MLT)

Integration method (d), explained earlier, is introduced into Equation 2.54 in the following form:

$$f(t) = \text{Real}\left[\left\{\frac{\exp(\alpha\tau)}{\pi}\right\} - 2\omega_0 \sum_{n=1}^{N} F\{(2n-1)\omega_0 - j\alpha\}\exp\{j(2n-1)\omega_0 t\}\right.$$
$$\left. \times \sin\frac{\{(2n-1)\pi/2N\}}{(2n-1)\pi/2N}\right] \qquad (2.60)$$

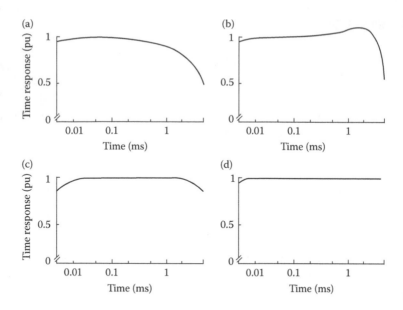

FIGURE 2.61
Results of unit step responses calculated by various methods of integration. (a) Tropezoidal-1,
(b) Tropezoidal-2, (c) Simpson's method, and (d) Odd-number sampling.

TABLE 2.8

Comparison of Accuracy (T_{max} = 5 ms)

Method	N	f_{max} (kHz)	T_1 (ms)	T_1/T_{max} (%)	T_2 (ms)	T_2/T_{max} (%)	T_3 (ms)	T_3/T_{max} (%)	Computation Time (s)
A	500	50	0.02	0.4	0.2	4	0.5	10	824
B	500	50	0.03	0.6	0.24	4.8	0.52	14	858
C	500	50	0.96	19.2	2.15	43.0	3.0	60	2393
D	250	49.9	4.97	99.4	5.0	100	5.0	100	215

T_1 = time for accuracy higher than 99.8%.
T_2 = time for accuracy higher than 99%.
T_3 = time for accuracy higher than 95%.
(except initial time)
Computer HITAC 8350 ($\approx$ 360/35).

2.6.2.5 Application of Fast Fourier Transform FFT: Fast Laplace Transform (FLT)

2.6.2.5.1 Principle and Algorithms of the FFT

The complex DFT is defined in the following form [42]:

$$F_n = \sum_{k=0}^{N-1} f_k \cdot \exp\left(\frac{-j2\pi kn}{N}\right), \quad n = 0, 1, \ldots, N-1 \qquad (2.61)$$

where

F_n is the nth coefficient of the DFT

f_k denotes the kth sample of the time series that consists of N samples

The inverse transform of Equation 2.61 is

$$f_k = \left(\frac{1}{N}\right) \sum_{n=1}^{N-1} F_n \cdot W^{kn}, \quad k = 0, 1, \ldots, N-1 \tag{2.62}$$

where

$$W = \exp\left(\frac{j2\pi}{N}\right) \tag{2.63}$$

f_k can be a complex number, but usually it is real in the field of electrical engineering, and F_n is almost always complex.

The principle and the algorithms of the FFT for the inverse DFT of Equation 2.62 are explained next.

Let us consider the case of $N = 8 = 2^3$. Then W becomes

$$W = \exp\left(\frac{j2\pi}{8}\right) = \cos\left(\frac{\pi}{4}\right) = j\sin\left(\frac{\pi}{4}\right) \tag{2.64}$$

Using $W^8 = 1$ and $W^r = W^s$, where $s = r$ mod 8, and also $W^2 = j$, $W^4 = -1$, $W^5 = -W$, etc., Equation 2.62 is rewritten in the following form:

$$
\begin{bmatrix} f_0 \\ f_1 \\ f_2 \\ f_3 \\ f_4 \\ f_5 \\ f_6 \\ f_7 \end{bmatrix} =
\begin{bmatrix}
1 & 1 & 1 & 1 & 1 & 1 & 1 & 1 \\
1 & W & W^2 & W^3 & W^4 & W^5 & W^6 & W^7 \\
1 & W^2 & W^4 & W^6 & W^8 & W^{10} & W^{12} & W^{14} \\
1 & W^3 & W^6 & W^9 & W^{12} & W^{15} & W^{18} & W^{21} \\
1 & W^4 & W^8 & W^{12} & W^{16} & W^{20} & W^{24} & W^{28} \\
1 & W^5 & W^{10} & W^{15} & W^{20} & W^{25} & W^{30} & W^{35} \\
1 & W^6 & W^{12} & W^{18} & W^{24} & W^{30} & W^{36} & W^{42} \\
1 & W^7 & W^{14} & W^{21} & W^{28} & W^{35} & W^{42} & W^{49}
\end{bmatrix}
\cdot
\begin{bmatrix} F_0 \\ F_1 \\ F_2 \\ F_3 \\ F_4 \\ F_5 \\ F_6 \\ F_7 \end{bmatrix}
$$

$$
=
\begin{bmatrix}
1 & 1 & 1 & 1 & 1 & 1 & 1 & 1 \\
1 & W & j & W^3 & -1 & -W & -j & -W^3 \\
1 & j & -1 & -j & 1 & j & -1 & -j \\
1 & W^3 & -j & W & -1 & -W^3 & j & -W \\
1 & -1 & 1 & -1 & 1 & -1 & 1 & -1 \\
1 & -W & j & -W^3 & -1 & W & -j & W^3 \\
1 & -j & -1 & j & 1 & -j & -1 & j \\
1 & -W^3 & -j & -W & -1 & W^3 & j & W
\end{bmatrix}
\cdot
\begin{bmatrix} F_0 \\ F_1 \\ F_2 \\ F_3 \\ F_4 \\ F_5 \\ F_6 \\ F_7 \end{bmatrix} \tag{2.65}
$$

Changing the columns and rearranging for (F_0, F_2, F_4, F_6) and (F_1, F_3, F_5, F_7), these matrices can be rewritten as

$$
[f_k] = \begin{bmatrix} 1 & 1 & 1 & 1 \\ 1 & j & -1 & -j \\ 1 & -1 & 1 & -1 \\ 1 & -j & -1 & j \\ 1 & 1 & 1 & 1 \\ 1 & j & -1 & -j \\ 1 & -1 & 1 & -1 \\ 1 & -j & -1 & j \end{bmatrix} \cdot \begin{bmatrix} F_0 \\ F_2 \\ F_4 \\ F_6 \end{bmatrix} + \begin{bmatrix} 1 & 1 & 1 & 1 \\ W & jW & -W & -jW \\ j & -j & j & -j \\ W^3 & -jW^3 & -W^3 & -jW^3 \\ -1 & -1 & -1 & -1 \\ -W & -jW & W & jW \\ -j & j & -j & j \\ -W^3 & jW^3 & W^3 & -jW^3 \end{bmatrix} \cdot \begin{bmatrix} F_1 \\ F_3 \\ F_5 \\ F_7 \end{bmatrix}
$$

$$(2.66)$$

These matrices include the following matrix:

$$
[T] = \begin{bmatrix} 1 & 1 & 1 & 1 \\ 1 & j & -1 & -j \\ 1 & -1 & 1 & -1 \\ 1 & -j & 1 & j \end{bmatrix}
$$

Defining matrix $[E_n]$ by

$$
\begin{bmatrix} E_0 \\ E_1 \\ E_2 \\ E_3 \end{bmatrix} = [T] \cdot \begin{bmatrix} F_0 \\ F_2 \\ F_4 \\ F_6 \end{bmatrix}, \quad \begin{bmatrix} E_4 \\ E_5 \\ E_6 \\ E_7 \end{bmatrix} = [T] \cdot \begin{bmatrix} F_1 \\ F_3 \\ F_5 \\ F_7 \end{bmatrix}
$$

Matrix $[f_k]$ is expressed in the following form:

$$
[f_k] = \begin{bmatrix} E_0 + E_4 \\ E_1 + WE_5 \\ E_2 + jE_6 \\ E_3 + W^3 E_7 \\ E_0 - E_4 \\ E_1 - WE_5 \\ E_2 - jE_6 \\ E_3 - W^3 E_7 \end{bmatrix}
$$

In matrix $[T]$, by interchanging the second and third columns, the following matrix is obtained:

$$[S] = \begin{bmatrix} 1 & 1 \\ 1 & -1 \end{bmatrix}$$

Using this orthogonal matrix, we can write

$$\begin{bmatrix} D_0 \\ D_1 \end{bmatrix} = [S] \begin{bmatrix} F_0 \\ F_4 \end{bmatrix}, \quad \begin{bmatrix} D_2 \\ D_3 \end{bmatrix} = [S] \begin{bmatrix} F_1 \\ F_5 \end{bmatrix}, \quad \begin{bmatrix} D_4 \\ D_5 \end{bmatrix} = [S] \begin{bmatrix} F_2 \\ F_6 \end{bmatrix}, \quad \begin{bmatrix} D_6 \\ D_7 \end{bmatrix} = [S] \begin{bmatrix} F_3 \\ F_7 \end{bmatrix} \quad (2.67)$$

The $[E_n]$ matrix is expressed in the following form:

$$\begin{bmatrix} E_0 \\ E_1 \\ E_2 \\ E_3 \end{bmatrix} = \begin{bmatrix} D_0 + D_4 \\ D_1 + jD_5 \\ D_0 - D_4 \\ D_1 - jD_5 \end{bmatrix}, \quad \begin{bmatrix} E_4 \\ E_5 \\ E_6 \\ E_7 \end{bmatrix} = \begin{bmatrix} D_2 + D_6 \\ D_3 + jD_7 \\ D_2 - D_6 \\ D_3 - jD_7 \end{bmatrix}$$

Therefore, $[f_k]$ can be obtained using the following procedure:

$$\{F_n\} \to \{D_n\} \to \{E_n\} \to \{f_k\} \quad (2.68)$$

Expressing the subscripts of each matrix by the binary code,

$$F_n = F(rqp)$$

then D_n from Equation 2.67 is given in the following form:

$$\left. \begin{array}{l} D(qp0) = F(0qp) + F(1qp) \\ D(qp1) = F(0qp) - F(1qp) \end{array} \right\} \quad q, \, p = 0, 1$$

Second, E_n is given by

$$\left. \begin{array}{l} E(p0\bar{r}) = D(0p\bar{r}) + D(1p\bar{r}) \\ E(p1\bar{r}) = D(0p\bar{r}) - D(1p\bar{r}) \end{array} \right\} \quad p = 0, 1, \quad \bar{r} = 0$$

$$\left. \begin{array}{l} E(p0\bar{r}) = D(0p\bar{r}) + W^2 D(1p\bar{r}) \\ E(p1\bar{r}) = D(0p\bar{r}) - W^2 D(1p\bar{r}) \end{array} \right\} \quad p = 0, 1, \quad \bar{r} = 1$$

Finally, f_k is obtained from these equations as follows:

$$f(oo\bar{r}) = E(ooo) + (-1)^{\bar{r}} W^0 E(ooo)$$

$$f(o1\bar{r}) = E(o1o) + (-1)^{\bar{r}} W^1 E(o11)$$

$$f(1o\bar{r}) = E(1oo) + (-1)^{\bar{r}} W^2 E(1o1)$$

$$f(11\bar{r}) = E(11o) + (-1)^{\bar{r}} W^{03} E(111)$$

The inverse Fourier transform of Equation 2.62 can be calculated in this manner. A more general description is given in many publications (see, for example, Reference 42).

2.6.2.5.2 Computation Time

The total number of calculation units by repeated application of the FFT becomes [6]

$$T_F = N(p_1 + p_2 + \cdots + p_n)$$

In the case of

$$p_1 = p_2 = \cdots = p_n = m$$

n being given by

$$n = \log_m N$$

the total number of calculation units becomes

$$T_F = N \cdot m \cdot \log_m N \tag{2.69}$$

The ratio between T_F and T_c is

$$\frac{T_F}{T_c} = \frac{(N\, m \log_m N)}{N^2} = \frac{m\, \log_m N}{N} = \left(\frac{m}{\log_2 m}\right)\left(\frac{\log_2 N}{N}\right) \tag{2.70}$$

Table 2.9 shows the value of $m/\log_2 m$. This becomes a minimum at $m = 3$. Table 2.10 shows T_c/T_F with $m = 2$. From the table, it is obvious that the FFT is highly efficient compared with the conventional Fourier transform.

2.6.2.5.3 Application of the FFT to MLT

The application of the FFT to the MLT in Equation 2.60 may be rather difficult compared with the application of the DFT because the integer term $(2k - 1)$ in the exponent of Equation 2.60 is an odd number [6]. Therefore, we need

TABLE 2.9

$m/\log_2 m$

(m)	(m/log₂m)
2	2.00
3	1.88
4	2.00
5	2.15
6	2.31
7	2.49
8	2.67
9	2.82
10	3.01

to modify this equation so that the integer term takes on sequential values of 1, 2, 3, 4... Thus, the following form is obtained, in which the FFT can be applied:

$$f_k = \text{Real}\left[\left\{\frac{\exp(\alpha k t_0)}{\pi}\right\}\cdot\exp\left(\frac{jk\pi}{N}\right)\right]\sum_N^{n=1} F_n\cdot\exp\left\{j\left(\frac{2\pi}{N}\right)(n-1)k\right\}$$

where

$$f_k = f(k\cdot t_0) = f(t), \quad t_0 = \frac{T}{N}, \quad \omega_0 = \frac{\pi}{T} \tag{2.71}$$

$$F_n = F\{(2n-1)\omega_0 - j\alpha\}\cdot\sin\frac{\{(2n-1)\pi/2N\}}{(2n-1)\pi/2N\cdot 2\omega_0}$$

TABLE 2.10

Theoretical Comparison of
Computation Time ($m = 2$)

N	T_c/T_F
32	3.2
64	5.2
128	9.1
256	16
512	28
1024	51
2048	92
4096	170
8192	315

```
      DIMENSION FV(N), FS(N)
      N2 = N/2
      AN = N
      JO = N2
      KEI = 1
      CJ = CMPLX(0., 1.)
      PAI = 3.14159
      CJP2 = CJ*2.*PAI/AN
      DW = 2.*PAI/TO
      DO 20 I = 1, N
      AI = FLOAT(1*2 - 1)2.
      WA = PAI*AI/AN
      SIGMA = SIN(WA)/WA
  20  FV(1) = FV(I)*SIGMA*DW
      DO 50 K = 1, M
      KE2 = KE1*2
      IL = 0
      DO 41 J = 1, N, KE2
      DO 40 I = 1, KE1
      IL = IL+1
      IR = IL+N2
      JW = (I - 1)*JO + 1
      AJW = JW-1
      FV1=FV(IL)
      FV2=FV(IR)*CEXP(CJP2*AJW)
      JL = J+I-1
      JR = L+KE1 FS(JL) = FV1 + FV2
  40  FS(JR) = FV1 + FV2
  41  CONTINUE
      KE1 = KE2
      JO = JO/2
      DO 45 I = 1, N
  45  FV(I) = FS(I)
  50  CONTINUE
      DO 60 I = 1, N
      AI = I-1
      T=TO*AI/AN/2.
      EA=EXP(ALFA*T)
      VT=REAL(FV/1)*EA/PAI
  60  WRITE T, VT
```

FIGURE 2.62
Complete program for computing the DLT of Equation 2.54 by the FFT method-subroutine FLT.

Figure 2.62 gives a complete program for computing the DLT in Equation 2.54 by the FFT. Figure 2.63 shows a calculation example in comparison with the exact solution.

2.6.3 Transient Simulation

A computer program for transient simulation by an FD method can be easily produced by a university student. Figure 2.64 illustrates its flow chart. The program is composed of three procedures [2,6], discussed next.

Time	EXACT SOL.	DLT SOL.	FLT SOL.
0.000E 00	0.196078E-01	0.119409E-01	0.119409E-01
0.500E-04	0.105274E 00	0.105109E 00	0.105109E 00
0.100E-03	0.159656E 00	0.159499E 00	0.159499E 00
0.150E-03	0.174684E 00	0.174562E 00	0.174562E 00
0.200E-03	0.152352E 00	0.152281E 00	0.152281E 00
0.250E-03	0.102884E 00	0.102846E 00	0.102846E 00
0.300E-03	0.412708E-01	0.412929E-01	0.412929E-01
0.350E-03	-0.167940E-01	-0.167490E-01	-0.167490E-01
0.400E-03	-0.585847E-01	-0.585381E-01	-0.585381E-01
0.450E-03	-0.768302E-01	-0.768006E-01	-0.768006E-01
0.500E-03	-0.706180E-01	-0.706177E-01	-0.706177E-01
0.550E-03	-0.447274E-01	-0.447603E-01	-0.447603E-01
0.600E-03	-0.775823E-02	-0.782012E-02	-0.782012E-02
0.650E-03	0.303518E-01	0.3028710E-01	0.302710E-01
0.700E-03	0.607726E-01	0.606862E-01	0.606862E-01
0.750E-03	0.776394E-01	0.775597E-01	0.775597E-01
0.800E-03	0.789786E-01	0.789150E-01	0.789150E-01
0.850E-03	0.666222E-01	0.665788E-01	0.665788E-01
0.900E-03	0.452667E-01	0.452425E-01	0.452425E-01
0.950E-03	0.210105E-01	0.209998E-01	0.209998E-01
0.100E-02	-0.227285E-03	-0.232818E-03	-0.232818E-03

Number of frequency samples = 128.
Maximum observation time = 1.28 ms.
Computation time : 1.0 by the DLT and 0.11 by the FLT.

FIGURE 2.63
Results calculated by the DLT and the FLT in comparison with the exact solution.

2.6.3.1 Definition of Variables

N is the number of frequency samples for $F(\omega)$ = number of time samples $f(t)$

DF is the frequency step; $FMAX = N*DF$ is the maximum frequency

$DT = 1/FMAX$ is the time step

$TMAX = N*DF = 1/DF$ is the observation time corresponding to the sampling theorem:

$$DW = 2\pi * DF, W = n * DW$$

$ALFA = 2\pi/TMAX$, $CJ = \exp(j\pi/2)$ is the symbol of the imaginary variable:

$$S = CJ * W + ALFA$$

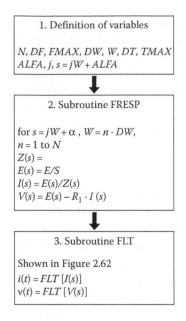

FIGURE 2.64
Flow chart of the FD method.

2.6.3.2 Subroutine to Prepare F(ω)

$F(\omega)$ containing N samples of frequency responses needs to be prepared. For example, let us obtain transient (time) responses of $v(t)$ in the RLC parallel circuit illustrated in Figure 2.65. Voltage $V(s)$ in the s-domain is given by

$$Y_1 = \frac{1}{R} + \frac{1}{sL} + sC$$

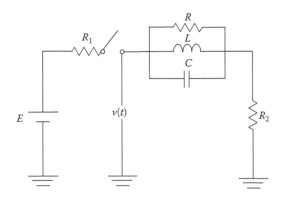

FIGURE 2.65
Switching of RLC parallel circuit. $R_1 = 500\ \Omega$, $R = 1000\ \Omega$, $L = 10$ mH, and $C = 1\ \mu$F. $R_2 = 10\ \Omega$, $E =$ Step function 1.

$$Z(s) = R_1 + \frac{1}{Y_1} + R_2$$

$$E(s) = \frac{E}{s}$$

$$I(s) = \frac{E(s)}{Z(s)}$$

$$V(s) = E(s) - R_1 \cdot I(s)$$

These frequency responses are produced for $n = 1$ to N, where

$$s = j\omega + \alpha, \quad \omega = n \cdot \omega_0$$

2.6.3.3 Subroutine FLT

The frequency responses $I(s)$ and $V(s)$ are sent to subroutine FLT given in Figure 2.62. Then, the FLT carries out the inverse Laplace transform and the time solutions are obtained. Figure 2.63 shows an example of a calculated result $v(t)$ in comparison with the accurate solution. Note that the accuracy of the FLT is quite high.

2.6.4 Remarks of the FD Method

The advantage of the FD method is that any frequency dependent effect is easy to handle as it is based on the frequency response of a transient to be solved. Thus, the frequency-dependent effect of a transmission line or cable, explained in Chapter 1, is very easily included in a simulation.

On the contrary, a sudden change in the time domain, such as switching, causes a difficulty because the change involves an initial condition problem that requires repeated time/frequency transforms. A nonlinear element (e.g., an arrester) requires a number of time/frequency transforms. Thus, the FD method is often used to check the accuracy of the time-domain method, such as the EMTP, on the frequency-dependent effect.

Appendix 2A

2A.1 Setup of the Field Test in Section 2.2.3.2

Section 2.2.3.2 introduces field test results on the single-circuit horizontal line and the double-circuit vertical line. This appendix introduces additional information regarding the test setup and test results on the former field test.

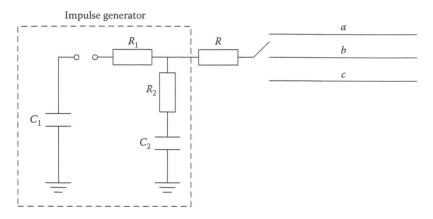

FIGURE 2A.1
Field test circuit with the IG.

The test line is a 500 kV transmission line of the Tokyo Electric Power Company. It is an untransposed single-circuit line with two GWs. The tower configuration and conductor data are already shown in Figure 1.22.

The test circuit is shown in Figure 2A.1. The test voltage was applied to the test line by an IG through series resistance R. The test voltages with various wave shapes were applied according to the standard wave shapes defined in the Japanese Electrotechnical Committee. Table 2A.1 shows circuit constants of the IG and the series resistance for each wave shape.

2A.2 Results of the Field Test

This section discusses the results of the field test introduced in Section 2A.1. The first result is the attenuation of the applied voltage at the receiving end. The attenuation is evaluated by the following definition and summarized in Table 2A.2:

$$\text{Attenuation } (\alpha) = \frac{V_r}{2V_s}$$

TABLE 2A.1

Circuit Constants of the Field Test

Wave-Shape (μs)	R (Ω)	C_1 (μF)	R_1 (Ω)	C_2 (μF)	R_2 (Ω)
1/23	415	0.1	–	–	550
1/420	415	1	–	–	–
1/4000	415	10	–	–	–
50/4000	207	10	208	0.1	20
200/4000	207	10	208	0.5	–

TABLE 2A.2

Attenuation of the Voltage Magnitude

Wave-shape (μs)	1/23	1/420	1/4000	50/4000	200/4000
Phase *a* energization	0.422	0.830	0.917	0.918	0.920
Phase *b* energization	0.424	0.830	0.938	0.934	0.925

where

V_s: magnitude of the voltage at the sending end

V_r: magnitude of the voltage at the receiving end

The attenuation is more significant when α is smaller.

Here, phase *a* is the left phase, and phase *b* is the middle phase. There are only minor differences between phase *a* energization and phase *b* energization with regard to the attenuation.

The results in Table 2A.2 show that the attenuation is highly dependent on the wave tail, and not on the wave front. The attenuation is more significant for wave shapes with a shorter wave tail. For the wave tail of 4000 μs, the difference of a wave front does not have a noticeable effect on the attenuation.

The second result is the length of a wave front of the voltage at the receiving end. The definition of the wave front is shown in Figure 2A.2. The voltage wave shape at the receiving end has two voltage spikes as explained in Section 1.6.4.5. The magnitude of the two voltage spikes is named as A_{f1} and A_{f2}, respectively. Two wave fronts T_{f1} and T_{f2} are defined using two lines that connect $0.3A_{fn}$ and $0.9A_{fn}$ ($n = 1, 2$).

Table 2A.3 shows the two wave fronts for phase *a* energization and phase *b* energization when the wave-shape of the applied voltage is 1/4000. The phase *b* energization has shorter wave fronts, as can be expected from the analytical calculation.

The last result is the comparison of field test results with the analytical calculation. Figure 2A.3 shows the comparison for phase *a* energization when

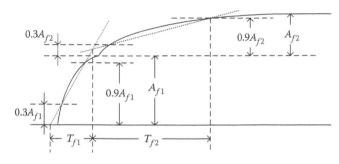

FIGURE 2A.2

Definition of the wave front.

TABLE 2A.3

Length of Wave-Front

	T_{f1} (µs)	T_{f2} (µs)
Phase *a* energization	6.9	71
Phase *b* energization	2.8	67

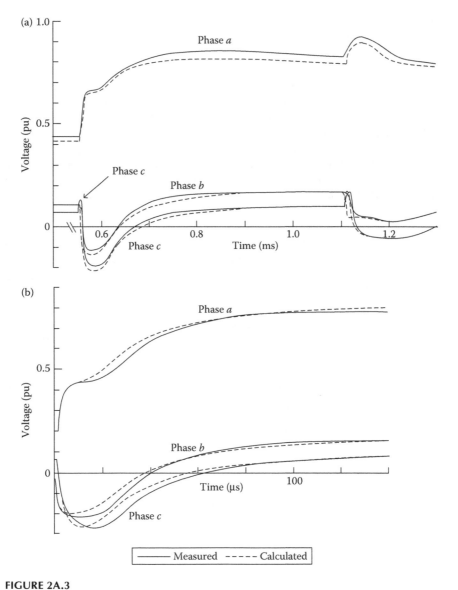

FIGURE 2A.3

Comparison of the measured and calculated results for phase *a* energization. (a) Sending-end voltages and (b) receiving-end voltages.

the wave-shape of the applied voltage is 1/4000. The analytical calculation agrees well with the field test results. The comparison for phase *b* energization is shown in Figure 2.7. Comparing with phase *b* energization in Figure 2.7, the wave front of the first voltage spike at the receiving end is longer as shown in Table 2A.3.

References

1. Ametani, A. 1987. *Power System Transient Analysis.* Kyoto, Japan: Doshisha University.
2. Ametani, A. 1990. *Distributed—Parameter Circuit Theory.* Tokyo, Japan: Corona Publishing Company.
3. Japanese Standard. 1994. High voltage testing (JEC-0102-1994). *IEE Japan.*
4. Ametani, A., T. Ono, Y. Honaga, and Y. Ouchi. 1974. Surge propagation on Japanese 500 kV untransposed transmission line. *Proc. IEE* 121 (2):136–138.
5. Ametani, A. and T. Ono. 1975. Surge propagation characteristics on an untransposed horizontal line. *IEE Jpn.* B-95 (12):591–598.
6. Ametani, A. 1972. The application of fast Fourier transform to electrical transient phenomena. *Int. J. Elect. Eng. Educ.* 10:277–287.
7. Ametani, A. et al. 1981. Wave propagation characteristics on an untransposed vertical twin-circuit line. *IEE Jpn.* B-101 (11):675–682.
8. Ametani, A., E. Ohsaki, and Y. Honaga. 1983. Surge characteristics on an untransposed vertical line. *IEE Jpn.* B-103 (2):117–124.
9. Jordan, C. A. 1934. Lightning computation for transmission line with overhead ground wires. *G. E. Rev.* 37 (4):180–186.
10. Bewley, L. V. 1951. *Traveling Waves on Transmission Systems.* New York: Wiley.
11. Wagner, C. F. 1956. A new approach to calculation of lightning performance of transmission lines. *AIEE Trans.* 75:1233–1256.
12. Lundholm, R., R. B. Finn, and W. S. Price. 1957. Calculation of transmission line lightning voltages by field concepts. *AIEE Trans.* 76:1271–1283.
13. Ozawa, J. et al. 1985. Lightning surge analysis in a multiconductor system for substation insulation design. *IEEE Trans. Power Appl. Syst.* 104:2244.
14. Kawamura, T. et al. 1987. A new approach of a lightning surge analysis in a power system. *IEE Japan.* Technical Report No. 244.
15. Kawamura, T. et al. 1989. Various parameters and the effects on lightning surges in a substation. *IEE Japan.* Technical Report No. 301.
16. CIGRE SC33-WG01. 1991. Guide to procedures for estimating lightning performance of transmission lines. *CIGRE Tech. Brochure.*
17. Kawamura, T. et al. 1992. A new method for estimating lightning surge in substations. *IEE Japan WG Report.* Technical Report No. 446.
18. IEEE WG. 1993. Estimating lightning performance of transmission lines, II— Update to analytical models. *IEEE Trans. Power Deliv.* 8:1254.
19. Kawamura, T. et al. 1995. An estimating method of a lightning surge for statistical insulation design of substations. *IEE Japan WG Report.* Technical Report No. 566.
20. IEEE Guide for Improving the Lightning Performance of Transmission Lines. 1997.

21. Ametani, A. et al. 2002. Power system transients and EMTP analyses. *IEE Japan WG Report*. Technical Report No. 872.
22. CRIPEI WG. 2003. Guide to transmission line protection against lightning. Report T72.
23. Ametani, A. and T. Kawamura. 2005. A method of a lightning surge analysis recommended in Japan using EMTP. *IEEE Trans. Power Deliv.* 20 (2):867–875.
24. Anderson, R. B. and A. J. Eriksson. 1980. Lightning parameters for engineering application. *Electra* 69:65.
25. Takami, J. and S. Okabe. 2007. Characteristic of direct lightning strokes to phase conductors of UHV transmission line. *IEEE Trans. Power Deliv.* 22 (1):537–546.
26. Study Committee of Lightning Protection Design. 1976. Lightning protection design guide-book for power stations and substations. *Central Research of Electric Power Industry (CRIEPI)* Report 175034.
27. Electric Research Association. 1988. Rationalization of insulation design. *ERA Report* 44 (3).
28. Ishii, M. et al. 1991. Multistory transmission tower model for lightning surge analysis. *IEEE Trans. Power Deliv.* 6 (3):1372.
29. Shindo, T. and T. Suzuki. 1985. A new calculation method of breakdown voltage-time characteristics of long air gaps. *IEEE Trans. Power Appl. Syst.* 104:1556.
30. Nagaoka, N. 1991. An archorn flashover model by means of a nonlinear inductance. *Trans. IEE Jpn.* B-111 (5):529.
31. Nagaoka, N. 1991. Development of frequency-dependent tower model. *Trans. IEE Jpn.* B-111:51.
32. Ametani, A., H. Morii, and T. Kubo. 2011. Derivation of theoretical formulas and an investigation on measured results of grounding electrode transient responses. *IEE Jpn.* B-131 (2):205–214.
33. Ueda, T., M. Yoda, and I. Miyachi. 1996. Characteristic of lightning surges observed at 77 kV substations. *Trans. IEE Jpn.* B-116 (11):1422.
34. Ametani, A. et al. 2002. Investigation of flashover phases in a lightning surge by new archorn and tower models. In *Proceedings of the IEEE PES T&D Conference 2002*, Yokohama, Japan, pp. 1241–1426.
35. Meyer, W. S. 1973. *EMTP Rule Book*, 1st edn. Portland, OR: BPA.
36. Nagaoka, N. and A. Ametani. 1988. A development of a generalized frequency-domain transient program—FTP. *IEEE Trans. Power Deliv.* 3 (4):1986–2004.
37. Nagaoka, N. and A. Ametani. 1986. Lightning surge analysis by means of a two-phase circuit model. *TIEE Jpn.* B-106 (5):403–410.
38. Day, S. J., N. Mullineux, and J. R. Reed. 1965. Developments in obtaining transient response using Fourier integrals. Pt. I: Gibbs phenomena and Fourier integrals. *Int. J. Elect. Eng. Educ.* 3:501–506.
39. Day, S. J., N. Mullineux, and J. R. Reed. 1966. Developments in obtaining transient response using Fourier integrals. Pt. II: Use of the modified Fourier transform. *Int. J. Elect. Eng. Educ.* 4:31–40.
40. Wedepohl, L. M. and S. E. T. Mohamed. 1969. Multi-conductor transmission lines: Theory of natural modes and Fourier integral applied to transient analysis. *Proc. IEE* 116:1553–1563.
41. Ametani, A. and K. Imanishi. 1979. Development of a exponential Fourier transform and its application to electrical transients. *Proc. IEE* 126 (1):51–59.
42. Cooley, J. 1967. What is the fast Fourier transform? *Proc. IEE* 55:1664–1674.

3

Transients on Cable Systems

3.1 Introduction

This chapter focuses on transient phenomena specifically related to cables. Transients on cable systems are characterized by the large charging capacities of cables and the presence of a metallic sheath around the phase conductor. Temporary overvoltages (TOVs), such as the overvoltage caused by system islanding and the resonance overvoltage observed on the cable system, contain low-frequency components due to large charging capacities. Because of the low frequency, namely low damping, these TOVs can be sustained for extended durations, posing challenges to the insulation performance of related equipment. This chapter introduces examples of such studies.

Other issues, such as the zero-missing phenomenon, the leading current interruption, and the cable discharge, also stem from the large charging capacities of cables. The effects of these issues on the cable system design are discussed in Section 3.5. The discussion includes countermeasures for the problems and suggestions for equipment selection.

Sheath bonding and grounding is another important issue in cable system design. Sheath overvoltage requires careful study not only to avoid failures of sheath voltage limiters (SVLs) and sheath interrupts, but also to ensure the safety of maintenance crews. Sections 3.2 and 3.3 cover all the major aspects of sheath bonding and grounding, thus providing a wide variety of information from fundamentals to applications. In addition, impedance calculations, wave propagation characteristics, and transient voltage behaviors discussed in these sections provide grounding for the transient phenomena discussed in the later sections.

3.2 Impedance and Admittance of Cable Systems

3.2.1 Single-Phase Cable

3.2.1.1 Cable Structure

The most significant difference between a cable and an overhead line is that a cable is generally composed of two conductors for one phase, while an overhead line is generally composed of one conductor. Thus, a three-phase cable consists of six conductors, while a three-phase overhead line consists of three conductors.

Figure 3.1 illustrates the cross section of a typical coaxial cable. The core conductor carries current in the same way the phase conductor of an overhead line does. The metallic sheath is grounded at both ends of the cable in order to shield the core current. Thus, the metallic sheath is often called the "shield."

3.2.1.2 Impedance and Admittance

The impedance and admittance of a single-phase cable are presented in matrix form because the cable contains two conductors:

$$[Z_i] = \begin{bmatrix} Z_{cc} & Z_{cs} \\ Z_{cs} & Z_{ss} \end{bmatrix}, \quad [Y_i] = \begin{bmatrix} Y_c & -Y_c \\ -Y_c & Y_s \end{bmatrix} \quad \text{for phase } i \tag{3.1}$$

Each element of the impedance matrix is composed of the cable internal impedance and the cable outer media (earth-return) impedance, as explained in Chapter 1 of this volume. In the overhead line case, the conductor internal impedance is composed of only one impedance (i.e., the outer surface

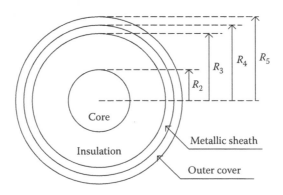

FIGURE 3.1
Cross section of a typical coaxial cable.

impedance of a conductor). The cable internal impedance consists of the following six components [1]:

1. Core outer surface impedance (same as the internal impedance of an overhead line)
2. Core-to-sheath insulator impedance
3. Sheath inner surface impedance
4. Mutual impedance between the sheath's inner and outer surfaces
5. Sheath outer surface impedance
6. Sheath outer insulator (the outer cover shown in Figure 3.1) impedance

It is quite clear that cable impedance is far more complicated than overhead line impedance [1,2].

The admittance matrix is expressed in the following form using the potential coefficient matrix:

$$[Y] = j\omega[C] = j\omega[P]^{-1} \quad \text{for one phase} \tag{3.2}$$

where

$$[P] = \begin{bmatrix} P_c & P_s \\ P_s & P_s \end{bmatrix} = \begin{bmatrix} P_{12} + P_{23} & P_{23} \\ P_{23} & P_{23} \end{bmatrix}$$

$$P_{12} = \frac{1}{2\pi\varepsilon_{i1}} \ln \frac{R_3}{R_2}$$

$$P_{23} = \frac{1}{2\pi\varepsilon_{i2}} \ln \frac{R_5}{R_4}$$

3.2.2 Sheath Bonding

Before we discuss the impedance and admittance of a three-phase cable, it is necessary to learn about sheath bonding. Underground cables that are longer than 2 km normally adopt cross-bonding to reduce sheath currents and to suppress sheath voltages at the same time [3]. Figure 3.2 shows a representative cross-bonding diagram of a cable. In the figure, one of the three sheath circuits is highlighted with a dotted line. Starting from the left termination, the sheath circuit goes along the phase *a* conductor in the first minor section, the phase *b* conductor in the second minor section, and the phase *c* conductor in the third minor section. Theoretically, the vector sum of the induced voltage of the sheath circuit in these three minor sections becomes zero when three phase currents in the phase conductors are balanced and the three minor sections are of the same length. This is why cross-bonding can reduce sheath currents and suppress sheath voltages at the same time.

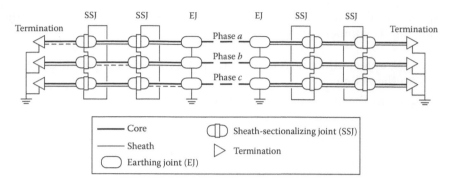

FIGURE 3.2
Example of a cross-bonding diagram of a cable.

If the lengths of the three minor sections are different, an imbalance in the induced voltages will result that causes sheath currents. However, when there are more than a few major sections, it is a common practice to design cross-bonding after considering the best balance for the induced voltage. This results in the so-called homogeneous nature of cable impedance [4,5].

For submarine cables, it is more common to adopt solid bonding due to the difficulty in constructing joints offshore, as shown in Figure 3.3. Hence, submarine cables have higher sheath currents compared to underground cables that are normally cross-bonded. In order to reduce the loss caused by higher sheath currents, the sheath conductors of submarine cables often have a lower resistance (i.e., a larger cross section).

Single-point bonding has an advantage in terms of reducing the sheath currents. The sheath current loss can be reduced virtually to zero by applying single-point bonding, as shown in Figure 3.4. However, it can only be applied to short cables or short cable sections due to a limitation in the acceptable sheath voltage. In order to prevent the sheath voltage from exceeding the limitation, SVLs are installed at the unearthed end of the sheath circuit. (The installation of SVLs is discussed in greater detail in Section 3.3.) Additionally, installing an earth continuity cable (ECC) is highly recommended in order to suppress sheath overvoltage.

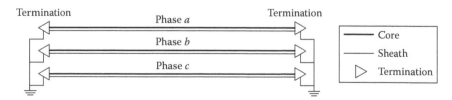

FIGURE 3.3
Solid bonding of a cable.

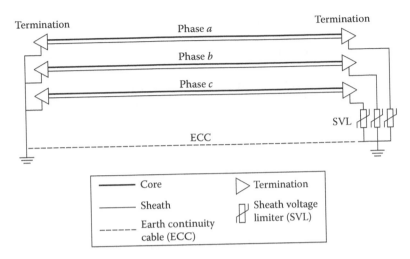

FIGURE 3.4
Single-point bonding of a cable.

Figure 3.5 shows an example in which single-point bonding is employed in a long cable. As discussed earlier, cross-bonding is adopted for the long cable. In the figure, the first three minor sections from the left termination compose one major section of cross-bonding. Since the number of minor sections is four, which is not a multiple of three, the fourth minor section from the left termination cannot become a part of cross-bonding. In this situation, single-point bonding is applied to the remaining minor section (as shown in Figure 3.5) as long as the sheath voltage allows it.

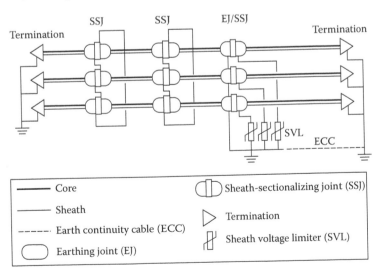

FIGURE 3.5
Single-point bonding as part of a cross-bonded cable.

This situation is often observed in actual installations, as the number of minor sections is not determined by the cross-bonding. Rather, it is determined to reduce the number of joints as much as possible as an aspect of cost consideration.

The joint labeled EJ/SSJ functions both as an earthing joint (EJ) and as a sheath-sectionalizing joint (SSJ). The left side of the joint is solidly grounded as in an earthing joint. The left and right sides of the joint are insulated as in a sheath-sectionalizing joint, and the right side of the joint is unearthed. Since the grounding resistance at the EJ/SSJ is normally much higher than the resistance at the termination (substation), this addition of the single-point-bonding section may significantly increase the zero-sequence impedance of the cable without the ECC.

3.2.3 Homogeneous Model of a Cross-Bonded Cable

3.2.3.1 Homogeneous Impedance and Admittance

Section 3.2.1 addressed the impedance and admittance of a single-phase cable. This section addresses the impedance and admittance of a cross-bonded three-phase cable and how 6×6 impedance and admittance matrices can be reduced to 4×4 matrices.

Figure 3.6 illustrates a major section of a cross-bonded cable. The bold solid line and the broken line express the core and sheath, respectively. The sheaths are grounded through grounding impedance Z_g at both sides of the major section. The core and sheath voltages V_k and V_k' and currents I_k and I_k' at the kth cross-bonded node are related as in the following equations:

$$\left(V_k'\right) = [R](V_k)$$
$$\left(I_k'\right) = [R](I_k) \tag{3.3}$$

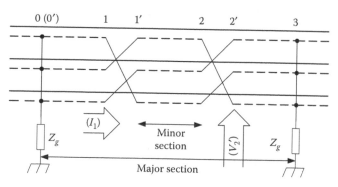

FIGURE 3.6
Major section of a cross-bonded cable.

$$(V_k) = \begin{pmatrix} (V_{kc}) \\ (V_{ks}) \end{pmatrix}, \quad (V_{kc}) = \begin{pmatrix} V_{kca} \\ V_{kcb} \\ V_{kcc} \end{pmatrix}, \quad (V_{ks}) = \begin{pmatrix} V_{ksa} \\ V_{ksb} \\ V_{ksc} \end{pmatrix} \qquad (3.4)$$

The second subscripts c and s denote the core and sheath, respectively, and the third subscripts, a, b, and c, express the phases. The other voltage and current vectors (V_k'), (I_k), and (I_k') have the same form as (V_k).

The sheath-sectionalizing joint is mathematically expressed by a rotation matrix $[R]$:

$$[R] = \begin{bmatrix} [U] & [0] \\ [0] & [R_{33}] \end{bmatrix}$$

$$[R_{33}] = \begin{bmatrix} 0 & 0 & 1 \\ 1 & 0 & 0 \\ 0 & 1 & 0 \end{bmatrix} \qquad (3.5)$$

where [0] and [U] denote a 3×3 null and unit matrix, respectively.

The rotation matrix has the following characteristics:

$$[R]^3 = [U], \quad [R]^2 = [R]_t = [R]^{-1} \qquad (3.6)$$

where the subscript t represents the transposed matrix.

When defining the voltage difference ΔV_{k-1} between nodes k–1 and k', the following equation is used:

$$(V_k) = (V_{k-1'}) + (\Delta V_k) \qquad (3.7)$$

The voltage difference between the major section ΔV (between nodes 0 and 3) is represented by

$$(\Delta V) = (V_3) - (V_0) \qquad (3.8)$$

From Equations 3.3 to 3.8, ΔV is expressed by ΔV_k ($k = 1, 2, 3$) in the following form:

$$\begin{aligned}
\Delta V &= (V_3) - (V_0) = (V_{2'}) + (\Delta V_3) - (V_0) = [R](V_2) + (\Delta V_3) - (V_0) \\
&= [R]\{(V_{1'}) + (\Delta V_2)\} + (\Delta V_3) - (V_0) = [R]\{[R](V_1) + (\Delta V_2)\} + (\Delta V_3) - (V_0) \\
&= [R]\{[R]\{(V_{0'}) + (\Delta V_1)\} + (\Delta V_2)\} + (\Delta V_3) - (V_0) \\
&= [R]^2 (V_{0'}) + [R]^2 (\Delta V_1) + [R](\Delta V_2) + (\Delta V_3) - (V_0) \qquad (3.9)
\end{aligned}$$

Voltage and current deviations are expressed by using the cable impedance [Z] and admittance [Y] in the following example:

$$(\Delta V_k) = -[Z]l_k(I_k), \quad (\Delta I_i) = -[Y]l_k(V_k) \tag{3.10}$$

where l_k is the length of the kth minor section.

The voltage difference between the terminals of the major section (ΔV) gives an equivalent impedance of a cross-bonded cable. It is obtained (3.13) by applying the following relations:

$$(I_{k-1'}) = (I_k) = [R](I_{k-1}) \tag{3.11}$$

$$(V_0) = [R]^2 (V_{0'}), \quad \because V_{0sa} = V_{0sb} = V_{0sc} = V_{0'sa} = V_{0'sb} = V_{0'sc} \tag{3.12}$$

$$
\begin{aligned}
(\Delta V) &= -[R]^2[Z]l_1(I_1) - [R][Z]l_2(I_2) - [Z]l_3(I_3) \\
&= -[R]^2[Z]l_1[R](I_{0'}) - [R][Z]l_2[R]^2(I_{0'}) - [Z]l_3[R]^3(I_{0'}) \\
&= -\left\{ [R]^2[Z]l_1[R] + [R][Z]l_2[R]^2 + [Z]l_3 \right\}(I_{0'})
\end{aligned} \tag{3.13}
$$

If the lengths of the minor sections are identical ($l_k = l$), an equivalent series impedance [Z'] can be obtained:

$$
\begin{aligned}
(\Delta V) &= -[Z']3l(I_{0'}) \\
[Z'] &= \frac{1}{3}([R]^2[Z][R] + [R][Z][R]^2 + [Z]) \\
&= \frac{1}{3}([R]_t[Z][R] + [R][Z][R]_t + [Z]) \\
&= \begin{bmatrix} [Z'_{cc}] & [Z'_{cs}] \\ [Z'_{cs}]_t & [Z'_{ss}] \end{bmatrix}
\end{aligned} \tag{3.14}
$$

The physical meaning of this equation can be explained using the following calculation:

$$
\begin{aligned}
[R]_t[Z][R] &= \begin{bmatrix} [U] & [0] \\ [0] & [R_{33}] \end{bmatrix}_t \begin{bmatrix} [Z_{cc}] & [Z_{cs}] \\ [Z_{cs}]_t & [Z_{ss}] \end{bmatrix} \begin{bmatrix} [U] & [0] \\ [0] & [R_{33}] \end{bmatrix} \\
&= \begin{bmatrix} [Z_{cc}] & [Z_{cs}] \\ [R_{33}]_t[Z_{cs}]_t & [R_{33}]_t[Z_{ss}] \end{bmatrix} \begin{bmatrix} [U] & [0] \\ [0] & [R_{33}] \end{bmatrix} \\
&= \begin{bmatrix} [Z_{cc}] & [Z_{cs}][R_{33}] \\ [R_{33}]_t[Z_{cs}]_t & [R_{33}]_t[Z_{ss}][R_{33}] \end{bmatrix}
\end{aligned} \tag{3.15}
$$

$$[R][Z][R]_t = \begin{bmatrix} [Z_{cc}] & [Z_{cs}][R_{33}]_t \\ [R_{33}][Z_{cs}]_t & [R_{33}][Z_{ss}][R_{33}]_t \end{bmatrix} \qquad (3.16)$$

where $[Z_{cc}]$, $[Z_{cs}]$, and $[Z_{ss}]$ are the submatrices of the cable impedance matrix. The submatrix for the cores $[Z_{cc}]$ remains unchanged based on the operation shown in Equation 3.14.

$$[Z'_{cc}] = [Z_{cc}] \qquad (3.17)$$

Equations 3.18 and 3.19 show that the operation to the submatrix for the mutual impedance between the cores and sheaths $[Z_{cs}]$ is averaging within the rows:

$$[Z'_{cs}] = \frac{1}{3}([Z_{cs}][R_{33}] + [Z_{cs}][R_{33}]_t + [Z_{cs}]) = \frac{1}{3}[Z_{cs}]([R_{33}] + [R_{33}]_t + [U]) \quad (3.18)$$

$$\frac{1}{3}([R_{33}] + [R_{33}]_t + [U]) = \frac{1}{3}\begin{bmatrix} 1 & 1 & 1 \\ 1 & 1 & 1 \\ 1 & 1 & 1 \end{bmatrix} \qquad (3.19)$$

The diagonal and mutual element (Z'_{ss} and Z'_{ssm}) of the submatrix for sheaths $[Z_{ss}]$ is the mean of the self- and mutual impedance of the sheath. The shape of the matrix is identical to that of a transposed overhead line:

$$\frac{1}{3}[R_{33}]_t[Z_{ss}][R_{33}] = \frac{1}{3}\begin{bmatrix} 0 & 1 & 0 \\ 0 & 0 & 1 \\ 1 & 0 & 0 \end{bmatrix}\begin{bmatrix} Z_{ss11} & Z_{ss12} & Z_{ss13} \\ Z_{ss12} & Z_{ss22} & Z_{ss23} \\ Z_{ss13} & Z_{ss23} & Z_{ss33} \end{bmatrix}\begin{bmatrix} 0 & 0 & 1 \\ 1 & 0 & 0 \\ 0 & 1 & 0 \end{bmatrix}$$

$$= \frac{1}{3}\begin{bmatrix} Z_{ss12} & Z_{ss22} & Z_{ss23} \\ Z_{ss13} & Z_{ss23} & Z_{ss33} \\ Z_{ss11} & Z_{ss12} & Z_{ss13} \end{bmatrix}\begin{bmatrix} 0 & 0 & 1 \\ 1 & 0 & 0 \\ 0 & 1 & 0 \end{bmatrix} = \frac{1}{3}\begin{bmatrix} Z_{ss22} & Z_{ss23} & Z_{ss12} \\ Z_{ss23} & Z_{ss33} & Z_{ss13} \\ Z_{ss12} & Z_{ss13} & Z_{ss11} \end{bmatrix}$$

$$(3.20)$$

$$\frac{1}{3}[R_{33}][Z_{ss}][R_{33}]_t = \frac{1}{3}\begin{bmatrix} 0 & 0 & 1 \\ 1 & 0 & 0 \\ 0 & 1 & 0 \end{bmatrix}\begin{bmatrix} Z_{ss11} & Z_{ss12} & Z_{ss13} \\ Z_{ss12} & Z_{ss22} & Z_{ss23} \\ Z_{ss13} & Z_{ss23} & Z_{ss33} \end{bmatrix}\begin{bmatrix} 0 & 1 & 0 \\ 0 & 0 & 1 \\ 1 & 0 & 0 \end{bmatrix}$$

$$= \frac{1}{3}\begin{bmatrix} Z_{ss13} & Z_{ss23} & Z_{ss33} \\ Z_{ss11} & Z_{ss12} & Z_{ss13} \\ Z_{ss12} & Z_{ss22} & Z_{ss23} \end{bmatrix}\begin{bmatrix} 0 & 1 & 0 \\ 0 & 0 & 1 \\ 1 & 0 & 0 \end{bmatrix} = \frac{1}{3}\begin{bmatrix} Z_{ss33} & Z_{ss13} & Z_{ss23} \\ Z_{ss13} & Z_{ss11} & Z_{ss12} \\ Z_{ss23} & Z_{ss12} & Z_{ss22} \end{bmatrix}$$

$$(3.21)$$

$$[Z'_{ss}] = \frac{1}{3}([R_{33}]_t[Z_{ss}][R_{33}] + [R_{33}][Z_{ss}][R_{33}]_t + [Z_{ss}]) = \begin{bmatrix} Z'_{sss} & Z'_{ssm} & Z'_{ssm} \\ Z'_{ssm} & Z'_{sss} & Z'_{ssm} \\ Z'_{ssm} & Z'_{ssm} & Z'_{sss} \end{bmatrix}$$

$$[Z'_{sss}] = \frac{1}{3}\sum_{i=1}^{3} Z_{ssii}$$

(3.22)

$$[Z'_{ssm}] = \frac{1}{3}\sum_{i=1}^{2}\sum_{j=i+1}^{3} Z_{ssij}$$

In this same manner, the equivalent admittance of a cross-bonded cable can be obtained from the current difference (ΔI):

$$(\Delta I) = -[Y']3l(V_0)$$

$$[Y'] = [R]^2[Y][R] + [R][Y][R]^2 + [Y] = [R]_t[Y][R] + [R][Y][R]_t + [Y]$$

(3.23)

The admittance matrix of the cable can be expressed as follows:

$$[Y] = \begin{bmatrix} [Y_{cc}] & -[Y_{cs}] \\ -[Y_{cs}] & [Y_{ss}] \end{bmatrix} = \begin{bmatrix} [Y_{cc}] & -[Y_{cc}] \\ -[Y_{cc}] & [Y_{ss}] \end{bmatrix}$$

(3.24)

$$[Y_{cc}] = j\omega \begin{bmatrix} C_{c1} & 0 & 0 \\ 0 & C_{c2} & 0 \\ 0 & 0 & C_{c3} \end{bmatrix}$$

(3.25)

$$[Y_{ss}] = j\omega \begin{bmatrix} C_{ss1} + C_{sm12} + C_{sm13} & -C_{sm12} & -C_{sm13} \\ -C_{sm12} & C_{ss2} + C_{sm12} + C_{sm23} & -C_{sm23} \\ -C_{sm13} & -C_{sm23} & C_{ss3} + C_{sm23} + C_{sm13} \end{bmatrix}$$

(3.26)

The core admittance submatrix $[Y_{cc}]$ is a diagonal matrix determined by the capacitances between the cores and the sheath, because a sheath encloses a core. The admittance submatrix of the cores for the cross-bonded cable is identical to the solidly bonded cable:

$$[Y'_{cc}] = [Y_{cc}]$$

(3.27)

Equations 3.18 and 3.19 show that the operation sheaths to the submatrix for the mutual admittance between the cores and sheaths $[Y_{cs}]$ is averaging within the rows:

$$[Y'_{cs}] = \frac{1}{3}([Y_{cs}][R_{33}] + [Y_{cs}][R_{33}]_t + [Y_{cs}]) = \frac{1}{3}[Y_{cs}]([R_{33}] + [R_{33}]_t + [U])$$

$$= j\omega\frac{1}{3}\begin{bmatrix} C_{c1} & C_{c1} & C_{c1} \\ C_{c2} & C_{c2} & C_{c2} \\ C_{c3} & C_{c3} & C_{c3} \end{bmatrix} \tag{3.28}$$

The sheaths diagonal and mutual element (Z'_{sss} and Z'_{ssm}) of the submatrix for the sheaths $[Z_{ss}]$ is the mean of the self- and mutual impedances of the sheath:

$$[Y'_{ss}] = \frac{1}{3}([R_{33}]_t[Y_{ss}][R_{33}] + [R_{33}[Y_{ss}][R_{33}]_t + [Y_{ss}]]) = \begin{bmatrix} Y'_{sss} & Y'_{ssm} & Y'_{ssm} \\ Y'_{ssm} & Y'_{sss} & Y'_{ssm} \\ Y'_{ssm} & Y'_{ssm} & Y'_{sss} \end{bmatrix}$$

$$Y'_{sss} = \frac{1}{3}\sum_{i=1}^{3} Y_{ssii} \tag{3.29}$$

$$Y'_{ssm} = \frac{1}{3}\sum_{i=1}^{2}\sum_{j=i+1}^{3} Y_{ssij}$$

3.2.3.2 Reduction of the Sheath

The lengths of the minor sections can have imbalances due to constraints on the locations of joints. The imbalances are designed to be as small as possible, since they increase sheath currents and raise sheath voltages. When a cable system has multiple major sections, the overall balance is considered for minimizing sheath currents. As a result, when a cable system has more than two major sections, sheath currents are generally balanced among three conductors, which allows for the reduction from three metallic sheaths to one conductor [4,5].

Since three-phase sheath conductors are short-circuited and grounded in every major section (as illustrated in Figure 3.6), the sheath voltages of the three phases are equal at each earthing joint. Assuming that the sheath currents are balanced among the three conductors, the sheath currents do not flow into the earth at each earthing joint:

$$\begin{aligned} V_{1sa} &= V_{1sb} = V_{1sc} \equiv V_{1s} \\ V_{4sa} &= V_{4sb} = V_{4sc} \equiv V_{4s} \\ I_{1sa} + I_{1sb} + I_{1sc} &\equiv I_{1s} \\ I_{4sa} + I_{4sb} + I_{4sc} &\equiv I_{4s} \end{aligned} \tag{3.30}$$

By applying connection matrix $[T]$, this equation can be rewritten as

$$(V_k) = [T]_t(V''_k), \quad (I''_k) = [T](I_k), \quad k = 0,3 \tag{3.31}$$

where

$$[T] = \begin{bmatrix} 1 & 0 & 0 & 0 & 0 & 0 \\ 0 & 1 & 0 & 0 & 0 & 0 \\ 0 & 0 & 1 & 0 & 0 & 0 \\ 0 & 0 & 0 & 1 & 1 & 1 \end{bmatrix} = \begin{bmatrix} [U] & [0] \\ (0)_t & (1)_t \end{bmatrix} \tag{3.32}$$

$$(0)_t = (0 \quad 0 \quad 0), \quad (1)_t = (1 \quad 1 \quad 1)$$

$$(V_k'') = \begin{pmatrix} (V_{kc}) \\ V_{ks} \end{pmatrix}, \quad (I_k'') = \begin{pmatrix} (I_{kc}) \\ I_{ks} \end{pmatrix} \tag{3.33}$$

From Equations 3.14, 3.23, and 3.31, the following relation is obtained:

$$\begin{aligned} (\Delta V'') &= -[Z'']3l(I_0'') \\ (\Delta I'') &= -[Y'']3l(V_0'') \end{aligned} \tag{3.34}$$

where

$$\begin{aligned} [Z''] &= \left([T][Z']^{-1}[T]_t \right)^{-1} \\ [Y''] &= [T][Y'][T]_t \end{aligned} \tag{3.35}$$

The impedance matrix $[Z'']$ and the admittance matrix $[Y'']$ are 4×4 matrices composed of three cores and a reduced single sheath.

Because the propagation mode of the cable can be expressed by both coaxial- and sheath-propagation mode, in a high-frequency region where skin depth is smaller than sheath thickness, the impedance matrix is composed of the following two submatrices:

$$[Z_{cc}] = [Z_{cd}] + [Z_{ss}] \tag{3.36}$$

$$[Z_{cs}] = [Z_{ss}] \tag{3.37}$$

where

$$[Z_{cd}] = \begin{bmatrix} Z_{cd} & 0 & 0 \\ 0 & Z_{cd} & 0 \\ 0 & 0 & Z_{cd} \end{bmatrix} \tag{3.38}$$

The reduced impedance in a high-frequency region becomes:

$$[Z''] = \left([T][Z']^{-1}[T]_t \right)^{-1} = \begin{bmatrix} [Z''_{cc}] & (Z''_{cs}) \\ (Z''_{cs})_t & Z''_{ss} \end{bmatrix} \tag{3.39}$$

The core impedances, including their mutual impedances, are identical to the original impedances:

$$[Z''_{cc}] = [Z'_{cc}] = [Z_{cc}] = [Z_{cd}] + [Z_{ss}] \tag{3.40}$$

The mutual impedance between the kth core and the reduced sheath Z''_{kj} is the average of the impedances between the core and the three-phase sheaths:

$$(Z''_{cs}) = \begin{pmatrix} Z''_{14} \\ Z''_{24} \\ Z''_{34} \end{pmatrix} \tag{3.41}$$

$$Z''_{k4} = \frac{1}{3}\sum_{j=1}^{3} Z_{sskj}$$

Finally, the sheath impedance is the average of all the elements of the original sheath impedance matrix:

$$Z''_{ss} = \frac{1}{9}\sum_{i=1}^{3}\sum_{j=1}^{3} Z_{ssij} \tag{3.42}$$

In the same manner, the reduced admittance matrix becomes:

$$[Y''_{cc}] = [Y'_{cc}] = [Y_{cc}] \tag{3.43}$$

$$(Y''_{cs}) = \begin{pmatrix} Y''_{14} \\ Y''_{24} \\ Y''_{34} \end{pmatrix} \tag{3.44}$$

$$Y''_{k4} = \sum_{j=1}^{3} Y_{cskj} = -j\omega C_{ck}$$

$$Y''_{ss} = \frac{1}{9}\sum_{i=1}^{3}\sum_{j=1}^{3} Y_{ssij} = j\omega\sum_{i=1}^{3} C_{ssii} \tag{3.45}$$

$$[Y''] = [T][Y'][T]_t = \begin{bmatrix} [Y''_{cc}] & (Y''_{cs}) \\ (Y''_{cs})_t & Y''_{ss} \end{bmatrix} \tag{3.46}$$

3.2.4 Theoretical Formula of Sequence Currents

The sequence impedance/current calculation of overhead lines is well-known and introduced in textbooks [2]. For underground cables, theoretical formulas are proposed for the cable itself [6–9]. However, in order to derive accurate theoretical formulas, it is necessary to consider the whole cable system, including sheath bonding, because the return current of an underground cable flows through both the metallic sheath and the ground. Until now, no formula existed for sequence impedances or currents that consider sheath bonding and sheath-grounding resistances at substations and earthing joints. As a result, it became a common practice to measure these sequence impedances or currents after installation, as it is considered difficult to predict these values in advance.

As mentioned earlier, it is a common practice for underground cable systems that are longer than approximately 2 km to cross-bond the metallic sheaths of three-phase cables to simultaneously reduce sheath currents and suppress sheath voltages [3]. Submarine cables, which are generally bonded solidly, are now gaining popularity due to the increase in offshore wind farms and cross-border transactions.

In this section, we derive theoretical formulas of the sequence currents for the majority of underground cable systems: that is, a cross-bonded cable that has more than two major sections. We also derive theoretical formulas for a solidly bonded cable considering the increased use of submarine cables.

3.2.4.1 Cross-Bonded Cable

3.2.4.1.1 Impedance Matrix

One cable system corresponds to six conductor systems composed of three cores and three metallic sheaths. As in the last section, the 6×6 impedance matrix of the cable system is represented by the following equation [1]:

$$[Z] = \begin{bmatrix} [Z_{cc}] & [Z_{cs}] \\ [Z_{cs}]^t & [Z_{ss}] \end{bmatrix} = \begin{bmatrix} [Z_{cc}] & [Z_{cs}] \\ [Z_{cs}] & [Z_{ss}] \end{bmatrix}$$

$$[Z_{cc}] = \begin{bmatrix} Z_{cc11} & Z_{cc12} & Z_{cc13} \\ Z_{cc12} & Z_{cc22} & Z_{cc23} \\ Z_{cc13} & Z_{cc23} & Z_{cc33} \end{bmatrix}, \quad [Z_{ss}] = \begin{bmatrix} Z_{ss11} & Z_{ss12} & Z_{ss13} \\ Z_{ss12} & Z_{ss22} & Z_{ss23} \\ Z_{ss13} & Z_{ss23} & Z_{ss33} \end{bmatrix}, \quad (3.47)$$

$$[Z_{cs}] = \begin{bmatrix} Z_{cs11} & Z_{cs12} & Z_{cs13} \\ Z_{cs12} & Z_{cs22} & Z_{cs23} \\ Z_{cs13} & Z_{cs23} & Z_{cs33} \end{bmatrix}$$

where
   c is the core
   s is the sheath
   t is the transpose

In Equation 3.47, cable phase a is assumed to be laid symmetrically to phase c and against phase b. The flat configuration and the trefoil configuration, which are typically adopted, satisfy this assumption.

By reducing the sheath conductors, the six-conductor system is reduced to a four-conductor system composed of three cores and one equivalent metallic sheath, as shown in Figure 3.7. The 4×4 reduced impedance matrix can be expressed as

$$
[Z''] =
\begin{bmatrix}
Z_{cc11} & Z_{cc12} & Z_{cc13} & Z''_{14} \\
Z_{cc12} & Z_{cc22} & Z_{cc23} & Z''_{24} \\
Z_{cc13} & Z_{cc23} & Z_{cc33} & Z''_{14} \\
Z''_{14} & Z''_{24} & Z''_{14} & Z''_{ss}
\end{bmatrix}
\tag{3.48}
$$

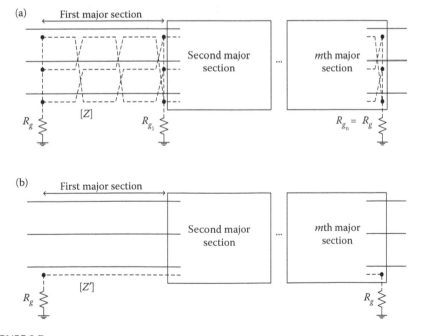

FIGURE 3.7
Cross-bonded cable and its equivalent model: (a) a cross-bonded cable system with m-major sections and (b) an equivalent four-conductor system.

Here, $Z''(4, j) = Z''(j, 4)$ can be calculated from the 6×6 impedance matrix Z shown in Equation 3.41. $Z''_{14} = Z''_{34}$ stands in the flat configuration and the trefoil configuration.

3.2.4.1.2 Zero-Sequence Current

The following equations are derived from Figure 3.8. Here, sheath grounding at earthing joints is ignored, but sheath grounding at substations can be considered through V_s:

$$(V_1) = [Z'](I_1) \tag{3.49}$$

where

$$(V_1) = \begin{pmatrix} E & E & E & V_s \end{pmatrix}^t$$

$$(I_1) = \begin{pmatrix} I_a & I_b & I_c & I_s \end{pmatrix}^t$$

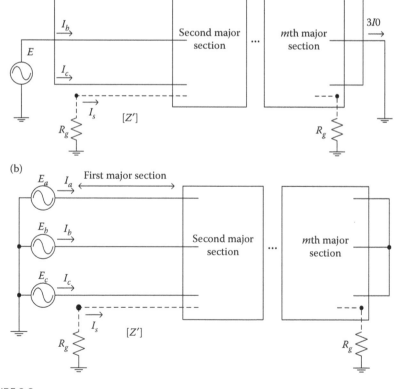

FIGURE 3.8
Setup for measuring sequence currents for a cross-bonded cable: (a) zero-sequence current and (b) positive-sequence current.

Figure 3.8a shows the setup for measuring the zero-sequence current for a cross-bonded cable.

Assuming the grounding resistance at substations R_g, the sheath voltage V_s can be obtained by:

$$V_s = -2R_g I_s \tag{3.50}$$

The following equations can be obtained by solving Equations 3.49 and 3.50:

$$
\begin{aligned}
I_a = I_c &= \frac{(Z_{22} - Z_{12})E}{\Delta_0} \\
I_b &= \frac{(Z_{11} - Z_{21})E}{\Delta_0}
\end{aligned}
\tag{3.51}
$$

where

$$
\begin{aligned}
\Delta_0 &= Z_{11}Z_{22} - Z_{12}Z_{21} \\
Z_{11} &= Z_{cc11} + Z_{cc13} - \frac{2Z_{14}''^2}{Z_{SR}''} \\
Z_{22} &= Z_{cc22} - \frac{Z_{24}''^2}{Z_{SR}''} \\
Z_{12} &= Z_{cc12} - \frac{Z_{14}''Z_{24}''}{Z_{SR}''}, \quad Z_{21} = 2Z_{12} \\
Z_{SR}'' &= Z_{ss}'' + 2R_g
\end{aligned}
$$

The zero-sequence current can be obtained from Equation 3.51 as follows:

$$I_0 = \frac{2I_a + I_b}{3} = \frac{E}{3\Delta_0}(Z_{11} + 2Z_{22} - 2Z_{12} - Z_{21}) \tag{3.52}$$

When three-phase cables are laid symmetrically to each other, the following equations are satisfied:

$$
\begin{aligned}
Z_{cc11} &= Z_{cc22} = Z_c, & Z_{ss11} &= Z_{ss22} = Z_s \\
Z_{cc12} &= Z_{cc13} = Z_m, & Z_{14}'' &= Z_{24}'' = Z_n
\end{aligned}
\tag{3.53}
$$

Using symmetrical impedances Z_c, Z_m, and Z_n in Equation 3.53, Z_{11}, Z_{22}, and Z_{12} can be expressed as

$$
\begin{aligned}
Z_{11} &= Z_c + Z_m - \frac{2Z_n^2}{Z_{SR}''} \\
Z_{22} &= Z_c - \frac{Z_n^2}{Z_{SR}''} \\
Z_{12} &= Z_m - \frac{Z_n^2}{Z_{SR}''}
\end{aligned}
\tag{3.54}
$$

Substituting Z_{11}, Z_{22}, and Z_{12} in Equations 3.51 and 3.52 by the symmetrical impedances will result in:

$$I_a = I_b = I_c \approx \frac{E}{\Delta 1}, \quad I_s \approx \frac{-3Z_n E}{Z''_{SR}\Delta_1}$$

$$I_b \approx \frac{E}{\Delta_1}$$

(3.55)

where $\Delta_1 = Z_c + 2Z_m - 3Z_n^2 / Z''_{SR}$.

3.2.4.1.3 Positive-Sequence Current

In Figure 3.8b, the equation $I_{sa} + I_{sb} + I_{sc} = 0$ is satisfied at the end of the cable line. The following equations are obtained since $V_s = 0$:

$$(V_1) = [E \quad \alpha^2 E \quad \alpha E \quad 0]^t$$

$$(I_1) = [I_a \quad I_b \quad I_c \quad I_s]^t$$

(3.56)

where $\alpha = \exp(j2\pi/3)$.

Solving Equation 3.56 for I_a, I_b, and I_c yields results in the following:

$$\begin{pmatrix} E \\ \alpha^2 E \\ \alpha E \end{pmatrix} = \begin{bmatrix} Z_{11} & Z_{12} & Z_{13} \\ Z_{12} & Z_{22} & Z_{12} \\ Z_{13} & Z_{12} & Z_{11} \end{bmatrix} \begin{pmatrix} I_a \\ I_b \\ I_c \end{pmatrix}$$

$$\therefore \begin{pmatrix} I_a \\ I_b \\ I_c \end{pmatrix} = \begin{bmatrix} Z_{11} & Z_{12} & Z_{13} \\ Z_{12} & Z_{22} & Z_{12} \\ Z_{13} & Z_{12} & Z_{11} \end{bmatrix}^{-1} \begin{pmatrix} E \\ \alpha^2 E \\ \alpha E \end{pmatrix}$$

$$= \frac{1}{\Delta} \begin{bmatrix} Z_{11}Z_{22} - Z_{12}^2 & Z_{12}(Z_{13} - Z_{11}) & Z_{12}^2 - Z_{13}Z_{22} \\ Z_{12}(Z_{13} - Z_{11}) & Z_{11}^2 - Z_{13}^2 & Z_{12}(Z_{13} - Z_{11}) \\ Z_{12}^2 - Z_{13}Z_{22} & Z_{12}(Z_{13} - Z_{11}) & Z_{11}Z_{22} - Z_{12}^2 \end{bmatrix} \begin{pmatrix} E \\ \alpha^2 E \\ \alpha E \end{pmatrix}$$

(3.57)

Here,

$$Z_{11} = Z_{cc11} - \frac{Z''^2_{14}}{Z''_{ss}}$$

$$Z_{22} = Z_{cc22} - \frac{Z''^2_{24}}{Z''_{ss}}$$

$$Z_{12} = Z_{cc12} - \frac{Z''_{14} Z''_{24}}{Z''_{ss}}$$

$$Z_{13} = Z_{cc12} - \frac{Z''^2_{14}}{Z''_{ss}}$$

The positive-sequence current is derived from Equation 3.57:

$$I_1 = \frac{1}{3}\left(I_a + \alpha I_b + \alpha^2 I_c\right) = \frac{E}{3\Delta_2}\{(Z_{11} - Z_{13})(Z_{11} + Z_{13} + 2Z_{12})$$
$$+ Z_{22}(2Z_{11} + Z_{13}) - 3Z_{12}^2\} \tag{3.58}$$

where $\Delta_2 = (Z_{11} - Z_{13})\{Z_{22}(Z_{11} + Z_{13}) - 2Z_{12}^2\}$.

When three-phase cables are laid symmetrically to each other, Equation 3.58 can be further simplified using Equation 3.53:

$$I_1 = \frac{E}{Z_c - Z_m} \tag{3.59}$$

3.2.4.2 Solidly Bonded Cable

3.2.4.2.1 Impedance Matrix

Figure 3.9 shows a sequence current measurement circuit for a solidly bonded cable. The following equations are given from the 6 × 6 impedance matrix in Equation 3.47 and Figure 3.9:

$$(E) = [Z_{cc}](I) + [Z_{cs}](I_s) \tag{3.60}$$

$$(V_s) = [Z_{cs}](I) + [Z_{ss}](I_s) = -2[R_g](I_s) \tag{3.61}$$

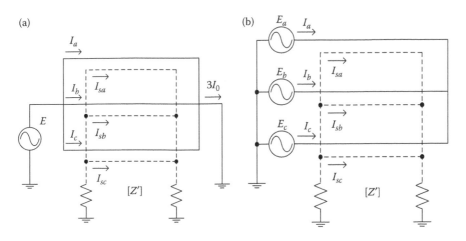

FIGURE 3.9
Setup for measuring sequence currents for a solidly bonded cable: (a) zero-sequence current and (b) positive-sequence current.

Here, $(I) = (I_a \, I_b \, I_a)^t$ is the core current and $(I_s) = (I_{sa} \, I_{sb} \, I_{sa})^t$ is the sheath current:

$$[R_g] = R_g \begin{bmatrix} 1 & 1 & 1 \\ 1 & 1 & 1 \\ 1 & 1 & 1 \end{bmatrix}$$

From Equation 3.61, sheath current (I_s) is found using:

$$(I_s) = -([Z_{ss}] + 2[R_g])^{-1} [Z_{cs}](I) \tag{3.62}$$

After eliminating the sheath current (I_s) in Equation 3.60, core current (I) can be derived as

$$(I) = ([Z_{cc}] - [Z_{cs}]([Z_{ss}] + 2[R_g])^{-1}[Z_{cs}])^{-1}(E) \tag{3.63}$$

3.2.4.2.2 Zero-Sequence Current

From Figure 3.9a, (E) and (I) are expressed as

$$(E) = [E \quad E \quad E]^t, \quad (I) = [I_a \quad I_b \quad I_a]^t \tag{3.64}$$

The core current (I) is obtained from Equations 3.63 and 3.64; and the zero-sequence current is calculated as $I_0 = (I_a + I_b + I_c)/3$.

Since the relationship $[Z_{cs}] \approx [Z_{ss}]$ generally stands, Equations 3.60 and 3.61 can be simplified to Equation 3.65 using Equation 3.53:

$$((E) - (V_s)) = ([Z_{cc}] - [Z_{ss}])(I) = (Z_c - Z_s)[U](I) \tag{3.65}$$

where $[U]$ is the 3×3 unit (identity) matrix.

Hence,

$$I_0 = I_a = I_b = I_c = \frac{1}{Z_c - Z_s}(E - V_s) \tag{3.66}$$

Using Equation 3.66, the core current (I) in Equation 3.61 can be eliminated, which yields

$$(V_s) = \frac{1}{Z_c - Z_s}[Z_{cs}]((E) - (V_s)) - [Z_{cs}](I_s) \tag{3.67}$$

After adding all three rows in Equation 3.67, the result is

$$3V_s = 3\frac{Z_s + 2Z_m}{Z_c - Z_s}(E - V_s) - \frac{Z_s + 2Z_m}{2R_g}V_s \tag{3.68}$$

After solving Equation 3.68 for V_s and then eliminating V_s from Equation 3.66, the zero-sequence current becomes

$$I_0 = \frac{6R_g + Z_s + 2Z_m}{6R_g(Z_c + 2Z_m) + (Z_c - Z_s)(Z_s + 2Z_m)}E \tag{3.69}$$

3.2.4.2.3 Positive-Sequence Current

From Figure 3.9b, (E) and (I) are expressed as

$$(E) = [E \quad \alpha^2 E \quad \alpha E]^t, \quad (I) = [I_a \quad I_b \quad I_c]^t \tag{3.70}$$

The core current (I) is obtained from Equations 3.63 and 3.70. Once the core current is solved for, the positive-sequence current can be calculated as

$$I_1 = (I_a + \alpha I_b + \alpha^2 I_c)/3.$$

Using Equation 3.65, the theoretical formula of the positive-sequence current simplifies to

$$I_1 = \frac{1}{3(Z_c - Z_s)}\left\{(E - V_s) + \alpha\left(\alpha^2 E - V_s\right) + \alpha^2(\alpha E - V_s)\right\} = \frac{E}{Z_c - Z_s} \tag{3.71}$$

Equation 3.71 shows that the positive-sequence current can be approximated by the coaxial mode current. It also shows that, in a manner similar to that of a cross-bonded cable, the positive-sequence current remains unaffected by the substation-grounding resistance R_g.

3.2.4.2.4 Example

Figure 3.10 shows the physical and electrical data of a 400-kV cable used for comparison. An existence of semiconducting layers introduces an error in the charging capacity of the cable. The relative permittivity of the insulation (XLPE) is converted from Equation 2.4 to 2.7, according to Equation 3.72, in order to correct the error and achieve a reasonable cable model [10]:

$$\varepsilon_r' = \frac{\ln(R_3/R_2)}{\ln(R_{so}/R_{si})}\varepsilon_r = \frac{\ln(61.40/32.60)}{\ln(59.50/34.10)} \cdot 2.4 = 2.729 \tag{3.72}$$

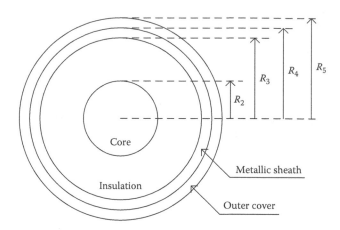

Core inner radius: 0.0 cm, $R_2 = 3.26$ cm, $R_3 = 6.14$ cm, $R_4 = 6.26$ cm, and $R_5 = 6.73$ cm
Core resistivity: 1.724×10^{-8} Ω m, metallic sheath resistivity: 2.840×10^{-8} Ω m,
and relative permittivity (XLPE, PE): 2.4

FIGURE 3.10
Physical and electrical data of the cable.

where
R_{si} is the inner radius of the insulation
R_{so} is the outer radius of the insulation

The total length of the cable is assumed to be 12 km. Figure 3.11 shows the layout of the cables. It is assumed that the cables are directly buried at a depth of 1.3 m with a separation of 0.5 m between the phases. Earth resistivity is set to 100 Ω m.

The calculation process in the case of a cross-bonded cable using the proposed formulas is shown as follows (the 6 × 6 impedance matrix Z is obtained using cable constants [1,11,12]):

$[Z']$ (upper: R, lower: X, unit: Ω)

0.71646353	0.59136589	0.59136589	0.5913705
8.44986724	6.28523815	5.76261762	6.63566685
0.59136589	0.71646353	0.59136589	0.5913705
6.28523815	8.44986724	6.28523815	6.80987369
0.59136589	0.59136589	0.71646353	0.5913705
5.76261762	6.28523815	8.44986724	6.63566685
0.5913705	0.5913705	0.5913705	0.83438185
6.63566685	6.80987369	6.63566685	6.63485268

Zero-sequence current
$\Delta_0 = -3.9605900 + j12.489448$
$Z_{11} = 4.0641578 + j2.2224336$

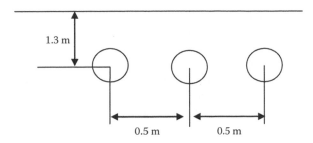

FIGURE 3.11
Layout of the cable.

$$Z_{22} = 2.1959039 + j2.1450759$$
$$Z_{12} = 2.0193420 + j0.1372637$$
$$Z_{21} = 4.0386840 + j0.2745275$$
$$I_0 \text{ (rms)} = 81.814700 - j31.778479$$

Positive-sequence current

$$\Delta_2 = -4.8574998 - j1.8394591$$
$$Z_{11} = 0.3670612 + j1.8221556$$
$$Z_{22} = 0.3801952 + j1.4707768$$
$$Z_{12} = 0.2482475 - j0.5159115$$
$$Z_{13} = 0.2419636 - j0.8650940$$
$$I_1 \text{ (rms)} = 14.118637 - j251.86277$$

Table 3.1 shows zero- and positive-sequence currents derived by the proposed formulas. Grounding resistances at substations are assumed to be 1 Ω. In the calculations, the applied voltage is set to $E = 1\text{kV}/\sqrt{3}$ (angle: 0°) and the source impedance is not considered. Here, sequence currents are determined according to the setups for measuring sequence currents shown in

TABLE 3.1

Comparison of Proposed Formulas with EMTP Simulations

	Zero Sequence		Positive Sequence	
	Amplitude (A)	Angle (deg.)	Amplitude (A)	Angle (deg.)
a. Cross-bonded cable				
Proposed formulas, Equations 3.8/3.14	124.1	−21.23	356.7	−86.79
b. Solidly bonded cable				
Proposed formulas, Equations 3.19, 3.20/3.26	124.8	−22.50	722.7	−49.08

Figures 3.8 and 3.9. The assumptions regarding the applied voltage and the source impedance match the condition in the actual setups for measuring sequence currents, since testing sets are generally used in such measurements.

The proposed formulas are known to have a satisfactory accuracy for planning and implementation studies. An acceptable level of error is introduced by the impedance matrix reduction discussed earlier. Owing to the matrix reduction, unbalanced sheath currents that flow into the earth at earthing joints are not considered in the proposed formulas.

Table 3.1 shows that the positive-sequence impedance is smaller for a solidly bonded cable than for a cross-bonded cable, and the positive-sequence current is larger for a solidly bonded cable. Because of this size differential, the return current flows only through the metallic sheath of the same cable and earth in the solidly bonded cable, whereas the return current flows through the metallic sheath of all three-phase cables in a cross-bonded cable $(Z_c - Z_m > Z_c - Z_s)$.

The impedance calculation in IEC 60909-2 assumes solid bonding. As a result, if the positive-sequence impedance of a cross-bonded cable is derived based on IEC 60909-2, it might be smaller than the actual positive-sequence impedance.

The phase angle of the zero-sequence current mentioned in Table 3.1 demonstrates that grounding resistance at substations in both cross-bonded and solidly bonded cables significantly affects the zero-sequence current. As a result, there is little difference in the zero-sequence impedance of the cross-bonded cable and the solidly bonded cable. The results indicate the importance of obtaining an accurate grounding resistance at the substations to derive accurate zero-sequence impedances of cable systems.

3.3 Wave Propagation and Overvoltages

3.3.1 Single-Phase Cable

3.3.1.1 Propagation Constant

As explained in Chapter 1, the evaluation of wave propagation-related parameters necessitates eigenvalue/eigenvector calculations. Because of the coaxial structure of a cable core and metallic sheath, the propagation-related parameters show the following characteristics when in a high-frequency region [2]:

1. Impedance matrix

    In a high-frequency region, the following relation is satisfied in Equation 3.1:

    $$Z_{cs} = Z_{ss} = Z_s$$

or

$$[Z_i] = \begin{bmatrix} Z_c & Z_s \\ Z_s & Z_s \end{bmatrix}, \quad Z_c = Z_{cc} \tag{3.73}$$

2. Voltage transformation matrix

$$(v) = [A]^{-1}(V), \quad [A]^{-1} = \begin{bmatrix} 0 & 1 \\ 1 & -1 \end{bmatrix} \tag{3.74}$$

where
   (v) is the modal voltage
   (V) is the actual phase voltage

3. Modal propagation constant
   The modal propagation constant γ is given in the following equation from the actual propagation constant matrix $[\Gamma]$ explained in Chapter 1:

$$[\gamma] = [A]^{-1}[\Gamma][A] \tag{3.75}$$

where $[\Gamma]^2 = [Z][Y]$.
   Considering Equation 3.74 with Equations 3.71 and 3.73:

$$[\gamma] = \begin{bmatrix} \gamma_e & 0 \\ 0 & \gamma_c \end{bmatrix} \tag{3.76}$$

where
   $\gamma_e = Z_s(Y_s - Y_c)$ is the earth-return mode (mode 1)
   $\gamma_c = Y_c(Z_s - Z_c)$ is the coaxial mode (mode 2)

4. Characteristic impedance
   The modal characteristic impedance $[z_0]$ is given as

$$[z_0] = \begin{bmatrix} z_{0e} & 0 \\ 0 & z_{0c} \end{bmatrix} \tag{3.77}$$

where
   $z_{0e} = Z_{0s}$ is the earth-return mode (mode 1)
   $z_{0c} = Z_{0c} - Z_{0s}$ is the coaxial mode (mode 2)

The actual characteristic impedance $[Z_0]$ is obtained from these equations in the following form:

$$[Z_0] = [A][z_0][B]^{-1} = \begin{bmatrix} Z_{0c} & Z_{0s} \\ Z_{0s} & Z_{0s} \end{bmatrix} \tag{3.78}$$

where $[B]^{-1} = [A]_t$ is the current transformation matrix.

From these equations, it should be clear that the coaxial mode current flows through the core conductor and returns through the metallic sheath in a high-frequency region. In fact, a communication signal cable and a measuring cable intentionally use coaxial mode propagation for signal transmission because the propagation characteristic is entirely dependent on the insulator between the core's outer surface and the sheath's inner surface. In such a case, propagation velocity c_c and characteristic impedance Z_{0c} of the coaxial mode are evaluated approximately by:

$$c_c = \frac{c_0}{\sqrt{\varepsilon_{i1}}}, \quad Z_{0c} = \frac{60}{\sqrt{\varepsilon_{i1}}} \ln \frac{R_3}{R_2} \tag{3.79}$$

3.3.1.2 Example of Transient Analysis

Figure 3.12 illustrates a circuit diagram of a single-phase coaxial cable. In the figure, the characteristic impedance of each section is defined as follows:

$$[Z_1] = \begin{bmatrix} R_0 & 0 \\ 0 & R_s \end{bmatrix}, \quad [Z_2] = \begin{bmatrix} Z_{0c} & Z_{0s} \\ Z_{0s} & Z_{0s} \end{bmatrix}, \quad [Z_3] = \begin{bmatrix} R_c & 0 \\ 0 & R_s \end{bmatrix} \tag{3.80}$$

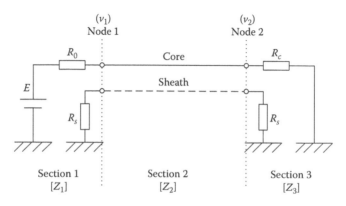

FIGURE 3.12
Circuit diagram of a single-phase coaxial cable.

where
  R_0 is the source impedance
  R_s is the sheath-grounding resistance
  R_c is the core-terminating resistance

The sheath-grounding resistance R_s generally ranges from 0.1 to 20 Ω depending on the earth resistivity and the grounding method. The source impedance is either the bus impedance or the transformer impedance. If the cable is connected to an overhead line, R_0 and R_c are the surge impedances of the overhead line. If the cable is extended beyond node 2, R_c is the core self-characteristic impedance Z_{0c} or the coaxial mode characteristic imped-ance z_{0c}.

As explained in Chapter 1, the refraction coefficient matrices at nodes 1 and 2 are given in the following forms:

$$[\lambda_{1f}] = \frac{2}{\Delta_1} \begin{bmatrix} Z_{0c}(Z_{0s} + R_s) - Z_{0s}^2 & R_0 Z_{0s} \\ R_s Z_{0s} & Z_{0s}(Z_{0c} + R_0) \end{bmatrix}$$

$$[\lambda_{2f}] = \frac{2}{\Delta_2} \begin{bmatrix} R_c(Z_{0s} + R_s) & -R_c Z_{0s} \\ -R_s Z_{0s} & R_s(Z_{0c} + R_c) \end{bmatrix} \tag{3.81}$$

where
$$\Delta_1 = (Z_{0c} + R_0)(Z_{0s} + R_s) - Z_{0s}^2$$
$$\Delta_2 = (Z_{0c} + R_c)(Z_{0s} + R_s) - Z_{0s}^2$$

Now, we consider the transient response of a coaxial cable to connect a PG to a circuit (that is, a current lead wire). Then, the following condition is given assuming that the receiving end of the core is open-circuited and the sheath is perfectly grounded:

$$R_0 = z_{0c}, \quad R_s = 0, \quad R_c = \infty \tag{3.82}$$

Going by this result, Equation 3.89 becomes

$$[\lambda_{1f}] = \begin{bmatrix} 1 & 1 \\ 0 & 2 \end{bmatrix}, \quad [\lambda_{2f}] = \begin{bmatrix} 2 & -2 \\ 0 & 0 \end{bmatrix} \tag{3.83}$$

Since the traveling wave voltage at node 1 is $E/2$ in Figure 3.12 from Thevenin's theorem (as explained in Chapter 1), the node 1 voltage (v_1) at $t = 0$ is calculated as

$$(v_1) = [\lambda_{1f}] \begin{pmatrix} \frac{E}{2} \\ 0 \end{pmatrix} = \begin{pmatrix} \frac{E}{2} \\ 0 \end{pmatrix} = (E_{12}) \quad \text{at } t = 0 \tag{3.84}$$

Traveling wave (E_{12}) from node 1 to node 2 is transformed into modal wave (e_{12}) as follows:

$$(e_{12}) = [A]^{-1}(E_{12}) = \begin{pmatrix} 0 \\ \frac{E}{2} \end{pmatrix} \tag{3.85}$$

Equation 3.85 indicates that a coaxial mode wave carries $E/2$ to the receiving end; that is, the cable works as a coaxial mode signal transfer system.

The coaxial mode wave arrives at node 2 at $t = t_a$. The wave is then transformed into an actual phase-domain wave (E_{2f}) as follows:

$$(E_{2f}) = [A](e_{12}) = \begin{bmatrix} 1 & 1 \\ 1 & 0 \end{bmatrix} \begin{pmatrix} 0 \\ \frac{E}{2} \end{pmatrix}$$

$$(v_2) = [\lambda_{2f}](E_{2f}) = \begin{pmatrix} E \\ 0 \end{pmatrix} \quad \text{at } t = t_a \tag{3.86}$$

The voltage E from the PG, therefore, appears at the open end of the coaxial cable at $t = t_a$.

3.3.2 Wave Propagation Characteristics

We will now discuss the wave propagation characteristics of a three-phase single-core cable. Figure 3.13 and Table 3.2 show a cross section and the parameters of a tunnel-installed cable.

Assuming that the tunnel is a pipe conductor, the propagation parameters are evaluated by using a pipe-type (PT) cable option of the EMTP cable constants.

Table 3.3a shows the calculated results of the impedance, the admittance, the modal attenuation constant, and the propagation velocity on the solidly bonded case; Table 3.3b shows the results on the cross-bonded case with the homogeneous model at frequency $f = 100$ kHz.

3.3.2.1 Impedance: R, L

In Table 3.3a-1, the top 3×3 matrix expresses a three-phase core impedance $[Z_{cc}]$, the top right and bottom left matrices are three-phase core-to-sheath $[Z_{cs}]$ and sheath-to-core $[Z_{sc}] = [Z_{cs}]_t$ impedances and the bottom 3×3 matrix expresses three-phase sheath impedances $[Z_{ss}]$. The upper line is

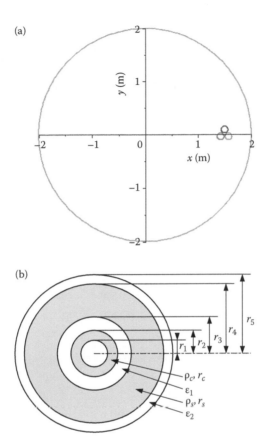

FIGURE 3.13
Tunnel-installed cable represented by a PT cable: (a) configuration of a three-phase single-core cable and (b) cross section of a single-core cable.

the resistance (Ω/km) and the lower line is the inductance (mH/km). The $[Z_{cc}]$ matrix of the solidly bonded cable in Table 3.3a-1 is the same as that of the cross-bonded cable in Table 3.3b-1. The impedance in the last column, $Z''_{4i} (= Z''_{i4})$, in Table 3.3b is given as the average of $[Z_{cs}]$ in Table 3.3a-1, as shown in Equation 3.41.

TABLE 3.2

Cable Parameters

r_1	0	ε_1	3.1
r_2	30.45 mm	ε_2	4.0
r_3	71.15 mm	μ_c	1.0
r_4	74.80 mm	μ_s	1.0
r_5	81.61 mm	ρ_c	$1.82 \times 10^{-8}\ \Omega$ m
ρ_e	100 Ω m	ρ_s	$2.83 \times 10^{-8}\ \Omega$ m

TABLE 3.3

Parameters of a Tunnel-Installed Cable at 100 kHz

(a) Solidly Bonded Cable

	Core-1	Core-2	Core-3	Sheath-1	Sheath-2	Sheath-3
a-1. Impedance [R] in (Ω/km) and [L] in (mH/km)						
R	83.150	81.660	82.120	82.200	81.660	82.120
L	1.172	0.801	0.800	0.989	0.801	0.800
R	81.660	82.690	81.890	81.660	81.740	81.890
L	0.801	1.172	0.800	0.801	0.990	0.800
R	82.120	81.890	83.620	82.120	81.890	82.670
L	0.800	0.800	1.171	0.800	0.800	0.989
R	82.200	81.660	82.120	82.200	81.660	82.120
L	0.989	0.801	0.800	0.989	0.801	0.800
R	81.660	81.740	81.890	81.660	81.740	81.890
L	0.801	0.990	0.800	0.801	0.990	0.800
R	82.120	81.890	82.670	82.120	81.890	82.670
L	0.800	0.800	0.989	0.800	0.800	0.989
a-2. Capacitance [C], in (nF/km)						
Core-1	191.0	0	0	−191.0	0	0
Core-2	0	191.0	0	0	−191.0	0
Core-3	0	0	191.0	0	0	−191.0
Sheath-1	−191.0	0	0	236.2	−19.0	−18.3
Sheath-2	0	−191.0	0	−19.0	235.4	−18.7
Sheath-3	0	0	−191.0	−18.3	−18.7	237.1
Mode	**1**	**2**	**3**	**4**	**5**	**6**
a-3. Current Transformation Matrix [T$_i$]						
Core-1	0	1.000	0	0	0	0
Core-2	1.000	0	0	0	0	0
Core-3	0	0	1.000	0	0	0
Sheath-1	0	−1.000	0	0.336	0.653	−0.229
Sheath-2	−1.000	0	0	0.227	−0.140	0.768
Sheath-3	0	0	−1.000	0.456	−0.503	−0.509
a-4. Modal Propagation Constant						
Attenuation (dB/km)	0.134	0.134	0.134	1.854	0.026	0.030
Velocity (m/μs)	169.6	169.6	169.6	220.3	287.7	288.1

(*Continued*)

TABLE 3.3 (*Continued*)

Parameters of a Tunnel-Installed Cable at 100 kHz

(b) Homogeneous Cable

	Core-1	Core-2	Core-3	Sheath
b-1. Impedance [R] in (Ω/km) and [L] in (mH/km)				
R	83.150	81.660	82.120	81.990
L	1.172	0.801	0.800	0.863
R	81.660	82.690	81.890	81.760
L	0.801	1.172	0.800	0.864
R	82.120	81.890	83.620	82.230
L	0.800	0.800	1.171	0.863
R	81.990	81.760	82.230	81.990
L	0.863	0.864	0.863	0.863
b-2. Capacitance [C], in (nF/km)				
Core-1	191.0	0	0	−191.0
Core-2	0	191.0	0	−191.0
Core-3	0	0	191.0	−191.0
Sheath	−191.0	−191.0	−191.0	596.7
Mode	**1**	**2**	**3**	**4**
b-3. Current Transformation Matrix [T_i]				
Core-1	−0.507	−0.339	0.333	0
Core-2	0.509	−0.335	0.333	−0.001
Core-3	−0.002	0.673	0.333	0.001
Sheath	0	0	−1.000	1.000
b-4. Modal Propagation Constant				
Attenuation (dB/km)	0.124	0.125	0.134	1.860
Velocity (m/μs)	118.8	118.8	169.6	220.5

The above results correspond to the fact that the cross-bonding acts as a transposition of an overhead transmission line, and the three sheath conductors are reduced to one equivalent conductor as explained in Section 3.2.3.2.

3.3.2.2 Capacitance: C

The capacitance matrix looks similar to the impedance matrix in Table 3.3a-2. The capacitance between the core and sheath C_{ck} of the homogeneous model is identical to that of the solidly bonded cable. The equivalent capacitance C''_{k4} of the cross-bonded cable in Table 3.3b-2 is given as the sum of the elements as shown in Equations 3.44 and 3.45.

3.3.2.3 Transformation Matrix

The role of the transformation matrix $[T_i]$ in Table 3.3a-3 and b-3, is to transform modal current (i) to phasor current (I), that is

$$(I) = [T_i](i) \tag{3.87}$$

In the solidly bonded cable, the first three modes (columns) shown in Table 3.3a-3 express coaxial-propagation modes (that is, the "core-to-sheath" mode [2]). The other modes (columns) correspond to one of the transformation matrices of an untransposed three-phase overhead line [2]. In this case, mode 4 expresses an earth-return mode and modes 5 and 6 correspond to aerial modes.

The reduced transformation matrix of a cross-bonded cable is shown in Table 3.3b-3. The composition of the top left 3×3 matrix (the first three modes) is similar to that of an overhead transmission line. The current of the third mode returns from the equivalent sheath instead of the earth. The fourth mode expresses the equivalent earth-return mode of the cross-bonded cable system.

3.3.2.4 Attenuation Constant and Propagation Velocity

Modes 1–3 in the solidly bonded cable shown in Table 3.3a-4 are coaxial modes and are the same as mode 3 in the homogeneous cross-bonded cable model. Although the attenuations of the inter-core modes (modes 1 and 2) shown in Table 3.3b-4 are almost identical to that of the coaxial mode of the solidly bonded cable, the velocities are lower. The velocity of the coaxial mode v_c is determined by the permittivity of the main insulator $\varepsilon_1 = 2.3$ shown in Table 3.2:

$$v_c = \frac{c_0}{\sqrt{\varepsilon_1}} = \frac{300}{\sqrt{2.3}} \approx 198 \text{ m}/\mu s \tag{3.88}$$

where c_0 is the speed of light.

The velocity converges to this value as the frequency increases.

The attenuation and velocity of the earth-return mode (mode 4) in both cable models are identical.

Because the cable is installed in a tunnel (that is, the cable is in air), the attenuation constant and the propagation velocity of mode 4 (earth return) and modes 5 and 6 (the first and second inter-sheath) in the solidly bonded cable show similar characteristics to those of an overhead line [1]. The attenuation constants of modes 5 and 6 are much smaller and the propagation velocity is much greater in the solidly bonded cable than those in the other modes.

3.3.3 Transient Voltage

Figures 3.14 and 3.15 show transient voltage waveforms at the sending-end core voltages and the first cross-bonding point of a cross-bonded

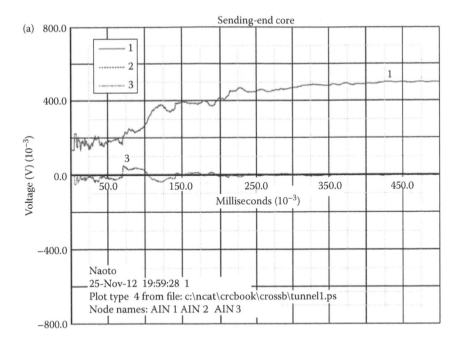

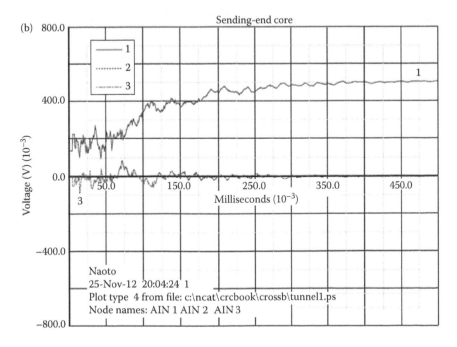

FIGURE 3.14
Calculated transient core voltages on a tunnel-installed cable: (a) cross-bonded, (b) mixed.
(*Continued*)

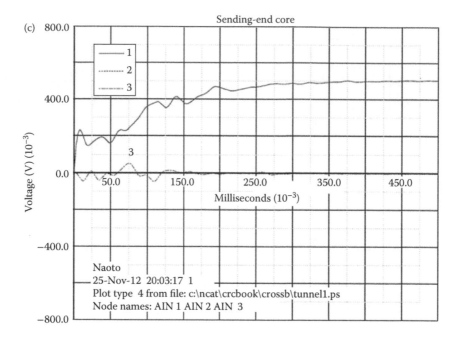

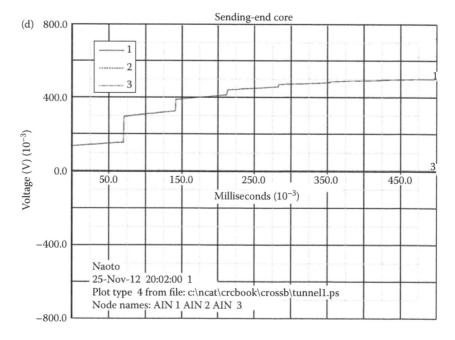

FIGURE 3.14 (Continued)
Calculated transient core voltages on a tunnel-installed cable: (c) homogeneous, and (d) solidly bonded.

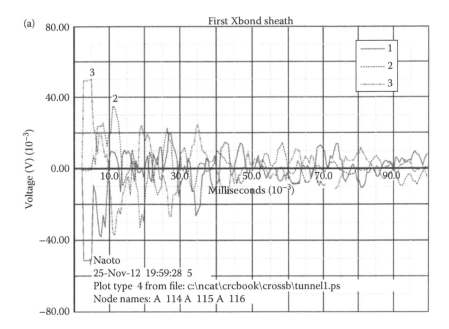

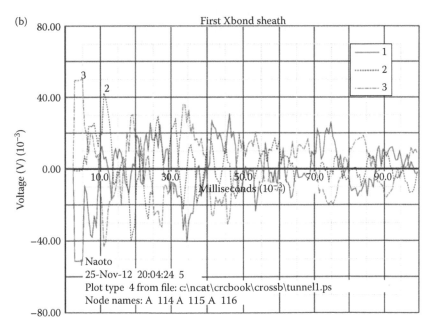

FIGURE 3.15
Calculated transient sheath voltages on a tunnel-installed cable: (a) cross-bonded and (b) mixed.

cable system with five major sections ($l_1 = l_2 = l_3 = 400$ m with total length $l = 5 \times 3 \times 0.4 = 6$ km) when a step voltage of 1 pu is applied to a sending-end core (phase a) through a resistor of $R_s = 200$ Ω. This resistor models a backward surge impedance. The physical parameters of the cable are shown in Figure 3.6 and Table 3.2. Each receiving-end core is grounded through a resistor of $R_e = 200$ Ω. The sheaths are short-circuited and grounded by a resistor of $R_s = 0.1$ Ω at both ends of each major section. The cable is represented by a constant parameter, Dommel's model.

The calculated results for minor sections exactly modeled by multiphase-distributed parameter lines are shown in Figures 3.14a and 3.15a. The induced voltages on the cores are observed in Figure 3.14a. The voltage is generated by the reflections at the cross-bonded points. After 70 μs, the voltage on the applied phase is gradually increased. The time is determined based on the round-trip time of the coaxial traveling wave:

$$T = \frac{2l_t}{v_c} = \frac{2 \times 6000}{169.6} = 70.8 \text{ μs} \tag{3.89}$$

where l_t and v_c denote the total cable length and the traveling velocity, respectively, of the coaxial mode shown in Table 3.3a-4 or b-4.

The time constant τ_1 of the voltage increase is determined by the sending-end resistance R_s, the terminating resistance R_e, and the total cable capacitance $C_c l_t$. The capacitance is obtained from Table 3.3a-2 or b-2.

$$\tau_1 = (R_s // R_e)C_c l = 100 \times 0.191 \times 6 \approx 110 \text{ μs} \tag{3.90}$$

The maximum sheath voltage at the first cross-bonded joint shown in Figure 3.15a becomes 0.05 pu, which is about 40% of the core voltage at the time. This voltage is generated by a reflection at the cross-bonded joint. This is an inherent characteristic of the cross-bonded cable. The high voltage (HV) is a key factor in the insulation design of a cross-bonded cable system.

Although the exact model of the cross-bonded cable is useful for the simulation of a simple grid, the simulation of a large-scale cable system with a cross-bonded cable is too complicated and difficult. The homogeneous model explained in Section 3.2.3 is comparatively simple and useful. Figures 3.14 and 3.15b illustrate the transient voltages when the first major section is expressed accurately and the other major sections are expressed by pi-equivalent circuits whose parameters are determined by the homogeneous model. By comparing the results from the exact model (a), a simplification is possible, providing sufficient information for an insulation design of the cable system. Figure 3.14c illustrates the result of a case in which all major sections are expressed by homogeneous pi-equivalent circuits. It is clear from the figure that the calculated result has enough accuracy for the simulation of the switching surge, although the sheath voltages at the cross-bonded joints cannot be obtained.

Figure 3.14d shows the transient response of the core voltage in a solidly bonded cable. It shows a stair-like waveform with a length of 70 µs. This length is determined by the round-trip time shown in Equation 3.89. Sheath voltages of the solidly bonded cable are much smaller than those of the cross-bonded cable. The results indicate that not all cross-bonded cables can be simplified by a solidly bonded cable from the viewpoint of not only the sheath voltages but also the core voltages.

3.3.4 Limitations of the Sheath Voltage

As mentioned in Section 3.2.2, the limitations of the sheath voltage are key for making decisions regarding sheath bonding and other cable system designs related to the sheath. There are two types of limitations in the sheath voltage: (1) continuous voltage limitation and (2) short-term voltage limitation.

Continuous voltage limitation is the limitation of the sheath voltage induced by the normal load flow in phase conductors without any faults. It is enforced by government or district regulations in many countries and differs in each area based on said regulations. This limitation was enforced for the safety of the maintenance crews who may come into contact with the sheath circuit. Even if this limitation is not enforced, utilities follow their own standards for continuous voltage limitation.

Since SVLs are designed not to be operated by continuous sheath voltages, continuous voltage limitation is maintained mainly by cable layouts, cross-bonding designs, and grounding resistances. The cable span length (minor length) is more often limited by restrictions in transportation, but can also be limited by a continuous voltage limitation.

A short-term voltage limitation is specified in IEC 62067 Annex G (informative) as impulse levels [13]. Considering short-term voltage limitation, the following phenomena are studied:

- SLG faults (external to the targeted major section)
- Three-phase faults (external to the targeted major section)
- Switching surges
- Lightning surges

When only power-frequency components are considered, SLG faults and three-phase faults are studied using theoretical formulas. Some utilities study SLG faults and three-phase faults using EMTP in order to consider transient components of the sheath voltage. Switching surges rarely become an issue for the sheath overvoltage.

Lightning surges have to be studied for a mixed overhead line/underground cable as shown in Figure 3.16. A lightning strike on the GW can propagate into the sheath circuit, since the transmission tower and cable sheath often share the grounding mesh or electrode at the transition site. The level

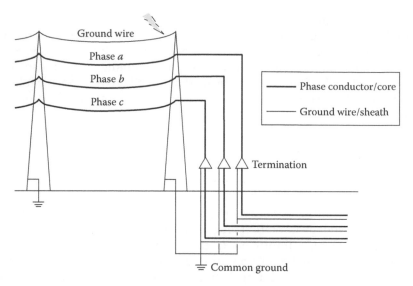

FIGURE 3.16
Lightning surge in a mixed overhead line/underground cable.

of the sheath overvoltage is highly dependent on the grounding resistance at the transition site. The space of the transition site is sometimes limited; it is necessary to achieve a low grounding resistance in order to lower the sheath overvoltage.

A BFO can occur when lightning strikes the GW. In addition, a lightning strike can directly hit the phase conductor due to shielding failure. In these cases, the lightning surge in the phase conductor can directly propagate into the cable core leading to sheath overvoltage. Since the lightning surge is not highly attenuated in this case, the voltage across the sheath interrupts at the first SSJ needs to be studied in addition to the sheath-to-earth voltage at the transition site.

Lightning surges are also studied when a limited number of feeders are connected to a substation together with a cable. In Figure 3.17, the substation has only two lines and one transformer considering the maintenance outage. The lightning surge on the overhead line can propagate into the cable core without significant attenuation depending on the substation layout.

3.3.5 Installation of SVLs

SVLs are installed at SSJs in order to suppress short-term sheath overvoltages. Figure 3.18 shows the connection of SVLs when the sheath circuit is cross-bonded in a link box. SVLs are often arranged in a star formation with their neutral points earthed. If study results show that the sheath overvoltage exceeds the TOV rating of SVLs, the ECC can be installed as shown in Figure 3.18. Other countermeasures include changing the neutral point from

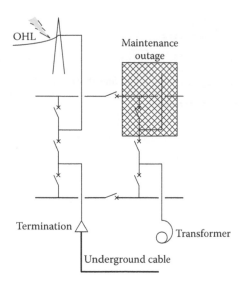

FIGURE 3.17
Example of a substation with a limited number of feeders.

solidly earthed to unearthed and changing the SVL connection from a star formation to a delta formation.

When the link box is not installed, SVLs are located immediately next to sheath-sectionalizing joints as shown in Figure 3.19. In this connection, SVLs are arranged in a delta formation. This formation has an advantage in

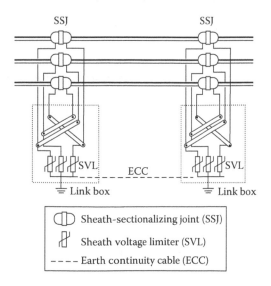

FIGURE 3.18
Connection of SVLs in a link box.

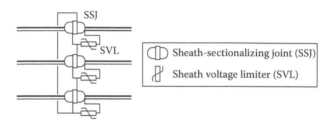

FIGURE 3.19
Connection of SVLs without a link box.

suppressing short-term sheath overvoltage, as bonding leads to SVLs can be much shorter than when using the link box.

3.4 Studies on Recent and Planned EHV AC Cable Projects

This section introduces recent and planned EHV AC cable projects and cable system transients studied for the projects. In order to compensate for the large charging capacity of EHV AC cables, shunt reactors are often installed together with these cables. The large charging capacity and large shunt reactors lower the natural frequency of the network which, at times, make it necessary to conduct resonance overvoltage studies. Load-shedding overvoltages and the zero-missing phenomena are the other power system transients specifically related to cable systems.

Similar to overhead line projects, switching transients such as cable energization, ground fault, and fault clearing are also studied for EHV AC cable projects as standard work. However, severe overvoltages related to these switching transients on cable systems have not been reported in the literature.

This section focuses on well-known long cable projects, which normally require shunt compensation, since the TOVs discussed can only be observed with these cables. Therefore, this section includes only cross-bonded land cables and submarine cables. It does not include short cables, typically installed inside power stations and substations, since power system transients specifically related to cable systems are normally not included in the study of short cables.

3.4.1 Recent Cable Projects

Table 3.4 lists long 500-/400-kV cables that are currently in operation. The number of long 400-kV cable projects is larger than the number of 500-kV cable projects [14–22]. The system voltage of 400 kV is adopted mainly based on the geographical area such as in Europe and the Middle East.

TABLE 3.4

Installed Long 500/400-kV Cables[a]

Location	Route Length (km)	Number of Circuits	Insulation	Commissioned in
500-kV Cables				
Vancouver, Canada				
Vancouver–Texada Island	30	2	SCFF[b]	1984
Texada Island–Mainland	8	2		
Mainland Japan–Shikoku	22.2	1	SCFF	1994
Island, Japan	22.2	1		2000
Tokyo, Japan	39.8	2	XLPE	2000
Shin-Toyosu Line				
Shanghai, China [14]	17	2	XLPE	2010
Moscow, Russia [15]	10.5	2	XLPE	2012
400-kV Cables				
Copenhagen, Denmark [16]				
Southern route	22	1	XLPE	1997
Northern route	12	1		1999
Spain–Morocco [17]	28	1	SCFF	1997
	31	1		2006
Berlin, Germany [18]				
Mitte–Friedrichshain	6.3	2	XLPE	1998
Friedrichshain–Marzahn	5.5	2		2000
Madrid, Spain [19]				
San Sebastian de los Reyes–Loeches–Morata Line	12.8	2	XLPE	2004
Jutland, Denmark [20]	14			
Trige–Nordjyllandsværket	(4.5 + 2.5 + 7.0)	2	XLPE	2004
London Ring [21]	20	1	XLPE	2005
Qatar	15	2	XLPE	2007
Istanbul, Turkey				
Ikitelly–Davutpasa–	12.8	1	XLPE	2007–2011
Yenibosna	7.2	1	XLPE	
London–West Ham	12.6	1	XLPE	2008
Rotterdam, Netherlands	12.5	1	XLPE	2010
Enecogen				
Dubai and Abu Dhabi, UAE	11.5	1	XLPE	2011
Mushrif–Al Mamzar				
Abu Dhabi, UAE [22]	13	2	XLPE	2011–2012
Sadiyat–ADST				
Doha, Qatar				
Umm Al Amad Super–Lusail Development Super 2	22	3	XLPE	2012–2013

[a] Include 500-/525-/550-kV cables and 380-/400-/420-kV cables.
[b] Self-contained fluid filled.

The world's first long 500-/400-kV cable was installed in Canada by BC Hydro in 1984 [23–25]. This 500-kV AC submarine cable is a double-circuit line that connects Vancouver Island to mainland Canada through Texada Island. The distance between Vancouver Island and Texada Island is approximately 30 km; the distance between Texada Island and mainland Canada is approximately 8 km. In between, the line has an overhead section on Texada Island. Shunt reactors totaling 1080 MVar were installed to compensate for the large charging capacity.

The longest 500-kV cable in the world, the Shin-Toyosu line, was installed in Japan by the Tokyo Electric Power Company in 2000. This double-circuit line has four 300-MVar shunt reactors for the compensation of the large charging capacity. This is the first 500-kV cable on which extensive power system transient studies are available in the literature [26,27]. In addition to ordinal switching transients, the overvoltage caused by system islanding, series resonance overvoltage, leading current interruption, and zero-missing phenomenon was also studied.

Here, we introduce the overvoltage caused by system islanding, studied on the Shin-Toyosu line. When one end of a long cable is open, a part of the network can be separated from the main grid and connected with the long cable, which can lead to severe overvoltage. Figure 3.20 illustrates the equivalent circuit where one end of the long cable is open due to a bus fault. A cable fault will not lead to overvoltage since it results in the removal of the long cable from the equivalent circuit.

From this equivalent circuit, the overvoltage caused by system islanding can be expressed using the following equations:

$$v(t) = V_m \sin \omega t - \frac{\omega}{\omega_0} V_m \sin \omega_0 t \tag{3.91}$$

$$V_m = \frac{E_m L}{L_0(1 - \omega^2 CL) + L}, \quad \omega_0 = \sqrt{\frac{1}{CL_0} + \frac{1}{CL}} \tag{3.92}$$

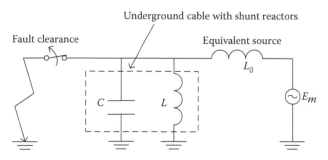

FIGURE 3.20
Equivalent circuit of the overvoltage caused by system islanding.

where

L_0 is the source impedance of the islanded system

E_m is the source voltage behind L_0

The charging capacity of the long cable and the inductance of the shunt reactors directly connected to the cable are expressed as C and L, respectively.

Equation 3.91 shows that the overvoltage contains two frequency components: the nominal frequency ω and the resonance frequency ω_0. Since the overvoltage is caused by the superposition of two frequency components, the resulting overvoltage is oscillatory and its level is often difficult to estimate before the simulation. The result of a simulation performed on the Shin-Toyosu line is shown in Figure 3.21.

Most of the 500-/400-kV cables shown in Table 3.4 are installed in highly populated areas, hence the route lengths are limited to 10–20 km. These cables are equipped with shunt reactors for the compensation of the charging capacity, but their unit size and the total capacity are not as large due to the shorter route lengths. For these cables, only studies such as the reactive power compensation, the design of the cable itself, and the laying method

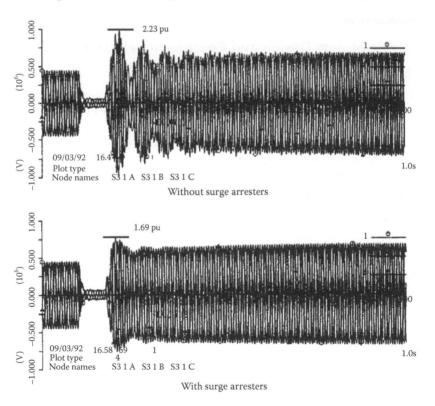

FIGURE 3.21

Example of an overvoltage caused by system islanding.

are discussed in the literature. Transient studies on these cables are not available.

3.4.2 Planned Cable Projects

Table 3.5 lists planned lengthy 400-kV cable projects that will be operational within a couple of years. There is no mention of planned 500-kV cable projects in any publicly available sources.

Various transient studies have been performed on the 400-kV cable that will connect Sicily to mainland Italy [28,29]. In addition to switching transients, these studies include the harmonic overvoltage caused by line energization, leading current interruption, and zero-missing phenomena. The studies identified the resonant condition at the second harmonic when the cable is energized from Sicily's side under a particular condition. The harmonic overvoltage caused by the resonant condition is avoided by the operational constraint.

3.5 Cable System Design and Equipment Selection

3.5.1 Study Flow

This section discusses the cable system design, except for overvoltage analysis and insulation coordination. The cable system design includes the selection and specification of the cable itself and the related equipment such as CBs and voltage transformers (VTs). The cable system design related to the sheath is discussed in Section 3.2.

During the planning stage, transmission capacity and reactive power compensation are normally studied. These studies mainly determine the cable

TABLE 3.5

Planned Long 400-kV Cables[a]

Location	Route Length (km)	Number of ckts	Insulation	Planned Operation
Sicily–Mainland Italy	38	2	SCOF	2013
Abu Dhabi Island, UAE [22]				
Sadiyat–ADST	13	1		2014
Bahia–Sadiyat	25	3	XLPE	2013
SAS Al Nakheel–Mahawi	21	2		2013
Mahawi–Mussafah	13	3		2013
Strait of Messina, Italy	42	2	XLPE	2015
Sorgente-Rizziconi				

[a] Include 380-/400-/420-kV cables.

route, the voltage level, the conductor size, and the amount and locations of shunt reactors.

When the transmission development plan is designed for the cable, the cable route will be studied further. One characteristic of cables, compared to overhead lines, is that the laying of the cables and soil conditions of the location affect planning studies in addition to the land availability. These factors affect the burial depth, soil thermal resistivity, and cable separation between phases, which may necessitate changes to the conductor size and the amount and locations of shunt reactors initially determined in the planning studies.

Figure 3.22 illustrates the study flow and the relationship of studies on the cable system design. In the figure, the boxes show study items, and the following bullets list the items that are mainly evaluated in these studies. The figure explains how the transmission capacity and the reactive power compensation studied during the planning stage affect the other studies conducted for the cables.

The amount of shunt reactors (that is, the compensation rate of a cable) is a key figure that has a major impact on the following studies. A compensation rate close to 100% is often preferred since it can eliminate the reactive power surplus created by the introduction of the cable. It also offers a preferable condition for the TOV but causes a severe condition for the zero-missing phenomenon. The negative effect on the zero-missing phenomenon is not a primary concern as there are countermeasures established for tackling this effect.

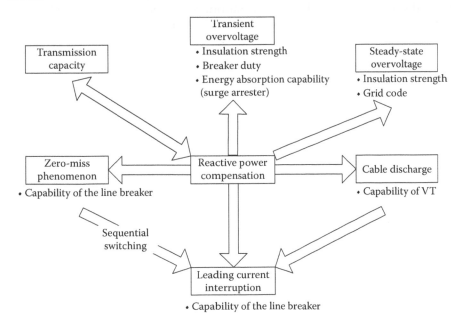

FIGURE 3.22
Study flow and the relationship of studies.

Usually, shunt reactors for 500-/400-kV cables are directly connected to the cables in order to mitigate the TOV when one end of the cable is opened. Shunt reactors for other cables are often connected to buses as the area compensation is applied at these voltage levels. When shunt reactors are connected to buses, the zero-missing phenomenon does not occur. In this case, however, the inductive VT connected to the cable needs to have enough discharge capability, and the line breaker needs to have sufficient leading current interruption capability.

3.5.2 Zero-Missing Phenomenon

A DC offset current (zero-missing current) appears when an EHV underground cable is energized with its shunt reactors [30–35]. In this case, an AC component of a charging current has an opposite phase angle to the AC component of a current flowing into the shunt reactors. If the compensation rate of the cable is 100%, the sum of these AC components becomes zero and only the DC component remains. Since the DC component decays gradually with time, it can take more than 1 s, depending on the compensation rate, before a current that flows through the line breaker crosses the zero point.

Figure 3.23 shows an example of current waveforms when an EHV cable is energized with its shunt reactors. In this example, the AC component of the energization current is very small, since the compensation rate is close to 100%. The simulation was run for 0.2 s, but the energization current did not cross the zero point during this duration. Since the zero-missing phenomenon is caused by a DC component of an energization current, it reaches its

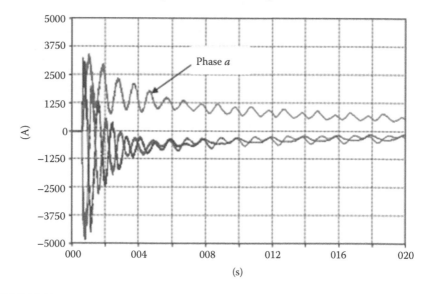

FIGURE 3.23
Zero-missing current in underground cable energization.

peak when the cable is energized and the maximum DC component is contained in the current. In order to realize this condition in phase *a*, the cable was energized at the zero-voltage point of phase *a*.

When a line breaker is used to interrupt this current without zero crossing, the arc current between the contacts cannot be extinguished within a couple of cycles and may continue for an extended duration. This extended duration may lead to the line breaker's failure depending on the amount of arc energy generated. The duration is mainly determined by the magnitude of the DC component and the relationship between the arc resistance and arc current inside the line breaker. Typical durations for EHV cables can be several hundreds of milliseconds in severe conditions.

The zero-missing phenomenon can theoretically be avoided by limiting the compensation rate to lower than 50%, but this is not the most common way of addressing the problem. Normally, a compensation rate near 100% is adopted (especially for 500-/400-kV cables) and the countermeasures listed in Table 3.6 are applied to the cables. All these countermeasures (except for Countermeasure (4)) have already been applied to the cable line in operation. Countermeasure (1) in particular has a number of application records to long EHV cable lines and is discussed in detail later in this section.

Countermeasure (3) will be applied to the 400-kV Sicily–mainland Italy cable [29]. This countermeasure can be implemented rather easily as long as a cable line is installed together with single-phase CBs and current differential relays. For this reason, Countermeasure (3) is more suited for EHV cable lines than HV cable lines.

In this countermeasure, the faulted phase is opened instantly and healthy phases are opened about 10 s later when the DC component has decayed

TABLE 3.6

Countermeasures of Zero-Missing Phenomenon

Countermeasures	Notes
1. Sequential switching	• Requires higher leading current interruption capability
	• Requires single-phase CB and current differential relay
2. Point-on-wave switching (synchronized switching)	• May cause higher-switching overvoltage
	• Requires single-phase CB
3. Delayed opening of healthy phases	• Requires single-phase CB and current differential relay
	• May not be possible to apply near generators
4. Breaker with preinsertion resistor	• May be necessary to develop a new CB (expensive)
5. Additional series resistance in the shunt reactor for energization	• Requires special control to bypass the series resistance after energization
6. Energize the shunt reactor after the cable	• Causes higher steady-state overvoltage and voltage step

enough. When this countermeasure is applied near a generator (especially when the cable line offers a radial path to the generator), it is necessary to evaluate the negative sequence current capability of the generator as Countermeasure (3) causes an unbalanced operation for a prolonged duration.

In Countermeasure (5), a resistance is connected in a series to shunt reactors when a cable line is energized. The resistance needs to be sized sufficiently for the DC component to decay fast enough. After the cable line is energized, the resistance is bypassed in order to reduce the losses. Considering the additional cost for the resistance, this countermeasure is more suited for HV cable lines than EHV cable lines.

Countermeasure (6) cannot always be applied; it is especially difficult to apply to long EHV cables due to their steady-state overvoltage.

3.5.2.1 Sequential Switching

Figure 3.24 shows a zero-missing current with an SLG fault in phase *b*. This assumes that a cable failure exists in phase *b* before energization, but it is not known to system operators until the cable is energized.

The zero-missing current is observed only in a healthy phase (phase *a*). The current in the faulted phase (phase *b*) crosses the zero point as it contains a large AC component due to the fault current. Hence, the line breaker of the faulted phase can interrupt the fault current.

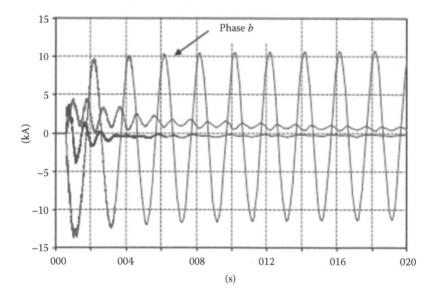

FIGURE 3.24
Zero-missing phenomenon with an SLG fault.

Figure 3.25 shows the time sequence of sequential switching when the cable line is energized from Substation A. In Step 1, the line breaker of phase *b* is opened at 60 ms after the fault and the fault is cleared by tripping this CB. Since the fault is already cleared by the opening of the phase *b* line breaker, some time can be allowed before opening the line breakers of other healthy phases.

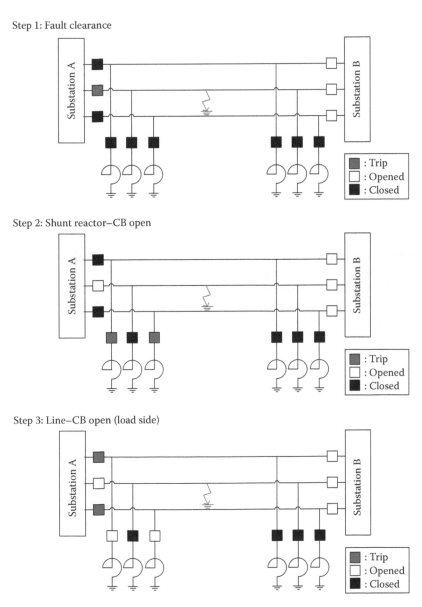

FIGURE 3.25
Time sequence of sequential switching.

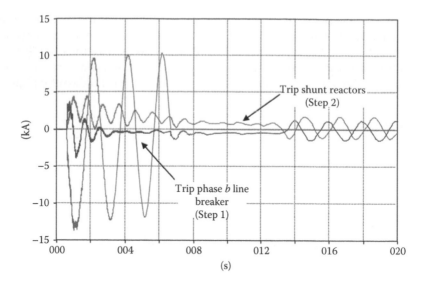

FIGURE 3.26
Zero-missing phenomenon with sequential switching.

In Step 2, shunt reactors are tripped before the line breakers of healthy phases. It is necessary to trip the shunt reactors of only healthy phases. At this time, it is not necessary or recommended to trip shunt reactors of the faulted phase since the current through shunt reactor breakers of the faulted phase does not cross the zero point.

It is recommended to trip at least half of the shunt reactors of healthy phases as shown in Figure 3.25 as the tripping will normally lower the compensation rate below 50%. The remaining shunt reactors will be useful in maintaining the charging current within the leading current interruption capability of the line breakers.

In Step 3, it is possible to open the line breakers of the healthy phases. Figure 3.26 shows that the current in the healthy phases contains the AC component and crosses the zero point after tripping the shunt reactors.

3.5.3 Leading Current Interruption

When the leading current is interrupted at current zero, it occurs at a voltage peak assuming that the current waveform is leading the voltage waveform by 90°. After the interruption, the voltage on the source side of the CB changes according to the system voltage; the voltage on the other side is fixed at peak voltage E as shown in Figure 3.27. The most severe overvoltage occurs during a restrike after half cycle when the voltage on the source side becomes $-E$. As the voltage difference between the source side and the other

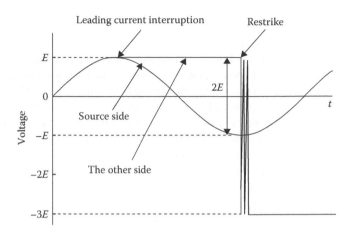

FIGURE 3.27
Overvoltage caused by leading current interruption and restrike.

TABLE 3.7

Leading Current Interruption Capability According to IEC 62271-100

Rated Voltage (kV)	Rated Capacitive-Switching Currents[a] (Cable) (A)
420	400
550	500

[a] Preferred values, voltage factor: 1.4 pu.

side is $2E$, the overvoltage can go as high as $-3E$. The restrike can be repeated to cause a very severe overvoltage.

The leading current interruption capability of CBs is specified in IEC 62271-100 (see Table 3.7) considering the severe overvoltage that it can cause.

When the charging capacity of a long EHV cable line is not compensated by shunt reactors that are directly connected to the cable, the leading current interruption capability requires careful attention [30]. Considering the typical capacitance of $0.2\ \mu F/km$, the maximum line length for a 400-kV cable line is limited approximately below 26 km without shunt reactors directly connected to the cable. Here, it is assumed that the leading current is interrupted at one end, and the other end is opened before the interruption.

Usually, long EHV cable lines are compensated by shunt reactors that are directly connected to the cable. When the compensation rate is high enough, the leading current interruption capability is not a concern. If sequential switching is applied to a cable line as a countermeasure to the zero-missing phenomenon, however, the tripping of shunt reactors will lower the compensation rate. This is the only situation that requires careful attention.

3.5.4 Cable Discharge

If a shunt reactor is directly connected to a cable, the cable line is discharged through the shunt reactor when it is disconnected from the network. In this case, the time constant of the discharge process is determined by the quality factor (Q factor) of the shunt reactor. Since the Q factor is around 500 for EHV shunt reactors, the time constant of the discharge process is around 8 min.

If the cable is disconnected from the network and energized again within a couple of minutes, a residual charge remains in the cable line, which can be highly dependent on the time separation between the disconnection and the re-energization. Under this condition, the re-energization overvoltage can exceed the switching impulse withstand voltage (SIWV) when the residual voltage has an opposite sign to the source voltage at the time of re-energization.

This is usually an issue for overhead lines since auto-reclose is applied to them. For cables, it is uncommon to apply auto-reclose as they may experience higher overvoltages because of their higher residual voltage. System operators should be aware that they need to wait for about 10 min (perhaps more to be conservative) before re-energizing a cable, though it is not common to re-energize a cable after a failure.

If a shunt reactor is not directly connected to the cable line, the cable line is discharged through inductive VTs. In this case, the discharge process will be completed within several milliseconds. The inductive VTs need to have enough discharge capability for a cable line to be operated (a) without a shunt reactor or (b) if all the shunt reactors are tripped by sequential switching. It takes several hours for the inductive VTs to dissipate heat after dissipating the cable charge. If the inductive VTs are required to dissipate the cable charge twice within several hours, the required discharge capability will be doubled.

3.6 EMTP Simulation Test Cases

PROBLEM 1

3.1. Assume that the sample cable in Section 3.2.4 is buried as a single-phase cable. Find the impedance and admittance matrices for the single-phase example cable using EMTP. Use the Bergeron model and calculate the impedance and admittance matrices at 1 kHz.

3.2. From the impedance and admittance matrices found in (1), find the phase constants for the earth-return mode and the coaxial mode using the voltage transformation matrix $A = \begin{bmatrix} 1 & 1 \\ 1 & 0 \end{bmatrix}$.

3.3. Find the propagation velocity for the coaxial mode and calculate the propagation time when the cable length is 12 km.

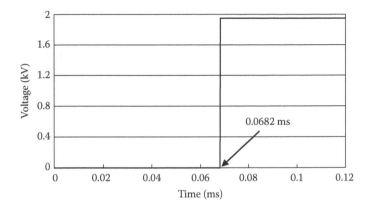

FIGURE 3.28
Propagation time of the 12-km cable obtained by EMTP simulation.

3.4. Using the cable data created in (1), find the propagation time for a 12-km cable with EMTP and compare it with the propagation time theoretically found in (3); assume that the sheath circuit is solidly grounded with zero-grounding resistance at both the ends.

Solution 1

3.1.
$$Z = \begin{bmatrix} 0.0010 + j0.0111 & 0.0010 + j0.0102 \\ 0.0010 + j0.0102 & 0.0010 + j0.0102 \end{bmatrix} (\Omega/m)$$

$$Y = \begin{bmatrix} j1.5068 \times 10^{-6} & -j1.5068 \times 10^{-6} \\ -j1.5068 \times 10^{-6} & j1.3095 \times 10^{-5} \end{bmatrix} (mho/m)$$

3.2. $\beta_e = 3.4446 \times 10^{-4}$ (Np/m): earth-return mode
$\beta_c = j3.5733 \times 10^{-5}$ (Np/m): coaxial mode

3.3. $c_c = 1.7584 \times 10^8$ (m/s), $t = 0.0682$ (ms)

3.4. A step voltage of 1.0 kV is applied at one end of the 12-km cable at 0 s. The coaxial mode arrives at the other end at 0.0682 ms. As shown in Figure 3.28, the propagation time found in EMTP exactly matches the time found in (3).

PROBLEM 2

Calculate zero- and positive-sequence currents using EMTP for the sample cable in Section 3.2.4. For this calculation, assume that the lengths of a minor section and a major section are 400 and 1200 m, respectively. As the total length of the cable is 12 km, the cable will have 10 major sections. Grounding resistance at earthing joints should be set to 10 Ω.

Solution 2

(a) Cross-Bonded Cable

	Zero-Sequence		Positive-Sequence	
	Amplitude (A)	Angle (deg.)	Amplitude (A)	Angle (deg.)
EMTP simulation	133.8	−21.42	356.4	−86.35
Proposed formulas, Equations 3.8/3.14	124.1	−21.23	356.7	−86.79

(b) Solidly Bonded Cable

	Zero-Sequence		Positive-Sequence	
	Amplitude (A)	Angle (deg.)	Amplitude (A)	Angle (deg.)
EMTP simulation	121.6	−21.80	694.9	−50.40
Proposed formulas, Equations 3.19, 3.20/3.26	124.8	−22.50	722.7	−49.08

References

1. Ametani, A. 1980. A general formulation of impedance and admittance of cables. *IEEE Trans. Power Appar. Syst.* PAS-99(3):902–910.
2. Ametani, A. 1990. *Distributed-Parameter Circuit Theory.* Tokyo, Japan: Corona Publishing Co.
3. CIGRE WG B1.19. 2004. General guidelines for the integration of a new underground cable system in the network. *CIGRE Tech. Brochure* 250.
4. Nagaoka, N. and A. Ametani. 1983. Transient calculations on crossbonded cables. *IEEE Trans. Power Appar. Syst.* PAS-102(4):779–787.
5. Ametani, A., Y. Miyamoto, and N. Nagaoka. 2003. An investigation of a wave propagation characteristic on a crossbonded cable. *IEEJ Trans. PE* 123(3):395–401 (in Japanese).
6. IEC/TR 60909-2 ed. 2.0. 2008. Short-circuit currents in three-phase a.c. systems—Part 2: Data of electrical equipment for short-circuit current calculations.
7. Central Station Engineers. 1964. *Electrical Transmission and Distribution Reference Book*, 4th edn. East Pittsburgh, PA: Westinghouse Electric Corporation.
8. Lewis Blackburn, J. 1993. *Symmetrical Components for Power Systems Engineering.* Boca Raton, FL: CRC Press.
9. Vargas, J., A. Guzman, and J. Robles. 1999. Underground/submarine cable protection using a negative-sequence directional comparison scheme. *26th Annual Western Protective Relay Conference*, Spokane, WA, October 25–28.
10. Gustavsen, B. 2001. Panel session on data for modeling system transients. Insulated cables. *Proceedings of the IEEE Power Engineering Society Winter Meeting*, Columbus, OH.
11. Ametani, A. 2009. On the impedance and the admittance in the EMTP cable constants/parameters. *European EMTP-ATP Users Group Meeting*, Delft, the Netherlands.

12. Scott-Meyer, W. 1982. *ATP Rule Book*. Can/Am EMTP User Group, USA.
13. IEC 62067 ed. 2.0. 2011. Power cables with extruded insulation and their accessories for rated voltages above 150 kV (*Um* = 170 kV) up to 500 kV (*Um* = 550 kV)—Test methods and requirements.
14. Dubois, D. 2007. Shibo 500 kV cable project—Shanghai—China. *Fall 2007 PES-ICC Meeting*. Scottsdale, AZ. Presentation available at: http://www.pesicc.org/iccwebsite/subcommittees/G/Presentations/Fall07/6.trans.Presentation_500kV_Shibo_3.pdf.
15. Kaumanns, J. 2012. The Skolkovo challenge: A 550 kV XLPE cable system with 138 joints put into operation within 17 month after order intake. *Proceedings of CIGRE Session 2012*, Paris, France.
16. Andersen, P. et al. 1996. Development of a 420 kV XLPE cable system for the metropolitan power project in Copenhagen. CIGRE Session 21-201.
17. JICABLE/WETS Workshop. 2005. Long insulated power cable links throughout the world. Reactive power compensation achievement. Results of the WETS'05 study.
18. Henningsen, C. G., K. B. Müller, K. Polster, and R. G. Schroth. 1998. New 400 kV XLPE long distance cable systems, their first application for the power supply of Berlin. CIGRE Session 21-109.
19. Granadino, R., M. Portillo, and J. Planas. 2003. Undergrounding the first 400 kV transmission line in Spain using 2.500 mm² XLPE cables in a ventilated tunnel: The Madrid "Barajas" Airport Project. JICABLE '03, A.1.2.
20. Argaut, P. and S. D. Mikkelsen. 2003. New 400 kV underground cable system project in Jutland (Denmark). JICABLE '03, A.4.3.
21. Sadler, S., S. Sutton, H. Memmer, and J. Kaumanns. 2004. 1600 MVA electrical power transmission with an EHV XLPE cable system in the underground of London. CIGRE Session B1-108.
22. Abu Dhabi Transmission & Despatch Company. 2011. Seven year electricity planning statement (2012–2018). Available at: http://www.transco.ae/media/docs.htm (accessed on June 6, 2013).
23. Crowley, E., J. E. Hardy, L. R. Horne, and G. B. Prior. 1982. Development programme for the design, testing and sea trials of the British Columbia Mainland to Vancouver Island 525 kV alternating current submarine cable link. CIGRE Session 21-10.
24. Foxall, R. G., K. Bjørløw-Larsen, and G. Bazzi. 1984. Design, manufacture and installation of a 525 kV alternating current submarine cable link from Mainland Canada to Vancouver Island. CIGRE Session 21-04.
25. Cherukupalli, S., G. A. Macphail, J. Jue, and J. H. Gurney. 2006. Application of distributed fibre optic temperature sensing on BC Hydro's 525 kV submarine cable system. CIGRE Session B1-203.
26. Momose, N., H. Suzuki, S. Tsuchida, and T. Watanabe. 1998. Planning and development of 500 kV underground transmission system in Tokyo Metropolitan Area. CIGRE Session 37-202.
27. Kawamura, T., T. Kouno, S. Sasaki, E. Zaima, T. Ueda, and Y. Kato. 2000. Principles and recent practices of insulation coordination in Japan. CIGRE Session 33-109.
28. Colla, L., S. Lauria, and F. M. Gatta. 2007. Temporary overvoltages due to harmonic resonance in long EHV cables. *International Conference on Power System Transients (IPST) 233*, Lyon, France.

29. Colla, L., M. Rebolini, and F. Iliceto. 2008. 400 kV ac new submarine cable links between Sicily and the Italian Mainland. Outline of project and special electrical studies. CIGRE Session C4-116.
30. Ohno, T. 2010. Operation and protection of HV cable systems in TEPCO. Global Facts, Trends and Visions in Power Industry, Swiss Chapter of IEEE PES. Presentation available on the web: http://www.ieee.ch/assets/Uploads/pes/downloads/1004/10042ohnoexperiencetepco.pdf.
31. Leitloff, V., X. Bourgeat, and G. Duboc. 2001. Setting constraints for distance protection on underground cables. In *Proceedings of the 2001 Seventh IEE International Conference on Developments in Power System Protection*, Amsterdam, the Netherlands.
32. Kulicke, B. and H. H. Schramm. 1980. Clearance of short-circuits with delayed current zeros in the Itaipu 550 kV-substation. *IEEE Trans. Power Appar. Syst.* PAS-99(4):1406–1414.
33. Michigami, T., S. Imai, and O. Takahashi. 1997. Theoretical background for zero-miss phenomenon in the cable network and field measurements. *IEEJ General Meeting* 1459, Tokyo, Japan (in Japanese).
34. Hamada, H., Y. Nakata, and T. Maekawa. 2001. Measurement of delayed current zeros phenomena in 500 kV cable system. *IEEJ General Meeting*, Nagoya, Japan.
35. Tokyo Electric Power Company. 2008. Joint feasibility study on the 400 kV cable line Endrup-Idomlund. Final Report.

4

Transient and Dynamic Characteristics of New Energy Systems

New energy, or the so-called green and sustainable energy, is gaining significance because of problems related to CO_2 in thermal power generation and nuclear waste in nuclear power generation. At the same time, the "smart grid" is becoming a very attractive research subject.

In this chapter, the transient and dynamic characteristics of a wind farm composed of many wind-turbine generators are described first. The model circuit of the wind farm, steady-state analysis, and transient calculations are described.

Next, the modeling of power-electronics circuit elements is described, and thermal calculations using the EMTP are explained.

Photovoltaic and wind power generation of energy necessitates energy storage using batteries. As an application example of a lithium-ion (Li-ion) battery, voltage-regulation equipment for a direct current (DC) railway system is developed based on EMTP simulations. EMTP simulation is explained in detail, and a comparison with the measured results is carried out. EMTP data lists are also given in this chapter.

4.1 Wind Farm

4.1.1 Model Circuit of a Wind Farm

Figure 4.1 illustrates a model circuit of a wind farm. The plant has ten generators with a capacity of 3 MW. Each generator is connected to a cable head through a step-up transformer, whose voltage ratio is 22/1 kV. The total capacity of the plant is 30 MW. The generated voltage is stepped up to 66 kV at Substation-L (S/S-L), and the plant is connected to a grid at S/S-K through a 12-km-long cable (Cable #1). The cable lengths in the plant are listed in Table 4.1. The parameters required for a circuit simulation by the EMTP are evaluated as follows.

The amplitude of the phase-to-ground voltage for the backward system is

$$V_m = \frac{\sqrt{2} \times 66}{\sqrt{3}} = 53.89 \,(\text{kV}) \tag{4.1}$$

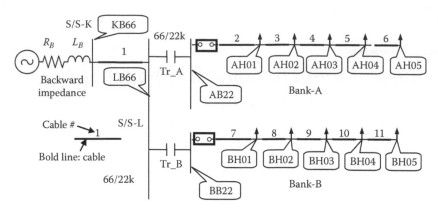

FIGURE 4.1
Circuit diagram of a wind farm.

The rated terminal voltage of the generator is

$$V_m = \frac{\sqrt{2} \times 1}{\sqrt{3}} = 0.81650 \; (\text{kV}) \tag{4.2}$$

The grid impedance $\% Z$ is assumed to be $j2.5\%$ (10-MVA base). The inductance L_B is obtained as follows:

$$Z_B = \frac{\% Z_B V^2}{100 P} = \frac{j2.5 \times 66^2}{100 \times 10} = j10.89 \; (\Omega)$$

$$L_B = \frac{Z_B}{j\omega} = \frac{Z_B}{j2\pi f} = \frac{j10.89}{j2\pi 50} = 34.664 \; (\text{mH}) \tag{4.3}$$

TABLE 4.1

Cable Lengths

Cable #	Size (mm²)	Length (km)
1	600	12.00
2	250	5.00
3	150	1.50
4	150	0.50
5	60	0.50
6	60	1.00
7	250	2.50
8	150	0.50
9	150	1.50
10	60	1.00
11	60	0.50

The capacity and impedance of every transformer installed in the substation S/S-L are 18 MVA and $j10\%$, respectively. Those of the step-up transformer are assumed to be 3.5 MVA and $j10\%$.

The transformer model installed in the EMTP requires leakage inductances and winding resistances. Although leakage inductances can be entered as winding data in theory, that of the secondary winding has to be nonzero in the EMTP. In this section, the winding resistances are combined and entered as the primary resistance. The leakage inductances are entered into a data column for the secondary winding.

The leakage inductance referred to the low-voltage side is obtained as follows:

$$Z_{TrAB} = \frac{\%ZV^2}{100P_{TrAB}} = \frac{j10 \times 22^2}{100 \times 18} = j2.6889\ (\Omega)$$

$$L_{TrAB} = \frac{Z_{TrB}}{j\omega} = \frac{Z_{TrB}}{j2\pi f} = 8.5590\ (\text{mH})$$

(4.4)

The winding resistance is assumed to be 1% of the leakage reactance, that is, the resistive component is almost neglected in this simulation. The resistance referred to the high-voltage side is

$$R_{TrAB} = \frac{Z_{TrAB}}{100}\left(\frac{V_1}{V_2}\right)^2 = \frac{2.6889}{100}\left(\frac{66}{22}\right)^2 = 0.242\ (\Omega)$$

(4.5)

The leakage inductance of the step-up transformer for each generator referred to the low-voltage side is obtained as follows:

$$Z_{TrG} = \frac{\%ZV^2}{100P_{TrAB}} = \frac{j10 \times 1^2}{100 \times 3.5} = j28.571\ (\text{m}\Omega)$$

$$L_{TrG} = \frac{Z_{TrB}}{j\omega} = \frac{Z_{TrB}}{j2\pi f} = 75.788\ (\mu\text{H})$$

(4.6)

The winding resistance is assumed to be 1% of the leakage reactance. The resistance referred to the high-voltage side is

$$R_{TrG} = \frac{Z_{TrAB}}{100}\left(\frac{V_1}{V_2}\right)^2 = \frac{0.028571}{100}\left(\frac{22}{1}\right)^2 = 0.13829\ (\Omega)$$

(4.7)

4.1.2 Steady-State Analysis

Voltage fluctuation caused by the charging current of cables is an important factor for the design of a wind farm. The voltage fluctuation can be simulated by steady-state analysis. In the analysis, the cable can be approximately expressed by a lumped parameter equivalent circuit. Since the steady-state behavior of a three-phase circuit is determined by its positive sequence component, the wind farm can be expressed by a single-phase circuit.

4.1.2.1 Cable Model

A three-phase cable system consisting of three single-core (SC) cables becomes a six-conductor circuit. If its sheath voltages can be neglected, the cable can be expressed by a three-phase circuit. The voltage drop due to the cable-series impedance is expressed by the following equation:

$$\Delta(V)_6 = \Delta \begin{pmatrix} V_{CA} \\ V_{CB} \\ V_{CC} \\ V_{SA} \\ V_{SB} \\ V_{SC} \end{pmatrix} = - \begin{bmatrix} Z_{11} & Z_{12} & \cdots & Z_{16} \\ Z_{21} & Z_{22} & \cdots & Z_{26} \\ \vdots & \vdots & \ddots & \vdots \\ Z_{61} & Z_{62} & \cdots & Z_{66} \end{bmatrix} \begin{pmatrix} I_{CA} \\ I_{CB} \\ I_{CC} \\ I_{SA} \\ I_{SB} \\ I_{SC} \end{pmatrix} \tag{4.8}$$

$$= -[Z]_{66}(I)_6$$

where the first subscripts C and S denote the core and the sheath, and the second subscripts A, B, and C indicate the phases. The voltage and current vectors in Equation 4.8 are defined as

$$(V_C) = \begin{pmatrix} V_{CA} \\ V_{CB} \\ V_{CC} \end{pmatrix}, \quad (I_C) = \begin{pmatrix} I_{CA} \\ I_{CB} \\ I_{CC} \end{pmatrix}$$

$$(V_S) = \begin{pmatrix} V_{SA} \\ V_{SB} \\ V_{SC} \end{pmatrix}, \quad (I_S) = \begin{pmatrix} V_{SA} \\ V_{SB} \\ V_{SC} \end{pmatrix} \tag{4.9}$$

$$(V)_6 = \begin{pmatrix} (V_C) \\ (V_S) \end{pmatrix}, \quad (I)_6 = \begin{pmatrix} (I_C) \\ (I_S) \end{pmatrix}$$

If the sheath voltages are negligible ($(V_S) = (0)$), the 6×6 impedance matrix $[Z]_{66}$ can be reduced to a 3×3 matrix $[Z]_{33}$.

$$\Delta(V)_6 = \Delta\begin{pmatrix}(V_C)\\(V_S)\end{pmatrix} \approx \Delta\begin{pmatrix}(V_C)\\(0)\end{pmatrix} = -[Z]_{66}\begin{pmatrix}(I_C)\\(I_S)\end{pmatrix}$$

$$\begin{pmatrix}(I_C)\\(I_S)\end{pmatrix} = -[Z]_{66}^{-1}\Delta\begin{pmatrix}(V_C)\\(0)\end{pmatrix}$$

$$= -\begin{bmatrix}[Z_{CC}] & [Z_{CS}]\\[Z_{SC}] & [Z_{SS}]\end{bmatrix}^{-1}\Delta\begin{pmatrix}(V_C)\\(0)\end{pmatrix} \tag{4.10}$$

$$= -\Big[\big([Z_{CC}]-[Z_{CS}][Z_{SS}]^{-1}[Z_{SC}]\big)^{-1}\Big]\Delta\begin{pmatrix}(V_C)\\(0)\end{pmatrix}$$

$$\Delta(V_C) = \big([Z_{CC}]-[Z_{CS}][Z_{SS}]^{-1}[Z_{SC}]\big)(I_C)$$

$$= -\big([Z_{CC}]-[Z_{CS}][Z_{SS}]^{-1}[Z_{CS}]_t\big)(I_C)$$

$$= -[Z]_{33}(I_C)$$

In the same manner, the admittance matrix of the line can be reduced to a 3×3 matrix:

$$\Delta(I)_6 = \Delta\begin{pmatrix}I_{CA}\\I_{CB}\\I_{CC}\\I_{SA}\\I_{SB}\\I_{SC}\end{pmatrix} = -\begin{bmatrix}Y_{11} & Y_{12} & \cdots & Y_{16}\\Y_{21} & Y_{22} & \cdots & Y_{26}\\\vdots & \vdots & \ddots & \vdots\\Y_{61} & Y_{62} & \cdots & Y_{66}\end{bmatrix}\begin{pmatrix}V_{CA}\\V_{CB}\\V_{CC}\\V_{SA}\\V_{SB}\\V_{SC}\end{pmatrix}$$

$$= -[Y]_{66}(V)_6 \tag{4.11}$$

$$\Delta(I)_6 = \Delta\begin{pmatrix}(I_C)\\(I_S)\end{pmatrix} = -[Y]_{66}\begin{pmatrix}(V_C)\\(V_S)\end{pmatrix} \approx -[Y]_{66}\begin{pmatrix}(V_C)\\(0)\end{pmatrix}$$

$$= -\begin{bmatrix}[Y_{CC}] & [Y_{CS}]\\[Y_{SC}] & [Y_{SS}]\end{bmatrix}\begin{pmatrix}(V_C)\\(0)\end{pmatrix} \tag{4.12}$$

$$\Delta(I_C) = -[Y_{CC}](I_C) = -[Y]_{33}(I_C)$$

For a steady-state analysis, a cable can be expressed by a single or a cascaded π-equivalent circuit instead of a distributed parameter line. In the EMTP, even if a cable is represented by a constant-parameter line model (Dommel's line model) or a frequency-dependent line model (Semlyen's or

TABLE 4.2

Technical Data for a Cable

		Example	
Nominal cross section of the conductor	S	600	mm²
Outer diameter of the conductor	$2r_2$	29.5	mm
Insulation thickness	d_i	10	mm
Thickness of the metallic sheath (screen)	d_s		
Thickness of the corrosion-proof layer	d_c	4	mm
Outer diameter of the cable	$2r_5$	67	mm
Conductor resistance	R_{dc}	30.8	mΩ/km

Marti's line model), the distributed parameter line is internally converted into a π-equivalent circuit and is passed to a steady-state analysis routine.

If the cable impedance and admittance per unit length are provided by the cable manufacturer, the data of the π-equivalent circuit can be easily calculated with the cable length. The cable impedance and admittance can be calculated by cable constants or cable parameters installed in the EMTP using the physical parameters of the cable. The parameters shown in Table 4.2 are generally provided by the manufacturer.

Although the cross section of the conductor S is given, the radius of the conductor r_2 is obtained from the outer diameter of the conductor $(S \neq \pi r_2^2)$. The inner radius of the metallic sheath r_3 is obtained from the radius of the conductor and the thickness of the insulator as $r_3 = r_2 + d_i$. The outer radius of the metallic sheath r_4 is obtained from the inner radius and the thickness of the sheath as $r_4 = r_3 + d_s$. If the sheath (screen) consists of metallic wires, its thickness is assumed to be the diameter of the wire screen. The outer radius of the cable r_5 can be directly obtained from the diameter of the cable. If the thickness of the metallic sheath is not given and the thickness of the corrosion-proof layer is given, the outer diameter of the metallic sheath is obtained from the cable radius and thickness as $r_4 = r_5 - d_s$.

The resistivity of the conductor ρ_c is obtained from the conductor resistance R_{dc} and the cross section of the conductor S $(\rho_c = R_{dc}\pi r_2^2 \neq R_{dc}S)$. In general, the resistivity is greater than that of the intrinsic resistivity of the conductor (e.g., copper: 1.8×10^{-8} Ω m) because of the gap within the stranded conductor. If the resistivity of the metallic sheath is not provided by the manufacturer, it is obtained by the resistance and its cross section is obtained in the same manner as the main conductor. The relative permittivities of the main insulator and the corrosion-proof layer are determined by their materials. For example, the relative permittivity of cross-linked polyethylene (XLPE) is 2.3. The permittivity of the corrosion-proof layer ranges widely. However, the value has no effect on the positive sequence impedance and admittance of the cable.

The data for cable impedance and admittance calculation using the EMTP are shown in List 4.1. Table 4.3 lists the calculated cable parameters of the

TABLE 4.3

Cable Parameters

Voltage (kV)	Size (mm²)	R (Ω/km)	L (mH/km)	C (μF/km)
66	600	0.059	0.182	0.247
22	250	0.143	0.509	0.232
22	150	0.202	0.490	0.191
22	60	0.388	0.619	0.139

TABLE 4.4

Cable Parameters for π-Equivalent Circuit

Cable #	Size (mm²)	Length (km)	R (Ω)	L (mH)	C (μF)	Voltage (kV)
1	600	4.00	0.235	0.726	0.989	66
1	600	4.00	0.235	0.726	0.989	66
1	600	4.00	0.235	0.726	0.989	66
2	250	5.00	0.713	2.544	1.159	22
3	150	1.50	0.303	0.735	0.287	22
4	150	0.50	0.101	0.245	0.096	22
5	60	0.50	0.194	0.310	0.070	22
6	60	1.00	0.388	0.619	0.139	22
7	250	2.50	0.357	1.272	0.579	22
8	150	0.50	0.101	0.245	0.096	22
9	150	1.50	0.303	0.735	0.287	22
10	60	1.00	0.388	0.619	0.139	22
11	60	0.50	0.194	0.310	0.070	22

XLPE cables. The parameter of the positive sequence is employed for the steady-state voltage simulation. If the cable parameters are provided by the cable manufacturer, parameter calculation by cable constants or cable parameters is not required. The model parameters are obtained from the cable parameters and length as shown in Table 4.4.

List 4.1: Data for Cable Parameter Calculation

```
1    BEGIN NEW DATA CASE
2    CABLE CONSTANTS
3    PUNCH
4    C TY] [SYS] [NPC] [EAR] [KMO] [ZFL] [YFL] [NPP] [NGD]
5        2    -1    3    0    1    2    2         3
6    C NP] [NCR] [IRS] [XMAJOR] [RSG] !
7        1    0    0      1000.    1.E-1B
8    C N1] [N2] [N3] [N4] [N4] [N5] [N6] [N7] [N8] [N9] [N10] [N11] [N12] [N13] [N14] [N15]
9        2    2    2
10   C    r1 ][    r2    ][    r3    ][    r4    ][    r5    ][    r6    ][    r7    ]
11   C       0.0 4.650E-03 1.165E-02 1.270E-02 1.500E-02 {22kV XLPE  60sqmm
```

```
12    C       0.0 7.350E-03  1.435E-02  1.550E-02  1.800E-02  {22kV XLPE 150sqmm
13    C       0.0 9.500E-03  1.650E-02  1.730E-02  2.000E-02  {22kV XLPE 250sqmm
14            0.0 1.475E-02  2.475E-02  2.950E-02  3.350E-02  {66 kV XLPE 600sqmm
15    C rohc  ][  muc   ][  mui1  ][  epsi1  ][  rohs  ][   mus  ][  mui2  ][  epsi2  ]
16       2.100E-08      1.0       1.0       2.3 2.100E-08       1.0       1.0       3.3
17    C   r1 ][     r2    ][    r3    ][    r4    ][    r5    ][    r6    ][    r7    ]
18    C       0.0 4.650E-03  1.165E-02  1.270E-02  1.500E-02  {22kV XLPE  60sqmm
19    C       0.0 7.350E-03  1.435E-02  1.550E-02  1.800E-02  {22kV XLPE 150sqmm
20    C       0.0 9.500E-03  1.650E-02  1.730E-02  2.000E-02  {22kV XLPE 250sqmm
21            0.0 1.475E-02  2.475E-02  2.950E-02  3.350E-02  {66kV XLPE 600sqmm
22    C rohc  ][  muc   ][  mui1  ][  epsi1  ][  rohs  ][   mus  ][  mui2  ][  epsi2  ]
23       2.100E-08      1.0       1.0       2.3 2.100E-08       1.0       1.0       3.3
24    C   r1 ][     r2    ][    r3    ][    r4    ][    r5    ][    r6    ][    r7    ]
25    C       0.0 4.650E-03  1.165E-02  1.270E-02  1.500E-02  {22kV XLPE  60sqmm
26    C       0.0 7.350E-03  1.435E-02  1.550E-02  1.800E-02  {22kV XLPE 150sqmm
27    C       0.0 9.500E-03  1.650E-02  1.730E-02  2.000E-02  {22kV XLPE 250sqmm
28            0.0 1.475E-02  2.475E-02  2.950E-02  3.350E-02  {66kV XLPE 600sqmm
29    C rohc  ][  muc   ][  mui1  ][  epsi1  ][  rohs  ][   mus  ][  mui2  ][  epsi2  ]
30       2.100E-08      1.0       1.0       2.3 2.100E-08       1.0       1.0       3.3
31    C Vert1  ][  Horiz1  ][  Vert2 ][  Horiz2  ][  Vert3  ][  Horiz3 ][  Vert4 ][ Horiz4 ]
32         0.50      -.15      0.50       0.0       0.50       .15
33    C    0.50      -0.3      0.50       0.0       0.50       0.3
34    C    rohe    ][     Freq.  ][DEC] [PNT][  DIST  ][  IPUN  ]
35           100.              50.                1000.
36    BLANK ending frequency data
37    $PUNCH
38    BLANK ending cable constant case
39    BEGIN NEW DATA CASE
40    BLANK
```

4.1.2.2 Charging Current

Table 4.5a shows the analytical charging currents of the cables

$$I = \frac{\omega C V}{\sqrt{3}} = \frac{2\pi f C V}{\sqrt{3}}$$ (4.13)

where
  V is the system voltage, which is 22 or 66 kV
  C is the capacitance of each cable
  f is the power frequency

The charging current of Bank-A in S/S-L is 6.984 A (0.266 MVA) and that of Bank-B is 4.672 A (0.178 MVA). Their ratio is almost identical to the ratio of the cable length. The total charging current of the system is 35.513 A (4.060 MVA), which is mainly determined by the cable connecting S/S-K and S/S-L. The calculated charging currents shown in Table 4.5b are slightly larger than those derived from the analytical calculation. The difference comes from the voltage increase due to charging, that is, the leading current. From the calculated voltage shown in Table 4.6, it can be seen that a voltage rise of 1% increases the charging current by 1%.

The calculation is carried out by the EMTP using the data shown in List 4.2.

TABLE 4.5

Charging Current

(a) Analytical Result

Cable#	Length (km)	Charging Current (A)	Charging Capacity (MVA)	Voltage (kV)
1	12.0	35.513	4.060	66
2	5.0	4.624	0.176	22
3	1.5	1.145	0.044	22
4	0.5	0.382	0.015	22
5	0.5	0.278	0.011	22
6	1.0	0.556	0.021	22
Subtotal Bank-A	8.5	6.984	0.266	22
7	2.5	2.312	0.088	22
8	0.5	0.382	0.015	22
9	1.5	1.145	0.044	22
10	1.0	0.556	0.021	22
11	0.5	0.278	0.011	22
Subtotal Bank-B	6.0	4.672	0.178	22
Total 22 kV		11.657	0.444	22
Total 66 kV		39.399	4.504	66

(b) Calculated Result

Node-M	Current (A)	Power (MW)	Reactive (MVA)
Subtotal Bank-A	7.08	0.00	−0.273
Subtotal Bank-B	4.73	0.00	−0.183
Total 66 kV	39.86	0.00	−4.609

TABLE 4.6

Node Voltages

Node	W/O Gen.	With Gen.
KB66_A	1.0114	0.9951
LB66_A	1.0118	0.9994
AB22_A	1.0133	0.9901
AH01_A	1.0136	1.0099
AH02_A	1.0136	1.0168
AH03_A	1.0136	1.0186
AH04_A	1.0136	1.0209
AH05_A	1.0136	1.0231
BB22_A	1.0128	0.9904
BH01_A	1.0129	1.0005
BH02_A	1.0129	1.0028
BH03_A	1.0129	1.0081
BH04_A	1.0129	1.0127
BH05_A	1.0129	1.0138

List 4.2: Voltage Distribution by Cable Charging

```
1    BEGIN NEW DATA CASE
2    C WER FREQUENCY                        [STATFR]
3    POWER FREQUENCY                          50.0
4    C FIX SOURCE
5    C    DT][ TMAX ][ XOPT ][ COPT ][EPSILN][TOLMAT]
6         0.0    0.0
7    C IOUT][ IPLOT][IDOUBL][KSSOUT][MAXOUT][ IPUN ][MEMSAV][ ICAT ][NENERG][IPRSUP]
8        100      1      1      3      1
9    C ----------------------------------------------------------- Backward impedance
10   C [BUS1][BUS2][BUS3][BUS4][ R ][ L ][ C ]                                      !
11     SOUR_AKB66PA                34.664    { ZB, 2.5%, 10 MVA }                    0
12   C -------------------------------------------------- Transformer A 66/22 kV
13   C TRANSFORMER [BUS3]         [Iste][Pste][BTOP][Rmag]                          !
14     TRANSFORMER                         LT66AA
15             9999
16   C [BUS1][BUS2]              [ Rk ][ Lk ][Volt]                                 !
17   1LB66AA                     0.242   0.0   66. { ! caution Y-Y }                0
18   2AB22_A                     0.08.5590   22. { 18 MVA, 10% }                    0
19   C TRANSFORMER  [BUS3]        [Iste][Pste][BTOP][Rmag]                          !
20     TRANSFORMER LT66AA                  LT66BA
21   1LB66BA                                                                        0
22   2BB22_A                                                                        0
23   C -------------------------------------------------- Stepup Transformer 22/1 kV
24   C TRANSFORMER [BUS3]         [Iste][Pste][BTOP][Rmag]                          !
25   TRANSFORMER                         AT01_A
26             9999
27   C [BUS1][BUS2]              [ Rk ][ Lk ][Volt]                                 !
28   1AH01_A                     .13829   0.0   22. { ! caution Y-Y 6.0%
29   2AG01_A                     .07579   1.0 { 3.5 MVA, 10% }
30     TRANSFORMER AT01_A                  AT02_A
31   1AH02_A
32   2AG02_A
33     TRANSFORMER AT01_A                  AT03_A
34   1AH03_A
35   2AG03_A
36     TRANSFORMER AT01_A                  AT04_A
37   1AH04_A
38   2AG04_A
39     TRANSFORMER AT01_A                  AT05_A
40   1AH05_A
41   2AG05_A
42     TRANSFORMER AT01_A                  BT01_A
43   1BH01_A
44   2BG01_A
45     TRANSFORMER AT01_A                  BT02_A
46   1BH02_A
47   2BG02_A
48     TRANSFORMER AT01_A                  BT03_A
49   1BH03_A
50   2BG03_A
51     TRANSFORMER AT01_A                  BT04_A
52   1BH04_A
53   2BG04_A
54     TRANSFORMER AT01_A                  BT05_A
55   1BH05_A
56   2BG05_A
57   C ----------------------------------------------------------- 22 kV CABLE
58   $VINTAGE, 1
59   C [BUS1][BUS2][BUS3][BUS4][    R11    ][    L11    ][    C11    ]
60   1KB66_AMB66_A               2.348088570E-01 7.260036593E-01 9.888645200E-01
61   1MB66_ANB66_A               2.348088570E-01 7.260036593E-01 9.888645200E-01
62   1NB66_ALB66_A               2.348088570E-01 7.260036593E-01 9.888645200E-01
63   1AH00_AAH01_A               7.134974733E-01 2.544378987E+00 1.158866150E+00
64   1AH01_AAH02_A               3.028793115E-01 7.347611338E-01 2.868727050E-01
```

```
65    1AH02_AAH03_A              1.009597705E-01  2.449203779E-01  9.562423500E-02
66    1AH03_AAH04_A              1.942236877E-01  3.095703631E-01  6.965880500E-02
67    1AH04_AAH05_A              3.884473753E-01  6.191407261E-01  1.393176100E-01
68    1BH00_ABH01_A              3.567487367E-01  1.272189493E+00  5.794330750E-01
69    1BH01_ABH02_A              1.009597705E-01  2.449203779E-01  9.562423500E-02
70    1BH02_ABH03_A              3.028793115E-01  7.347611338E-01  2.868727050E-01
71    1BH03_ABH04_A              3.884473753E-01  6.191407261E-01  1.393176100E-01
72    1BH04_ABH05_A              1.942236877E-01  3.095703631E-01  6.965880500E-02
73    $VINTAGE,  0
74    BLANK ending BRANCH
75    C [BUS1] [BUS2] MEASURING                                                      !
76       AB22_AAH00_A                                 MEASURING            1
77       BB22_ABH00_A                                 MEASURING            1
78       LB66_ALB66AA                                 MEASURING            1
79       LB66_ALB66BA                                 MEASURING            1
80       KB66PAKB66_A                                 MEASURING            1
81    BLANK ending SWITCH
82    C ------------------------------------------------------------- 66 kV system
83    C [BUS1] [] [  AMP.  ] [  FREQ.  ] [  ANGLE  ]          [ TSTART ] [ TSTOP ]
84    14SOUR_A  53.88878E3      50.0       -90.0                          -1.
85    BLANK ending Source
86    C [BUS1]  [BUS2]  [BUS3]  [BUS4]  [BUS5]  [BUS6]  [BUS7]  [BUS8]  [BUS9]  [BUSA]  [BUSB]  [BUSC]  [BUSD]
87       KB66_ALB66_A
88       AB22_A
89       AH01_AAH02_AAH03_AAH04_AAH05_A
90       BB22_A
91       BH01_ABH02_ABH03_ABH04_ABH05_A
92    BLANK ending OUTPUT Specification
93    BLANK ending PLOT
94    BEGIN NEW DATA CASE
95    BLANK
```

4.1.2.3 Load-Flow Calculation

It can be easily estimated that the voltage distribution within a wind farm depends on the operation of the generators. The effective power of each generator is mainly determined by the phase difference, and the reactive power is determined by the amplitude of the voltage. However, the correct values cannot be obtained by a linear calculation. The EMTP has a load-flow feature called "FIX SOURCE." In this section, an example is shown assuming that every generator operates at its rated capacity and the reactive power is controlled to be zero.

The key word "FIX SOURCE" should be inserted before the miscellaneous data (uncomment Line 4 in List 4.2). The source data expressing the generators (List 4.3) are inserted before Line 85 in List 4.2. The amplitudes and angles of the sources are automatically corrected by the FIX SOURCE routine. The entered values are used as initial values of the nonlinear calculation in the FIX SOURCE routine. In general, each amplitude of the source is specified as the amplitude of the phase-to-ground voltage given in Equation 4.2. The phase angle is given by the angle of the source of the backward grid specified by Line 84 in List 4.2, because the system is expressed by a single-phase system, in this case. If the system is expressed by a three-phase system, the phase shift by a transformer winding, that is, Y-Δ or Δ-Y connection should be taken into account.

List 4.3: Source Data for Generators

```
1   C [BUS1][] [  AMP.   ] [  FREQ.  ] [  ANGLE  ]        [ TSTART ] [  TSTOP ]
2   14AG01_A    8.1650E+2     50.0       -90.0              -1.
3   14AG02_A    8.1650E+2     50.0       -90.0              -1.
4   14AG03_A    8.1650E+2     50.0       -90.0              -1.
5   14AG04_A    8.1650E+2     50.0       -90.0              -1.
6   14AG05_A    8.1650E+2     50.0       -90.0              -1.
7   14BG01_A    8.1650E+2     50.0       -90.0              -1.
8   14BG02_A    8.1650E+2     50.0       -90.0              -1.
9   14BG03_A    8.1650E+2     50.0       -90.0              -1.
10  14BG04_A    8.1650E+2     50.0       -90.0              -1.
11  14BG05_A    8.1650E+2     50.0       -90.0              -1.
12  BLANK ending Source
```

The generator powers have to be present in the data shown in List 4.4, followed by a BLANK line for terminating the source data (after Line 85 in List 4.2). The power is one-third of the generated power, because the system is expressed by a single-phase model, in this case. The power of the 3-MW generator is specified by Line 3 of the data shown in List 4.4.

List 4.4: Additional Data for FIX SOURCE

```
1   BLANK ending Source
2   C [BUS1] [BUS2] [BUS3] [    Pk     ] [    Qk     ] [ VMIN  ] [  VMAX  ] [THMI] [THMA]
3       AG01_A              1.000E+6         0.0
4       AG02_A              1.000E+6         0.0
5       AG03_A              1.000E+6         0.0
6       AG04_A              1.000E+6         0.0
7       AG05_A              1.000E+6         0.0
8       BG01_A              1.000E+6         0.0
9       BG02_A              1.000E+6         0.0
10      BG03_A              1.000E+6         0.0
11      BG04_A              1.000E+6         0.0
12      BG05_A              1.000E+6         0.0
13  C        [NNNOUT] [NITERA] [NFLOUT] [NPRINT] [RALCHK] [CFITEV] [CFITEA] [VSCALE] [KTAPER]
14                       1      6000              1     0.001    0.005    0.10              2
```

The miscellaneous data for the load-flow calculation (Line 14 in List 4.4) should be specified, followed by the generator powers. The calculated result is shown in List 4.5. The outputs of the generators are converged to the specified data within 1% error.

List 4.5: Calculated Load Flow

```
Exit the load flow iteration loop with counter NEKITE=1366. If no warning on the
preceding line, convergence was attained.
  Row  Node  Name     Voltage magnit    Degrees    Real power P    Reactive power
    2    36  AG01_A   8.23225827E+02   -75.33539   9.99043389E+05  7.08178870E+01
    3    37  AG02_A   8.28950371E+02   -75.04581   9.99017237E+05  6.16391826E+01
    4    38  AG03_A   8.30382307E+02   -74.97387   9.99010832E+05  5.93467166E+01
    5    39  AG04_A   8.32251107E+02   -74.91679   9.99005868E+05  5.62211791E+01
    6    40  AG05_A   8.34117694E+02   -74.85980   9.99000958E+05  5.31057736E+01
    7    41  BG01_A   8.15497500E+02   -75.93822   9.99146319E+05  7.91157970E+01
    8    42  BG02_A   8.17424393E+02   -75.83980   9.99137898E+05  7.62948678E+01
    9    43  BG03_A   8.21754221E+02   -75.61970   9.99119250E+05  6.99421381E+01
   10    44  BG04_A   8.25520356E+02   -75.50343   9.99109641E+05  6.41407115E+01
   11    45  BG05_A   8.26461175E+02   -75.47438   9.99107270E+05  6.26963934E+01
```

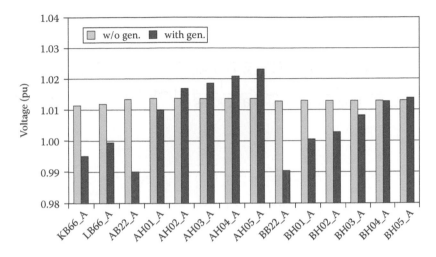

FIGURE 4.2
Calculated steady-state voltage.

In general, the load-flow calculation requires a long computational time. If the initial voltages and angles are specified by the data shown in List 4.5, the time of the subsequent calculations will be fairly reduced.

Figure 4.2 illustrates a calculated result of cable energization, that is, when all generators are disconnected (w/o gen., Table 4.6) and when the generators are operated at their rated capacity (with gen.). Voltage increase in the case without generators is due to the leading current for cable charging. The minor voltage differences within the plant indicate that the voltage is increased by the cable between the grid and the substation S/S-K. When generators are operated, each voltage increases as the distance between the substation and the generator increases. This is due to the voltage drop caused by cable impedance.

4.1.3 Transient Calculation

Figure 4.3 illustrates the calculated transient responses when the CB at Bank-B is closed for charging the cables at Bank-B while the generators at Bank-A are fully operated. The initial conditions are determined by the load-flow feature of the EMTP by specifying each output power of the Bank-A generator to be 3 MW, that is, 1 MW for each phase. Although the effect of the closing on the 66 kV bus voltage is minor, the maximum bus voltage of Bank-B is increased to 2 pu and oscillates with a frequency of 1.6 kHz. The oscillation is principally generated by a resonance between the leakage inductance of the transformer L_{TrB} shown in Equation 4.4 and the total cable capacitance of Bank-B shown in Table 4.4 (cable No. 7–11).

$$f = \frac{1}{\pi\sqrt{L_{TrB}C}}$$

$$= \frac{1}{2\pi\sqrt{(8.559 + 8.559)\,\text{mH} \times (0.579 + 0.096 + 0.287 + 0.139 + 0.070)\,\mu\text{F}}}$$

$$= \frac{1}{2\pi\sqrt{17.118\,\text{mH} \times 1.171\,\mu\text{F}}} = \frac{1}{0.6290\,\text{ms}} = 1.590\,(\text{kHz}) \qquad (4.14)$$

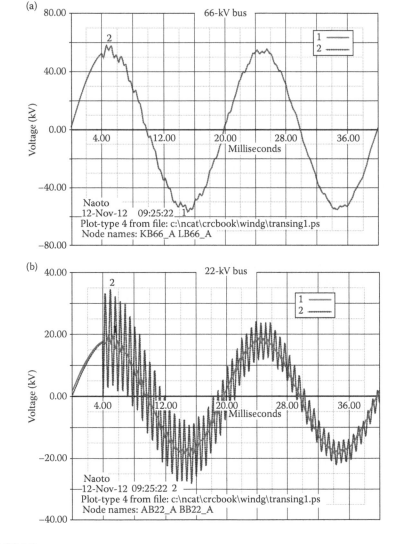

FIGURE 4.3
Calculated transient voltage: (a) 66 kV bus voltages and (b) 22 kV bus voltages.   (*Continued*)

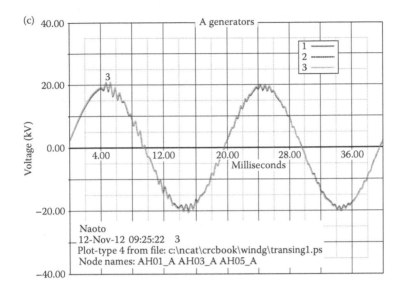

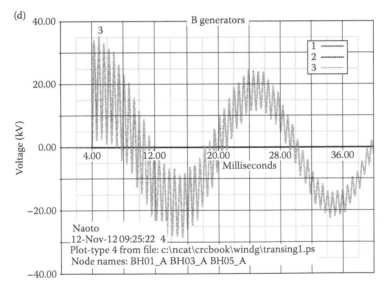

FIGURE 4.3 (Continued)
Calculated transient voltage: (c) Bank-A voltages and (d) Bank-B voltages.

4.2 Power-Electronics Simulation Using the EMTP

The simulation of a switching circuit is important while designing equipment using the power-electronics technique. For the simulation of an electronics circuit, numerical simulators specialized in the circuit, such as simulation program with integrated circuit emphasis (SPICE), are widely used. Although they have accurate semiconductor models and high functionality to simulate the behavior of the electronics circuit, they have no model of power-system equipment, such as accurate multiphase transmission line models, synchronous machines, etc. The programs cannot be applicable to power-system analysis involving power-electronics equipment, such as an inverter.

The solutions are the following: (1) expansion of the electronics simulator to power-system analysis by developing some modes of power apparatuses and (2) expansion of the power-system simulator, such as the EMTP, by developing some modes of semiconductor devices. In this section, the latter method is employed [1–8].

Models of a bipolar transistor and of a metal oxide semiconductor field-effect transistor (MOSFET), along with simulation techniques using the EMTP are explained with EMTP data in this section. The techniques of these devices are applicable to model an insulated-gate bipolar transistor (IGBT). Transient analysis of controlled systems (TACS) or MODELS (see Section 1.8.3) installed in the EMTP are indispensable for modeling. These features were originally developed for modeling a control circuit of a power system. They can be applicable to express the characteristics of the semiconductor because their functionality and generality are quite high. In general, the required accuracy of the semiconductor model is lower than that of an electronics-circuit simulator for power-system simulation. The model should be as simple as possible if the accuracy requirement of the power-system simulation is satisfied.

4.2.1 Simple Switching Circuit

Figure 4.4 illustrates a logical inverter (NOT) circuit using a bipolar transistor. The source voltage V_{CC} and the collector resistance R_C are assumed to be 5 V and 4.7 kΩ, respectively.

The collector current I_C becomes

$$I_C = \frac{V_{CC}}{R_C} = \frac{5\,(\text{V})}{4.7\,(\text{k}\Omega)} \approx 1\,(\text{mA}) \tag{4.15}$$

If the current gain of the transistor h_{FE} is 100, the base current should be greater than 10 μA:

$$I_B \geq \frac{I_C}{h_{FE}} = \frac{1\,(\text{mA})}{100} = 10\,(\mu\text{A}) \tag{4.16}$$

An input signal is applied by a signal source with an internal impedance of 50 Ω (R_{sig}). The base current I_B is obtained from the amplitude of the input voltage V_{Sigout} and base resistance R_{b1}:

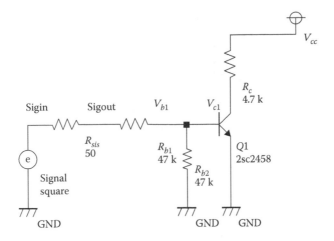

FIGURE 4.4
Switching circuit.

$$I_B \approx \frac{V_{Sigin} - V_{BE}}{R_{Sig} + R_{b1}} = \frac{V_{Sigout} - V_{BE}}{R_{b1}}, \quad I_B \gg \frac{V_{BE}}{R_{b2}} \tag{4.17}$$

Equation 4.17 gives the maximum base resistance R_{b1}:

$$R_{b1} = \frac{V_{Sigout} - V_{BE}}{I_B} \leq \frac{V_{Sigout} - V_{BE}}{I_C / h_{FE}} = \frac{(5 - 0.7)\,(V)}{1\,(mA)/100} = 430\,(k\Omega) \tag{4.18}$$

In this section, base resistance R_{b1} is assumed to be 47 kΩ. R_{b2} is required for discharging the charge remaining in the transistor. The resistance is 47 kΩ.

4.2.2 Switching Transistor Model

In this section, a simulation method for a basic switching circuit using the EMTP is explained. The technique can be applied to a power-switching device.

The base–emitter characteristic can be expressed by a nonlinear resistor model installed in the EMTP. Although both TYPE-92 and TYPE-99 models accept point-by-point data expressing its current–voltage characteristic, only the true nonlinear resistor model (TYPE-92) is suitable for the simulation from the viewpoint of stability. This characteristic is easily obtained from the data sheet of the transistor.

4.2.2.1 Simple Switch Model

The simplest switching transistor model of the EMTP is the TACS-controlled switch model (TYPE-13) illustrated in Figure 4.5.

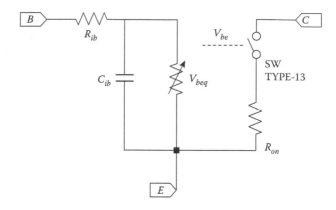

FIGURE 4.5
Simplest switching transistor model.

If the saturation voltage between the collector and the emitter is not negligible, a resistor R_{on} is inserted in series with the switch. The resistor R_{ib} and the capacitor C_{ib} represent a base-spreading resistance and a base input capacitance, respectively.

The "OPEN/CLOSE" signal of the TYPE-13 switch is synthesized within the TACS from the base–emitter voltage of the transistor. Figure 4.6 shows the flow chart of the control signal. List 4.6 and Figure 4.7 show the input data for the simple model. The EMTP data are given as a text file, that is, by character user interface (CUI). ATP-Draw was developed for data input by a graphical user interface (GUI). Although the latter method is easy to use and to grasp the configuration of the circuit, the circuit parameters cannot be obtained from the graphics. In this section, the data are explained using the original data format, CUI.

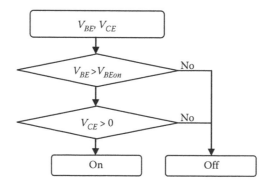

FIGURE 4.6
Flow chart for the simplest switching transistor model.

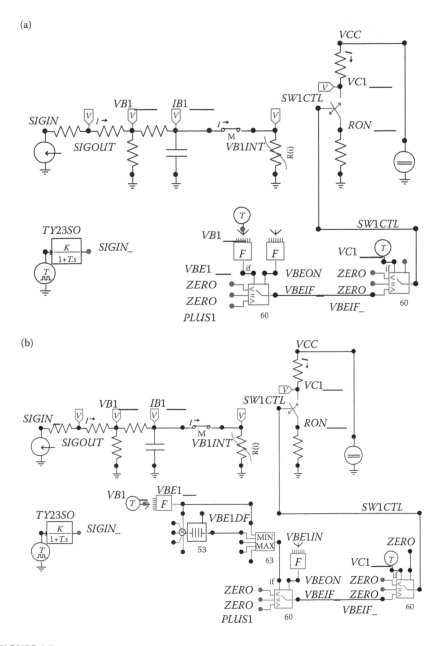

FIGURE 4.7
EMTP data for a simple switching circuit (ATP-Draw, GUI): (a) without delay and (b) with delay.

List 4.6: EMTP Data for Simple Switching Circuit (Conventional Format, CUI)

```
 1    BEGIN NEW DATA CASE { ---------------------------------------------- C2458NOT1.dat
 2    C Logical NOT (Inverter) circuit simulation using 2SC2458
 3    C   DT][ TMAX ][ XOPT ][ COPT ] [EPSILN] [TOLMAT]
 4      40.E-09 60.E-06
 5    C  IOUT] [IPLOT] [IDOUBL] [KSSOUT] [MAXOUT] [IPUN] [MEMSAV] [ICAT] [NENERG] [IPRSUP]
 6        100      2      1       0       1      0       0       1
 7    TACS HYBRID
 8    C ================================================================ Signal source model
 9    C [VAR1]      [  Amp.  ] [   T(s)  ] [width(s)]                       [ T-START] [ T-END ]
10    23TY23SO          5.0   25.00E-06  12.50E-06
11    C [NAME]   +[ IN1]  +[ IN2]  +[ IN3]  +[ IN4]  +[ IN5]  [GAIN] [F-LO] [F-HI] [N-LO] [N-HI]
12     1SIGIN_  +TY23SO
13         1.0
14         1.0  0.050E-06
15    C  D0  ][  D1  ][  D2  ][   D3   ][   D4   ][  D5   ][  D6  ][  D7   ]
16    C ================================================================================
17    C [NODE]
18    90VB1___   { Base voltage
19    90VC1___   { Collector voltage
20    C VE1___   { Emitter voltage
21    C [NAME]  =------------------------ FREE FORMAT -------------------------------
22    99VBE1__  = VB1___  { -VE1___ }   { Vbe =Vb - Ve
23    99VCE1__  = VC1___  { -VE1___ }   { Vce =Vc - Ve
24    C
25    99VBEON_  = 0.62                  { threshold Vbe on-voltage
26    C $DISABLE { ==============================-============== comment out for simple model
27    C [NAME]60 + [ IN1]  +[IN2]  +[ IN3]                      [CNST]        [SIG1] [SIG2]
28    99VBEIF_60+ ZERO    +ZERO   +PLUS1                                     VBE1__ VBEON_
29    98SW1CTL60+ ZERO    +ZERO   +VBEIF_                                    VCE1__ ZERO
30    C $ENABLE  { ============================================ comment out for simple model
31    $DISABLE { ============================================== comment out for delay model
32    C [NAME]53+[ IN1]   +[ IN2]  +[ IN3]  +[ IN4]  +[ IN5]  [hist] [FDel] [MaxD] [NDel] [Nhis]
33    99VBE1DF53+VBE1__          {transport_delay, tdoff}         1.0E-6
34    C [NAME]63+[ IN1]   +[ IN2]  +[ IN3]  +[ IN4]  +[ IN5]  [CNTL]
35    99VBE1IN63+VBE1__   +VBE1DF         {instantaneous max}      1.0
36    C [NAME]60+[ IN1]   +[ IN2]  +[ IN3]                    [CNST]         [SIG1] [SIG2]
37    99VBEIF_60+ZERO    +ZERO   +PLUS1                                    VBE1INVBEON_
38    98SW1CTL60+ZERO    +ZERO   +VBEIF_                                    VCE1__ ZERO
39    $ENABLE  { ============================================== comment out for delay model
40    BLANK ENDING TACS
41    C  [BUS1] [BUS2] [BUS3] [BUS4] [NFLS]      4444.                                        !
42    92VB1INT                                   4444.   {Rbe}
43    C       R-lin          ][       V-flash     ][       V-zero      ]
44                   0.0                      0.0                     0.0
45    C       Current        ][       Voltage     ]
46              1.00000E-06                  -1.00
47                    0.00                   0.00
48              2.37704E-07              0.539007092
49              8.80783E-07              0.567375887
50              2.35231E-06              0.588652482
51              4.52805E-06              0.602836879
52              8.71621E-06              0.617021277
53              2.04206E-05              0.635460993
54              5.10802E-05              0.655319149
55              0.000105998              0.673758865
56              0.000309901              0.723404255
57              0.000731080              0.763120567
58                     9999
59    C SIGOUT VB1___  IB1___  VB1INT VC1___   VCC
60    C    +--Rb1--+--Rib--+---+       +--RC--+
61    C       |       |    |   |       SW      |
62    C      Rb2    Cib Reqb   |       Vcc     |
63    C       |       |    |   |      Ron      |
64    C    +-------+------+---+------+------ + GND (Emitter)
65    C [BUS1] [BUS2] [BUS3] [BUS4] [ R ][  L ][  C ]                                          !
```

```
66    SIGIN_SIGOUT              50.
67    SIGOUTVB1___             47.E3                      {Rb1}                                          1
68    VB1___                  47.E3                      {Rb2}
69    VCC VC1___               4.7E3                     {RC}                                           1
70    RON___                  10.0                       {Ron}
71    VB1___IB1___            50.0                       {Rib}
72    IB1___                   0.0          20.E-6       {Cib}
73    BLANK ENDING BRANCH CARDS
74    C [BUS1] [BUS2]                                        MEASURING                               !
75    IB1___VB1INT         {Base current sensor}            MEASURING                              1
76    C [BUS1] [BUS2]                                        CLOSED              [CLMP]    !!
77    13VC1___RON___       {Collecter-Emitter }                                  SW1CTL     0
78    BLANK ENDING SWITCH CARDS
79    60SIGIN_             {Signal voltage}
80    C [BUS1][][ AMP. ]                                            [ TSTART ][  TSTOP ]
81    11VCC         5.0     {Power supply}
82    BLANK ENDING SOURCE CARDS
83    SIGOUTVB1INTVC1___VB1___IB1___
84    BLANK ENDING OUTPUT SPECIFICATION CARDS
85    C !![H][ST]END][MIN]MAX][BUS1][BUS2][BUS3][BUS4][ HEADING LABEL][ VERTICAL AXIS]
86    145 5. 10. 60. -10. 10.SIGOUTVB1___VC1___                               VOLTAGE (V)
87    195 5. 10. 60.         SIGOUTVB1___              BASE CUR                CURRENT (A)
88    BLANK ENDING PLOT CARDS
89    BEGIN NEW DATA CASE
90    BLANK
```

The data of the EMTP are divided into two parts: a TACS or MODELS part for the controlling circuit and an electrical part.

At first, a square-wave signal source, with an amplitude of 5 V and a frequency of 40 kHz, is defined in Line 10 in List 4.6, just after "TACS HYBRID" declaration. The signal is defined by the TYPE-23 built-in source and a first-order transfer function (s-block) (Lines 12–14) to represent its rise and fall times. The output "SIGIN_" is sent to the electrical part of the EMTP and is expressed as a voltage source by a TACS-controlled source (TYPE-60, Line 79).

The base and collector voltages (VB1___and VC1___) in the electrical part are sent to the TACS using TYPE-90 TACS sources (Lines 18 and 19). In this case, the voltages are identical to the base–emitter and the collector–emitter voltage (VBE1__and VCE1__) because the emitter is directly grounded. If a circuit has an emitter resistor, that is, the emitter voltage is different from zero, the definitions should be modified by subtraction of the emitter voltage, VE1___, from the base and collector voltages, respectively (Lines 20, 22, and 23).

The threshold voltage V_{Beon} in Figure 4.6 can be obtained from the I_B–V_{BE} characteristic provided by the manufacturer or by an experimental result. The threshold value (Line 25) is determined as the voltage where the base current becomes the minimum base current given in Equation 4.16.

The IF-Devices (Device 60) of the TACS (Lines 28 and 29) are used for the logical judgments illustrated in the flow chart (Figure 4.6). The output SW1CTL is used for the "OPEN/CLOSE" signal of the TYPE-13 switch in the electrical part (Line 77).

The next commented-out data by $DISABLE and $ENABLE (Lines 31–39) are for a delay model, which will be described in Section 4.2.2.2.

The nonlinear characteristic between the base and emitter of the transistor is expressed by a nonlinear resistor R_{beq}. This characteristic can be expressed by a TYPE-92 ZnO arrester model installed in the EMTP (Lines 42–58). The current–voltage characteristic is almost identical to that of a diode, and is given by point-by-point data.

The resistors and the capacitor illustrated in Figures 4.4 and 4.5 are specified as RLC branches (Lines 66–72). In this simulation, the base-spreading resistance R_{ib} and the base input capacitance C_{ib} are assumed to be 50 Ω and 20 pF, respectively.

The MEASURING switch between nodes IB1___ and VB1INT is used for detecting the base current (Line 75). The TYPE-13 switch expresses the switching operation of the transistor (Line 77). The "OPEN/CLOSE" signal SW1CTL is defined in the TACS (Line 29).

The TYPE-60 source (SIGIN_) expresses the signal source and the TYPE-11 source is for the power source, V_{CC} (Lines 79 and 81).

Figures 4.8 and 4.9 illustrate the measured and the calculated results of the switching circuit. The switching operation can be simply expressed by the simple model, although the time delay at the turn off cannot be reproduced by the model. The model is accurate enough if the switching frequency is much lower than the transition frequency f_T of the transistor.

4.2.2.2 Switch with Delay Model

The accuracy of the simple switch model decreases as the frequency of the signal source is increased due to the delay of the transistor. The turn-off delay is generally greater than the turn-on delay. The turn-off delay is easily included into the model using a pulse-delay device (Device 53) and an

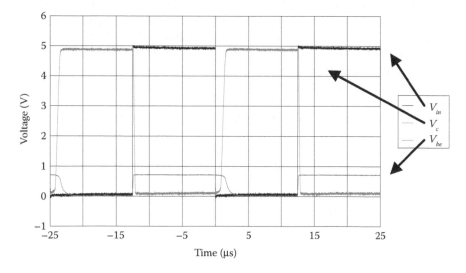

FIGURE 4.8
Measured result.

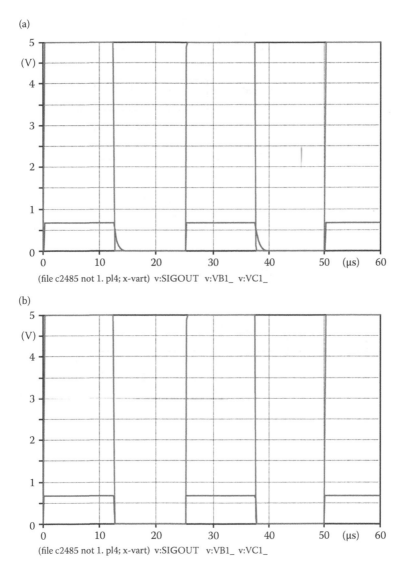

FIGURE 4.9
Calculated result by a simple switch model: (a) $C_{ib} = 2\,0$ pF and (b) $C_{ib} = 0$.

instantaneous-maximum device (Device 63) of the TACS. Figure 4.10 and List 4.7 (Lines 33–38 in List 4.6) illustrate the control algorithm.

The transistor model represented by a switch cannot express the fall time of the collector–emitter voltage V_{CE}, although it reproduces the switching delay time (Figure 4.11). The fall time is quite important for the thermal design of a switching circuit. A more generalized transistor model can be expressed by nonlinear resistances as shown in Figure 4.12.

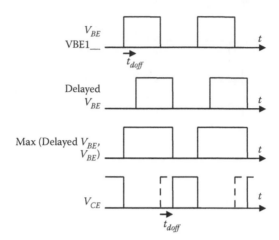

FIGURE 4.10
Block diagram of an off-delay representation.

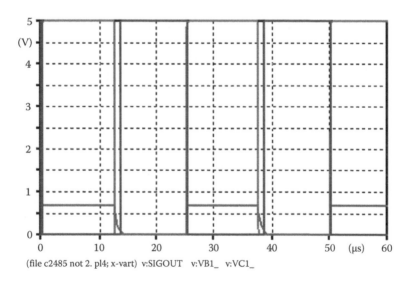

(file c2485 not 2. pl4; x-vart) v:SIGOUT   v:VB1_   v:VC1_

FIGURE 4.11
Calculated result with the turn-off delay.

List 4.7: Turn-Off Delay by Device 53

```
1  C [NAME]53+[IN1]  +[IN2]  +[IN3]  +[IN4]+[IN5][hist][FDel][MaxD][NDel][Nhis]
2  99VBE1DF53+VBE1__           {transport_delay, tdoff}   1.0E-6
3  C [NAME]63+[IN1]  +[IN2]  +[IN3]  +[IN4]+[IN5]         [CNTL]
4  99VBE1IN63+VBE1__+VBE1DF       {instantaneous max}  1.0
5  C [NAME]60+[IN1]  +[IN2]  +[IN3]               [CNST]        [SIG1][SIG2]
6  99VBEIF_60+ZERO   +ZERO   +PLUS1                        VBE1INVBEON_
7  98SW1CTL60+ZERO   +ZERO   +VBEIF_                       VCE1__ZERO
```

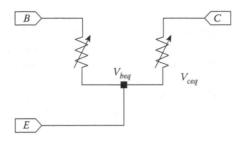

FIGURE 4.12
Nonlinear resistor model of a transistor.

The nonlinear characteristic between the collector and emitter cannot be expressed by any conventional nonlinear resistor model, such as the TYPE-92 resistor, because its characteristic depends not only on its terminal voltage V_{CE} but also on the base current I_B. The resistor R_{ceq} is expressed by the TYPE-91 TACS-controlled resistor. Its resistance is calculated in the TACS according to the collector–emitter voltage V_{CE}, base current I_B, and transient characteristic of a transistor. This modeling technique is explained in Section 4.2.3.

4.2.3 Metal Oxide Semiconductor Field-Effect Transistor

MOSFET is widely used as a switching device for high-frequency operations. The drain current of the MOSFET is controlled by its gate-source voltage V_{GS}. Generally, MOSFET comes with a built-in reverse diode. Two MOSFET models are introduced in Sections 4.2.3.1 and 4.2.3.2.

4.2.3.1 Simple Model

The simplest model of a MOSFET with a reverse diode can be expressed using a switch and a diode as illustrated in Figure 4.13. The switch status

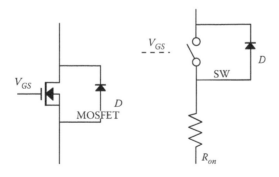

FIGURE 4.13
MOSFET simple model.

is controlled by the gate-source voltage V_{GS}. If the voltage is greater than its threshold voltage V_P, the switch is closed. The control signal is easily produced by the TACS. If the gate-source voltage V_{GS} is smaller than the threshold voltage V_P, the MOSFET acts as a diode for bypassing its reverse current. The model, which consists of a switch and a diode, can be simply expressed by a Type-11 switch (diode) model of the EMTP. The diode model has an "OPEN/CLOSE" signal for controlling the switch status. If the signal is positive, the switch is closed as long as the signal is active. If the signal is zero, the switch acts as an ideal diode. The series-connected resistor R_{on} expresses the on-resistance of the MOSFET.

However, the model cannot be used from the ATP-Draw, because there is no input column for the "OPEN/CLOSE" signal. The switch and the diode have to be separately entered. In this case, the switch and the diode are required to have their own on-resistance, because the EMTP cannot handle parallely connected switches.

4.2.3.2 Modified Switching Device Model

The transistor model described in Section 4.2.3.1 is simple and useful for low-frequency switching circuits. However, the accuracy of the model is reduced as the switching frequency increases. A modified MOSFET model is illustrated in Figure 4.14.

The input (gate-source) circuit of a MOSFET can be simply illustrated by a capacitor C_{iss}. The output (drain-source) circuit is expressed by an equivalent resistor R_{dseq} and a capacitor C_{oss}. The resistance of the equivalent resistor R_{dseq} is controlled by the gate-source voltage, V_{GS}. The static coupling between the gate and the drain is expressed by the reverse transfer capacitance C_{rss}.

Figure 4.15 illustrates an example of the drain-current versus gate-source voltage (I_D–V_{GS}) characteristic. The characteristic should be expressed as accurately as possible for a precise simulation. A function approximation of the characteristic is useful for this purpose.

In a high-voltage region ($V_{GS} \geq V_{th}$), the I_D–V_{GS} characteristic can be expressed by the following linear equation:

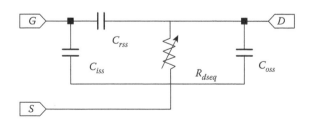

FIGURE 4.14
MOSFET model.

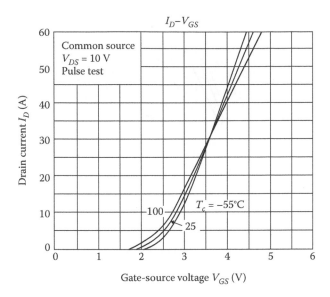

FIGURE 4.15
I_D–V_{GS} characteristic of MOSFET 2SK2844.

$$I_D = a_h V_{GS} + b_h, \quad V_{GS} \geq V_{th} \tag{4.19}$$

In a low-voltage region ($V_{GS} < V_{th}$), the characteristic is approximated by a quadratic function

$$I_D = a_l (V_{GS} - V_p)^2, \quad V_p < V_{GS} < V_{th} \tag{4.20}$$

where V_p is the threshold voltage ($I_D = 0$ at $V_{GS} = V_p$).

The coefficient a_l is obtained by a condition in which both curves are in contact at $V_{GS} = V_{th}$:

$$a_l = -\frac{a_h^2}{4(a_h V_P + b_h)} \tag{4.21}$$

$$V_{th} = -\left(V_p + \frac{2b_h}{a_h}\right) \tag{4.22}$$

Figure 4.16 illustrates a drain-current versus drain-source voltage (I_D–V_{DS}) characteristic. Although the saturating characteristic has no significant effect on switching operations, on-resistance is an important factor. Resistance is expressed as the inverse of the slope of the I_D–V_{DS} characteristic in a low-voltage region.

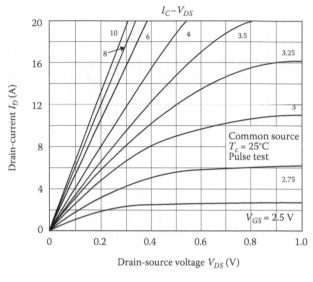

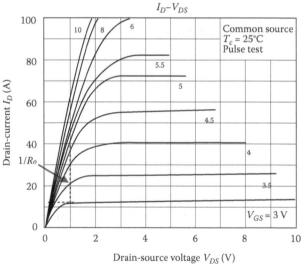

FIGURE 4.16
I_D–V_{DS} characteristic of MOSFET 2SK2844.

The on-resistance taken from the data sheet of the MOSFET is shown in Table 4.7.

It is clear from Table 4.7 and Figure 4.16 that the on-resistance depends on the gate-source voltage and it decreases as the voltage increases. The resistance can be approximated by the following equation:

$$R_{on} = R_{on0}V_{GS}^{-R_{on\tau}} \tag{4.23}$$

TABLE 4.7

Drain Source on Resistance R_{on}

	Typ.	Max.	
$V_{GS} = 4\ V, I_D = 18\ A$	26	35	mΩ
$V_{GS} = 10\ V, I_D = 18\ A$	16	20	mΩ

If the saturation on the I_D–V_{DS} characteristic has to be expressed, it can be approximated by the following function involving an exponential function:

$$I_D = I_{D\max}\left\{1 - \exp\left(-\frac{V_{DS} - R_{on}I_D}{V_\tau}\right)\right\} \tag{4.24}$$

where $I_{D\max}$ is the maximum drain current given by Equation 4.19 or 4.20.

Table 4.8 shows the parameters of 2SK2844 MOSFET for its static characteristics.

The transient (dynamic) characteristics of a MOSFET are determined by capacitors and by the behavior of carriers in the device. In general, the capacitance is larger than that of a bipolar transistor. Table 4.9 shows the capacitances taken from the data sheet of 2SK2844 MOSFET.

The capacitance is not negligible in high-frequency switching operations. In most cases, the transient overvoltage generated in a switching circuit is caused by resonance between the capacitors and stray inductors in the switching circuit.

TABLE 4.8

Model Parameters

a_1	12.2 A/V²	a_h	28.3 A/V
V_p	1.93 V	b_h	−71.1 A
V_{lh}	3.09 V	V_τ	0.15 V
R_{on0}	60 mΩ	$R_{on\tau}$	0.59

TABLE 4.9

Typical Capacitance

Input capacitance	C_{iss}	980 pF
Reverse-transfer capacitance	C_{rss}	270 pF
Output capacitance	C_{oss}	580 pF

$V_{DS} = 10\ V, V_{GS} = 0\ V$, and $f = 1\ MHz$.

TABLE 4.10

Typical Switching Time

Rise time	t_r	14 ns
Turn-on time	t_{on}	23 ns
Fall time	t_f	64 ns
Turn-off time	t_{off}	190 ns

The switching characteristic of MOSFET is expressed by the parameters shown in Table 4.10. Although the physical behavior of a MOSFET is too complicated for an EMTP simulation, the operational characteristic can be reproduced with satisfactory accuracy from the viewpoint of the numerical simulation of a power system including power-electronics apparatuses. A simple representation method of the dynamic characteristic is proposed in this section. Figure 4.17 illustrates the signals used for representing the transient-switching characteristic of a MOSFET. The gate-source voltage V_{GS} is delayed by t_{on} and by t_{off} using a transport delay device (Device 53). The rise time t_r is expressed by an s-block $F(s)$ from the delayed signals:

$$F(s) = \frac{1}{1+s\tau} \tag{4.25}$$

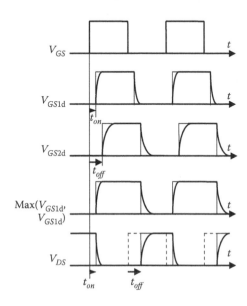

FIGURE 4.17
Control signal for R_{dseq}.

The time constant τ is obtained as a solution of the following equations:

$$0.1 = 1 - \exp\left(-\frac{t_1}{\tau}\right)$$

$$0.9 = 1 - \exp\left(-\frac{t_1 + t_r}{\tau}\right) = 1 - \exp\left(-\frac{t_1}{\tau}\right)\exp\left(-\frac{t_r}{\tau}\right) \tag{4.26}$$

The time constant τ becomes

$$\tau = \frac{t_r}{\ln(9)} = 0.455 t_r \tag{4.27}$$

In the same manner, the time constant for the fall time t_f is obtained.

An instantaneous-maximum device (Device 63) gives an equivalent gate-source voltage from the deformed signals ($V_{GS1}d$ and $V_{GS2}d$).

4.2.3.3 Simulation Circuit and Results

Figure 4.18 illustrates a switching circuit using MOSFET 2SK2844, and List 4.8 and Figure 4.19 show the input data for an analysis using the switch (diode) model. A square-wave voltage, whose amplitude is 10 V and the frequency is 200 kHz (Lines 10–14, and 63 in List 4.8), is applied to the gate through resistors R_{sig} and R_{g1} (Lines 48 and 49 in List 4.8). A voltage source (V_{cc}) of 15-V amplitude is applied to the drain through resistor R_d of 15 Ω (Lines 54 and 65 in List 4.8). Thus, the drain current becomes 1 A. The sum of the stray inductance of the drain resistor R_d and the source V_{cc} is 1.6 μH (Line 54 in List 4.8).

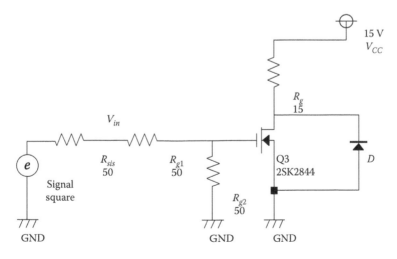

FIGURE 4.18
Switching circuit.

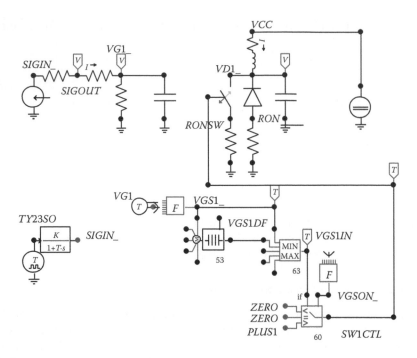

FIGURE 4.19
MOSFET switching circuit (switch model, ATP-Draw, GUI).

List 4.8: MOSFET Switching Circuit (Switch Model, Conventional Format, CUI)

```
1   BEGIN NEW DATA CASE { -------------------------------------- k2844notsw.dat
2   C Switching circuit simulation (device MOSFET 2SK2844)
3   C    DT][ TMAX ][ XOPT ][ COPT ][EPSILN][TOLMAT]
4      2.E-09   50.E-6
5   C IOUT][ IPLOT][IDOUBL][KSSOUT][MAXOUT][ IPUN ][MEMSAV][ ICAT ][NENERG][IPRSUP]
6      100      2      1      0      1      0      0      1
7   TACS HYBRID
8   C ====================================================== Signal source model
9   C [VAR1]     [ Amp. ][ T(s) ][width(s)]                   [ T-START][ T-END ]
10  23TY23SO            10.0 5.000E-06 2.500E-06   { 200 kHz, square wave
11  C [NAME]   +[IN1]+[IN2]+[IN3]+[IN4]+[IN5] [GAIN][F-LO][F-HI][N-LO][N-HI]
12  1SIGIN_   +TY23SO
13        1.0
14        1.0 0.010E-06                      { freq. characteristic of signal source
15  C   D0    ][  D1   ][  D2   ][  D3   ][  D4   ][  D5   ][  D6   ][  D7   ]
16  C =============================================================
17  C [NODE]   ================================================ voltage sensors
18  90VG1___   { Gate voltage
19  90VD1___   { Drain voltage
20  C =============================================================
21  C [NAME] =-------------------- FREE FORMAT ---------------------
22  99VGS1__  = VG1___         { VGS =VG - VS, source is grounded (VS =0)
23  99VDS1__  = VD1___         { VDS =VD - VS, source is grounded (VS =0)
24  99NZERO   = 1.E-9          {small non-zero value
```

```
25  99R1ONM_ =16.E-3          {minimum on resistance}
26  99VGSON_ =1.930434783     {threshold VGS on-voltage
27  C $DISABLE { comment out for simple sw model
28  C ================================= MOS-FET=sw (diode) model without delay
29  C [NAME]60+[IN1] +[IN2] +[IN3]                    [CNST]        [SIG1][SIG2]
30  99SW1CTL60+ZERO  +ZERO  +PLUS1                                 VGS1__VGSON_
31  C $ENABLE { comment out for simple sw model
32  $DISABLE { comment out for sw with delay model
33  C ================================= MOS-FET=sw (diode) model with delay
34  C [NAME]53+[IN1]+[IN2]+[IN3]+[IN4]+[IN5] [hist][FDel][MaxD][NDel][Nhis]
35  99VGS1DF53+VGS1__        {transport_delay, tdr}      .25E-6
36  C [NAME]63+[IN1]+[IN2] +[IN3]+[IN4]+[IN5]            [CNTL]
37  99VGS1IN63+VGS1__ +VGS1DF         {instantaneous max}   1.0
38  99SW1CTL60+ZERO   +ZERO +PLUS1                                 VGS1INVGSON_
39  $ENABLE { comment out for sw with delay model
40  BLANK ENDING TACS
41  C SIGOUT VG1___           VD1___   VCC
42  C +--Rg1---+-----------+      +--RD--+
43  C          |           |      |      |
44  C        Rg2         Cgs     Rds    Vcc
45  C          |           |      |      |
46  C +------- +-----------+-------+----- + GND (Emitter)
47  C [BUS1] [BUS2] [BUS3] [BUS4] [ R  ] [ L ][ C  ]                            !
48    SIGIN_SIGOUT                50.          {Rsig}
49    SIGOUTVG1___                50.          {Rg1}                            1
50    VG1___                      50.          {Rg2}                            1
51    VG1___                        .98E-3     {Cgs}
52  C VG1___VD1___                 .27E-3      {Cgd}
53    VD1___                        .58E-3     {Cds}
54    VCC___VD1___               15.1.6E-3     {Rd}                             1
55  C [BUS1] [BUS2] [BUS3] [BUS4] [ R  ] [ L ][ C  ]                            !
56    RON___                      30.E-3
57  BLANK ENDING BRANCH CARDS
58  C[BUS1] [BUS2]                                   MEASURING                  !
59    P1DS__HSINK1                                   MEASURING
60  C[BUS1] [BUS2]                                   CLOSED        [CLMP]      !!
61  11RON___VD1___           {Source-Drain}                        SW1CTL       0
62  BLANK ENDING SWITCH CARDS
63  60SIGIN_                 {Signal voltage}
64  C [BUS1]  []  [AMP.]                                    [ TSTART ][ TSTOP ]
65  11VCC___          15.0    {Power supply}
66  BLANK ENDING SOURCE CARDS
67    SIGOUTVB1INTVD1___VG1___HSINK1
68  BLANK ENDING OUTPUT SPECIFICATION CARDS
69  C !![H][ST]END][MIN]MAX][BUS1][BUS2][BUS3][BUS4][HEADING LABEL][VERTICAL AXIS]
70    145 1. 5.0 15.          SIGOUTVG1___VD1___                  VOLTAGE (V)
71  BLANK ENDING PLOT CARDS
72  BEGIN NEW DATA CASE
73  BLANK
```

The commented-out data by $DISABLE and $ENABLE (Lines 32–39 in List 4.8) express the turn-off delay of the MOSFET. The technique is identical to that of the switching transistor explained in Section 4.2.2.2.

List 4.9 and Figure 4.20 show input data for the switching circuit using the nonlinear resistor model. In these data, the dynamic on-resistance (Equation 4.23) and the saturation characteristic (Equation 4.24) are neglected because they might have minor effects on the result. The data also include a thermal model, described in Section 4.2.4.

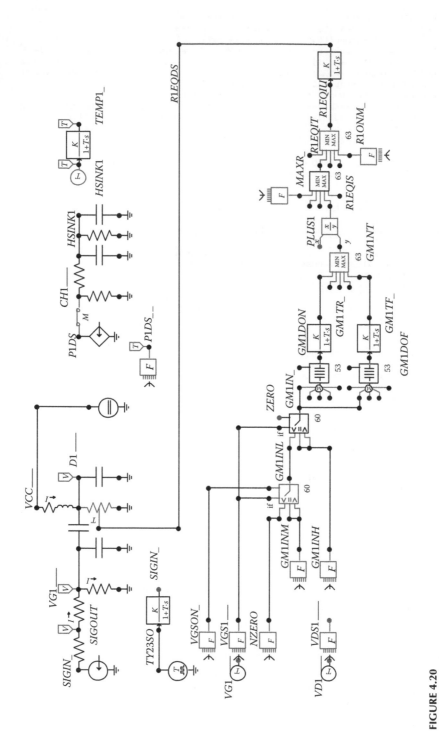

FIGURE 4.20
MOSFET switching circuit (nonlinear model with heat sink, GUI).

List 4.9: MOSFET Switching Circuit (Nonlinear Model with Heat Sink, CUI)

```
1     BEGIN NEW DATA CASE { ------------------------------------k2844NOTnonl.dat
2     C Switching circuit simulation (device MOSFET 2SK2844)
3     C    DT] [ TMAX ] [ XOPT ] [ COPT ] [EPSILN] [TOLMAT]
4      2.E-09  50.E-6
5     C IOUT] [ IPLOT] [IDOUBL] [KSSOUT] [MAXOUT] [ IPUN ] [MEMSAV] [ ICAT ] [NENERG] [IPRSUP]
6        100       2       1        0       1         0         0         1
7     TACS HYBRID
8     C ==================================================== Signal source model
9     C [VAR1]   [  Amp.  ] [  T(s) ] [width(s)]              [ T-START] [ T-END ]
10    23TY23SO   10.0 5.000E-06 2.500E-06   { 200kHz, square wave
11    C [NAME]  +[ IN1] +[ IN2] +[ IN3] +[ IN4] +[ IN5] [GAIN] [F-LO] [F-HI] [N-LO] [N-HI]
12    1SIGIN_   +TY23SO
13        1.0
14        1.0 0.010E-06                  { freq. characteristic of signal source
15    C  D0  ] [   D1   ] [   D2  ] [   D3   ] [   D4   ] [   D5   ] [   D6  ] [   D7  ]
16    C ======================================================================
17    C [NODE]   ========================================= voltage sensors
18    90VG1___    { Gate voltage
19    90VD1___    { Drain voltage
20    C ======================================================================
21    C [NAME]  = --------------------- FREE FORMAT ------------------------
22    99VGS1__  = VG1___                { VGS =VG - VS, source is grounded (VS =0)
23    99VDS1__  = VD1___                { VDS =VD - VS, source is grounded (VS =0)
24    99NZERO   = 1.E-9                 {small non-zero value
25    99R1ONM_  = 16.E-3                {minimum on resistance}
26    99VGSON_  = 1.930434783           {threshold VGS on-voltage
27    C =========================== MOS-FET =nonlinear resistor model based on gm
28    99GM1INM  = 12.15776308*(VGS1__-VGSON_)**2/VDS1__  {Id-Vgs in small Vgs region
29    99GM1INH  = (28.3*VGS1__-71.1)/VDS1__              {Id-Vgs in large Vgs region
30    C [NAME]60+[ IN1]  +[ IN2]  +[ IN3]        [CNST]        [SIG1] [SIG2]
31    99GM1INL60+NZERO+GM1INM+GM1INM                        VGS1__ VGSON_
32    99GM1IN_60+GM1INL+GM1INH +GM1INH           3.0943     VGS1__ ZERO
33    C [NAME]53+[ IN1] +[ IN2] +[ IN3] +[ IN4] +[ IN5] [hist] [FDel] [MaxD] [NDel] [Nhis]
34    99GM1DON53 + GM1IN_    {transport_delay, ton}    23.E-9
35     1GM1TR_  +GM1DON
36        1.0
37        1.0  6.37E-09    { 0.455*tr }
38    99GM1DOF53 + GM1IN_    {transport_delay, toff}    .19E-6
39     1GM1TF_  +GM1DOF
40        1.0
41        1.0 29.12E-09    { 0.455*tf }
42    C [NAME] 63+[ IN1] +[ IN2] +[ IN3] +[ IN4] +[ IN5]      [CNTL]
43    99GM1INT63 GM1TR_ GM1TF_                 {instantaneous max} 1.0
44    99R1EQIS =1/GM1INT                       {eq. R between D & S}
45    C
46    99MAXR__ =10000.
47    99R1EQIT63 MAXR__ R1EQIS    {instantaneous max}   -1.0
48    99R1EQIU63 R1ONM_ R1EQIT    {instantaneous max}   1.0
49    1R1EQDS+R1EQIU
50        1.0
51        1.0 0.020E-06
52    C [NAME] 63+[ IN1] +[ IN2] +[ IN3] +[ IN4] +[ IN5]      [CNTL]
53    99ID1EST = VDS1__/R1EQDS    {estimated collector current}
54    C $ENABLE { comment out for nonlinear model
55    C ============================================= Temperature calculation
56    99P1DS__   = VDS1__**2/R1EQDS
57    90HSINK1
```

```
58    1TEMP1_ +HSINK1
59       1.0
60       1.0    5.0E-6
61    C [VAR1][VAR2][VAR3][VAR4][VAR5][VAR6][VAR7][VAR8][VAR9][VARA][VARB][VARC][VARD]
62    33R1EQDS
63    33ID1EST
64    33GM1TR_GM1TF_GM1IN_GM1INT
65    33P1DS__HSINK1TEMP1_
66    BLANK ENDING TACS
67    C SIGOUT VG1___                        VD1___    VCC
68    C    +--Rg1---+---------+              +--RD--+
69    C            |         |               |      |
70    C           Rg2       Cgs             Rds    Vcc
71    C            |         |               |      |
72    C    +--------+---------+----------+------+ GND (Emitter)
73    C [BUS1][BUS2][BUS3][BUS4][ R ][ L ][ C ]                              !
74     SIGIN_SIGOUT              50.
75     SIGOUTVG1___              50.                {Rg1}                     1
76     VG1___                    50.                {Rg2}                     1
77     VG1___                              .98E-3  {Cgs}
78    C $DISABLE
79     VG1___VD1___                        .27E-3  {Cgd}
80    C $ENABLE
81     VD1___                              .58E-3  {Cds}
82     VCC___VD1___              15.1.6E-3          {Rd}                      1
83    C [BUS1][BUS2]TACS [BUS4]                                              !
84    91VD1___          TACS R1EQDS                 {Rce}                     4
85    C ===================================== Thermal model for TO220 package
86     CH1___HSINK1          2.08 { K/W channel to case,    Thermal resistance
87     CH1___               83.30 { K/W channel to ambient, Thermal resistance
88     HSINK1                      .16663 { J/K *10^6 1/40 Thermal capacity
89    $DISABLE { comment out to take into account the heat-sink
90     HSINK1                      .01666 { J/K *10^7 1/40 Thermal capacity
91    C ======================================================= Heat-Sink model
92    C [BUS1][BUS2][BUS3][BUS4][ R ][ L ][ C ]                              !
93     HSINK1                17.30 { K/W case to ambient,   Thermal resistance
94     HSINK1                      .66652 { J/K *10^7      Thermal capacity
95    C =========================================================================
96    $ENABLE { comment out to take into account the heat-sink
97    BLANK ENDING BRANCH CARDS
98    C [BUS1][BUS2]                                    MEASURING            !
99     P1DS__HSINK1                                     MEASURING
100   BLANK ENDING SWITCH CARDS
101   60SIGIN_                   {Signal voltage}
102   C [BUS1][][ AMP. ]                                  [ TSTART ][ TSTOP ]
103   11VCC___         15.0      {Power supply}
104   C
105   60P1DS__-1                 {Thermal
106   BLANK ENDING SOURCE CARDS
107    SIGOUTVB1INTVD1___ VG1___HSINK1
108   BLANK ENDING OUTPUT SPECIFICATION CARDS
109   C !![H][ST]END][MIN]MAX][BUS1][BUS2][BUS3][BUS4][ HEADING LABEL][ VERTICAL AXIS]
110    145 1. 5.0 15.         SIGOUTVG1___VD1___           VOLTAGE (V)
111    195 1. 5.0 15.         TACS R1EQDS     EQUIV. R     RESISTANCE (OHM)
112    195 1. 5.0 15. -1.0 1.0TACS ID1ESTVCC___VD1___ ID        CURRENT (A)
113    195 1. 5.0 15.         BRANCH          CONDUCTANCE  CONDUCTANCE (S)
114                           TACS GM1TR_TACS GM1TF_TACS GM1IN_TACS GM1INT
115    195 1. 5.0 15.         TACS P1DS__     POWER             POWER (W)
116    195 5. 0.0 50.         TACS HSINK1TACS TEMP1_ TEMP. RISE TEMPERATURE (K)
117   BLANK ENDING PLOT CARDS
118   BEGIN NEW DATA CASE
119   BLANK
```

Figure 4.21a illustrates a measured result of the switching circuit. High-frequency oscillations are observed on the drain voltage just after the MOSFET is turned off. Its frequency is determined by the stray inductance of the drain circuit and the capacitance of the MOSFET. The oscillation is induced on the gate voltage through the transfer capacitance C_{rss}.

Figure 4.21b shows the results obtained by the simplified switch model. The amplitude of the high-frequency oscillation is far greater than that of

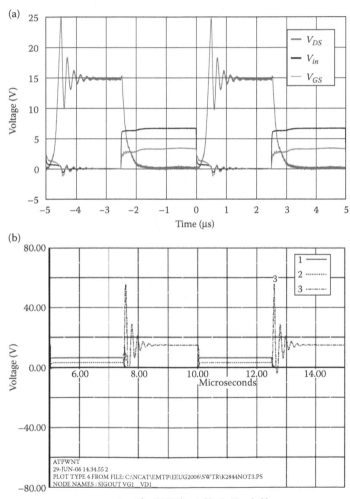

1 μs/div, 20 V/div, 1: V_{in}, 2: V_{GS}, 3: V_{DS}

FIGURE 4.21
Responses of the switching circuit: (a) measured result and (b) calculated result by the switch model. *(Continued)*

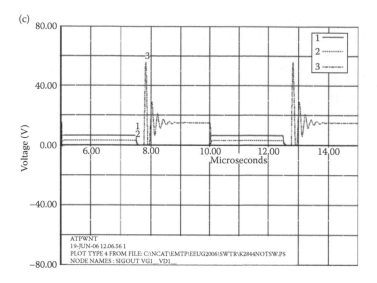

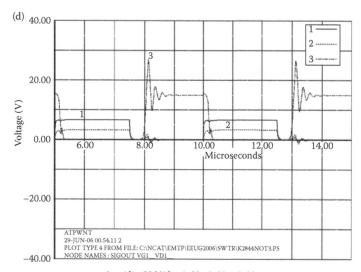

1 μs/div, 20 V/div, 1: V_{in}, 2: V_{GS}, 3: V_{DS}

FIGURE 4.21 (Continued)
Responses of the switching circuit: (c) calculated result by the switch model (with delay) and (d) calculated result by the proposed model.

the measured result, because the model cannot correctly represent the resistance within the MOSFET, which attenuates the oscillation. The switching delay is not represented in this case, because the Type-11 switch is directly controlled by the gate-source voltage V_{GS}. If the delayed signal explained in Section 4.2.2.1 is used as the control signal, the delay could be approximately introduced. In the simulation case, the transfer capacitor C_{rss} is neglected for a stable calculation. If high-frequency oscillations are induced on the gate-source voltage at around the threshold voltage V_p through the transfer capacitor, the switching operation becomes unstable.

Figure 4.21c illustrates the results obtained using the switch model with the turn-off delay. Even if the switching delay is included into the simulation, the accuracy is not improved.

Figure 4.21d illustrates the results obtained using the accurate model. The high-frequency oscillation as well as the switching delay is accurately reproduced. The loss of the MOSFET reduces the oscillation.

4.2.4 Thermal Calculation

The EMTP has been widely used for estimating transient overvoltages on a power system. In a power-electronics field, the prediction of temperature rise and overvoltages in the switching circuit is one of the important goals of numerical simulations. The result provides valuable information for designing power-electronics apparatuses.

The thermal equation is analogous to the electrical-circuit equation. Electrical power consumption P corresponds to a current source. Static heat-transfer properties are usually specified using a thermal resistance R_θ that defines a relation between heat flow per unit time Q and temperature difference $\theta = T - T_0$:

$$Q = \frac{1}{R_\theta}\theta \tag{4.28}$$

The thermal capacitance C_θ is specified to model the dynamical properties of the heat transfer:

$$Q = C_\theta \frac{d\theta}{dt} \tag{4.29}$$

An equivalent equation can be derived using electrical quantities instead of thermal quantities. The relations between electrical and thermal quantities are given in Table 4.11.

TABLE 4.11

Relations between Thermal and Electrical Quantities

Thermal Quantity	Electrical Quantity
Heat flux Q	Current I
Temperature difference θ	Voltage V
Thermal resistance R_θ	Resistance R
Thermal capacitance C_θ	Capacitance C

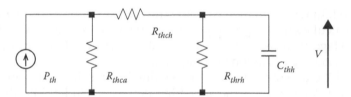

FIGURE 4.22
Thermal-equivalent circuit.

The equivalent voltage V is proportional to the difference between an absolute temperature T and a reference temperature T_0. Usually, T_0 is selected to represent ambient temperature. The equivalent current source I transfers heat to the thermal circuit, and its value is proportional to power dissipation P in an electrical component.

The time constant of a thermal-equivalent circuit is usually far greater than that of an electrical circuit. An accelerating coefficient a_t is introduced to compress the difference. Equation 4.30 is transformed into an "accelerated domain":

$$t = a_t t' \tag{4.30}$$

$$Q = C_\theta \frac{d\theta}{dt} = \frac{C_\theta}{a_t} \frac{d\theta}{dt'} \tag{4.31}$$

In the accelerated domain, the thermal capacitance is inversely proportional to the accelerating coefficient a_t.

Figure 4.22 illustrates a thermal-equivalent circuit for a switching device. The resistor R_{thca} expresses the heat resistance between the channel of the MOSFET and the ambient air, and the resistor R_{thch} expresses the heat resistance between the channel and the package. The resistor R_{thrh} expresses the heat radiation from the package to the ambient air. The resistance is generally determined by a heat sink. The capacitor C_{thh} corresponds to the thermal capacitance of the heat sink. If dynamic heat characteristics are not important, the capacitor is negligible. The current source P_{th} represents power dissipation within the channel. The voltage across the capacitor V expresses the temperature rise of the MOSFET.

In this section, two cases are investigated: (a) without the heat sink and (b) with the heat sink. Tables 4.12 and 4.13 show the circuit parameters used

TABLE 4.12

Thermal Characteristics of 2SK2844 MOSFET

Characteristics	Symbol	Max	Unit
Channel to case	R_{thch}	2.08	°C/W
Channel to ambient	R_{thca}	83.3	°C/W

TABLE 4.13

Thermal Characteristics of Heat Sink

Characteristics	Symbol	Max	Unit
Thermal resistance	R_{thha}	17.3	°C/W
Surface area	S_h	42	cm²
Weight	M_h	7.6	g
Specific heat of aluminum		0.877	J/g°C
Thermal capacity	C_{ths}	6.67	J/°C

in this section. Figure 4.23 illustrates the results calculated by the proposed model. The voltage across the thermal capacitor with a sawtooth oscillation is smoothed by a first-order s-block with a time constant of 5 μs. Figure 4.23a shows the result when the heat sink is removed and the acceleration coefficient a_t of 10^6 is applied. Although the maximum observation time T_{max} is 50 μs, the smoothed waveform expresses the change in temperature up to 50 s (= $T_{max} \times a_t$). The result of the heat sink is obtained with $a_t = 10^7$ and is illustrated in Figure 4.23b.

The power consumption of the MOSFET is determined by its switching loss, because its on-resistance is small enough for the application. The periodic power consumption causes the oscillations in the results. The oscillations are not observed in practical situations. The difference is caused by the acceleration introduced for saving computational time. The error can be easily suppressed by a smoothing filter. The calculated results show that the temperature rise is 20°C when the heat sink is removed. The measured surface temperature of the MOSFET is 46.2°C when the ambient temperature is 25.7°C. The difference is 20.5°C, which agrees with the calculated result. Figure 4.23b shows that the heat sink reduces the temperature rise to 3°C and the temperature converges at about 500 s (= $T_{max} \times a_t$).

4.3 Voltage-Regulation Equipment Using Battery in a DC Railway System

4.3.1 Introduction

A numerical simulation method of a voltage regulator using Li-ion battery in a DC railway system using the EMTP with TACS is proposed in this section. A couple of TACS-controlled resistors are used for representing each line resistance within a feeding section for expressing a train's operation. In addition, a power compensator, which regulates the line voltage, can be expressed by the functions installed in the TACS. The calculated result of

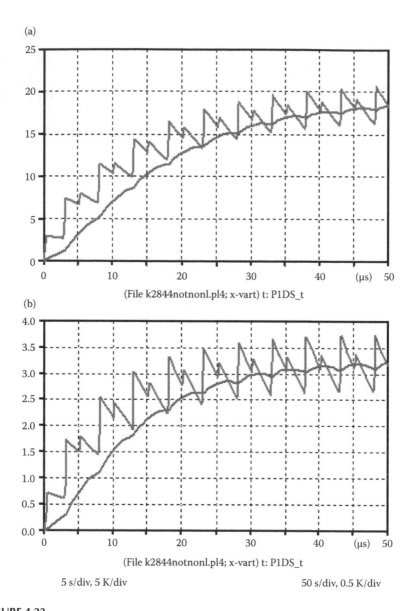

(a)

(File k2844notnonl.pl4; x-vart) t: P1DS_t

(b)

(File k2844notnonl.pl4; x-vart) t: P1DS_t

5 s/div, 5 K/div   50 s/div, 0.5 K/div

FIGURE 4.23
Calculated results of temperature rise: (a) without heat sink ($a_t = 10^6$) and (b) with heat sink ($a_t = 10^7$).

the system including the voltage regulator agrees well with the measured result of a practical train system. The proposed method indicates the optimal installing position and capacity of the compensator. The numerical simulation using the EMTP enables a computer-aided design of feeding circuits of a DC train system.

The voltage distribution along the feeding line of a DC railway system is determined by output voltages of substations and line voltage drops. Voltage drops are proportional to the train's current, feeding line resistance, and line length between the substation and the train, that is, the train's position. The line voltage is lowered by the current of a powering train and is increased by the regenerative current of a braking train. In recent years, line-voltage fluctuation of the feeding system has become larger because the train current is increased for increasing the transportation capacity.

The line voltage should be kept at its rating voltage as much as possible. The voltage drop caused by train operation is proportional to the distance between the substation and the train. Some apparatuses for stabilizing the line voltage have been proposed. A substation with a voltage regulator, which consists of thyristors and a controller, is effective for compensating the voltage drop caused by the powering train.

Recently, power compensators with some kind of electrical storage device have been developed to stabilize voltage and to increase the efficiency of DC train systems [9]. The regenerative energy is stored in Li-ion batteries or electrical double-layer capacitors. Power compensators can be installed on the ground or in a train. A verification test of a prototype of the storage system installed on the ground is currently being carried out [10–13]. The compensator releases the stored energy to powering trains and has the capability to store the regenerative energy of a braking train.

To design an efficient feeding system including the voltage regulator and/or the various compensators, an estimation of the voltage and current distributions is indispensable. Numerical simulation is one of the solutions for estimation. For an accurate simulation of the train-feeding system, the movement of trains has to be taken into account. Thus, the line resistance should be expressed by time-dependent resistors for expressing the train operation. A flexible modeling capability is also required for a circuit-simulation program for expressing various characteristics of compensators.

The EMTP has been widely used as a standard transient-analysis program in the field of power-system simulation. TACS, which is installed in the EMTP as a modeling tool of the control system, is suitable for the simulation of a train-feeding system. The line resistance taking into account the moving train is expressed by a TACS-controlled resistor. In addition, the power compensators can be expressed by the functions installed in the TACS.

4.3.2 Feeding Circuit

Figure 4.24 illustrates the feeding circuit investigated in this chapter. The circuit has five substations. The substations S/S_2 and S/S_3 have conventional voltage regulators, which consist of thyristors. A power compensator is installed between these substations. In the figure, the line length is expressed by l_k and the train position is represented by distance l_t from the substation S/S_1 as a function of time. The line lengths are shown in Table 4.14.

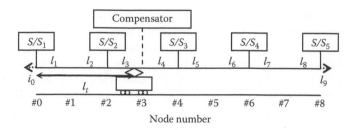

FIGURE 4.24
Feeding circuit.

TABLE 4.14

Distance between Substations

Section	$l_{1,2}$	$l_{3,4}$	$l_{5,6}$	$l_{7,8}$
Distance	6.3 km	7.4 km	5.05 km	4.95 km

These values are taken from a practical feeding system. The total length of the system is 47.4 km. The resistance of the feeding line R_l, including the rail resistance, is assumed to be 40 mΩ/km.

The rated voltage of the system V_r is 1500 V, and the capacity of the substation is assumed to be 4000 kW. The rated current of the substation I_r is

$$I_r = \frac{P_r}{V_r} = \frac{4000\,(\text{kW})}{1.5\,(\text{kV})} \approx 2.7\,(\text{kA}) \tag{4.32}$$

Figure 4.25 illustrates a numerical model of the single section of the feeding circuit. The substation is expressed by a series circuit of a diode D_m, an internal resistor R_m, and a voltage source E_m.

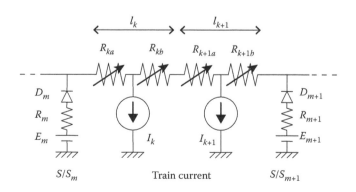

FIGURE 4.25
Feeding-circuit model.

If the backward impedance, that is, the voltage-fluctuation ratio of the substation, is 9% (= Z_{pu}), the internal voltage E_m of the substation model becomes

$$E_m = 1.09V_r = 1.09 \times 1500 = 1635 \text{ (V)} \tag{4.33}$$

The internal resistance R_m is obtained from the backward impedance:

$$R_m = \frac{Z_{pu}V_r^2}{P_r} = \frac{0.09 \times 1.5^2}{4} = 0.051 \text{ (}\Omega\text{)} \tag{4.34}$$

If a voltage regulator is installed into a substation, the substation can be simply modeled with a small internal resistor $R_m = 1$ mΩ and an ideal source E_m of 1690 V, which is the target voltage of the voltage regulator.

A diode is inserted in series for preventing the reverse current of substations S/S_2, S/S_3, and S/S_4. There are no diodes in the substations at both ends (S/S_1 and S/S_5) to approximately express the currents flowing out from the model sections (i_0 and i_9 in Figure 4.24) as reverse currents.

The train operation is modeled by the current sources and nonlinear resistors illustrated in Figure 4.25. Train position l_t is obtained from train velocity $v(t)$ using an integrator (1/s) illustrated in Figure 4.26:

$$l_t(t) = \int_0^t v(t)\, dt \tag{4.35}$$

The comparator in the figure determines the section where the train operates at the time. The comparator is represented by a nonlinear function (Device 56) and a truncation function TRUNC() installed in the TACS.

The resistances of the feeding line are determined by Equation 4.36. These resistances, R_{ka} and R_{kb}, are modeled by TACS-controlled resistors (Type-91) as time-varying resistors:

$$R_{ka}(t) = \left(l_t(t) - \sum_{n=1}^{k-1} l_n \right) R_l$$

$$\tag{4.36}$$

$$R_{kb}(t) = \left(\sum_{n=1}^{k} l_n - l_t(t) \right) R_l$$

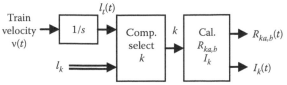

FIGURE 4.26
Train model.

The movement of the train is modeled by current sources $I_k(t)$ with these nonlinear resistors. A current source in the section, where the train is running, is activated, and the other sources are deactivated

$$I_k(t) = I_t(t)$$
$$I_n(t) = 0 \qquad (n \neq k)$$

(4.37)

where $I_t(t)$ is the train current.

4.3.3 Measured and Calculated Results

4.3.3.1 Measured Results

Figure 4.27 illustrates the measured feeding line voltages at node #4 and the output current of the substation S/S_3 for Cases a and b shown in Table 4.15. The voltage regulators installed in substations S/S_2 and S/S_3 are turned on in Case-a, and the regulator in S/S_2 at node #2 only operates in Case-b.

A pulse-like current waveform is observed in both the results. The time region where a high current is observed indicates that the train is powering, and the low-current region indicates that the train is coasting. The difference between the current waveforms in Figure 4.27a and b mainly comes from the variations in train operation. The base current of about 100 A observed in Figure 4.27a expresses the power sent to trains operating in the other sections.

If the voltage regulator at observation node #4 is turned on, the feeding line voltage is stabilized by the regulator and the output voltage of the substation becomes 1690 V (Figure 4.27a). The maximum current of the substation is 2.1 kA. The current linearly increases as the train comes close to the substation and decreases after the train passes the substation, indicated with a broken line. The decreasing rate is greater than the increasing rate because the length between the substations on the left-hand side of the substation S/S_3 ($l_3 + l_4$) is greater than that on the right-hand side ($l_5 + l_6$). The slope is determined by the resistances between the substations.

Figure 4.27b illustrates the result when the voltage regulator at observation node #4 is turned off. The output voltage fluctuates due to the train current even if the other regulator operates. The maximum current is 1.56 kA, which is smaller by 25% compared to the previous result. The maximum current depends on the voltage regulator as well as the train operation. No current is observed during coasting, although a current of 100 A flows when the voltage regulator at the substation operates. The feeding line voltage at the coasting operation is about 1.64 kV.

4.3.3.2 Calculated Results of the Conventional System

Investigation on the electrical characteristics of the feeding system using numerical simulations is helpful rather than that using measurements,

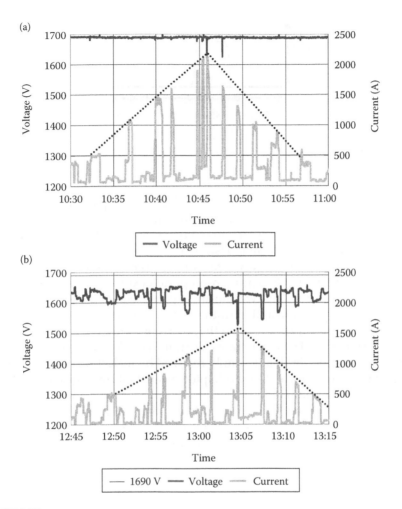

FIGURE 4.27
Measured feeding line voltages at node #4 and output current of substation S/S_3: (a) with voltage regulators (Case-a) and (b) with a single voltage regulator (Case-b).

TABLE 4.15

Circuit Conditions

	Voltage Regulator	Compensator
Case-a	S/S_2 and S/S_3 (#2 and #4)	No
Case-b	S/S_2 (#2)	No
Case-c	No	No
Case-d	No	#3

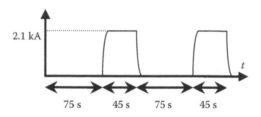

FIGURE 4.28
Train-current waveform.

because voltage and current are affected by many parameters and their simultaneous measurement is quite difficult. Some numerical simulations by the proposed method are carried out using the EMTP in this section.

The train-current waveform is assumed as illustrated in Figure 4.28, in the simulations. Its amplitude, duty ratio, and period are assumed to be 2.1 kA, 5/8, and 120 s, respectively. The turn-off period expresses that the train runs in coasting operation. This waveform is easily generated using a square-wave source (Type-23) installed in the TACS. The rise and fall times of the current are assumed to be 2.2 s and they are represented by an s-block with a time constant of 1 s. The regenerative current is neglected in this simulation. If the current is required, it is simply realized by including negative current pulses into the waveform. In this chapter, a constant train speed of 54 km/h is assumed in the simulations.

Figure 4.29 illustrates the calculated results at node #4 (S/S_3). It is clear from Figure 4.29a that there is no voltage fluctuation, if the conventional voltage regulator installed in S/S_3 operates. The maximum current is 2.1 kA and is identical to the maximum train current. The minimum current is 65 A, which is observed while the train is coasting. This result expresses that substation S/S_3 feeds power to trains running in the other sections. No pulse current is observed before 0:15, because the current while the train is running in the first section (between nodes #0 and #2) is fed by S/S_1 and S/S_2, which has a voltage regulator. On the contrary, pulse currents are observed after 0:43 when the train is running in the last section (between nodes #6 and #8), because the feeding line voltage at node #4 (S/S_3) is kept high by the voltage regulator and no voltage regulator is installed in the section on the right (S/S_4 and S/S_5).

It is clear from Figure 4.29b that the maximum current of the substation is reduced from 2.1 to 1.63 kA (−22%), if the voltage regulator at S/S_3 is turned off. The remaining current (0.47 kA) is fed from S/S_2, which has a voltage regulator. No current is observed while in the coasting periods, although some current is observed in the previous result. There is no pulse current observed before 0:15, because substation S/S_2 with the voltage regulator feeds to the train running in the first stage. On the contrary, small pulse currents are observed after 0:43, that is, the train is running in the last section

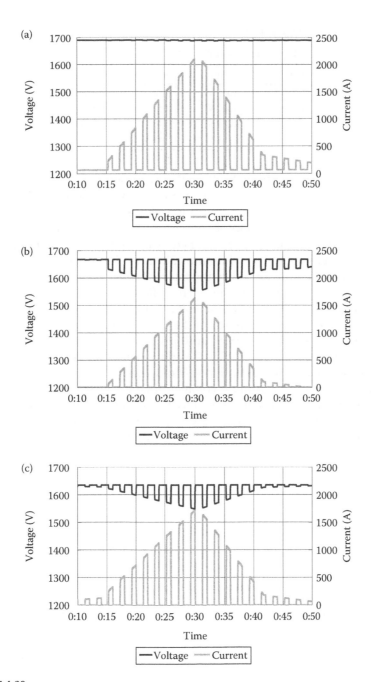

FIGURE 4.29
Calculated results of a conventional system: (a) with voltage regulators (Case-a), (b) with a single voltage regulator (Case-b), and (c) without voltage regulators (Case-c).

TABLE 4.16

Minimum Voltages

Node #	#3	#4 (S/S_3)	#5
Case-a	1386 V	1688 V	1433 V
Case-b	1340 V	1553 V	1379 V
Case-c	1284 V	1548 V	1376 V
Case-d	1451 V	1548 V	1376 V

(between nodes #6 and #8). Substation S/S_3 feeds a minor current to the train running in the next section, if the next substation has no voltage regulator.

The voltage while the train is coasting is 1.67 kV, which is greater than the open-circuited voltage of 1.64 kV shown in Equation 4.33. This result also shows that the feeding line voltage is maintained by the voltage regulator operating at the adjoining substation. The minimum voltage is 1.55 kV, and is determined by the voltage drop mainly caused by the internal impedance of substation S/S_3.

Figure 4.29c indicates that the current flowing from substation S/S_3 is increased by a disconnection of the voltage regulator at S/S_2. The maximum current is 1.72 kA. The difference in the train current (0.38 kA = 2.1–1.72) is fed by adjoining substations (S/S_2 and S/S_4). For the same reason, the pulse currents are observed before 0:15 in Case-b and after 0:43 in Case-c. The minimum voltage is almost identical to that of Case-b.

Table 4.16 shows the minimum voltages obtained by simulations. Nodes #3 and #5 denote halfway points of the feeding sections. The minimum voltage in the table is observed at node #3 in Case-c (without voltage regulators). The voltage at node #3 (1.28 kV) is lower than that at node #5 (1.38 kV), because the line resistance between substations S/S_2 and S/S_3 is higher than that between S/S_3 and S/S_4. The minimum voltage at the substation is 1.55 kV, and is almost identical to the result of Case-b. If the regulator at node #4 (S/S_3) is also turned on (Case-a), the middle-point voltages at nodes #3 and #5 are increased. Comparisons between the results of Cases b and c and between the results of Cases a and b indicate that the voltage regulator is effective in increasing the voltage of the feeding line connected to the regulator.

4.3.3.3 Calculated Results with Power Compensator

A power compensator has been developed for the DC railway system to compensate the voltage drop by the line resistance [10–13]. The power compensator installed on the ground is composed of some parallely connected units, and each unit consists of a bidirectional DC/DC converter and a Li-ion battery bank. Table 4.17 shows the specifications of the unit. The battery bank consists of 182 cells connected in series. The capacity of a cell is 60 Ah and the maximum discharging current is 600 A. In this simulation, the number of units N_u is assumed to be 8. The maximum power of the compensator is 3145 kW (= 393×8), which is 79% of that of the substation.

TABLE 4.17

Specifications of a Power Compensator

Maximum battery bank voltage	746 V
Nominal battery bank voltage	655 V
Minimum battery bank voltage	564 V
Number of cells in a battery bank	182
Battery capacity	60 Ah
Maximum discharging current	600 A (10 C)
Maximum discharging capacity	393 kW
Internal resistance of a battery cell	0.8 mΩ

The information of the train operation cannot be accessed by the ground-based compensator. The operational characteristic is determined according to the line voltage at the installed point as shown in Figure 4.30. The vertical axis is scaled in the charging current into a battery bank installed in a unit of the compensator. Because the voltage conversion ratio of the DC/DC converter γ is 2.3 ($\approx$ 1500/655), the maximum injecting current to the feeding line becomes 130 A (= 300/2.3 = I_{Bd}/γ) per unit.

All conventional voltage regulators are turned off in the following simulation. Figure 4.31 illustrates the result of the proposed model [10,11], when the compensator is installed at node #3 where the minimum voltage is observed in Case-c. Pulse currents are injected into the feeding line from the compensator at node #3 as illustrated in Figure 4.31a, and the current is determined by the control characteristic illustrated in Figure 4.30. The maximum injected current I_{imax} is about 1 kA, which is calculated from the maximum discharge current I_{Bd} (300 A) illustrated in Figure 4.30, voltage conversion ratio γ, and the number of units N_u installed in the compensator:

$$I_{imax} = N_u \frac{I_{Bd}}{\gamma} = 8 \frac{300}{2.3} = 8 \times 130 = 1.0 \text{ (kA)} \qquad (4.38)$$

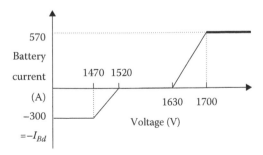

FIGURE 4.30
Control characteristic of a compensator unit.

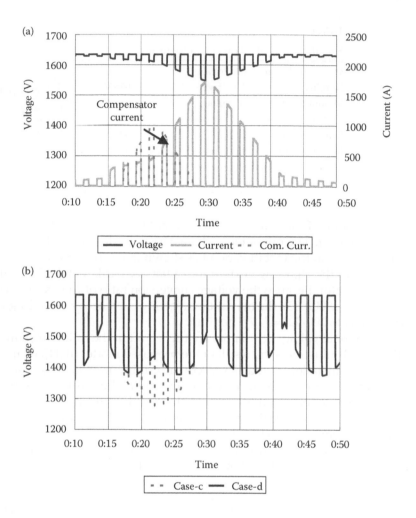

FIGURE 4.31
Calculated results with a compensator: (a) with a compensator (Case-d) and (b) pantograph voltage.

A comparison between the results shown in Figures 4.29c and 4.31a indicates that the minimum voltage at node #4 and the maximum current flowing from substation S/S_3 are almost identical. The voltage and current wave forms in the period from 0:17 to 0:27 are however slightly different from the results in the case without the compensator. Voltage drop in and current flowing from the substation are reduced by the compensator.

Figure 4.31b shows a comparison of the pantograph voltages. The figure clearly indicates that the voltage fed to the train is improved by the compensator. The improvement is also observed in Table 4.16.

The numerical simulation shows that the parameters used in this chapter are optimal from the viewpoint of voltage equalization. If further reduction of the feeding voltage fluctuation is required, the maximum discharging current (I_{Bd}) should be increased and also the compensators should be installed in the other sections.

4.4 Concluding Remarks

Numerical simulations of a wind farm using the EMTP are explained in this chapter. Voltage increase due to the charging current of the cables is easily obtained by the EMTP's steady-state analysis routine. The load-flow calculation option of the EMTP called "FIX SOURCE" enables an estimation of the steady-state behavior of the wind farm, which has plural generators. These techniques are applicable to a simulation of conventional grids.

Simulation models of a switching transistor and a MOSFET are also explained in this chapter. The model parameters of the device are easily obtained from a data sheet supplied by its manufacturer or from a simple experiment without complicated physical parameters of the semiconductor. The proposed model also enables temperature estimation. The accuracy of the models is satisfactory for the design of a switching circuit, such as a DC/DC converter and an inverter.

A numerical simulation model of a train-feeding system for the EMTP is proposed in this chapter. The feature of TACS installed in the EMTP is suitable for the simulation. A TACS-controlled nonlinear resistor is used for representing the movement of the train. The calculated results of the system including the voltage regulator agree with the measured results of a practical train system. This proves the accuracy of the proposed method. Both the voltage regulator and the power compensator installed for stabilizing the line voltage are effective for regulating the feeding line voltage. If the compensator is installed in the middle of a feeding section, its effectiveness is greater than that of the conventional voltage regulator. For an optimal feeding system design, the proposed simulation method is useful to confirm the effectiveness and to determine the parameters of the control characteristics. Although conventional regulators are effective in increasing the feeding line voltage, they cannot decrease the voltage rise caused by the regenerative brake because a substation generally has no reverse power flow capability. The line voltage has to be kept below the maximum rating voltage of the feeding system. The voltage rise caused by the regenerative brake is proportional to the distance between the braking and powering train, which consumes the regenerated power. The proposed simulation model is also useful in the analysis of voltage rise, because the model is capable of taking the regenerative current into account.

These techniques are useful for numerical simulations in the new energy system and for expanding the application of the EMTP in various other fields.

References

1. Thompson, M. T. 2006. *Intuitive Analog Circuit Design*. Oxford, UK: Newnes.
2. Nagaoka, N. 1988. Large-signal transistor modeling using the ATP version of EMTP. *EMTP News* 1(3):1–7.
3. Nagaoka, N. and E. Yamamoto. 1988. Numerical analysis of transistor circuit by EMTP, Part-1 Switching circuit. *Record of the 1988 Kansai-section joint convention of Institute of Electrical Engineers*, Japan, G1-11.
4. Nagaoka, N. and E. Yamamoto. 1988. Numerical analysis of transistor circuit by EMTP, Part-2 Amplifier. *Record of the 1988 Kansai-section joint convention of Institute of Electrical Engineers*, Japan, G1-12.
5. Nagaoka, N. and Y. Kimura. 1990. Numerical analysis of FET circuit by EMTP, *Record of the 1988 Kansai-section joint convention of Institute of Electrical Engineers*, Japan, G3-18.
6. Nagaoka, N. and Y. Kimura. 1990. Development of numerical model of switching FET. *Proceedings of the 28th Convention of Science and Engineering Research Institute of Doshisha University*, Science and Engineering Research Institute of Doshisha University, Kyoto, Japan.
7. Nose, N., A. Sugiyama, I. Okada, N. Nagaoka, and A. Ametani. 1998. Transistor model including temperature dependent characteristic for EMTP. National convention record, IEEJ 709.
8. Nagaoka, N. and A. Ametani. 2006. Semiconductor device modeling and circuit simulations by EMTP. Joint Convention of Power Engineering and Power System Engineering IEEJ, PE-06-74/PSE-06-74.
9. Nagaoka, N. 2006. Chapter 3 DC/DC converter, *Textbook of EEUG 2006 Course*, pp. III-1-III-20.
10. Nagaoka, N. et al. 2006. Power compensator using lithium-ion battery for DC railway and its simulation by EMTP. *63rd IEEE Vehicular Technology Conference*, Conf. CD, 6P-9.
11. Nagaoka, N., M. Sadakiyo, N. Mori, A. Ametani, S. Umeda, and J. Ishii. 2006. Effective control method of power compensator with lithium-ion battery for DC railway system. *Proceedings of the 41st International Universities Power Engineering Conference*, Newcastle upon Tyne, United Kingdom, pp. 1067–1071.
12. Umeda, S., J. Ishii, and H. Kai. 2004. Development of hybrid power supply system for DC electric railways. JIASC04 3-84.
13. Umeda, S., J. Ishii, N. Nagaoka, H. Oue, N. Mori, and A. Ametani. 2005. Energy storage of regenerated power on DC railway system using lithium-ion battery. *Proceedings of the 2005 International Power Electronics Conference*, Niigata, Japan, pp. 455–460.

5

Numerical Electromagnetic Analysis Methods and Their Application in Transient Analyses

5.1 Fundamentals

5.1.1 Maxwell's Equations

Before going into NEA methods, let us look at Maxwell's equations, which are the fundamentals of electromagnetics. They are stated in the time domain as

$$\nabla \times E(r,t) = -\frac{\partial B(r,t)}{\partial t} \tag{5.1}$$

$$\nabla \times H(r,t) = \frac{\partial D(r,t)}{\partial t} + J(r,t) \tag{5.2}$$

$$\nabla \cdot D(r,t) = \rho(r,t) \tag{5.3}$$

$$\nabla \cdot B(r,t) = 0 \tag{5.4}$$

where
  $E(r, t)$ is the electric field
  $H(r, t)$ is the magnetic field
  $D(r, t)$ is the electric-flux density
  $B(r, t)$ is the magnetic-flux density
  $J(r, t)$ is the conduction current density
  $\rho(r, t)$ are the volume charge density, each at space point r and at time t

Equation 5.1 represents Faraday's law, and Equation 5.2 represents Ampere's law. Equations 5.3 and 5.4 represent Gauss's laws for electric and magnetic fields, respectively.

In the frequency domain (FD), Maxwell's equations are expressed as

$$\nabla \times E(r) = -j\omega B(r) \tag{5.5}$$

$$\nabla \times H(r) = J(r) + j\omega D(r) \tag{5.6}$$

$$\nabla \cdot D(r) = \rho(r) \tag{5.7}$$

$$\nabla \cdot B(r) = 0 \tag{5.8}$$

where
  j is the imaginary unit
  ω is the angular frequency

The relations between the electric- and magnetic-flux densities and the electric and magnetic fields are given as

$$D(r) = \varepsilon E(r) = \varepsilon_r \varepsilon_0 E(r) \tag{5.9}$$

$$B(r) = \mu H(r) = \mu_r \mu_0 H(r) \tag{5.10}$$

where
  ε_0 is the permittivity of the vacuum (8.854×10^{-12} F/m)
  μ_0 is the permeability of the vacuum ($4\pi \times 10^{-7}$ H/m)
  ε_r is the relative permittivity of the medium
  μ_r is the relative permeability of the medium
  ε is the permittivity of the medium
  μ is the permeability of the medium

In a conductive medium, the following relation between the electric field and the conduction current density, known as Ohm's law, is fulfilled:

$$J(r) = \sigma E(r) \tag{5.11}$$

where σ is the conductivity of the medium.

5.1.2 Finite-Difference Time-Domain Method

The finite-difference time-domain (FDTD) method [1] is one of the most frequently used techniques in electromagnetics. It involves space–time discretization of the whole working space and the finite-difference approximation to Maxwell's differential equations. For analyzing the electromagnetic response of a structure in an unbounded space using the FDTD method, an absorbing boundary condition, which suppresses unwanted reflections from the surrounding boundaries that truncate the unbounded space, needs to be applied. To avoid numerical instabilities or spurious resonances, a time increment, or step Δt, needs to be determined following the Courant–Friedrichs–Lewy (CFL) criterion [2]: $\Delta t < \Delta s/\sqrt{3}c$ in three-dimensional (3-D) computations, where Δs is the (cubic) cell-side length and c is the speed of light.

The advantages of this method are: (1) it is based on a simple procedure in electric- and magnetic-field computations, making its programming relatively easy; (2) it is capable of treating complex geometrical shapes and inhomogeneities; (3) it is capable of incorporating nonlinear effects and components; and (4) it can yield wideband data from one run with the help of a time-to-frequency transforming tool.

The disadvantages are: (1) it is inefficient compared with the method of moments (MoM); (2) it cannot deal with oblique boundaries that are not aligned with the Cartesian grid; and (3) it would require a complex procedure for analyzing the electromagnetic response of dispersive materials (materials with frequency-dependent constitutive constants).

The method requires that the whole working space be divided into cubic or rectangular cells. The cell size should not exceed one-tenth of the wavelength corresponding to the highest frequency in the excitation. The electromagnetic field components E_x, E_y, E_z, H_x, H_y, and H_z are located in each cell in the 3-D Cartesian coordinate system, as shown in Figure 5.1. Time-updating equations for electric and magnetic fields are derived next.

Considering Equations 5.9 and 5.11, Ampere's law (Equation 5.2) is rewritten as

$$\nabla \times H^{n-(1/2)} = \varepsilon \frac{\partial E^{n-(1/2)}}{\partial t} + J^{n-(1/2)} = \varepsilon \frac{\partial E^{n-(1/2)}}{\partial t} + \sigma E^{n-(1/2)} \qquad (5.12)$$

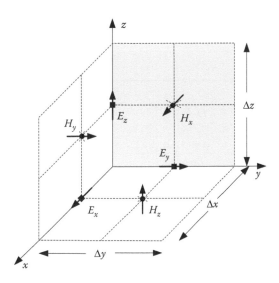

FIGURE 5.1
FDTD cell with x-, y-, and z-directed vectors of electric and magnetic fields.

where $n - (1/2)$ is the present time step. If the time-dependent, partial-differential term in Equation 5.12 is approximated by its central finite difference, Equation 5.12 is expressed as

$$\varepsilon \frac{\partial E^{n-(1/2)}}{\partial t} + \sigma E^{n-(1/2)} \approx \varepsilon \frac{E^n - E^{n-1}}{\Delta t} + \sigma \frac{E^n + E^{n-1}}{2} \approx \nabla \times H^{n-(1/2)} \quad (5.13)$$

where Δt is the time increment. Note that $E^{n-(1/2)}$ in Equation 5.13 is approximated by its average value, $(E^n + E^{n-1})/2$.

Rearranging Equation 5.13, the updated equation for the electric field at time step n from its value E^{n-1} at the previous time step and magnetic-field rotation $H^{n-(1/2)}$ at the previous half time step is obtained as follows:

$$E^n = \left(\frac{1 - (\sigma \Delta t / 2\varepsilon)}{1 + (\sigma \Delta t / 2\varepsilon)} \right) E^{n-1} + \left(\frac{\Delta t / \varepsilon}{1 + (\sigma \Delta t / 2\varepsilon)} \right) \nabla \times H^{n-(1/2)} \quad (5.14)$$

$\nabla \times H$ is given by

$$\nabla \times H = \begin{vmatrix} i & j & k \\ \dfrac{\partial}{\partial x} & \dfrac{\partial}{\partial y} & \dfrac{\partial}{\partial z} \\ H_x & H_y & H_z \end{vmatrix} = i \left(\frac{\partial H_z}{\partial y} - \frac{\partial H_y}{\partial z} \right) + j \left(\frac{\partial H_x}{\partial z} - \frac{\partial H_z}{\partial x} \right) + k \left(\frac{\partial H_y}{\partial x} - \frac{\partial H_x}{\partial y} \right)$$

$$(5.15)$$

where i, j, and k are x-, y-, and z-directed unit vectors, respectively.

From Equations 5.14 and 5.15, the updated equation for E_x^n at the space point $(i + (1/2), j, k)$, for example, is expressed as

$$E_x^n \left(i + \frac{1}{2}, j, k \right)$$

$$= \left(\frac{1 - (\sigma \Delta t / 2\varepsilon)}{1 + (\sigma \Delta t / 2\varepsilon)} \right) E_x^{n-1} \left(i + \frac{1}{2}, j, k \right)$$

$$+ \left(\frac{\Delta t / \varepsilon}{1 + (\sigma \Delta t / 2\varepsilon)} \right) \left[\frac{\partial H_z^{n-(1/2)}(i + (1/2), j, k)}{\partial y} - \frac{\partial H_y^{n-(1/2)}(i + (1/2), j, k)}{\partial z} \right]$$

$$\approx \left(\frac{1 - (\sigma \Delta t / 2\varepsilon)}{1 + (\sigma \Delta t / 2\varepsilon)} \right) E_x^{n-1} \left(i + \frac{1}{2}, j, k \right)$$

$$+ \left(\frac{\Delta t / \varepsilon}{1 + (\sigma \Delta t / 2\varepsilon)} \right) \left[\begin{array}{c} \dfrac{H_z^{n-(1/2)}(i + (1/2), j + (1/2), k) - H_z^{n-(1/2)}(i + (1/2), j - (1/2), k)}{\Delta y} \\[3mm] - \dfrac{H_y^{n-(1/2)}(i + (1/2), j, k + (1/2)) - H_y^{n-(1/2)}(i + (1/2), j, k - (1/2))}{\Delta z} \end{array} \right]$$

$$(5.16)$$

where the spatial, partial-differential terms in the first equation are approximated by their central finite differences. In the same manner, updated equations for E_y^n and E_z^n can be derived.

Considering Equation 5.10, Faraday's law (Equation 5.1) is rewritten as

$$\nabla \times \boldsymbol{E}^n = -\mu \frac{\partial \boldsymbol{H}^n}{\partial t} \tag{5.17}$$

where n is the present time step. If the time-dependent, partial-differential term in Equation 5.17 is approximated by its central finite difference, Equation 5.17 can be expressed as

$$\mu \frac{\partial \boldsymbol{H}^n}{\partial t} \approx \mu \frac{\boldsymbol{H}^{n+(1/2)} - \boldsymbol{H}^{n-(1/2)}}{\Delta t} \approx -\nabla \times \boldsymbol{E}^n \tag{5.18}$$

Rearranging Equation 5.18, the updated equation for the magnetic field at time step $n + (1/2)$ from its value $\boldsymbol{H}^{n-(1/2)}$ at the previous time step and electric-field rotation $\boldsymbol{E}^n$ at the previous half time step is obtained as follows:

$$\boldsymbol{H}^{n+(1/2)} = \boldsymbol{H}^{n-(1/2)} - \frac{\Delta t}{\mu} \nabla \times \boldsymbol{E}^n \tag{5.19}$$

From Equations 5.19 and 5.15, the updated equation for $H_x^{n+(1/2)}$ at the location $(i, j + 1/2, k + (1/2))$, for example, is expressed as follows:

$$H_x^{n+(1/2)}\left(i, j+\frac{1}{2}, k+\frac{1}{2}\right) = H_x^{n-(1/2)}\left(i, j+\frac{1}{2}, k+\frac{1}{2}\right) - \frac{\Delta t}{\mu}\left[\begin{array}{c} \dfrac{\partial E_z^n(i, j+(1/2), k+(1/2))}{\partial y} \\[2mm] -\dfrac{\partial E_y^n(i, j+(1/2), k+(1/2))}{\partial z} \end{array}\right]$$

$$\approx H_x^{n-(1/2)}\left(i, j+\frac{1}{2}, k+\frac{1}{2}\right)$$

$$-\frac{\Delta t}{\mu}\left[\begin{array}{c} \dfrac{E_z^n(i, j+1, k+(1/2)) - E_z^n(i, j, k+(1/2))}{\Delta y} \\[2mm] -\dfrac{E_y^n(i, j+(1/2), k+1) - E_y^n(i, j+(1/2), k)}{\Delta z} \end{array}\right] \tag{5.20}$$

where the spatial, partial-differential terms in the first equation are approximated by their central finite differences. In the same manner, updated equations for $H_y^{n+(1/2)}$ and $H_z^{n+(1/2)}$ can be derived.

The updated equations for $E_x^n, E_y^n, E_z^n, H_x^{n+(1/2)}, H_y^{n+(1/2)}$, and $H_z^{n+(1/2)}$ are summarized in the following equations:

$$
E_x^n\left(i+\frac{1}{2},j,k\right)
$$

$$
= \frac{1-(\sigma(i+(1/2),j,k)\Delta t/2\varepsilon(i+(1/2),j,k))}{1+(\sigma(i+(1/2),j,k)\Delta t/2\varepsilon(i+(1/2),j,k))} E_x^{n-1}\left(i+\frac{1}{2},j,k\right)
$$

$$
+ \frac{\Delta t/\varepsilon(i+(1/2),j,k)}{1+(\sigma(i+(1/2),j,k)\Delta t/2\varepsilon(i+(1/2),j,k))}
\left[
\begin{array}{c}
\dfrac{H_z^{n-(1/2)}(i+(1/2),j+(1/2),k)}{} \\
\dfrac{-H_z^{n-(1/2)}(i+(1/2),j-(1/2),k)}{\Delta y} \\
H_y^{n-(1/2)}(i+(1/2),j,k+(1/2)) \\
\dfrac{-H_y^{n-(1/2)}(i+(1/2),j,k-(1/2))}{\Delta z}
\end{array}
\right]
$$

$$
\tag{5.21}
$$

$$
E_y^n\left(i,j+\frac{1}{2},k\right)
$$

$$
= \frac{1-(\sigma(i,j+(1/2),k)\Delta t/2\varepsilon(i,j+(1/2),k))}{1+(\sigma(i,j+(1/2),k)\Delta t/2\varepsilon(i,j+(1/2),k))} E_y^{n-1}\left(i,j+\frac{1}{2},k\right)
$$

$$
+ \frac{\Delta t/\varepsilon(i,j+(1/2),k)}{1+(\sigma(i,j+(1/2),k)\Delta t/2\varepsilon(i,j+(1/2),k))}
\left[
\begin{array}{c}
H_x^{n-(1/2)}(i,j+(1/2),k+(1/2)) \\
\dfrac{-H_x^{n-(1/2)}(i,j+(1/2),k-(1/2))}{\Delta z} \\
H_z^{n-(1/2)}(i+(1/2),j+(1/2),k) \\
\dfrac{-H_z^{n-(1/2)}(i-(1/2),j+(1/2),k)}{\Delta x}
\end{array}
\right]
$$

$$
\tag{5.22}
$$

$$
E_z^n\left(i,j,k+\frac{1}{2}\right)
$$

$$
= \frac{1-(\sigma(i,j,k+(1/2))\Delta t/2\varepsilon(i,j,k+(1/2)))}{1+(\sigma(i,j,k+(1/2))\Delta t/2\varepsilon(i,j,k+(1/2)))} E_z^{n-1}\left(i,j,k+\frac{1}{2}\right)
$$

$$
+ \frac{\Delta t/\varepsilon(i,j,k+(1/2))}{1+(\sigma(i,j,k+(1/2))\Delta t/2\varepsilon(i,j,k+(1/2)))}
\left[
\begin{array}{c}
H_y^{n-(1/2)}(i+(1/2),j,k+(1/2)) \\
\dfrac{-H_y^{n-(1/2)}(i-(1/2),j,k+(1/2))}{\Delta x} \\
H_x^{n-(1/2)}(i,j+(1/2),k+(1/2)) \\
\dfrac{-H_x^{n-(1/2)}(i,j-(1/2),k+(1/2))}{\Delta y}
\end{array}
\right]
$$

$$
\tag{5.23}
$$

$$H_x^{n+(1/2)}\left(i, j+\frac{1}{2}, k+\frac{1}{2}\right)$$

$$= H_x^{n-(1/2)}\left(i, j+\frac{1}{2}, k+\frac{1}{2}\right)$$

$$- \frac{\Delta t}{\mu(i, j+(1/2), k+(1/2))}\left[\begin{array}{c} \dfrac{E_z^n(i, j+, k+(1/2)) - E_z^n(i, j, k+(1/2))}{\Delta y} \\ - \dfrac{E_y^n(i, j+(1/2), k+1) - E_y^n(i, j+(1/2), k)}{\Delta z} \end{array}\right] \quad (5.24)$$

$$H_y^{n+(1/2)}\left(i+\frac{1}{2}, j, k+\frac{1}{2}\right)$$

$$= H_y^{n-(1/2)}\left(i+\frac{1}{2}, j, k+\frac{1}{2}\right)$$

$$- \frac{\Delta t}{\mu(i+(1/2), j, k+(1/2))}\left[\begin{array}{c} \dfrac{E_x^n(i+(1/2), j, k+1) - E_x^n(i+(1/2), j, k)}{\Delta z} \\ - \dfrac{E_z^n(i+1, j, k+(1/2)) - E_z^n(i, j, k+(1/2))}{\Delta x} \end{array}\right] \quad (5.25)$$

$$H_z^{n+(1/2)}\left(i+\frac{1}{2}, j+\frac{1}{2}, k\right)$$

$$= H_z^{n-(1/2)}\left(i+\frac{1}{2}, j+\frac{1}{2}, k\right)$$

$$- \frac{\Delta t}{\mu(i+(1/2), j+1/2, k)}\left[\begin{array}{c} \dfrac{E_y^n(i+1, j+(1/2), k) - E_y^n(i, j+(1/2), k)}{\Delta t} \\ - \dfrac{E_x^n(i+(1/2), j+1, k) - E_x^n(i, j+(1/2), j, k)}{\Delta y} \end{array}\right] \quad (5.26)$$

By updating Equations 5.21 through 5.26 at every point, transient electric and magnetic fields throughout the working space are obtained.

For analyzing lightning surges on power systems, it is necessary to appropriately represent thin wires, which include overhead transmission-line conductors, distribution-line conductors, and steel frames of towers and buildings. Noda and Yokoyama [3] have found that a thin wire in air has an equivalent radius of $r_0 = 0.23\Delta s$ (Δs is the side length of cubic cells used in FDTD simulations), when the electric field along the axis of the thin wire is set to zero in an orthogonal and uniform Cartesian grid for FDTD simulations. They further showed that a thin wire with an arbitrary radius r could be equivalently represented by placing a zero-radius wire in an artificial rectangular prism, coaxial with the thin wire, with a cross-sectional area of $2\Delta s \times 2\Delta s$ and modified permittivity $m\varepsilon_0$ and permeability μ_0/m given by

$$m\varepsilon_0, \quad \frac{\mu_0}{m}, \quad m = \frac{\ln(1/0.23)}{\ln(\Delta s/r)} \tag{5.27}$$

For example, in representing a thinner wire with radius r ($<r_0$), the permittivity of its enclosed cells is decreased and the permeability is increased. This modification increases the characteristic impedance of the wire, but the speed of the electromagnetic wave in the enclosed artificial cells remains equal to the speed of light.

Lumped voltage and current sources are represented as follows. A lumped voltage source in the z direction at the space point $(i, j, k + (1/2))$, which generates a time-varying voltage $V_z(n\Delta t)$, is represented by

$$E_z^n\left(i, j, k + \frac{1}{2}\right) = \frac{V_z(n\Delta t)}{\Delta z} \tag{5.28}$$

A lumped current source in the z direction at the space point $(i, j, k + (1/2))$, which generates a time-varying current $I_z[(n - 1/2)\Delta t]$, is realized by

$$
\begin{aligned}
&E_z^n\left(i, j, k + \frac{1}{2}\right) \\
&= \frac{1 - \sigma(i, j, k + (1/2))\Delta t/2\varepsilon(i, j, k + (1/2))}{1 + \sigma(i, j, k + (1/2))\Delta t/2\varepsilon(i, j, k + (1/2))} E_z^{n-1}\left(i, j, k + \frac{1}{2}\right) \\
&\quad + \frac{\Delta t/\varepsilon(i, j, k + (1/2))}{1 + \sigma(i, j, k + (1/2))\Delta t/2\varepsilon(i, j, k + (1/2))}
\left[
\begin{array}{c}
\dfrac{H_y^{n-(1/2)}(i+(1/2), j, k+(1/2)) - H_y^{n-(1/2)}(i-(1/2), j, k+(1/2))}{\Delta x} \\[4pt]
- \dfrac{H_x^{n-(1/2)}(i, j+(1/2), k+(1/2)) - H_x^{n(1/2)}(i, j-(1/2), k+(1/2))}{\Delta y}
\end{array}
\right] \\
&\quad - \frac{\Delta t/\varepsilon(i, j, k + (1/2))}{1 + \sigma(i, j, k + (1/2))\Delta t/2\varepsilon(i, j, k + (1/2))} \cdot \frac{1}{\Delta x \Delta y} I_z[(n - 1/2)\Delta t]
\end{aligned}
\tag{5.29}
$$

Voltage and current sources in x- and y-directions can be represented in a similar manner.

The representation of lumped elements, a resistor R, a capacitor C, and an inductor L, are described here. For voltage $V_z[(n - 1/2)\Delta t]$ across a resistor R in the z direction, the current flowing through it, $I_z[(n - 1/2)\Delta t]$, has the relation

$I_z[(n-1/2)\Delta t] = V_z[(n-1/2)\Delta t]/R \approx \left(E_z^n + E_z^{n-1}\right)\Delta z/2R$. If this relation is substituted into Equation 5.29, the following updated equation for a z-directed resistor located at the space point $(i, j, k+1/2)$ is obtained:

$$
E_z^n\left(i, j, k+\frac{1}{2}\right)
$$

$$
= \frac{1-(\Delta t \Delta z/2R\varepsilon(i,j,k+(1/2))\Delta x \Delta y)}{1+(\Delta t \Delta z/2R\varepsilon(i,j,k+(1/2))\Delta x \Delta y)} E_z^{n-1}\left(i,j,k+\frac{1}{2}\right)
$$

$$
+ \frac{\Delta t/\varepsilon(i,j,k+1/2)}{1+(\Delta t \Delta z/2R\varepsilon(i,j,k+1/2)\Delta x \Delta y)}
\begin{bmatrix}
\dfrac{H_y^{n-(1/2)}(i+(1/2)j,k+(1/2))}{-H_y^{n-(1/2)}(i-(1/2),j,k+(1/2))}{\Delta x} \\[2ex]
\dfrac{H_y^{n-(1/2)}(i,j+(1/2),k+(1/2))}{-H_x^{n-(1/2)}(i,j-(1/2),k+(1/2))}{\Delta y}
\end{bmatrix}
$$

$$(5.30)$$

For voltage V_z $[(n-1/2)\Delta t]$ across a capacitor C in the z direction, the current flowing through it, I_z $[(n-1/2)\Delta t]$, has the relation $I_z[n-1/2\Delta t] = C \, \partial V_z[(n-1/2)\Delta t]/\partial t \approx C\left(E_z^n - E_z^{n-1}\right)\Delta z/\Delta t$. If this relation is substituted into Equation 5.29, the following updated equation for a z-directed capacitor at the space point $(i, j, k+(1/2))$ is obtained:

$$
E_z^n\left(i, j, k+\frac{1}{2}\right)
$$

$$
= E_z^{n-1}\left(i, j, k+\frac{1}{2}\right)
$$

$$
+ \frac{\Delta t/\varepsilon(i,j,k+1/2)}{1+(C \, \Delta z/\varepsilon(i,j,k+1/2)\Delta x \Delta y)}
\begin{bmatrix}
\dfrac{H_y^{n-(1/2)}(i+(1/2),j,k+(1/2))}{-H_y^{n-(1/2)}(i-(1/2),j,k+(1/2))}{\Delta x} \\[2ex]
\dfrac{H_x^{n-(1/2)}(i,j+(1/2),k+(1/2))}{-H_x^{n-(1/2)}(i,j-(1/2),k+(1/2))}{\Delta y}
\end{bmatrix}
$$

$$(5.31)$$

For voltage V_z $[(n-1/2)\Delta t]$ across an inductor L in the z direction, the current flowing through it, I_z $[(n-1/2)\Delta t]$, has the relation $I_z[(n-1/2)\Delta t] = \int V_z dt/L \approx \Delta z \Delta t \sum_{m=1}^{(n-1)} E_z^m /L$. If this relation is substituted into

Equation 5.29, the following updated equation for a z-directed inductor at the space point $(i, j, k + 1/2)$ is obtained:

$$E_z^n\left(i,j,k+\frac{1}{2}\right) = E_z^{n-1}\left(i,j,k+\frac{1}{2}\right)$$

$$+ \frac{\Delta t}{\varepsilon(i,j,k+1/2)} \left[\begin{array}{c} H_y^{n-(1/2)}(i+(1/2),j,k+(1/2)) \\ \dfrac{-H_y^{n-(1/2)}(i-(1/2),j,k+(1/2))}{\Delta x} \\ H_x^{n-(1/2)}(i,j+(1/2),k+(1/2)) \\ -\dfrac{-H_x^{n-(1/2)}(i,j-(1/2),k+(1/2))}{\Delta y} \end{array} \right]$$

$$- \frac{\Delta z(\Delta t)^2}{\varepsilon(i,j,k+1/2)L\Delta x\Delta y} \sum_{m=1}^{n-1} E_z^m\left(i,j,k+\frac{1}{2}\right) \qquad (5.32)$$

Lumped elements in x and y directions can be represented in a similar manner.

5.1.3 Method of Moments

The MoM [4] is also frequently employed in transient electromagnetic computations. This method is based on an electric-field integral equation in either a frequency or time domain that relates the induced current on a conductor to the incident electric field. Only the conducting structure to be analyzed has to be represented as a combination of short cylindrical segments.

The advantages of this method are: (1) it is computationally more efficient than the FDTD method; (2) it requires no absorbing boundary condition; (3) it can represent oblique conductors easily without any staircase approximation; (4) it is capable of considering dispersive materials in the FD MoM; and (5) it is capable of incorporating nonlinear effects and components in the time-domain MoM.

Its disadvantages are: (1) it cannot deal with complex boundaries, compared with the FDTD method; (2) in the time-domain MoM, it would require a complex procedure for considering dispersive materials; and (3) in the FD MoM, it would be impossible to incorporate nonlinear effects and components.

Next we will look at an electric-field integral equation in the FD, and then the corresponding electric-field integral equation in the time domain.

The electric field at the space point r is generally expressed in the FD as

$$E(r) = -j\omega A(r) - \nabla\varphi(r) \qquad (5.33)$$

where

$A(r)$ is the magnetic vector potential

$\varphi(r)$ is the electric scalar potential

If the Lorenz gauge given as

$$\nabla \cdot A(r) = -j\omega\mu\varepsilon\varphi(r) \tag{5.34}$$

is applied to Equation 5.33, the following wave equations for each potential are obtained:

$$\nabla^2 A(r) + \omega^2\mu\varepsilon A(r) = -\mu J(r) \tag{5.35}$$

$$\nabla^2\varphi(r) + \omega^2\mu\varepsilon\varphi(r) = -\frac{\rho(r)}{\varepsilon} \tag{5.36}$$

Each of these wave equations, excited by a source at the location r' in the form of Dirac delta function $\delta(r, r')$, can be expressed as

$$\nabla^2 g(r,r') + k^2 g(r,r') = -\delta(r,r') \tag{5.37}$$

If the source is assumed to be located at the origin $r' = 0$, Equation 5.37 can be rewritten in the spherical coordinate as

$$\frac{1}{r^2}\frac{\partial}{\partial r}\left(r^2 \frac{\partial}{\partial r} g(r)\right) + k^2 g(r) = -\delta(r) \tag{5.38}$$

The solution of Equation 5.38 is expressed as $g(r) = e^{-jkr}/(4\pi r)$. Thus, the solution at the location r for the arbitrary source point r' is expressed as

$$g(r,r') = \frac{1}{4\pi}\frac{e^{-jk|r-r'|}}{|r-r'|} \tag{5.39}$$

On the basis of the superposition principle, the potential generated by arbitrary sources can be written as

$$\varphi(r) = \int_V \frac{\rho(r')}{\varepsilon} g(r,r')dV \tag{5.40}$$

$$A(r) = \int_V \mu J(r')g(r,r')dV \tag{5.41}$$

The total electric field E at the space point r is the sum of the incident electric field E^i and the scattered electric field E^s. This relation can be written as

$$E(r) = E^i(r) + E^s(r) \qquad (5.42)$$

Also, the total electric field follows Ohm's law, which is

$$E(r) = E^i(r) + E^s(r) = \frac{J(r)}{\sigma} \qquad (5.43)$$

where σ is the conductivity of the material of interest.

If Equations 5.40 and 5.41 are substituted into Equation 5.33, the scattered electric field E^s at the space point r due to the current J at the space point r' is expressed as

$$E^s(r) = -j\omega A(r) - \nabla\varphi(r) = -j\omega\mu \int_V g(r,r')J(r')dV - \frac{\nabla}{\varepsilon}\int_V g(r,r')\rho(r')dV \qquad (5.44)$$

If Equation 5.44 is substituted into Equation 5.43, the relation of the incident electric field to the induced current density and volume charge density is obtained as

$$E^i(r) = \frac{J(r)}{\sigma} + j\omega\mu \int_V g(r,r')J(r')dV + \frac{\nabla}{\varepsilon}\int_V g(r,r')\rho(r')dV \qquad (5.45)$$

The volume charge density ρ is related to the current density J via the charge conservation equation, which is as follows:

$$\nabla \cdot J(r') = -j\omega\rho(r') \qquad (5.46)$$

If Equation 5.46 is substituted into Equation 5.45, the following form is obtained:

$$E^i(r) = \frac{J(r)}{\sigma} + j\omega\mu \int_V g(r,r')J(r')dV - \frac{\nabla}{j\omega\varepsilon}\int_V g(r,r')\nabla \cdot J(r')dV \qquad (5.47)$$

When a perfect conductor is analyzed, the tangential component of the total electric field on the conductor's surface is zero. Therefore, the following relation is fulfilled on the conductor surface:

$$n \times [E^i(r) + E^s(r)] = n \times \frac{J(r)}{\sigma} = 0 \qquad (5.48)$$

Furthermore, the current and charge are distributed only on the surface of a perfect conductor. Thus, Equation 5.48 is rewritten as

$$n \times E^i(r) = -n \times E^s(r)$$

$$= n \times \left[j\omega\mu \int_S g(r,r')J_s(r')dS - \frac{\nabla}{j\omega\varepsilon} \int_S g(r,r')\nabla \cdot J_s(r')ds \right] \quad (5.49)$$

where
  n is a unit-normal vector on the conductor surface
  J_s is the surface current density

When the radius of a perfectly conducting wire is much smaller than the wavelength of interest, the current *I* and charge *q* could be assumed to be confined to the wire axis, as shown in Figure 5.2. This assumption is called thin-wire approximation. With this assumption, the electric-field integral equation, Equation 5.47 or Equation 5.49, for a perfectly conducting thin wire in air, is simplified to:

$$E^i(r) = j\omega\mu_0 \int_C I(r')g(r,r')s'ds' - \frac{\nabla}{j\omega\varepsilon_0} \int_C g(r,r')\nabla \cdot [I(r')s']ds' \quad (5.50)$$

where
  s is the unit tangential vector along the wire surface *C(r)*
  s' is the unit tangential vector on the wire axis

The incident electric field, which is tangential to the wire surface and parallel with the wire axis, is given as a dot product $s \cdot E^i$, as follows:

$$s \cdot E^i(r) = j\omega\mu_0 \int_C s \cdot s'I(r')g(r,r')ds' - s \cdot \frac{\nabla}{j\omega\varepsilon_0} \int_C g(r,r')\nabla \cdot [I(r')s']ds' \quad (5.51)$$

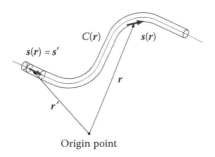

FIGURE 5.2
Thin-wire approximated conductor.

or

$$s \cdot E^i(r) = j\omega\mu_0 \int_C s \cdot s' I(r') g(r,r') ds' - \frac{1}{j\omega\varepsilon_0} \int_C \frac{\partial[I(r')g(r,r')]}{\partial s \partial s'} ds'$$

$$= \frac{j\eta}{k} \int_C \left(k^2 s \cdot s' - \frac{\partial}{\partial s \partial s'} \right) I(r') g(r,r') ds' \qquad (5.52)$$

where $k = \omega\,(\mu_0\varepsilon_0)^{1/2}$ and $\eta = (\mu_0/\varepsilon_0)^{1/2}$.

Next we examine an electric-field integral equation in the time domain. Electric scalar potential and magnetic vector potential are expressed in the time domain as follows:

$$\varphi(r,t) = \frac{1}{4\pi\varepsilon} \int_V \frac{\rho(r',t')}{R} dV \qquad (5.53)$$

$$A(r,t) = \frac{\mu}{4\pi} \int_V \frac{J(r',t')}{R} dV \qquad (5.54)$$

where
$R = |r - r'|$
$t' = t - R/v$
v is the speed of the electromagnetic wave in the medium of interest

The electric field at the space point r is expressed as

$$E(r,t) = -\frac{\partial A(r,t)}{\partial t} - \nabla\varphi(r,t) \qquad (5.55)$$

If Equations 5.53 and 5.54 are substituted into Equation 5.55, the scattered electric field E^s at the space point r due to the current J at the space point r' is expressed as

$$E^s(r,t) = -\frac{\mu}{4\pi} \int_V \frac{1}{R} \cdot \frac{\partial J(r',t')}{\partial t'} dV - \frac{\nabla}{4\pi\varepsilon} \int_V \frac{\rho(r',t')}{R} dV \qquad (5.56)$$

The volume charge density in Equation 5.56 is evaluated by the following continuity equation:

$$\rho(r',t') = -\int_{-\infty}^{t'} \nabla \cdot J(r',\tau)\, d\tau \qquad (5.57)$$

When a perfect conductor is analyzed, the current and charge are distributed only on the surface of the perfect conductor and the tangential component of the total electric field on the conductor's surface is zero. Therefore, the following relation is fulfilled on the conductor's surface:

$$n \times E^i(r,t) = -n \times E^s(r',t) = n \times \left[\frac{\mu}{4\pi} \int_S \frac{1}{R} \cdot \frac{\partial J_s(r',t')}{\partial t'} dS + \frac{\nabla}{4\pi\varepsilon} \int_S \frac{\rho_s(r',t')}{R} dS \right] \quad (5.58)$$

where ρ_s is the surface charge density, which is evaluated by

$$\rho_s(r',t') = -\int_{-\infty}^{t'} \nabla \cdot J_s(r',\tau) d\tau \quad (5.59)$$

The electric-field integral equation in the time domain for a perfectly conducting thin wire in air is obtained similarly to that in the FD, as follows:

$$s \cdot E^i(r,t) = \frac{\mu_0}{4\pi} \int_C s \cdot s' \frac{1}{R} \cdot \frac{\partial I(r',t')}{\partial t'} ds' + s \cdot \frac{\nabla}{4\pi\varepsilon_0} \int_C \frac{q(r',t')}{R} ds' \quad (5.60)$$

where

$$q(r',t') = -\int_{-\infty}^{t'} \frac{\partial I(r',\tau)}{\partial s'} d\tau$$

The last term in Equation 5.60 is converted into

$$\nabla \left(\frac{q(r',t')}{R} \right) = \frac{\partial}{\partial R} \left(\frac{q(r',t')}{R} \right) \frac{R}{R} = \left(q \frac{\partial(1/R)}{\partial R} + \frac{1}{R} \frac{\partial q}{\partial R} \right) \frac{R}{R}$$
$$= \left(-\frac{1}{R^2} q + \frac{1}{R} \frac{\partial q}{\partial s'} \right) \frac{R}{R} = \left(-\frac{1}{R^2} q - \frac{1}{R} \frac{\partial q}{c \partial t'} \right) \frac{R}{R} \quad (5.61)$$

With the continuity equation in one-dimensional form, which is given as

$$\frac{\partial q}{\partial t'} = -\frac{\partial I}{\partial s'} \quad (5.62)$$

the last term is expressed as

$$\nabla \left(\frac{q(r',t')}{R} \right) = \left(-\frac{1}{R^2} q + \frac{1}{Rc} \frac{\partial I}{\partial s'} \right) \frac{R}{R} \quad (5.63)$$

If Equation 5.63 is substituted into Equation 5.60, the electric-field integral equation in the time domain for a thin-wire perfect conductor is obtained as

$$s \cdot E^i(r,t) = \frac{\mu_0}{4\pi} \int_C \left[s \cdot s' \frac{1}{R} \cdot \frac{\partial I(r',t')}{\partial t'} + c \frac{s \cdot R}{R^2} \frac{\partial I(r',t')}{\partial s'} - c^2 \frac{s \cdot R \, q(r',t')}{R^3} \right] ds' \quad (5.64)$$

In solving Equation 5.52 or Equation 5.64, a mathematical function for approximating the distribution of current along the wire axis is employed. The function is usually expressed as a linear combination of basis functions having unknown coefficients, and the unknown coefficients are evaluated numerically. Equation 5.52 can be written as follows:

$$L(f) = E \quad (5.65)$$

where
L is a linear operator
E is the excitation function
f is the unknown current function

The unknown function $f(x)$ can be expanded as

$$f(x) = \sum_{n=1}^{N} a_n f_n(x) \quad (5.66)$$

where
a_n is an unknown coefficient
$f_n(x)$ is a known basis function that is illustrated in Figure 5.3

If Equation 5.66 is substituted into Equation 5.65, the following equation is obtained:

$$\sum_{n=1}^{N} a_n L(f_n(x)) \approx E \quad (5.67)$$

To solve Equation 5.67, the dot product of weight function f_m is applied to Equation 5.67. Then the following equation is obtained:

$$\sum_{n=1}^{N} a_n \langle f_m, L(f_n) \rangle \approx \langle f_m, E \rangle \quad (5.68)$$

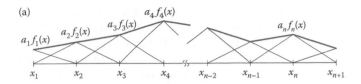

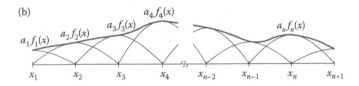

FIGURE 5.3
MoM typical basis functions for approximating the distribution of current along the wire axis. (a) Piecewise triangular function and (b) piecewise sinusoidal function.

Equation 5.68 can be written in matrix form as

$$Za = b \qquad (5.69)$$

with

$$Z_{mn} = \langle f_m, L(f_n) \rangle \qquad (5.70)$$

and

$$b_{mn} = \langle f_m, E \rangle \qquad (5.71)$$

where
Z_{mn} is an element of matrix Z at row m and column n
b_m is an element of vector b at row m

The unknown coefficients of the current function are obtained by solving Equation 5.69.

The numerical electromagnetic code (NEC) [5,6], the widely used computer program, is based on the MoM in the FD. The thin-wire time domain (TWTD) code [7] is based on the MoM in the time domain.

5.2 Applications

5.2.1 Grounding Electrodes

The role of grounding electrodes is to effectively drain fault currents into the soil and thereby mitigate the damage of installations of telecommunication systems and electrical power systems. Thus, the performance of such a system is influenced by the transient characteristics of its grounding electrodes. It is therefore important to study these characteristics. Recently, NEA methods have been successfully applied in analyzing transient responses.

Tanabe [8] analyzed the transient response of a vertical grounding electrode of 0.5 m × 0.5 m × 3 m, shown in Figure 5.4, using the FDTD method [1]. For FDTD computations, the conductor system shown in the figure is accommodated in a working volume of 27.5 m × 61 m × 55 m, which is divided uniformly into cubic cells of 0.25 m × 0.25 m × 0.25 m. The working volume is surrounded by six planes of Liao's second-order absorbing boundary condition [9] to minimize unwanted reflections. The time increment is set to 0.481 ns, which is determined following the CFL criterion [3]. The conductivity, relative permittivity, and relative permeability of the ground are set to $\sigma = 1.9$–2.7 mS/m (based on the low-frequency measurement), $\varepsilon_r = 50$, and $\mu_r = 1$, respectively. Figure 5.5 shows the FDTD-computed voltage and current waveforms for a vertical grounding electrode and the corresponding measured waveforms [8]. The FDTD-computed waveforms are in good agreement with the corresponding measured ones.

Tanabe et al. [10] studied the transient response of a horizontally placed, square-shaped grounding electrode of 7.5 m × 7.5 m, buried 0.5 m deep, using the FDTD method. For FDTD computations, the conductor system, consisting of a 50 m-long, horizontal voltage reference wire, a 26.25 m-long horizontal current lead wire, and the grounding electrode, is accommodated in a working volume of 83.75 m × 67.5 m × 30 m, which is divided uniformly into cubic cells of 0.25 m × 0.25 m × 0.25 m. The working volume is surrounded by six planes of Liao's second-order absorbing boundary condition. The time increment is set to 0.481 ns, which is determined on the basis

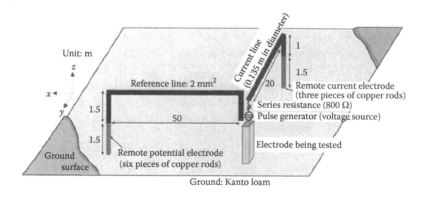

FIGURE 5.4
Configuration of a 3 m-long vertical grounding electrode and its auxiliary wires for the measurement of its surge response. (Tanabe, K., Novel method for analyzing dynamic behavior of grounding systems based on the finite-difference time-domain method, *IEEE Power Eng. Rev.* 21(9):55–577. © 2001 IEEE.)

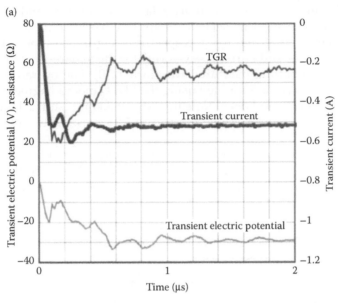

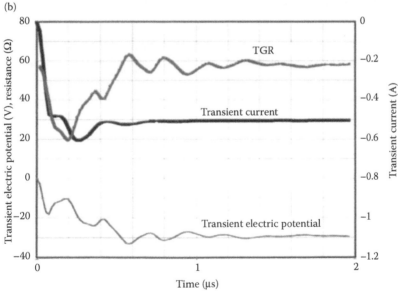

FIGURE 5.5
(a) Measured waveforms of voltage and current for the 3 m-long vertical grounding electrode and (b) the corresponding FDTD-computed waveforms. (Tanabe, K., Novel method for analyzing dynamic behavior of grounding systems based on the finite-difference time-domain method, *IEEE Power Eng. Rev.* 21(9):55–57. © 2001 IEEE.)

of the CFL criterion. The conductivity, relative permittivity, and relative permeability of the ground are set to $\sigma = 3.8$ mS/m (based on the low frequency measurement), $\varepsilon_r = 50$, and $\mu_r = 1$, respectively. Figure 5.6 shows the FDTD-computed voltage and current waveforms for the square-shaped electrode and the corresponding measured waveforms [10]. The overall waveforms of voltage and current computed using the FDTD method agree reasonably well with the measured ones. Note that Miyazaki and Ishii [11] reproduced the measured waveforms, shown in Figure 5.6a, reasonably well using the MoM in the FD.

Ala et al. [12] considered soil ionization around a grounding electrode in their FDTD computations. The ionization model is based on the dynamic soil-resistivity model of Liew and Darveniza [13]. Figure 5.7 shows the

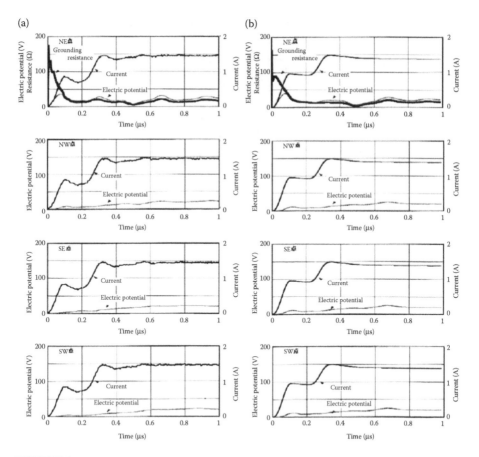

FIGURE 5.6
(a) Measured waveforms of voltage at each corner of a horizontally placed 7.5 × 7.5 m square-shaped grounding electrode and injected current and (b) the corresponding FDTD-computed waveforms. (Reprinted from Tanabe, K. et al., *IEEJ Trans. Power Energy*, 123(3), 358–367, 2003. With permission from IEEJ.)

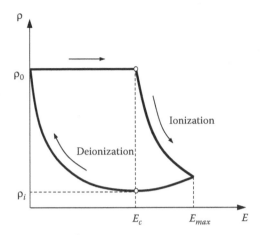

FIGURE 5.7
Resistivity profile in the dynamic model proposed by Liew and Darveniza [13] and employed by Ala et al. [12] for FDTD computations.

resistivity profile in the dynamic model proposed by Liew and Darveniza [13] and employed by Ala et al. [12]. In the model, the resistivity of each soil-representing cell is controlled by the instantaneous values of the electric field and time. When the instantaneous value of the electric field E at a soil-representing cell is lower than the critical electric field E_c, the resistivity ρ is equal to its steady-state value ρ_0:

$$\rho = \rho_0 \tag{5.72}$$

When E at a soil-representing cell exceeds the critical electric field E_c, ρ begins to decrease with time, as follows:

$$\rho = \rho_0 \exp\left(-\frac{t}{\tau_1}\right) \tag{5.73}$$

where
  t is the time defined so that $t = 0$ at the instant of $E = E_c$
  τ_1 is the ionization time constant

This decreasing resistivity with time represents the soil ionization process. When E at a cell in the ionized soil region falls below E_c, ρ begins to increase with time, as follows:

$$\rho = \rho_i + (\rho_0 - \rho_i)\left[1 - \exp\left(-\frac{t}{\tau_2}\right)\right]\left(1 - \frac{E}{E_c}\right)^2 \tag{5.74}$$

where

ρ_i is the minimal value reached by the ionization process
t is the time defined so that $t = 0$ at the instant of $E = E_c$
τ_2 is the deionization time constant

This increasing resistivity with time from ρ_i to ρ_0 represents the deionization process of the soil.

Figure 5.8 shows a 0.61 m-long, vertical grounding rod buried in the homogeneous soil of resistivity $\rho_0 = 50$ Ω-m, relative permittivity $\varepsilon_r = 8$, and relative permeability $\mu_r = 1$, to be analyzed using the FDTD method. The vertical grounding rod is energized by a lumped current source whose other terminal is connected to four auxiliary grounding electrodes via overhead wires. The current source generates a current with a magnitude of approximately 3.5 kA and a rise time of approximately 5 μs, as shown in Figure 5.9a. The working volume is divided uniformly into 61 mm × 61 mm × 61 mm cubic cells and is surrounded by six planes of an absorbing boundary condition to minimize unwanted reflections. The equivalent radius of the vertical grounding rod is 14 mm (= 0.23Δs = 0.23 × 61 mm) [3,14].

Figure 5.9b shows the waveform of voltage at the top of the vertical grounding rod, computed using the FDTD method with the soil ionization model. In this computation, $E_c = 110$ kV/m, $\tau_1 = 2.0$ μs, and $\tau_2 = 4.5$ μs were employed for the soil ionization model. Figure 5.9b also shows the voltage waveform computed without the soil ionization model. The peak voltage computed with the soil ionization model is about 40% less than that computed without the soil ionization model.

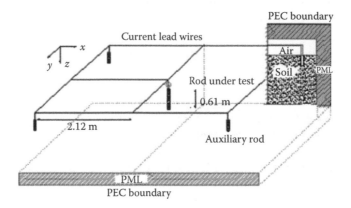

FIGURE 5.8
Configuration of a 0.61 m-long vertical grounding electrode and its auxiliary wires for the measurement of the electrode's surge response. (Reprinted from Ala, G. et al., *IET Sci. Meas. Technol.*, 2(3), 134–145, 2008. With permission from IET.)

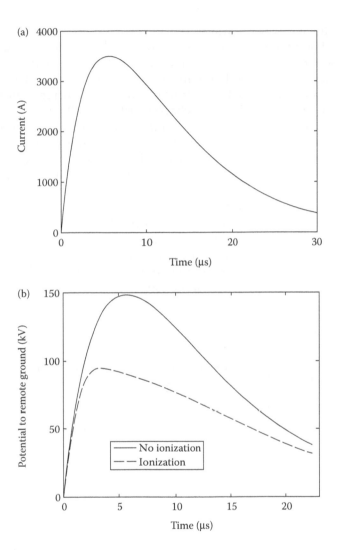

FIGURE 5.9
Waveform of current injected into the top of the 0.61 m-long vertical grounding rod and wave-forms of the voltage at the top of the grounding rod, computed using the FDTD method with and without the soil ionization model. (a) Current injected and (b) voltages computed with and without soil ionization. (Reprinted from Ala, G. et al., *IET Sci. Meas. Technol.*, 2(3), 134–145, 2008. With permission from IET.)

5.2.2 Transmission Towers

Lightning overvoltages in overhead power transmission systems are mainly caused by back flashovers (BFOs) of tower insulations. The electromagnetic field around a transmission tower when it is hit by lightning changes dynamically, while electromagnetic waves make several round trips between the shield wire and the ground.

During this interval, the waveforms of insulator voltages exhibit complex variations. For a tall structure such as an extra-high-voltage, double-circuit transmission tower, the contribution of the tower-surge characteristic to the insulator voltages becomes dominant because the travel time of a surge along the tower is comparable to the rise time of a lightning current. Therefore, it is important to investigate the surge characteristics of tall towers.

Mozumi et al. [15], using the MoM in the time domain [7], computed voltages across insulators of a 500-kV, double-circuit transmission-line tower with two overhead ground wires located above a flat, perfectly conducting ground when the top of the tower is struck by lightning, resulting in a BFO across the insulator of one phase. In order to analyze BFOs using the MoM in the time domain, they incorporated a flashover model developed by Motoyama [16] into the MoM. For computation purposes, the lightning return-stroke channel is represented by a vertical, perfectly conducting wire in air. The lightning channel and the tower are excited by a lumped voltage source in series, with a 5 kΩ lumped resistance inserted between them. Figure 5.10 shows the structure of the tower to be analyzed. This conductor

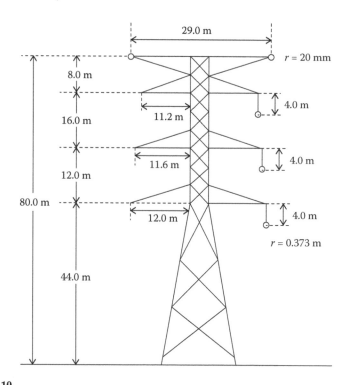

FIGURE 5.10
Structure of a 500-kV transmission-line tower analyzed using the MoM in the time domain. (Mozumi, T. et al., Numerical electromagnetic field analysis of archorn voltages during a back flashover on a 500 kV twin-circuit line, *IEEE Trans. Power Deliv.* 18(1):207–213. © 2003 IEEE.)

system is divided into thin, cylindrical segments of about 4 m in length. The time increment is set to 20 ns.

Figure 5.11 shows waveforms of insulator voltages computed using the MoM in the time domain and using the electromagnetic transients program (EMTP) [17] when an upper-phase BFO occurs for a current having a magnitude of 150 kA and a rise time of 1 μs. Note that, in the EMTP computation, a multistory transmission-line tower model [18] is employed and the characteristic impedance of the top of the tower is set to 245 Ω and the impedance of the bottom is set to 180 Ω. MoM-computed waveforms are reasonably well reproduced by the corresponding EMTP-computed waveforms.

Noda [19], using the FDTD method [1], computed voltages across insulators of a 500-kV, double-circuit transmission-line tower, located on flat ground with a conductivity of 10 mS/m, when the top of the tower is struck by lightning. In his computation, the lightning return-stroke channel is represented by a vertical, perfectly conducting wire with an additional distributed series inductance of 10 μH/m. The resultant speed of the current wave propagating along the wire is 0.33c. The lightning channel and the tower are excited by a lumped current source inserted between them. Figure 5.12 shows the structure of the tower analyzed using the FDTD method. This conductor system is accommodated in a working volume of 250 m × 250 m × 150 m, which is divided uniformly into cubic cells of 1 m × 1 m × 1 m. The working volume is surrounded by six planes of Liao's second-order absorbing boundary condition [9] to minimize unwanted reflections.

Figure 5.13 shows waveforms of insulator voltages computed using the FDTD method and using the EMTP when a ramp current with a magnitude of 1 A and a rise time of 1 μs is injected. Note that in Noda's EMTP computation, a new circuit model for a tower [19] is employed, with the characteristic

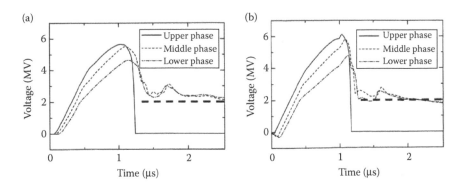

FIGURE 5.11
Waveforms of insulator voltages computed using the MoM in the time domain and using EMTP, in the case of an upper-phase BFO for a current with a magnitude of 150 kA and a rise time of 1 μs. (a) TWTD and (b) EMTP. (Mozumi, T. et al., Numerical electromagnetic field analysis of archorn voltages during a back flashover on a 500 kV twin-circuit line, *IEEE Trans. Power Deliv.* 18(1):207–213. © 2003 IEEE.)

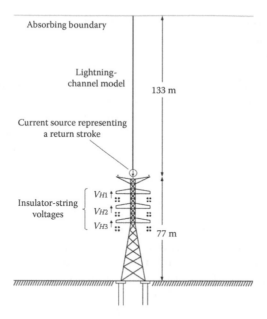

FIGURE 5.12
Structure of a 500-kV transmission-line tower analyzed using the FDTD method. (Reprinted from Noda, T., *IEEJ Trans. Power Energy*, 127(2), 379–388, 2007. With permission from IEEJ.)

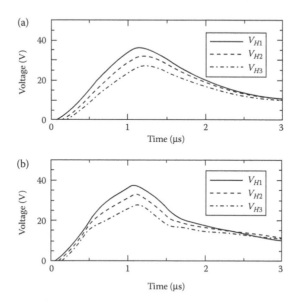

FIGURE 5.13
Waveforms of insulator voltages computed using FDTD and EMTP methods, in the case of a ramp current with a magnitude of 1 A and a rise time of 1 μs is injected. (a) FDTD and (b) EMTP. (Reprinted from Noda, T., *IEEJ Trans. Power Energy*, 127(2), 379–388, 2007. With permission from IEEJ.)

impedance set to 192 Ω. FDTD-computed waveforms are reasonably well reproduced by the corresponding EMTP-computed waveforms. Also, Noda [19] showed that his FDTD-computed waveforms of the tower-top voltage and the tower current for a similar tower agree reasonably well with the corresponding measured waveforms.

5.2.3 Distribution Lines: Lightning-Induced Surges

In order to optimize ways to protect telecommunication and power distribution lines from lightning, one needs to know voltages that can be induced on overhead wires by lightning strikes to the ground or to nearby grounded objects. NEA methods have recently been employed to analyze lightning-induced voltages on overhead telecommunication and power distribution lines.

Using the MoM in the FD [2], Pokharel et al. [20] reproduced lightning-induced voltages on an overhead horizontal wire of radius 0.25 mm, length 25 m, and height 0.5 m, which were measured by Ishii et al. [21]. Figure 5.14 shows the configuration of Ishii et al.'s small-scale experiment. In the experiment, the vertical return-stroke channel is represented by a coiled wire along which a current wave propagates upward at a speed of approximately 125 m/μs. The close (to the simulated channel) end of the overhead horizontal wire is either terminated in a 430-Ω resistor or left open, and the remote end is terminated in a 430-Ω resistor. The lightning-induced voltages at both ends of the wire are measured using voltage probes with 20 pF input

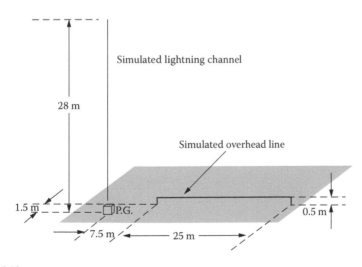

FIGURE 5.14
Configuration of a small-scale experiment for measuring lightning-induced voltages. (Ishii, M., Michishita, K., and Hongo, Y., Experimental study of lightning-induced voltage on an overhead wire over lossy ground, *IEEE Trans. Electromagn. Compat.* 41(1):39–45. © 1999 IEEE.)

capacitance. Figure 5.15 shows MoM-computed and measured waveforms of induced voltages. Note that in the MoM computations the lightning channel is represented by a vertical wire with 1 Ω/m series-distributed resistance and 3 μH/m series-distributed inductance, and the ground conductivity and its relative permittivity are set to σ = 0.06 S/m and ε_r = 10, respectively. The conductor system is modeled as a combination of cylindrical segments that are either 1 m or 0.5 m long. Computation is carried out over the frequency range of 195.3 kHz to 50 MHz with an increment step of 195.3 kHz. This corresponds to the time interval from 0 to 5.12 μs with a time increment

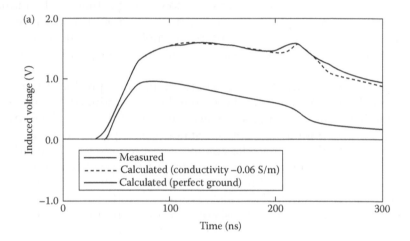

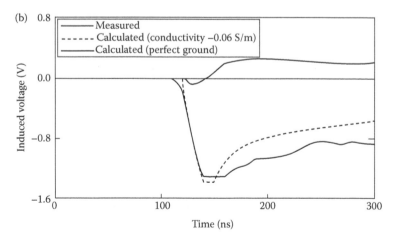

FIGURE 5.15

MoM-computed and measured waveforms of lightning-induced voltages at both ends of an overhead wire; each end is terminated with a 430-Ω resistor. (a) Voltages at the close end and (b) voltages at the remote end. (Pokharel, R. K., Ishii, M., and Baba, Y., Numerical electromagnetic analysis of lightning-induced voltage over ground of finite conductivity, *IEEE Trans. Electromagn. Compat.* 45(4):651–656. © 2003 IEEE.)

of 10 ns. In Figure 5.15, MoM-computed voltage waveforms for the case of a perfectly conducting ground are also shown for reference. Owing to the finitely conducting ground, the polarity of the remote-end voltage is opposite to that of the close end.

Note that Ishii et al. [21] reproduced lightning-induced voltages measured in their experiment with Agrawal et al.'s field-to-wire electromagnetic coupling model [22], and Baba and Rakov [23] reproduced the same measured waveforms using the FDTD method [1].

Using the MoM in the time domain [7], Pokharel and Ishii [24] computed lightning-induced voltages on a 500 m long overhead wire, which is located above a flat, perfectly conducting ground. The overhead wire is terminated with a 540-Ω resistor at each end and is connected to the ground in the middle of the wire via a surge arrester. The surge arrester is represented by a nonlinear resistor whose characteristics are shown in Figure 5.16. Figure 5.17 shows the computed waveforms of lightning-induced voltage in the middle of the wire with or without the surge arrester and the arrester current. Note that the simulated lightning channel is located 100 m away from the middle of the overhead wire, the magnitude of the lightning current is set to 10 kA, and its rise time is set to 1 μs.

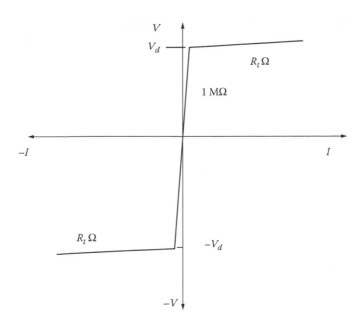

FIGURE 5.16

Approximate voltage versus current characteristics of nonlinear resistance representing a surge arrester, employed by Pokharel and Ishii [24] in their computations using the MoM in the time domain. V_d is set to 30 kV. (Pokharel, R. K. and Ishii, M., Applications of time-domain numerical electromagnetic code to lightning surge analysis, *IEEE Trans. Electromagn. Compat.* 49(3):623–631. © 2007 IEEE.)

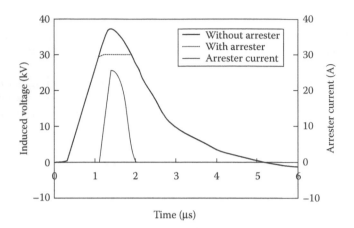

FIGURE 5.17
Waveforms of lightning-induced voltage in the middle of a 500 m–long overhead wire with or without a surge arrester and the arrester current computed using the MoM in the time domain. (Pokharel, R. K. and Ishii, M., Applications of time-domain numerical electromagnetic code to lightning surge analysis, *IEEE Trans. Electromagn. Compat.* 49(3):623–631. © 2007 IEEE.)

Using the same time-domain method, Moini et al. [25] performed a computation of lightning-induced voltages on vertically arranged and horizontally arranged multiphase conductors above a flat, perfectly conducting ground.

5.2.4 Transmission Lines: Propagation of Lightning Surges in the Presence of Corona

When the overhead shield wire of a transmission line is struck by lightning, corona discharge occurs on the wire. Corona discharge around a shield wire reduces its characteristic impedance and increases the coupling between the shield wire and phase conductors. The reduced characteristic impedance of the shield wire results in a lesser tower current, and the increased coupling to the phase conductors increases their voltages. As a result, corona discharge leads to reduced insulator voltages. Also, it distorts the wave fronts of propagating lightning-surge voltages. Thus, it is important to consider corona effects when computing lightning surges on transmission lines and designing ways to protect lines from lightning surges.

Thang et al. [26] proposed a simplified corona discharge model for FDTD computations. They represent the radial progression of corona streamers from energized wires by the radial expansion of the cylindrical conducting region. The critical electric field E_0 on the surface of a cylindrical wire of radius r_0 for the initiation of corona discharge is given by Hartmann's [27] equation, which is reproduced as follows:

$$E_0 = m \cdot 2.594 \times 10^6 \left(1 + \frac{0.1269}{r_0^{0.4346}}\right) (V/m) \tag{5.75}$$

where m of Equation 5.75 is the coefficient depending on the wire surface conditions. Note that this coefficient was not employed by Hartmann, but was later introduced by Guillier et al. [28]. Since radial electric-field computation points closest to the wire are located not at $0.23\Delta x$ and $0.23\Delta z$ (which are equal to the equivalent wire radius) from the wire axis, but at $0.5\Delta x$ and $0.5\Delta z$, they assume that corona streamers start emanating from the wire when the radial electric field at $0.5\Delta x$ (and $0.5\Delta z$) exceeds $0.46E_0$ ($=E_0 \times 0.23\Delta x/0.5\Delta x$). They set the critical background electric field necessary for streamer propagation (which determines the maximum extent of the radially expanding corona region) for positive (E_{cp}) and negative (E_{cn}) polarities as follows:

$$\left.\begin{matrix} E_{cp} = 0.5 \; (MV/m) \\ E_{cn} = 1.5 \; (MV/m) \end{matrix}\right\} \tag{5.76}$$

The corona radius r_c was obtained using the analytical expression (5.77), based on E_c (0.5 or 1.5 MV/m, depending on polarity; see Equation 5.76) and the FDTD-computed charge per unit length (q). Then the conductivity of the cells located within r_c was set to $\sigma_{cor} = 20$ or 40 µS/m:

$$E_c = \frac{q}{2\pi\varepsilon_0 r_c} + \frac{q}{2\pi\varepsilon_0 (2h - r_c)} (V/m) \tag{5.77}$$

The simulation of corona discharge implemented in the FDTD procedure is summarized as follows:

1. If the FDTD-computed electric field E_{zb}^n, at time step n and at a point located below and closest to the wire (at $0.5\Delta z$ from the wire axis shown in Figure 5.18a), exceeds $0.46E_0$, the conductivity of $\sigma_{cor} = 20$ or 40 µS/m is assigned to x- and z-directed sides of the four cells closest to the wire.

2. The radial current I^n per unit length of the wire at $y = j\Delta y$ from the excitation point at time step n is evaluated by numerically integrating the radial conduction and displacement current densities as follows:

$$
\begin{aligned}
I^n(j\Delta y) = &\; \sigma\left[\left(E_{xl}^n + E_{xr}^n\right)\Delta z + \left(E_{za}^n + E_{zb}^n\right)\Delta x\right]\Delta y \\
&+ \varepsilon_0 \left[\begin{matrix} \left(\dfrac{E_{xl}^n - E_{xl}^{n-1}}{\Delta t} + \dfrac{E_{xr}^n - E_{xr}^{n-1}}{\Delta t}\right)\Delta z \\[2ex] + \left(\dfrac{E_{za}^n - E_{za}^{n-1}}{\Delta t} + \dfrac{E_{zb}^n - E_{zb}^{n-1}}{\Delta t}\right)\Delta x \end{matrix}\right] \Delta y
\end{aligned} \tag{5.78}
$$

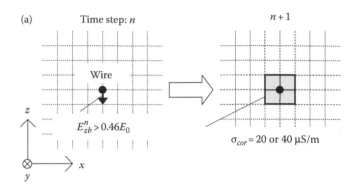

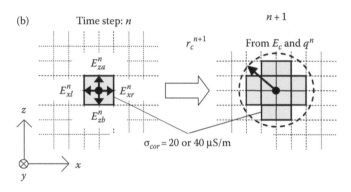

FIGURE 5.18
FDTD representations of (a) the inception of corona discharge at the wire surface and (b) radial expansion of corona discharge. (Thang, T. H. et al., FDTD simulation of lightning surges on overhead wires in the presence of corona discharge, *IEEE Trans. Electromagn. Compat.* 54(6):1234–1243. © 2012 IEEE.)

where E_{xl}, E_{xr}, E_{za}, and E_{zb} are the radial electric fields closest to the wire shown in Figure 5.18b.

The total charge (charge deposited on the wire and the emanated corona charge) per unit length of the wire at $y = j\Delta y$ from the excitation point at time step n is calculated as follows:

$$q^n(j\Delta y) = q^{n-1}(j\Delta y) + \frac{I^{n-1}(j\Delta y) + I^n(j\Delta y)}{2}\Delta t \qquad (5.79)$$

From q^n, yielded by Equation 5.79, and E_c, given by Equation 5.76, the corona radius r_c^{n+1} at time step $n+1$ is calculated using Equation 5.77. The conductivity of $\sigma_{cor} = 20$ or $40\ \mu S/m$ is assigned to x- and z-directed sides of all cells located within r_c^{n+1}.

Figure 5.19a shows a 3-D view of a 12.65 mm-radius, 1.4 km-long, overhead, horizontal, perfectly conducting wire located 22.2 m above the ground

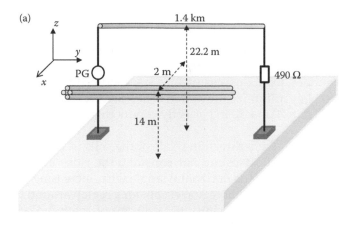

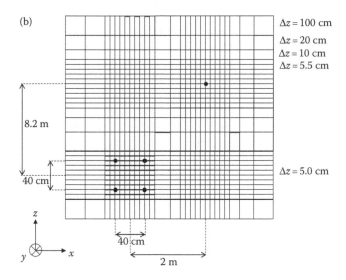

FIGURE 5.19
(a) 3-D and (b) cross-sectional views of a horizontal single wire of radius 12.65 mm and length 1.4 km, located 22.2 m above the ground with a conductivity of 10 mS/m, and a four-conductor bundle of length 1.4 km, located 14 m above the ground and horizontally 2 m away from the single wire [30]. One end of the single wire is energized by a lumped voltage source and the other end is connected to the ground via a 490-Ω resistor. (Thang, T. H. et al., A simplified model of corona discharge on an overhead wire for FDTD computations, *IEEE Trans. Electromagn. Compat.* 54(3):585–593. © 2012 IEEE.)

with a conductivity of 10 mS/m and a 1.4 km-long bundled perfect conductor (four conductors in the bundle) located 14 m above the same ground and horizontally 2 m away from the single wire. This configuration represents one of Inoue's experiments [29]. The radius of each conductor of the bundle is 11.5 mm and the distance between the conductors is 0.4 m. One end of the single wire is energized by a lumped voltage source and the other end is connected to the ground via a 490-Ω (matching) resistor. For FDTD computations, this conductor system is accommodated in a working volume of 60 m × 1460 m × 80 m, which is divided nonuniformly into rectangular cells and is surrounded by six planes of Liao's second-order absorbing boundary condition [9] to minimize unwanted reflections. At each ground connection point, a 20 m × 20 m × 10 m perfectly conducting grounding electrode is employed.

The side length in the y-direction of all of the cells is 1 m (constant). Cell sides along the x- and z-axes are not constant; the sides are 5.5 cm in the vicinity (220 × 220 cm) of the horizontal single wire, increasing gradually (to 10, 20, and 100 cm) beyond that region except for a region around the bundled conductor, and 5 cm in the vicinity (80 × 80 cm) of the bundled conductor except for a region around the horizontal single wire, increasing gradually (to 10, 20, and 100 cm) beyond that region, as shown in Figure 5.19b. The equivalent radius of the horizontal single wire used in this experiment is $r_0 \approx 12.65$ mm (= $0.23\Delta x = 0.23\Delta z = 0.23 \times 5.5$ cm), which is equal to those used in the corresponding experiment of Inoue [29]. The time increment was set to $\Delta t = 1.75$ ns.

Figure 5.20 shows waveforms of a positive surge voltage at $d = 0, 350, 700$, and 1050 m from the energized end of the horizontal single wire above the ground whose conductivity is 10 mS/m, computed using the FDTD method for corona region conductivity $\sigma_{cor} = 40$ μS/m [30]. The critical electric field for corona onset on the wire surface was set at $E_0 = 2.4$ MV/m (for $m = 0.5$). The corresponding measured waveforms [29] are also shown in this figure. FDTD-computed waveforms agree reasonably well with the corresponding measured ones. Both FDTD-computed and measured waveforms of surge voltage suffer from distortion, which becomes more significant when the applied peak voltage and propagation distance are increased.

The maximum corona radii for positive peak voltages of 1580, 1130, and 847 kV are 66, 44, and 27.5 cm, respectively.

Figure 5.21 shows FDTD-computed waveforms of surge voltages without considering corona discharge for 847 kV positive voltage application. The measured waveforms [29] with corona discharge are also shown in this figure. In the absence of corona, the FDTD-computed surge voltages suffer little distortion with propagation and significantly differ from the corresponding measured waveforms with corona discharge.

Figure 5.22 shows FDTD-computed waveforms of induced voltages at $d = 700$ and $d = 1050$ m on a 1.4 km-long, horizontal, four-conductor bundle, which is located horizontally 2 m away from the energized horizontal wire and 14 m above the ground [30]. The corresponding measured waveforms [29] are also shown in Figure 5.22. The computed waveforms of voltages induced on the bundled conductor also agree fairly well with the corresponding measured waveforms.

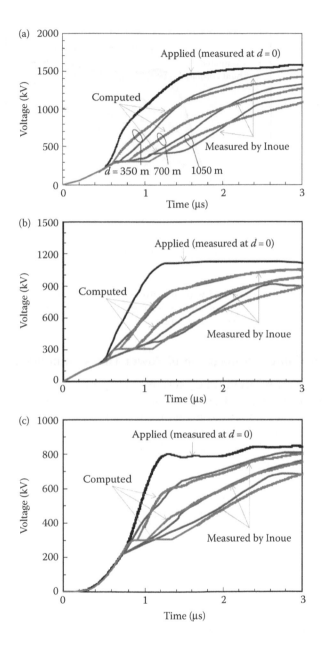

FIGURE 5.20
FDTD-computed (for $\sigma_{cor} = 40\ \mu S/m$ and $E_0 = 2.4\ MV/m$) and measured waveforms of surge voltage at $d = 0$, 350, 700, and 1050 m from the energized end of the 12.65 mm-radius, 1.4 km-long horizontal wire, located 22.2 m above the ground with a conductivity of 10 mS/m [30]. The applied voltage is positive and $E_{cp} = 0.5\ MV/m$. The applied peak voltages are (a) 1580 kV, (b) 1130 kV, and (c) 847 kV. (Thang, T. H. et al., A simplified model of corona discharge on an overhead wire for FDTD computations, *IEEE Trans. Electromagn. Compat.* 54(3):585–593. © 2012 IEEE.)

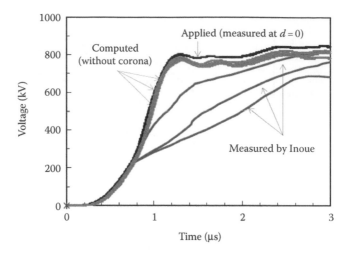

FIGURE 5.21
Same as Figure 5.20c, but computed without corona discharge. (Thang, T. H. et al., A simplified model of corona discharge on an overhead wire for FDTD computations, *IEEE Trans. Electromagn. Compat.* 54(3):585–593. © 2012 IEEE.)

5.2.5 Power Cables: Propagation of Power Line Communication Signals

Power line communication (PLC) systems use power distribution lines and cables for data communication in frequency ranges up to 30 MHz. Within a power cable, semiconducting layers are usually incorporated between the core conductor of the power cable and the insulating layer and between the insulating layer and the sheath conductor. Because power cables are not designed for effectively transmitting the PLC signals, they might attenuate significantly along the cables due to the presence of semiconducting layers.

Okazima et al. [31] investigated the propagation characteristics of a PLC signal of frequency 30 MHz along a single-core power cable with two 3 mm–thick, semiconducting layers using the FDTD method [1] in the two-dimensional (2-D) cylindrical-coordinate system. Figure 5.23 shows a 130 m–long, single-core power cable analyzed using the FDTD method. The radius of the core conductor is 5 mm and the inner radius of the sheath conductor is 25 mm. The core and sheath conductors are perfectly conducting. Figure 5.23 also shows a 14 mm–thick insulating layer with semiconducting layers of 3-mm thickness on both inner and outer surfaces. The relative permittivity ε_r of the insulating layer and of each semiconducting layer is set to $\varepsilon_r = 3$. The conductivity of each semiconducting layer is set to a value ranging from $\sigma = 10^{-5}$ to 10^5 S/m. Note that this power cable is rotationally symmetric around its axis and has a circular cross section. It is represented without the staircase-approximated contour in the FDTD method using a 2-D cylindrical-coordinate system with a working space of 130 m × 27 mm rectangle, contoured by the thick black line shown in Figure 5.23. At one end of

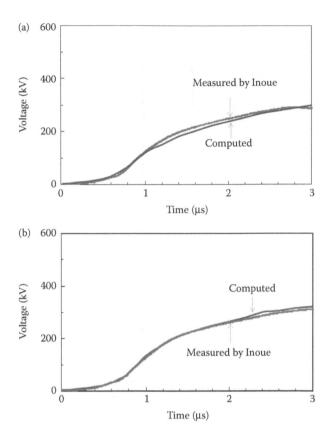

FIGURE 5.22
FDTD-computed (for $\sigma_{cor} = 40$ µS/m and $E_0 = 2.4$ MV/m) and measured waveforms of voltage induced on a nearby four-conductor bundle at $d = 700$ and 1050 m, located 14 m above the ground with a conductivity of 10 mS/m [30]. The applied voltage is positive and $E_{cp} = 0.5$ MV/m. The applied peak voltage is 1580 kV. (a) $d = 700$ m and (b) $d = 1050$ m. (Thang, T. H. et al., A simplified model of corona discharge on an overhead wire for FDTD computations, *IEEE Trans. Electromagn. Compat.* 54(3):585–593. © 2012 IEEE.)

the cable, a 10 V, positive half-sine pulse of frequency $f = 30$ MHz is applied between the core and sheath conductors. The other end of the cable model is terminated using Liao's second-order absorbing boundary. The working space of 130 m × 27 mm for the FDTD computation is divided into 1 × 1 mm square cells. The time increment is set to 2.3 ps.

Figure 5.24 shows waveforms of the voltage between the core and sheath conductors of the power cable at different distances of 20, 40, 60, 80, and 100 m from the excitation point. Figure 5.25 shows how σ affects the magnitude of the voltage between the core and sheath conductors at a distance that is 100 m from the excitation point. It can be observed from Figures 5.24 and 5.25 that the magnitude of the voltage pulse decreases when the propagation distance is increased in all cases considered. However, the dependence of the

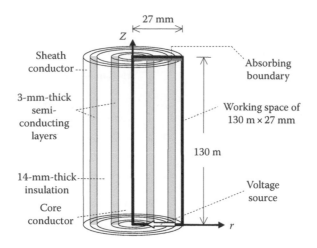

FIGURE 5.23
A 130 m-long, single-core power cable with semiconducting layers analyzed using the FDTD method. The 130 m × 27 mm rectangle space, contoured by a thick black line, is the actual working space for the present FDTD computations in the 2-D cylindrical-coordinate system. (Okazima, N. et al., Propagation characteristics of power line communication signals along a power cable having semiconducting layers, *IEEE Trans. Electromagn. Compat.* 52(3):756–759. © 2010 IEEE.)

signal attenuation on σ is not monotonic; the attenuation is significant around $\sigma = 10^{-3}$ and 10^3 S/m, while it is not significant when σ is lower than 10^{-5} S/m or σ is around 1 S/m. When $\sigma = 10^{-3}$ and 10^3 S/m, dispersion is also marked. Therefore, it is quite difficult to conduct PLC signals in a power system cable with semiconducting layers that have conductivity of approximately $\sigma = 10^{-3}$ or 10^3 S/m, but there are more possibilities if $a \leq 10^{-5}$ S/m or $\sigma = 1$ S/m.

The signal attenuation around $\sigma = 10^{-3}$ S/m is caused by the capacitive charging and discharging of the semiconducting layers in a radial direction. For $\sigma = 10^3$ S/m, axial current propagation in the semiconducting layers is the dominant cause of attenuation. These are quantified as follows.

The time constant τ of each semiconducting layer is given by

$$\tau = CR = \frac{2\pi\varepsilon_0\varepsilon_r}{\ln(r_2/r_1)} \frac{\ln(r_2/r_1)}{2\pi\sigma} = \frac{\varepsilon_0\varepsilon_r}{\sigma} \tag{5.80}$$

where
$C = 2\pi\varepsilon_r\varepsilon_0/\ln(r_2/r_1)$ is the per-unit length capacitance of each semiconducting layer
$R = \ln(r_2/r_1)/(2\pi\sigma)$ is its radial direction per-unit length resistance
r_2 is the outer radius of the semiconducting layer
r_1 is its inner radius
ε_0 is the permittivity of the vacuum
ε_r is the relative permittivity of the semiconducting layer
σ is its conductivity

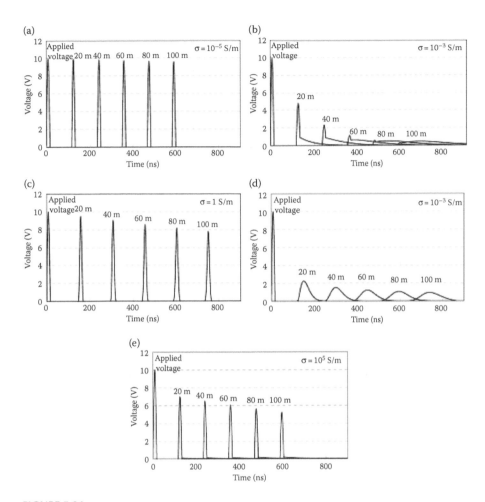

FIGURE 5.24
Waveforms of the voltage between the core and sheath conductors at different distances from the excitation point when a 30 MHz and 10 V half-sine pulse is injected and the semiconducting-layer conductivity σ is (a) 10^{-5}, (b) 10^{-3}, (c) 1, (d) 10^3, and (e) 10^5 S/m. (Okazima, N. et al., Propagation characteristics of power line communication signals along a power cable having semiconducting layers, *IEEE Trans. Electromagn. Compat.* 52(3):756–759. © 2010 IEEE.)

For $\varepsilon_r = 3$ and $\sigma = 10^{-3}$ S/m, the time constant is $\tau = 27$ ns. Charging and discharging processes in the radial direction of the semiconducting layer with $\sigma = 10^{-3}$ S/m will have a strong effect on a 30-MHz signal with 17-ns half-cycle, leading to significant attenuation and distortion. Note that in this condition, the magnitude of the radial conduction current across the semiconducting layer is close to that of the radial displacement current. In other words, the conductance of the semiconducting layer is close to its susceptance.

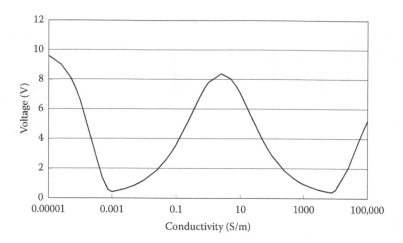

FIGURE 5.25
Dependence of the magnitude of the voltage between the core and sheath conductors at a distance of 100 m from the excitation point on the semiconducting-layer conductivity σ when a 30 MHz and 10 V half-sine pulse is injected. (Okazima, N. et al., Propagation characteristics of power line communication signals along a power cable having semiconducting layers, *IEEE Trans. Electromagn. Compat.* 52(3):756–759. © 2010 IEEE.)

At a high conductivity of σ = 10^3 S/m, the depth d of penetration for an electromagnetic wave of frequency f into a medium of conductivity σ and permeability μ_0 is relevant for loss calculations. This depth is given by

$$d = \frac{1}{\sqrt{2\pi f \sigma \mu_0}} \qquad (5.81)$$

For $\mu_0 = 4\pi \times 10^{-7}$ H/m, $f = 30$ MHz, and σ = 10^3 S/m, Equation 5.81 yields $d = 2$ mm, which is close to the thickness of the semiconducting layer (3 mm). Therefore, most of the axial current flows in the semiconducting layers rather than on the core and sheath conductor surfaces. This results in significant signal attenuation and dispersion.

5.2.6 Air-Insulated Substations

In order to estimate the voltage level that substations can withstand from lightning impulses, lightning overvoltages that would be generated in 3-D complex-structure substations need to be known.

Using the FDTD method [1], Watanabe et al. [32] computed surge voltages on an air-insulated substation. They applied an impulse voltage to the substation and compared the FDTD-computed voltage waveforms with the corresponding waveforms measured on a 1/10-scale model, shown in Figure 5.26. Figure 5.27 shows the plan view of the FDTD model and the reduced-scale experimental model. An impulse voltage is applied to the terminal of

FIGURE 5.26
1/10-scale model of an air-insulated substation. (Reprinted from Watanabe, T. et al., The measurement and analysis of surge characteristics using miniature model of air insulated substation, Paper presented at *IPST 2005*, Montreal, Quebec, Canada, 2005. With permission from IPST.)

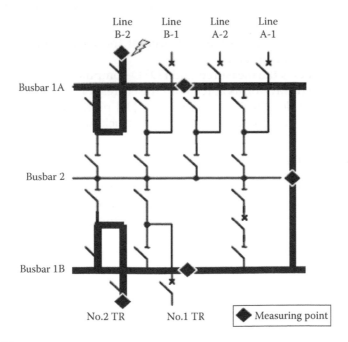

FIGURE 5.27
Plan view of the small-scale model and the FDTD simulation model: An impulse voltage is applied to the terminal of line B-2, and surge voltages are measured at the voltage application point and at the No. 2 transformer. (Reprinted from Watanabe, T. et al., The measurement and analysis of surge characteristics using miniature model of air insulated substation, Paper presented at *IPST 2005*, Montreal, Quebec, Canada, 2005. With permission from IPST.)

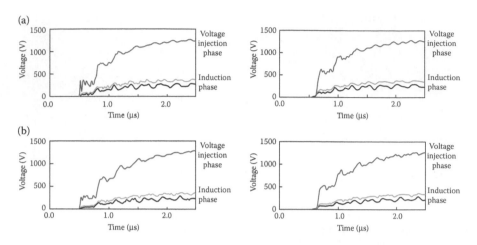

FIGURE 5.28
Measured and FDTD-computed waveforms: The plots on the left show voltages at the voltage application point, and the plots on the right show voltages at the No. 2 transformer. (a) Measured voltages and (b) FDTD-computed voltages. (Reprinted from Watanabe, T. et al., The measurement and analysis of surge characteristics using miniature model of air insulated substation, Paper presented at *IPST 2005*, Montreal, Quebec, Canada, 2005. With permission from IPST.)

line B-2, and surge voltages are measured at the voltage application point and at the No. 2 transformer.

Figure 5.28 shows the measured and FDTD-computed voltage waveforms. FDTD-computed waveforms agree with the corresponding measured waveforms. Note that de Oliveira and Sobrinho [33] performed a similar FDTD computation for an air-insulated substation struck by lightning.

5.2.7 Wind-Turbine Generator Towers

Wind-turbine generator towers are frequently struck by lightning. In order to optimize ways to protect wind-turbine generator systems from lightning, it is important to know the mechanism of lightning overvoltages generated in the systems.

Yamamoto [34] and Yamamoto et al. [35] investigated lightning-protection methods for wind-turbine generator systems using the FDTD method [1] experimentally with a small-scale model of a wind-turbine generator tower struck by lightning. Figure 5.29 shows a 3/100-scale model of a 50 m-high wind-turbine tower with 25 m-long blades, one of which is connected to a lightning channel-representing, vertical current lead wire or the core conductor of a vertical coaxial cable. The grounding resistance is set to 9.4 Ω. In the FDTD simulation, this conductor system is accommodated in a working volume of 6 m × 5 m × 7.5 m, which is divided uniformly into cubic cells of 25 mm × 25 mm × 25 mm. The working volume is surrounded by six planes of Liao's second-order absorbing boundary condition [9]. V_{11} to V_{14} in Figure 5.30

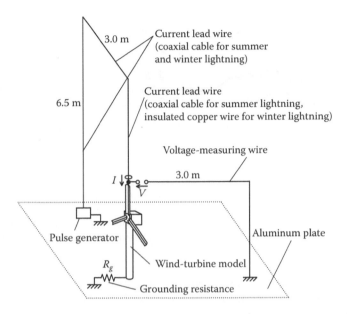

FIGURE 5.29
3-D view of a 3/100-scale model of a wind-turbine generator tower, blades, nacelle, current lead wire, and a voltage reference wire. (Reprinted from Yamamoto, K., A study of overvoltages caused by lightning strokes to a wind turbine generation system, PhD thesis, Doshisha University, Kyoto, Japan, 2007. With permission.)

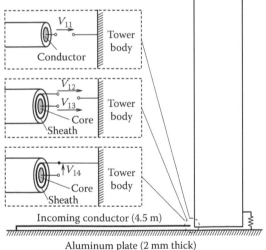

FIGURE 5.30
Configuration of incoming conductors to the wind-turbine tower base for measuring and computing overvoltages between the incoming conductors and equipment at the tower base. (Reprinted from Yamamoto, K., A study of overvoltages caused by lightning strokes to a wind turbine generation system, PhD thesis, Doshisha University, Kyoto, Japan, 2007. With permission.)

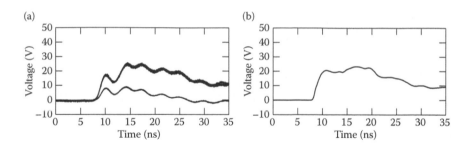

FIGURE 5.31
Voltage differences between the incoming conductor and equipment in the tower base. The grounding resistance is 9.4 Ω and the wave front of the injected current is 4 ns. (a) Voltage $V_{11}-V_{14}$ of the measured results and (b) voltage V_{11} of FDTD-computed results. (Reprinted from Yamamoto, K., A study of overvoltages caused by lightning strokes to a wind turbine generation system, PhD thesis, Doshisha University, Kyoto, Japan, 2007. With permission.)

represent the voltage differences between the incoming conductor connected to a distant point and equipment at the tower base. The voltage difference is generated by the voltage rise at the tower base. This voltage difference might lead to an overvoltage between the power line and the power converter or transformer installed inside the tower. Figure 5.31 shows the measured and FDTD-computed voltages for an injected current with a rise time of 4 ns. The FDTD-computed waveforms agree well with the corresponding measured ones.

References

1. Yee, K. S. 1966. Numerical solution of initial value problems involving Maxwell's equations in isotropic media. *IEEE Trans. Antenn. Propag.* 14(3):302–307.
2. Courant, R., K. Friedrichs, and H. Lewy. 1965. On the partial difference equations of mathematical physics. *IBM Res. Develop.* 11(2):215–234.
3. Noda, T. and S. Yokoyama. 2002. Thin wire representation in finite difference time domain surge simulation. *IEEE Trans. Power Deliv.* 17(3):840–847.
4. Harrington, R. F. 1968. *Field Computation by Moment Methods.* New York: Macmillan Co.
5. Burke, G. and A. Poggio. 1980. Numerical electromagnetic code (NEC)-method of moment. Technical document 116. Naval Ocean Systems Center, San Diego, CA.
6. Burke, G. 1992. Numerical electromagnetic code (NEC-4)-method of moment. UCRL-MA-109338. Lawrence Livermore National Laboratory, Livermore, CA.
7. Van Baricum, M. and E. K. Miller. 1972. TWTD—A computer program for time-domain analysis of thin-wire structures. UCRL-51-277. Lawrence Livermore Laboratory, Livermore, CA.

8. Tanabe, K. 2001. Novel method for analyzing dynamic behavior of grounding systems based on the finite-difference time-domain method. *IEEE Power Eng. Rev.* 21(9):55–57.

9. Liao, Z. P., H. L. Wong, B. P. Yang, and Y. F. Yuan. 1984. A transmission boundary for transient wave analysis. *Sci. Sin.* A27(10):1063–1076.

10. Tanabe, K., A. Asakawa, M. Sakae, M. Wada, and H. Sugimoto. 2003. Verifying the computational method of transient performance with respect to grounding systems based on the FD–TD method. *IEEJ Trans. Power Energy* 123(3):358–367 (in Japanese).

11. Miyazaki, S. and M. Ishii. 2005. Analysis of transient response of grounding system based on moment method. Paper presented at *ISH 2005*, Beijing, China.

12. Ala, G., P. L. Buccheri, P. Romano, and F. Viola. 2008. Finite difference time domain simulation of earth electrodes soil ionisation under lightning surge condition. *IET Sci. Meas. Technol.* 2(3):134–145.

13. Liew, A. C. and M. Darveniza. 1974. Dynamic model of impulse characteristics of concentrated earth. *Proc. IEE* 121(2):123–135.

14. Baba, Y., N. Nagaoka, and A. Ametani. 2005. Modeling of thin wires in a lossy medium for FDTD simulations. *IEEE Trans. Electromagn. Compat.* 47(1):54–60.

15. Mozumi, T., Y. Baba, M. Ishii, N. Nagaoka, and A. Ametani. 2003. Numerical electromagnetic field analysis of archorn voltages during a back-flashover on a 500 kV twin-circuit line. *IEEE Trans. Power Deliv.* 18(1):207–213.

16. Motoyama, H. 1996. Development of a new flashover model for lightning surge analysis. *IEEE Trans. Power Deliv.* 11(2):972–979.

17. Scott-Meyer, W. 1977. *EMTP Rule Book.* Portland, OR: Bonneville Power Administration.

18. Ishii, M. et al. 1991. Multistory transmission tower model for lightning surge analysis. *IEEE Trans. Power Deliv.* 6(3):1327–1335.

19. Noda, T. 2007. A tower model for lightning overvoltage studies based on the result of an FDTD simulation. *IEEJ Trans. Power Energy* 127(2):379–388 (in Japanese).

20. Pokharel, R. K., M. Ishii, and Y. Baba. 2003. Numerical electromagnetic analysis of lightning-induced voltage over ground of finite conductivity. *IEEE Trans. Electromagn. Compat.* 45(4):651–656.

21. Ishii, M., K. Michishita, and Y. Hongo. 1999. Experimental study of lightning-induced voltage on an overhead wire over lossy ground. *IEEE Trans. Electromagn. Compat.* 41(1):39–45.

22. Agrawal, A. K., H. J. Price, and H. H. Gurbaxani. 1980. Transient response of multiconductor transmission lines excited by a nonuniform electromagnetic field. *IEEE Trans. Electromagn. Compat.* 22(2):119–129.

23. Baba, Y. and V. A. Rakov. 2006. Voltages induced on an overhead wire by lightning strikes to a nearby tall grounded object. *IEEE Trans. Electromagn. Compat.* 48(1):212–224.

24. Pokharel, R. K. and M. Ishii. 2007. Applications of time-domain numerical electromagnetic code to lightning surge analysis. *IEEE Trans. Electromagn. Compat.* 49(3):623–631.

25. Moini, R., B. Kordi, and M. Abedi. 1998. Evaluation of LEMP effects on complex wire structures located above a perfectly conducting ground using electric field integral equation in time domain. *IEEE Trans. Electromagn. Compat.* 40(2):154–162.

26. Thang, T. H. et al. 2012. FDTD simulation of lightning surges on overhead wires in the presence of corona discharge. *IEEE Trans. Electromagn. Compat.* 54(6):1234–1243.
27. Hartmann, G. 1984. Theoretical evaluation of Peek's law. *IEEE Trans. Ind. Appl.* 20(6):1647–1651.
28. Guillier, J. F., M. Poloujadoff, and M. Rioual. 1995. Damping model of travelling waves by corona effect along extra high voltage three phase lines. *IEEE Trans. Power Deliv.* 10(4):1851–1861.
29. Inoue, A. 1983. Study on propagation characteristics of high-voltage traveling waves with corona discharge. CRIEPI Report 114 by Central Research Institute of Electric Power Industry Report (in Japanese).
30. Thang, T. H. et al. 2012. A simplified model of corona discharge on an overhead wire for FDTD computations. *IEEE Trans. Electromagn. Compat.* 54(3):585–593.
31. Okazima, N., Y. Baba, N. Nagaoka, A. Ametani, K. Temma, and T. Shimomura. 2010. Propagation characteristics of power line communication signals along a power cable having semiconducting layers. *IEEE Trans. Electromagn. Compat.* 52(3):756–759.
32. Watanabe, T., K. Fukui, H. Motoyama, and T. Noda. 2005. The measurement and analysis of surge characteristics using miniature model of air insulated substation. Paper presented at *IPST 2005*, Montreal, Quebec, Canada.
33. de Oliveira, R. M. S. and C. L. S. S. Sobrinho. 2009. Computational environment for simulating lightning strokes in a power substation by finite-difference time-domain method. *IEEE Trans. Electromagn. Compat.* 51(4):995–1000.
34. Yamamoto, K. 2007. A study of overvoltages caused by lightning strokes to a wind turbine generation system. PhD thesis, Doshisha University, Kyoto, Japan (in Japanese).
35. Yamamoto, K., T. Noda, S. Yokoyama, and A. Ametani. 2009. Experimental and analytical studies of lightning overvoltages in wind turbine generator systems. *Elec. Power Syst. Res.* 79(3):436–442.

6

Electromagnetic Disturbances in Power Systems and Customer Homes

6.1 Introduction

Since the mid-1980s, a number of analog control circuits have been replaced by digital control circuits in power stations and substations in Japan. This is the case for any industrial product, such as automobiles and home appliances, due to the advancement of digital circuit technologies. However, electromagnetic compatibility (EMC) environments are becoming a significant problem for digital circuits, which are sensitive to high-frequency electromagnetic waves such as switching surges (SS) and lightning surges (LS).

Consequently, in 1990 the Japanese Electrotechnical Research Association began a 10-year investigation of digital control circuit disturbances experienced by Japanese utilities in generator stations and substations [1]. There was a total of 330 disturbances, one-third of which were protective relays. Protective relays made up the majority of the disturbances (details of these protective relay disturbances are presented in References 2 and 3).

Similarly, electromagnetic disturbances that affect users have become a significant problem, and a number of surveys have been conducted by utility companies and public organizations related to home appliances [4–9]. For example, it was reported that there were more than 1000 cases of damage to home appliances over the course of 1 year for one Japanese utility company.

This chapter summarizes the disturbances experienced in Japan, including the disturbed equipment, incoming surge routes, and characteristics of the disturbances [1,10]. Results of these disturbances in power system operations, such as countermeasures and costs, are also explained, and some case studies are presented.

Additionally, this chapter includes measurement and simulation results to describe how LSs due to nearby activity and communication lines can strike customers.

Finally, an analytical method is discussed for calculating the induced voltage in a pipeline or communication line from a power line.

6.2 Disturbances in Power Stations and Substations

6.2.1 Statistical Data of Disturbances

6.2.1.1 Overall Data

In total, 330 cases of disturbances occurred in substations and generator stations over the course of 10 years, beginning in 1990 [1]. Table 6.1 classifies the disturbances by (a) troubled equipment and (b) causes. As shown in Table 6.1a, one-third of the disturbances affected protection equipment, a quarter affected telecontrol equipment, and another quarter affected control equipment. Table 6.1b shows that two-thirds of the disturbances were caused by LSs, one-sixth by SSs in the main circuits (high-voltage side), and one-twelfth by SSs in DC circuits that are part of control circuits (low-voltage side).

TABLE 6.1

Total Number of Disturbances and Failures Collected over 10 Years

(a) Disturbed Equipment

Equipment	Number	Equipment	Number
Protection	105	Control[a]	73
Telecontrol	73	Measuring	49
Supervisory control	5	Automatic processing	2
Communication	15	Others	8
		Total	330

(b) Causes

Basic Cause	Type of Surges	Number	Subtotal
Lightning	LS	220	220 (0.72)
Main-circuit switching operation	DS SS	21	47 (0.15)
	CB SS	24	
	Capacitor bank SS	2	
DC circuit switching operation	DC circuit SS	21	21 (0.07)
Others	Fault surge	2	19 (0.06)
	CPU switching noise	2	
	Welder noise	2	
	Not clear	13	
Total			307[b] (ratio 1.0)

Source: Working Group (Chair: Agematsu, S.) Japanese Electrotechnical Research Association, Technologies of countermeasure against surges on protection relays and control systems, *ETRA Report*, 57(3), 2002 (in Japanese).

DS, disconnector; CB, circuit breaker; SS, switching surge; LS, lightning surge.

[a] Control: control board, station power circuit board, and generator sequencer.

[b] Communication and others in Table 6.1a deleted.

The disturbances recorded in Table 6.1 that were caused by LSs (220 cases) and main-circuit SSs (47) were noticed in the following ways:

1. Malfunction with no reaction of equipment—LS: fifty-six cases, SS: twenty-one cases
2. Indication of abnormal supervision (alarms, etc.)—LS: 124 cases, SS: eighteen cases
3. Routine maintenance—LS: twenty cases, SS: three cases
4. Other—LS: twenty cases, SS: six cases

The 307 cases of disturbed control equipment recorded in Table 6.1 were installed as follows:

1. Centralized (control room): 275 cases (89.6%)
2. Dispersed: twenty-one cases (6.8%)
3. Unknown: eleven cases (3.6%)

This shows that electromagnetic disturbances were far less frequent in the dispersed control equipment than in the centralized equipment. While a surge can cause many disturbances, the dispersed control equipment limits the incoming surge route. Furthermore, there are more incoming surge routes for centralized equipment than for dispersed control equipment.

6.2.1.2 Disturbed Equipment

The details of the disturbed equipment listed in Table 6.1a are summarized in Table 6.2. According to Table 6.2, more than 80% of the equipment disturbed by lightning are protection, control, and telecommunication circuits, and another 15% are measuring circuits and indicators. These circuits should be well protected from LSs, and the incoming routes, as discussed in Section 6.2.1.3, should be designed to reduce surges. Table 6.2 also shows that more than 70% of the equipment disturbed by SSs are protection and measuring/indicator circuits, but only a few disturbances occur in the control and telecommunication circuits. The difference might be caused by a characteristic of an SS, which is different from that of an LS. Table 6.2b also clearly shows that the number of disturbances in digital control equipment is much greater than that in analog equipment; the ratio is 4:1 for SSs (see Table 6.2c), whereas the ratio is 1:1 for LSs (see Table 6.2b). The rate of disturbances in mechanical control circuits reached nearly 10%, which suggests that an LS can cause a disturbance to any type of control equipment due to its high overvoltage and current, whereas an SS causes more disturbances in digital equipment due to its high frequency.

This observation suggests that overvoltage/current reduction (using arresters and surge absorbers) is more effective against LSs, while reducing frequency (using surge capacitors) is effective against SSs.

TABLE 6.2

Details of Disturbed Equipment

(a) Disturbed Equipment

	Protection	Control	Telecom[a]	Measuring Indicator	Centralized Superv./Control	Automatic Process	Total (Ratio)
LS	66 (0.30)	64 (0.29)	50 (0.23)	33 (0.15)	5 (0.02)	2 (0.01)	220 (1.0)
SS	24 (0.51)	3 (0.06)	9 (0.06)	11 (0.23)	0	0	47 (1.0)

(b) Types of Control Equipment Disturbed by LS

	Mechanical	Analog	Digital	Unknown	Total (Ratio)
All data	20 (0.09)	79 (0.36)	73 (0.33)	48 (0.22)	220 (1.0)
Protection	2 (0.03)	26 (0.45)	21 (0.37)	9 (0.15)	58 (1.0)

(c) Types of Control Equipment Disturbed by SS

	Analog	Digital	Unknown	Total (Ratio)
All data	9 (0.19)	36 (0.76)	2 (0.05)	47 (1.0)
Protection	3 (0.11)	22 (0.89)	0	25 (1.0)

(d) Disturbed Elements

	IC Board	Aux. Relay	Lamp, Fuse, Switch	Source	Analog Input	Transmit.	Digital Input
Number	154	36	29	29	15	20	35

(e) Details of IC Board Disturbances

	Digital Process	Digital Input	Analog Input	Source	Relay	Arrester	Wiring Terminal	Measuring	Analog	Others	Unknown	Total
Number	37	35	20	15	14	10	8	10	6	24	6	154

[a] Distant supervision.

TABLE 6.3

Surge Incoming Routes Resulting in Disturbances

	Control Cable	DC Power	Communication	VT	CT
LS	40	36	24	15	2
SS	20	8	2	5	2
	Direct to Panel	**Others**	**Subtotal**	**Unknown**	**Total**
LS	4	8	130	90	220
SS	0	1	38	9	47

VT, measuring voltage transformer; CT, current transformer.

Table 6.2d shows that nearly half of the disturbed equipment are integrated circuit (IC) boards; of these, digital processors and input circuits have the highest risk of disturbance.

6.2.1.3 Incoming Surge Routes

Table 6.3 summarizes incoming surge routes that were determined from corresponding disturbances. It is not easy to find the incoming route of a surge; however, the most likely route for an LS and/or SS is a control cable. Communication circuits and voltage transformers (VT) are also at high risk of being incoming surge routes for LSs. Table 6.3 suggests a position where a surge protector can be installed. From the table and References 11 and 12, the incoming route of a surge can be drawn, as shown in Figure 6.1.

6.2.2 Characteristics of Disturbances

6.2.2.1 Characteristics of LS Disturbances

Table 6.4 categorizes the isokeraunic level (IKL) of substations in which disturbances were found. Table 6.2 indicates that the number of disturbances tends to be proportional to the IKL for substations of 154 kV and below. However, no correlation to the IKL is observed in substations of 187 kV and above, because countermeasures against LSs are performed for these substations.

Table 6.5 shows the types of disturbances caused by LSs. Table 6.5a shows that more than 70% of the disturbances are permanent—that is, they lead to the complete breakdown of equipment and thus result in a permanent halt/lock and malfunction.

6.2.2.2 Characteristics of SS Disturbances

One of the main causes of electromagnetic disturbances in control circuits of gas-insulated substations (GIS) is an SS due to a disconnect switch (DS)—or occasionally a circuit breaker (CB)—operation. Because of the complex

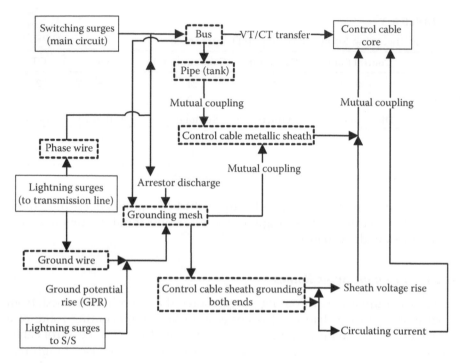

FIGURE 6.1
Incoming routes of SSs and LSs via a control cable.

combination of short gas-insulated buses and lines in the GIS, multiple reflections and refractions of traveling waves at the boundaries between the buses and the lines generate a high-frequency surge. This surge invades low-voltage control circuits via a capacitor voltage transformer and a current transformer (CT) and results in malfunction and occasionally insulation failure of digital elements of the control circuits [13,14].

Table 6.6 shows the relationship between operating CBs/DSs and disturbances. The total number of GISs is 11,102, and the total number of non-GISs is 27,456; the ratio of the disturbances in the GISs is 0.0017, which is higher than that in the non-GISs, 0.00087.

TABLE 6.4

Number of Disturbances and the Ratio Categorized by IKL

IKL	Higher Than 20	Less Than 20
187 kV/above	9 (0.056)	7 (0.043)
154 kV/below	108 (0.671)	37 (0.230)
Total	117	44

TABLE 6.5

Types of Disturbances Due to LSs

(a) Aspects of Disturbances

Permanent (Breakdown)	Nonrepetitive	Unknown	Total
155 cases	63	2	220

(b) Types of Disturbances Due to LSs

Type	Malfunction	Not Operated	Erroneous Display	Halt/Lock	Others	Unknown	Total
Number	27	21	20	116	29	7	220

(c) Disturbed Elements

	IC Board	Aux. Relay	Lamp	DC Converter	Others	Relay Element	Wiring	Arrester	Meter	Unknown	Total
Permanent	67	29	25	17	16	13	5	4	0	2	178[a]
Nonrepetitive	35	2	1	8	4	—	—	4	2	7	63

[a] Includes multiple disturbances.

TABLE 6.6

Number of Disturbances Due to SSs

Cause	Operating	Number of Disturbances			
		A	B	Unclear	Subtotal
DS	DS in GIS	4	4	2	10
	Non-GIS	5	5	1	11
CB	CB in GIS	2	5	2	9
	Non-GIS	9	1	3	13
	Unknown		1	1	2
Capacitor bank	Non-GIS		1		1
	Unknown			1	1
Subtotal		20	15 (2)	8 (2)	43 (4)

The total number of GISs: 11,102; non-GISs: 27,456. A: control/protection circuits of the operating CB/DS. B: control/protection circuits independent of the operating CB/DS. () for unknown and capacitor bank switching.

The disturbances are categorized as follows:

Malfunction 15 (7), malfunction (not operated) 3 (2), halt/lock 14 (11), erroneous indication 8 (3), others 6 (1), unknown 1 (1), total 47 (25) cases, the number in () for protection relays.

The impact of disturbances is summarized as follows:

DS operation: seven permanent failures, fourteen temporary failures

CB operation: seven permanent failures, seventeen temporary failures

Figure 6.2 summarizes the overall results of the voltage–frequency relationship for SSs at CT secondary circuits measured in thirteen different GISs

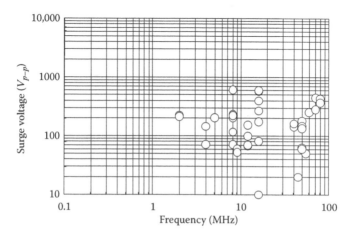

FIGURE 6.2
Voltage–frequency characteristic of GIS surges.

in Japan (fifty-eight cases). The figure shows that the frequency of the SSs ranges from 2 to 80 MHz and the peak-to-peak voltage ranges from 10 to 600 V. It is important to note that no frequency component from 20 to 40 MHz is observed. Thus, two average values of the frequency exist: approximately 10 and 60 MHz. The results suggest that it is necessary to revise the oscillating frequency of a test wave in the existing International Electrotechnical Commission standard 61,000-4-12 [15]. The average of the peak-to-peak voltage is 100–200 V.

Figure 6.3 shows surge voltages at various parts of control circuits: (1) VT secondary; (2) CT secondary; (3) source circuit DC 110 V P-E; and (4) CB control (pallet) circuit. In general, the figure depicts the following trend of the voltage amplitude:

CT secondary > VT secondary > source circuit > CB control circuit.

This trend makes sense as the CT and VT circuits are connected directly to the main (high-voltage) circuit of a GIS, and the number of turns of the CT is less than that of the VT.

Figure 6.4 shows the frequency of surge voltages due to a DS operation (*o*) and a CB operation (*x*). The frequency due to the DS operation tends to be higher than that of the CB operation.

Generally, a DS produces a surge voltage with a higher frequency than that produced by a CB, because the operating speed of the DS is slow and

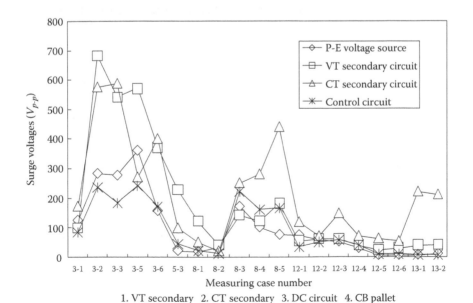

FIGURE 6.3
Voltages at various parts of control circuits.

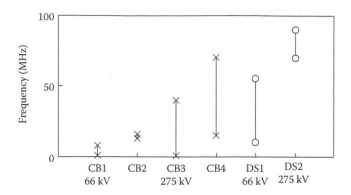

FIGURE 6.4
Frequency of surge voltages due to CBs and DSs.

the polarity of the source-side voltage can become opposite to that of the disconnected side. Thus, a discharge across the poles occurs due to a sufficiently high voltage across the poles. Such a phenomenon cannot occur in the CB operation, because its operating speed is high and the voltage across the poles is reduced by a stray capacitance. Also, the circuit length is shorter in the DS operation than in the CB operation.

6.2.2.3 SSs in DC Circuits

1. *Basic characteristics:* twenty-one total cases; thirteen cases due to SSs within DC circuits.
2. *Disturbed equipment:*
    a. *Types of equipment:* nine digital, three analog, and one unknown; thirteen total cases. It is clear that there are more disturbances (three times as many) in digital equipment than in analog equipment. The ratio (3:1) is similar to that due to main-circuit SSs (thirty-six digital, nine analog) in Section 6.2.1.2.
    b. *Voltage class:* All disturbances are in the voltage class of 154 kV and below because of countermeasures to reduce surges in a higher voltage class.
    c. *Disturbed equipment:* seven telecommunication, three protection, and three control. There are more disturbances in telecommunication equipment because it is connected to CB control circuits and auxiliary relays, which are the sources of DC SSs.
    d. *Relationship to standards:* The seven disturbances in the telecommunication equipment resulted from either switching overvoltages that exceeded the voltage defined in the standard or the inclusion of parts not defined in the standard [15,16].

e. *Disturbed elements:* six IC base plates, three auxiliary relays, one lamp, one source circuit, two others, and thirteen total. This ratio of disturbed elements is similar to that of the disturbances due to main-circuit SSs, excluding auxiliary relays, which are characteristic of DC SSs.

3. *Types of disturbances:*

a. Types of disturbances: three malfunctions, one malfunction (not operated), three halts/locks, one erroneous indication, four others, one unknown, and thirteen total.

b. Aspects of disturbances: three permanent, ten nonrepetitive.

4. *Surge voltage and frequency:*

a. Surge voltage: 3–3.6 kV.

b. Surge frequency: lower than the frequency of main-circuit SSs.

6.2.3 Influence, Countermeasures, and Costs of Disturbances

This section investigates the disturbances that influenced power system operation. While some of the disturbances caused no trouble in power system operation, one control circuit element did break down. Therefore, the total number of disturbances is not the same as that explained in Section 6.2. In fact, forty-three of the total 330 cases listed in Table 6.1a resulted in disturbances in power system operation.

6.2.3.1 Influence of Disturbances on Power System Operation

The types of disturbances for each type of equipment are categorized in Table 6.7a. It is clear from the table that halts and locks (freezes) of equipment make up about half of the total disturbances, and those including malfunctions (not operated too) constitute 70% of the total disturbances. These disturbances result in troubles of power supply and system operation as shown in Figure 6.5. For example, more than 10% of the control circuit disturbances cause power system troubles. As long as a control equipment disturbance is contained within a control circuit, the disturbance does not cost much. However, if it affects power system operation, it becomes an urgent and significant problem. More than one-third of control equipment disturbances affect generation, power supply, and system operation at the same time.

The causes and incoming routes of the disturbances listed in Table 6.7a are categorized in Table 6.7b. While an LS is more likely to cause disturbances in power generation and supply, an SS can also result in these disturbances. It makes sense that SS disturbances occur only in substations. The incoming routes for surges are through control circuits (29%) and CTs/VTs and others (40%), as shown in Table 6.7b. It is important to consider these routes while establishing countermeasures.

TABLE 6.7

Influences of Control Circuit Disturbances on Power Systems

(a) Types of Disturbances of Each Control Equipment

Equipment	Halt/Lock	Malfunc.	Not Operat.	Others	Total
Protection	48 (0.46)	25 (0.24)	8 (0.08)	24 (0.22)	105 (1.0)
Control	31 (0.42)	12 (0.16)	11 (0.15)	19 (0.27)	73 (1.0)
Telecom	42 (0.58)	2 (0.03)	3 (0.04)	26 (0.35)	73 (1.0)
Measure/indicat.	20 (0.41)	13 (0.27)	4 (0.08)	12 (0.24)	49 (1.0)
Total	141 (0.47)	52 (0.17)	26 (0.09)	81 (0.27)	300 (1.0)

(b) Causes and Incoming Routes of Power System Disturbances

Total	55 (Overlapped 12)	Generation Power Supply 32 (Overlapped 2)	Syst. Operation 23 (Overlapped12)
Type of surge	Lightning	25	14
	Switching	4	6
	Others	3	3
Place	Generator station	18	—
	Substation	14	23
Incoming route	Control circuit	9	7
	Signal transmission	6	3
	Source circuit	5	3
	Others (CT/VT, etc.)	12	10
Cause	Protection	13	15
	Control	10	3
	Measure/indicat.	7	4
	TC/distant control	2	1

(): ratio.

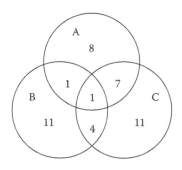

FIGURE 6.5
Power system disturbances resulting from Table 6.10a: (A) generation: seventeen cases; (B) power supply: seventeen cases; and (C) system operation: twenty-three cases.

6.2.3.2 Conducted Countermeasures

Countermeasures were established for 153 cases (51%) of the total 279 disturbances listed in Figure 6.6. There were twenty-eight cases for which it is unclear whether countermeasures or repairs were used.

Countermeasures were taken for thirty-nine cases (85%) of main-circuit SSs. The high percentage is primarily due to repetitive disturbances. The maximum switching overvoltage in the main circuit can be predicted by a numerical simulation, and the incoming route can also be estimated. Countermeasures against SSs involve the modification of software used for digital control equipment. For example, software that refreshes the contents of stored memory is installed to counteract malfunction of the central processing unit (CPU) board, and recovery processing is added to protect against freezing of a keyboard controller.

Ninety-four countermeasures were taken against disturbances caused by LSs, only 43% of the total disturbances (220 cases); this percentage is relatively low because disturbances due to lightning are mostly permanent and involve the breakdown of the disturbed circuit (element). Thus, 118 circuits are replaced by a new one or repaired, as shown in Figure 6.6. Remind some countermeasures and repair are overlapped.

6.2.3.3 Cost of Countermeasures

6.2.3.3.1 Flow of Countermeasures

The first stage in the flow of countermeasures is restoration of the disturbed equipment. Only when the restoration is not effective, the manufacturer

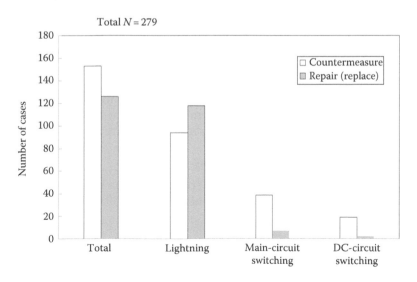

FIGURE 6.6
Countermeasures carried out.

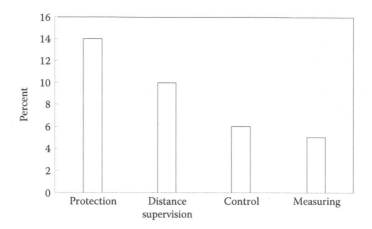

FIGURE 6.7
Ratio of the third-stage countermeasures to the total number of disturbances.

is involved. The second stage occurs when (1) similar disturbances occur repeatedly, (2) the disturbances have a significant impact on equipment, or (3) the cause and incoming route of the surge are unclear. The countermeasure is investigated by both the utility company and manufacturer. The third stage involves estimating the number of similar disturbances and/or which disturbances appear to occur repeatedly. Occasionally, the third stage results in the revision of an existing standard [16], a summary of which is provided in Appendix 6A.1. Of the 153 countermeasures, twenty-five cases reached the third stage. Figure 6.7 shows the ratio of the number of third-stage countermeasures to the total number of disturbances. The third-stage countermeasures were taken against equipment that significantly affected power system operation. Many of the countermeasures involved IC boards of CPUs, digital input/output circuits, and analog-to-digital conversion circuits.

Restoration was frequently needed for the disturbances caused by lightning, while countermeasures were normally taken for disturbances caused by main-circuit switching due to their repetitive nature.

6.2.3.3.2 *Manpower and Cost of Countermeasures*

It appears that the average manpower spent was 50 man-days per countermeasure. The average cost per countermeasure was $10,000 (see Figure 6.8).

6.2.4 Case Studies

This section presents case studies on the experienced disturbances [1].

Each case study is categorized by cause and includes the following items:

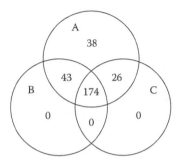

FIGURE 6.8
Manpower and material costs for utilities and manufacturers: (A) utilities, (B) manufacturers, and (C) material costs.

1. Simulation of disturbances: (a) the voltage class of a generator station or a substation in which the disturbance was found; (b) the cause of the disturbance (i.e., LS or SS); and (c) the impact of the disturbance, such as the breakdown of a diode or the freezing of control equipment
2. Disturbed equipment
3. Incoming surge routes to the disturbed equipment
4. Countermeasures taken against the disturbance
5. Investigation and analysis of the disturbance if necessary

6.2.4.1 Case No. 1

When lightning strikes the tower nearest to a substation, the diode that protects the relay of CB enclosure equipment against surges (52X1 in Figure 6.9a) is broken, and the auxiliary relay contact is melted. Also, the auxiliary relay coil for indication of the telecontrol equipment (79SX1 in Figure 6.9a) breaks. The following outline corresponds to the template provided in Section 6.2.4 and presents the details of this case study:

1. (a) 77 kV, (b) lightning, (c) DC short circuit, erroneous indication.
2. 77-kV line CB enclosure equipment.
    a. Diode for surge protection broken.
    b. Auxiliary relay contact melted.
    c. Telecontrol equipment's auxiliary relay coil for indication broken.
3. Lightning to the tower nearest to the substation: A lightning current flowed into a ground mesh and induced a surge voltage in a control cable. A differential mode voltage was induced in the diode.

(a)

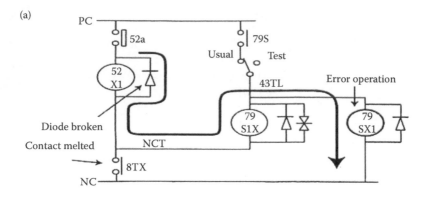

(b)

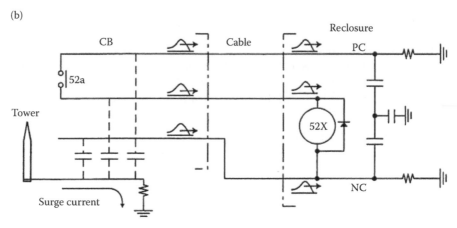

(c)

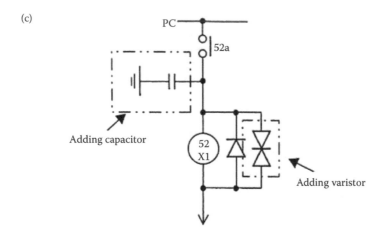

FIGURE 6.9
A disturbed circuit for Case No. 1: (a) disturbed circuit, (b) flow of the surge voltage, and (c) countermeasures.

TABLE 6.8

Recommended Values of Lightning Currents Flowing into a Ground Mesh

Lightning to	S/S Tower	GW	Tower 1	2nd	3rd
Distance from S/S		150 m	300 m	600 m	900 m
Mesh current	10.9 kA	6.6 kA	2.8 kA	1.7 kA	1.4 kA

Standard lightning current waveform 1.2/50 µs, estimated wave front of above 0.5 µs.

4. Countermeasures:

   a. Varistor installation for auxiliary relays in reclosure.

   b. Installation of independent molded case circuit breakers for DC circuits in reclosure.

5. Reclosing terminal.

Assuming that a lightning current of 30 kA strikes a tower of a 66-kV system, the current flowing into the mesh is estimated as shown in Table 6.8.

A lightning current of 6 kA that strikes the ground wire (GW) (see Table 6.8) and flows into a ground mesh would result in a surge voltage at the reclosing terminal, which is estimated in Table 6.9 as

CVV cable case 0.24 V/A × 6 kA = 1,440 V

CVVS cable case (with metallic sheath) 0.036 V/A × 6 kA = 220 V

The voltage exceeds the surge strength of the broken diode that was manufactured in 1975.

TABLE 6.9

Induced Voltage to a Control Cable Due to Lightning

Distance from Applied Node	Number of Cables	Induced Voltage (V: Peak Value)	
		Current-Applied Node	End Side
0 m (A)	8	16,007	11,082
	4	12,717	5,813
14 m (B)	4	3,179	2,634

Model of substation ground mesh.

Lightning current: 10.9 kA, 1.2/50 µs, CVV cable: 8 mm²

A lightning overvoltage induced in a relay terminal (i.e., at the end of a control cable) is given in Table 6.9. The figure included in Table 6.9 is a schematic diagram of an experiment for which the results are listed in the table. With a cable length of 200 m, the voltage at the relay terminal is estimated to be 11,082 V = 2.42 V/A × 10.9 kA × 0.5/1.2 μs under the worst conditions in an experiment. However, when considering the real length of the cable, 50 m (see Figure 6.9b), the estimated voltage at the relay terminal becomes

$$11,082 \text{ V} \times 150/200 \text{ m} = 8422 \text{ V}$$

6.2.4.2 Case No. 2

Figure 6.10 shows a circuit configuration in a GIS for Case No. 2.

CB1 and CB2 were opened in the substation where Line 1 and Line 2 were charged from the other end in Figure 6.10. When DS1 (or DS2) was opened (or closed), relay (for 66 kV line protection equipment) malfunction occurred, and no CB was tripped. The surge was transferred through the control cable via a CT. As a countermeasure, the CVV control cable (without a metallic shield) for the CT circuits was replaced by a CVVS control cable (with a metallic shield) with grounded ends. Then, the frequency of the malfunction decreased. Ferrite cores were also installed at the secondary circuit of an internal auxiliary transformer for the CT. No malfunctions occurred after this countermeasure.

Figure 6.11 shows the test results of surge propagation along a 201 m CVVS cable with both terminals of the metallic shield grounded [1]. The test results show that the surge transferred through a CT was weakened and the oscillating frequency decreased during propagation along the CVVS cable, and the surge at the relay terminal became low enough to cause a disturbance, as expected in the countermeasure.

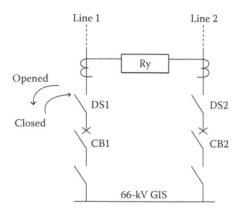

FIGURE 6.10
Circuit configuration for Case No. 2: 66/6.6 kV distribution substation.

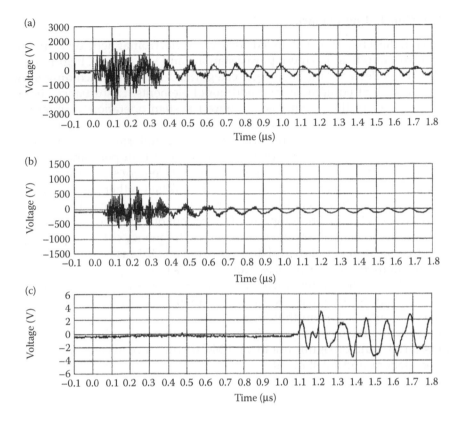

FIGURE 6.11
A test result of surge propagation along a CVVS control cable: (a) CT secondary terminal, (b) GIS control box, and (c) relay terminal.

6.2.4.3 Case No. 3

The following outline provides the details for Case No. 3:

1. (a) 275 kV, (b) DC SS, and (c) flicker of light-emitting diodes (LEDs) for system-state indicator and measurement
2. Numerical control equipment (6.6 kV distribution system)
3. a. 6.6 kV vacuum circuit breaker (VCB) closing
   b. DC SS appeared in the DC 110-V circuit (700 V/0-peak, 10–30 kHz)
   c. LEDs connected to the DC 5-V circuits were affected by the DC SS
4. a. Separation of DC 110-V circuit from the DC 5-V circuit
   b. Replacement of wires used in the DC 5-V circuit by twist-pair type
   c. Installation of noise filters in the DC 5 V power circuit

6.2.5 Concluding Remarks

Thus far, this chapter has detailed 10 years of electromagnetic disturbances in control circuits in Japanese power stations and substations, beginning in 1990. The collected data have been categorized by cause, incoming surge route, disturbed equipment, and additional elements. Disturbance characteristics, case studies, the impact on power system operation, countermeasures, and costs were also presented.

Average manpower and material cost per countermeasure totaled 50 man-days and $10,000, respectively. LSs often result in permanent failure, including the breakdown or burnout of a control element such as diodes (155 permanent failures out of 220 cases), which makes it rather easy to identify the disturbance. On the contrary, SSs tend to be nonrepetitive disturbances, such as freezes (thirty-one temporary failures out of forty-five cases), because the surge overvoltage is low but the oscillating frequency of the surge is high. Although the number of disturbances due to LSs (220 cases) is much higher than the number of disturbances caused by SSs (sixty-eight cases including the DC circuit), the nonrepetitive nature of SSs makes it difficult to identify the disturbances; this suggests that such a disturbance may have severe consequences, such as system shutdown.

Considering the increase in digital control equipment and compact substations, an SS could be significant for an EMC disturbance in control circuits in power stations and substations.

6.3 Disturbances in Customers' Home Appliances

Insulation design and coordination of high-voltage systems are quite effective, and the systems are well protected against various voltage surges. However, low-voltage distribution and service systems for customers are *not* well protected against overvoltages, and disturbances in customers' home appliances have often been reported [4–9]. The increasing use of digital appliances makes their protection from overvoltages and electromagnetic disturbances an important issue.

6.3.1 Statistical Data of Disturbances

Figure 6.12 shows statistical data for the number of damaged home electric appliances (HEAs) from a Japanese utility company from 1987 to 1991 and 1996 to 2006 [8,9]. The data were based on information from more than 2,000 monitors. There were 21–34 thunderstorm days per year from 1987–1991

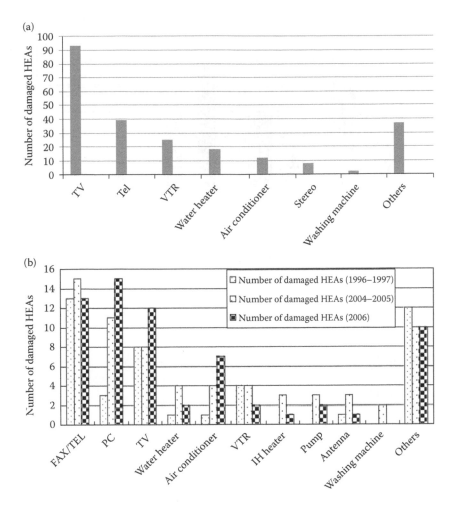

FIGURE 6.12
Number of damaged HEAs: (a) 1987–1991 and (b) 1996–1997, 2004–2005, and 2006.

(included in Figure 6.12a), 15–23 thunderstorm days per year in 1996–1997, and 24 thunderstorm days in 2004–2005 (included in Figure 6.12b).

Between 1987 and 1991, 228 HEAs were damaged in 129 houses: 49% TV/ video antennas, 18% communication equipment, and 16% grounding. In the data from 1996 to 1997, 32% of the damaged appliances were antennas, 40% communication equipment, and 30% grounding. The results indicate that the protection of TV/video players against lightning was improved. However, because of the increasing number of digital circuits used in communication equipment and their weakness and sensitivity to LSs, the ratio of damaged

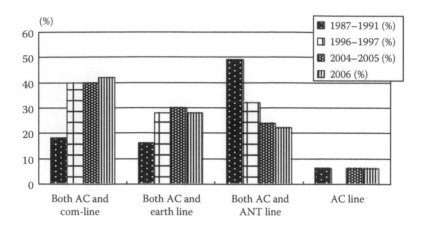

FIGURE 6.13
Ratio of the connecting circuits of damaged HEAs.

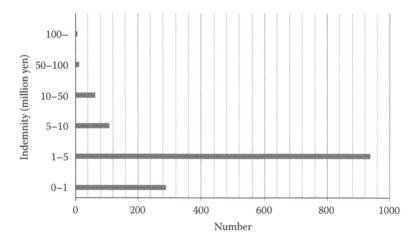

FIGURE 6.14
Expenses (Insurance) Paid For Damages. (From Ieej Wg. 2002. *Ieej Tech. Report* 902.)

communication equipment increased. Figure 6.13 shows the ratio of the connecting circuits of damaged HEAs.

Figure 6.14 shows the compensation paid to customers by an insurance company [17]. The figure shows that 82 of the 1,417 cases from 1987 to 2000 paid more than 10 million Japanese yen.

6.3.2 Breakdown Voltage of Home Appliances

6.3.2.1 Testing Voltage

A voltage waveform with 1.2/50 µs (wave front T_f = 1.2 µs, wave tail T_t = 50 µs) and a current waveform with 8/20 µs are used as lightning impulses to test the withstand voltage of home appliances or electronic equipment in general. The impulse voltage is mainly used for high-impedance equipment, and its maximum value is taken as 6 kV. On the other hand, the impulse current is mainly for low-impedance equipment, and its peak value is taken as 3 kA [18,19].

In addition, ring waves of 6 kV/500 A with 0.5 µs/100 kHz and of 0.6 kV/120 A with 1.5 µs/5 kHz are recommended by the Institute of Electrical and Electronics Engineers for insulation tests of home appliances [4,18].

As for an induced LS in a communication line, an impulse voltage with 10/200 µs is used [18], and an impulse voltage of 1.5 kV with 10/700 µs is widely recommended for lightning impulse tests of home communication equipment [20].

The testing voltages of home appliances surveyed in this chapter are summarized in Table 6.10a. In general, the testing voltage of a home appliance ranges from 1 to 1.5 kV [21].

6.3.2.2 Breakdown Test

An experiment was conducted to determine the actual breakdown voltage of home appliances. Impulse voltages of about 1.0/23 µs were applied to home appliances, and the test results are recorded in Table 6.10b. As shown in the table, the breakdown voltages of the home appliances are greater than 5 kV for line-to-ground and greater than 7 kV for line-to-line, although the recommended testing voltage is about 1.5 kV. The test results agree with those obtained in Reference 4, where the breakdown voltages were found to be 4–6 kV.

6.3.3 Surge Voltages and Currents to Customers Due to Nearby Lightning

6.3.3.1 Introduction

This section investigates incoming paths of LSs to customers/users due to nearby lightning [12,22]. Four incoming paths exist: (1) a low-voltage distribution line (feeder) through a distribution pole, (2) a telephone line, (3) a TV antenna, and (4) the grounding electrode of the customer's electrical equipment. A lightning strike and an induced lightning voltage to the distribution line result in path (1). A lightning strike to a telephone line and an induced lightning voltage result in path (2), and the same applies to path (3) when lightning hits a TV antenna. A lightning strike to the ground, wood, or a distribution pole near a customer/user results in a ground potential rise (GPR) as the lightning current flows into the ground [23,24]. In path (4), the current causes a potential rise to the grounding electrode of the customer's equipment.

Based on the measured results, modeling of electrical elements related to these paths is developed for an LS simulation, and electromagnetic transients

TABLE 6.10

Withstand Voltages of Home Appliances

(a) Testing Voltages Surveyed from Standards and Guides

		Rated Voltage	Voltage	Applied Time
Electric equipment	Incandescent lamp	~30	500	1 min
	Discharge lamp	~150	1000	
		~300	1500	
		~1000	2E + 1000	
Home appliances	Air conditioner	100 V	AC 1000 V	1 min
		200 V	AC 1500 V	
	Microwave oven	100 V	AC 1000 V	1 min
		200 V	AC 1500 V	
Information and communication appliance	Telephone	100 V		
	Fax			
	PC	100 V		
Audiovisual	TV	100 V	AC 1000 V	1 min
	Video			

(b) Experimental Results of Breakdown Voltages

	Breakdown Voltage (kV)	
	Line-to-Ground	Line-to-Line
Ventilating fan	6.0	8.0
Electric rice cooker	5.0	20.0
Refrigerator	11.0	13.0
Electric fan 1	7.0	7.0
Electric fan 2	8.0	7.0
Electric fan 3	8.0	7.0
Cassette deck	8.0	12.0
Hair dryer	10.0	7.0

program (EMTP) simulations are demonstrated in the model. The simulation results are compared with the measured results, and the accuracy of the modeling method is discussed.

6.3.3.2 Model Circuits for Experiments and EMTP Simulations

6.3.3.2.1 Experimental Conditions

Kansai Electric Power Co. (KEPCO) has conducted experiments to investigate lightning currents that flow into a house [25,26]. An impulse current, which represents a lightning current, is injected into an antenna, distribution pole, and the ground from an impulse current generator (IG, maximum voltage 3 MV, maximum current 40 kA). An impulse current from 500 to 5,000 A is applied to the following:

1. An antenna (see Figure 6.15a)
2. A distribution pole (see Figure 6.15b)
3. A structure near a house (see Figure 6.15c)
4. A messenger wire of a telephone line (see Figure 6.15d)

At a KEPCO test site, three distribution poles and two poles of National Telephone and Telecommunication (NTT) company were constructed, and 6600/220/110 V distribution lines with a pole transformer were installed as depicted in Figure 6.15. A telephone line and a messenger wire were also installed. A model house was built near the pole transformer, and a feeder line from the transformer led into the house. In the house, model circuits of an air conditioner and a fax machine were installed, as shown in Figure 6.16, and included a surge protection device (SPD) and surge arrester (air conditioner, operating voltage 2670 V; fax, 2800 V; and NTT SPD, 500 V). The poles, the transformer, the telephone line (NTT), and the air conditioner (home appliance) were grounded individually, as in shown Figure 6.16.

Table 6.11 summarizes the measured results of maximum voltages and currents through the grounding resistances.

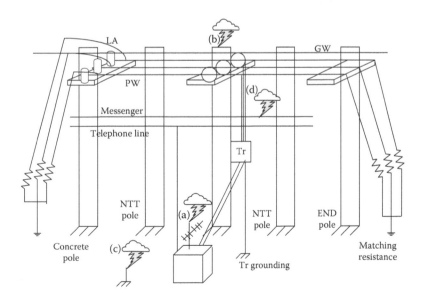

FIGURE 6.15
Lightning and its path to a house. Lightning to (a) an antenna, (b) a pole, (c) the ground, and (d) a telephone (messenger) line.

6.3.3.2.2 *Modeling for EMTP Simulations*

The distribution line, the pole, and the home appliances in the house shown in Figure 6.16 can be represented by horizontal and vertical distribution line models and lumped parameter circuits [27,28]. The grounding electrodes of the pole, the telephone line SPD, and the home appliances, if grounded, are modeled by a combination of a distributed line and a lumped parameter circuit to simulate the transient characteristic [29]. However, this section adopts a simple resistance model with a resistance value taken from the

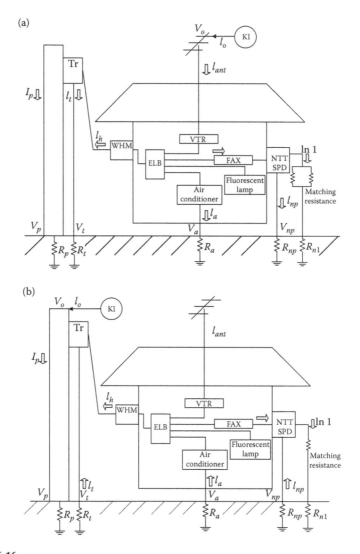

FIGURE 6.16
Experimental circuit and lightning current path: (a) lightning to antenna and (b) lightning to a distributed line.
(*Continued*)

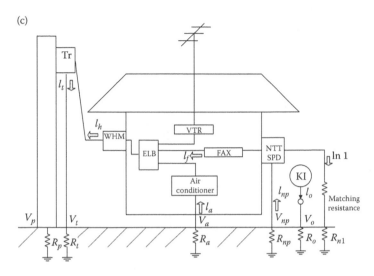

(c)

FIGURE 6.16 (Continued)
Experimental circuit and lightning current path: (c) lightning to the ground.

experiments discussed in References 29 and 30, as the vertical grounding electrode used for a home appliance is short and the transient period is much shorter in the phenomenon investigated in this chapter. A protection device (PD) is installed in a home appliance, and the NTT SPD is represented by a time-controlled switch prepared in the EMTP [31].

A model circuit for an EMTP simulation is shown in Figure 6.17. In the figure, Z_p is the surge impedance of a distribution pole, which is represented by a lossless distributed line with a propagation velocity of 300 m/µs. The surge impedance is evaluated by the following formula of a vertical conductor [27]:

$$Z_p = 60\left\{\ln\left(\frac{h_p}{r_p}\right) - 1\right\}(\Omega) \tag{6.1}$$

where
  h_p is the height of the pole
  r_p is the radius of the pole

$r_p = 17$–$19\ \Omega$ is the grounding resistance of the pole. The value is given by a utility company [25,26]. Tr is a pole-mounted distribution transformer, which is represented by an ideal transformer with the voltage ratio of 6600:110 V and stray capacitances.

In Figure 6.17, TRN-TRNB is a grounding lead of the transformer represented by a lossless distributed line (the surge impedance is explained in Reference 32). R_t is the grounding resistance of the transformer grounding lead, and its values by utility are provided in Table 6.11 [25,26]. PW represents a phase wire of a distribution line and is modeled by a lossless

TABLE 6.11

Test Conditions and Results

	Case												
	A				B				C1				C2
Appl. Current I_0 (A)	306	739	1247	1660	602	1183	1977	2416	676	1532	2815	4134	2771
Grounding resistance (Ω)													
R_p	17	17	17	17	19	19	19	19	19	19	19	19	19
R_t	89	89	89	89	75	75	75	75	85	85	85	85	85
R_a	830	830	830	830	100	100	100	100	140	140	140	140	–
R_{np}	900	900	900	900	300	300	300	300	380	380	380	380	380
R_{n1}	153	153	153	153	150	150	150	150	100	100	100	100	100
Voltage (kV)													
V_o	11.6	25.9	41.5	56.1	10	19.3	31.5	37.6	55.4	88	122	146.3	122
V_t	3.8	7.7	12.9	17.5	3.8	5.6	21.2	24.6	–	–	2.5	–	2.5
V_a	–	–	–	–	0.8	5.1	19.6	22.8	1.8	2.2	3.1	4.4	3.1
V_{np}	–	–	–	–	1.3	2.9	16.7	19.1	1.3	2.9	5	6.3	0.8
V_p	3.5	7.5	12.2	16	–	–	–	–	–	–	–	–	–
Current (A)													
I_a	–	–	–	–	0	44	177	206	0	27.5	58.2	83.1	55.7
I_{np}	–	–	–	–	3.6	5.9	52	68	13.2	31.2	56.6	80.2	–
I_t	34.4	72	120	159	3.4	–27	174	197	–	–	–	–	–
I_p	247	500	888	1150	–	–	–	–	–	–	–	–	–
I_f	–	–	–	–	–	–	–	–	0	0	0	0	31.8

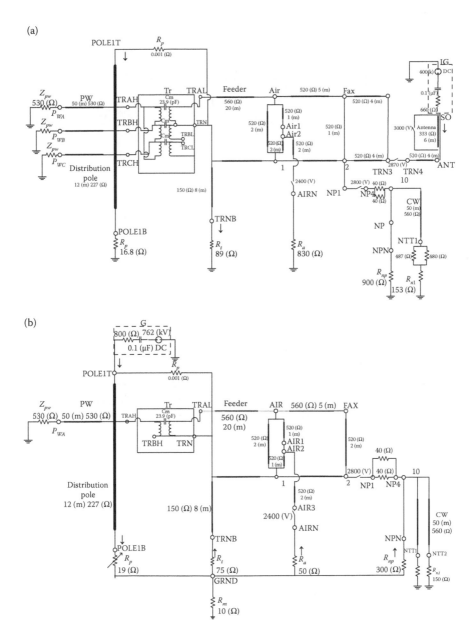

FIGURE 6.17
Model circuits for EMTP simulations: (a) Lightning to an antenna, (b) lightning to a pole. PW:
power line; CW: communication (NTT) line. *(Continued)*

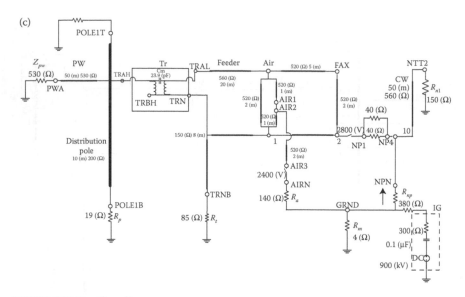

FIGURE 6.17 (Continued)
Model circuits for EMTP simulations: (c) lightning to a ground.

distributed line with the surge impedance $Z_{pw} = 530\ \Omega$. The other end of PW1 is terminated by the matching impedance Z_{pw}.

TRAL-FAX and TRAL-HOME1 are a feeder line from the pole transformer to the house. For the transformer, A represents phase A, and N indicates neutral. The standard steady-state AC voltage between phase A and neutral is 114 V in Japan. When lightning struck the antenna, case A, a flashover between the transformer grounding lead and the steel of the distribution pole was observed. The flashover is represented by short-circuiting the transformer and the pole by resistance R_p in Figure 6.17a.

The feeder line is represented by a lossless distributed line with a surge impedance of 560 Ω and a velocity of 300 m/μs. R_a is the grounding resistance of an air conditioner (the value is provided in Table 6.11). Z_a represents the air conditioner expressed by a lead wire inductance of 1 μH and by a time-controlled switch representing a surge arrester that operates when the voltage exceeds 2670 V in parallel with the resistance. Z_b represents a fax machine. It is expressed in the same manner as Z_a, except the other terminal of Z_b is connected to a telephone line through an NTT SPD, a surge arrester valve with an operating voltage of 500 V represented by a time-controlled switch. The telephone line is represented by a lossless distributed line with a surge impedance of 560 Ω and a propagation velocity of 200 m/μs; the other end is terminated by a resistance matching a grounding resistance of 153 Ω.

An IG is the impulse current source used in the experiment. It is represented by a charged capacitance and a resistance [25,26]. R_m is the mutual grounding impedance between various grounding electrodes, and a value of 2–10 Ω is used [24,30].

6.3.3.3 Experimental and Simulation Results

6.3.3.3.1 Experimental Results

Figures 6.18 through 6.20 show the experimental and simulation results in the case of lightning to (a) an antenna, (b) a pole, and (c) the ground. From the experimental results, the following observations were made:

1. Lightning to antenna: When lightning strikes the antenna of a house (see Figure 6.16a), the lightning current flows out (1) to a distribution

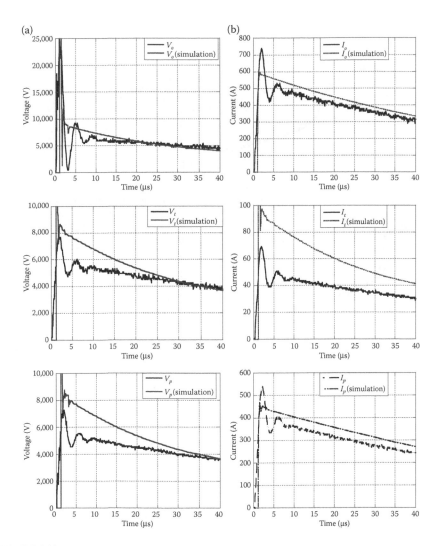

FIGURE 6.18
Experimental and simulated results in the case of lightning to an antenna: (a) voltage and (b) current.

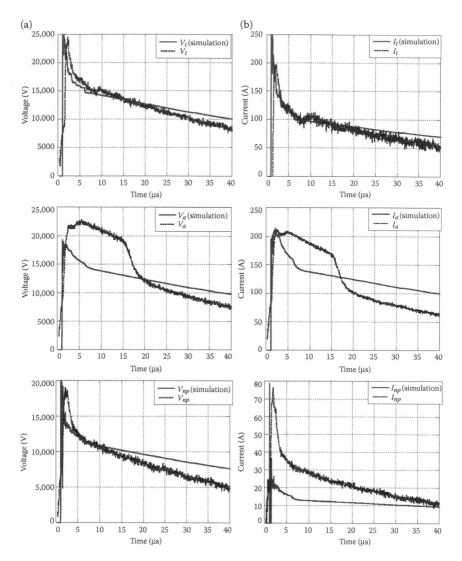

FIGURE 6.19

Experimental and simulation results in the case of lightning to a pole: (a) voltage and (b) current.

line I_1 (I_n, I_t) through feeders in the house, and (2) to the ground I_2 (I_a, I_{np}) through the grounding electrodes of home appliances. The ratio of I_1 and I_2 is dependent on the grounding resistances of (1) the house feeder to the distribution line, (2) the house to the telephone line, and (3) the grounding electrodes of the home appliances. In the experimental results (for example, see Figure 6.18 with the applied current $I_0 = 739$ A), about 85% of the current that strikes the antenna

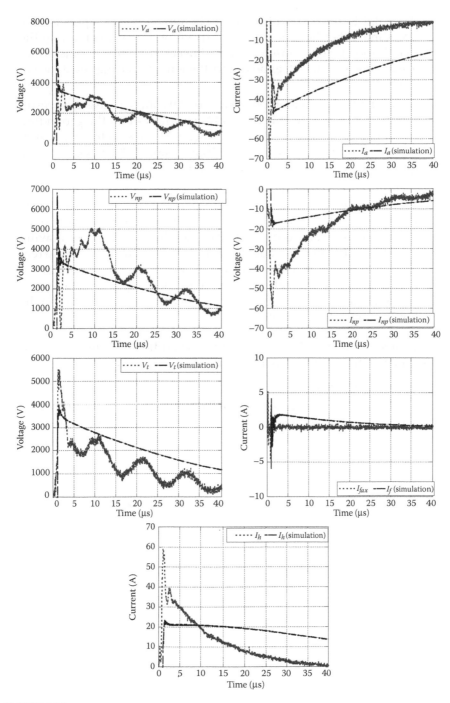

FIGURE 6.20

Experimental and simulation results in the case of lightning to the ground.

flows out to the distribution line. The remaining 15% presumably flows into other houses. A large current of 543 A flowing into the distribution pole is caused by a flashover between the transformer grounding lead, the steel pole, and the low grounding resistance of the pole.

2. Lightning to distribution pole: Figure 6.19 shows a measured result when I_0 is 2,416 A. In this case, 8.1% of I_0 flows into the grounding resistance R_t of the neutral transformer, 8.5% flows into the grounding resistance R_a of the air conditioner, and 2.8% flows into the grounding resistance R_{np} of the telephone line SPD.

3. Lightning to grounding: Part of the applied current I_0 flows into the air conditioner ($I_{a\ max} = 58.2$ A for $I_0 = 2815$ A) and the NTT SPD ($I_{np\ max} = 56.6$ A) in Figure 6.20 (case C1). The current flowing into the air conditioner flows out to a distribution line. The current flowing into the NTT SPD flows out through the telephone line. No current flows into the fax machine because the operating voltage of 2,800 V of the surge arrester within the fax machine is higher than the voltage across it (i.e., the voltage difference between the feeder line in the house and the telephone line).

Compared with case C2, which has no air conditioner grounding, the currents flowing into the house increased rapidly.

6.3.3.3.2 Simulation Results

Table 6.12 summarizes the simulation conditions and results, and Figures 6.18 through 6.20 show a comparison of the simulated voltage and current waveforms and the measured results. From the figures, the following observations were made:

1. In Figure 6.18, when lightning struck the antenna, the simulation results of transient voltages and currents—except the wave front of the applied current and voltage at the top of the pole and the voltage at the primary winding of the transformer—were similar to the measured results.

2. When lightning struck the top of the pole (as shown in Figure 6.19), voltage differences between the top and the bottom of the pole and at the air conditioner grounding electrode were smaller than those of the measured results. Otherwise, the simulation results agreed with the measured results.

3. For lightning that struck the ground near a house, the simulation results in Figure 6.20 show close agreement with the measured results, and thus the proposed approach to use mutual impedance to protect equipment against a GPR is proved effective.

TABLE 6.12

Simulation Results Corresponding to Table 6.11

	Case												
	A				B				C1				C2
Appl. Current I_0 (A)	306	739	1247	1660	602	1183	1977	2416	676	1532	2815	4134	2771
Voltage (kV)													
V_0	87.2	174.3	261.4	348.5	16.2	30.6	49.6	60	50.8	81.24	121.9	162.4	121.9
V_t	17.6	35.2	52.8	70.5	13.6	25.6	41.5	50.2	1.7	2.7	4	5.3	3.4
V_a	—	—	—	—	5.9	11.2	18.2	22	4.3	6.9	103.3	137.8	103.8
V_{np}	—	—	—	—	9.8	18.4	29.9	36.2	2.9	4.6	6.9	9.3	6.5
V_p	22.4	44.8	67.1	89.5	—	—	—	—	—	—	—	—	—
Current (A)													
I_a	—	—	—	—	61	114.9	186.5	225.6	−19.8	−31.3	−47.6	−62.6	−53.9
I_{np}	—	—	—	—	23.7	44.7	72.5	87.8	−8	−12.7	−19	−25.4	—
I_t	198	395	593	791	143	269.3	437.2	528.8	—	—	—	—	—
I_p	240	480	721	961	189.4	356.7	579.2	700.5	—	—	—	—	—
I_f	—	—	—	—	—	—	—	—	−2.4	−3.9	−6	−7.9	7.3

However, further improvement of the overall simulation method is needed to achieve quantitative agreement with the measured results. For example, an SPD has to be carefully represented based on its circuit and nonlinear characteristics. Also, grounding impedance should be modeled based on its transient characteristics rather than the simple resistance used in this section.

6.3.3.4 Concluding Remarks

This section has shown the experimental and EMTP simulation results of an LS at a house. The simulation results agree qualitatively with the experimental results, and thus the simulation models in the chapter are adequate. Because measurements were calculated during different time periods over the course of 3 years, experimental conditions such as the soil resistivity, the voltage probe used, and the reference voltage line for each measurement varied, and these differences were not considered in the simulations. Some oscillations observed in the measured results are presumably caused by mutual coupling between the measuring wires and feeder lines in the experiments. Also, a grounding electrode may couple with other electrodes. Further improvements for modeling the experimental circuits are required to achieve greater accuracy.

LS voltages to a customer/user and lightning currents flowing out to distribution lines can be determined from the experimental and simulation results. The results should assist in the protection coordination of SPDs for customers/users and telephone lines and investigate the necessity of home appliance grounding.

6.3.4 LS Incoming from a Communication Line

6.3.4.1 Introduction

Many digital appliances, such as PCs, are used at home. Thus, protection of such appliances against LSs is essential. However, the characteristics of an LS to devices connected to communication lines are not well investigated. Therefore, incoming LSs have been investigated; installation of a PD is recommended for significant LSs [33]. In this section, the characteristics of PDs and incoming LSs are examined through experiments and finite-difference time-domain (FDTD) simulations [34]. Also, the optimum method to protect appliances and the adequacy of the FDTD simulation are discussed.

6.3.4.2 Protective Device

6.3.4.2.1 Experiment

Figure 6.21a shows the experimental circuit. A step-like voltage from a pulse generator (PG) is applied to a discharge tube through a 3D 2-V cable and a resistor. The circuit is grounded by an aluminum plate.

(a)

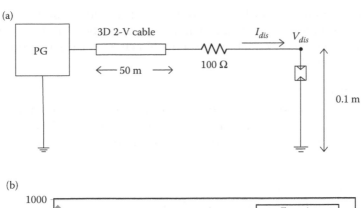

(b)

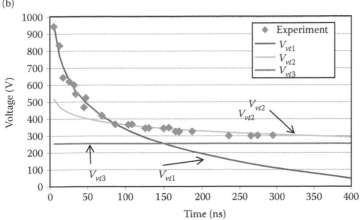

(c)

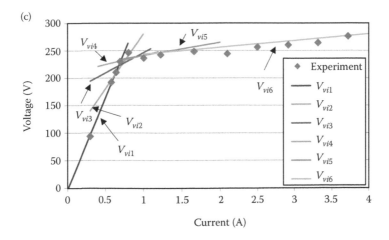

FIGURE 6.21
Experimental circuit and results for PD characteristics: (a) experimental circuit for a PD characteristic, (b) measured results of the V–t characteristic, and (c) measured results of the V–I characteristic.

Figure 6.21b and c show the measured results of the V–t and the V–I characteristics of PDs and their approximate curves. The V–t characteristic is approximated by three equations, and the V–I characteristic is approximated by six equations.

6.3.4.2.2 Simulation

Figure 6.22 shows a simulation circuit. The analytical space of an FDTD simulation is 31 cm × 21 cm × 23 cm, and the cell size is 1 cm. The time step is 19 ps, and an absorbing boundary condition of Berenger's perfectly matched layer is used [34].

The model circuit of the PD is developed based on measured results. Until the PD flashes over, its performance follows the V–t characteristic, and its resistance is 1 MΩ. First, the threshold voltage is calculated based on the V–t characteristic in Figure 6.21b. When the PD's voltage reaches the threshold voltage, it flashes over and the calculation based on the V–t characteristic is shifted to that based on the V–I characteristic in Figure 6.21c. The value of the variable resistance is the gradient of the linear approximation indicated as V_{vi1} to V_{vi6} in Figure 6.21c. For example, the resistance is 333 Ω corresponding to V_{vi1} when $V_{dis} = 180$ V. The resistance is 12.4 Ω corresponding to V_{vi6} when $V_{dis} = 250$ V.

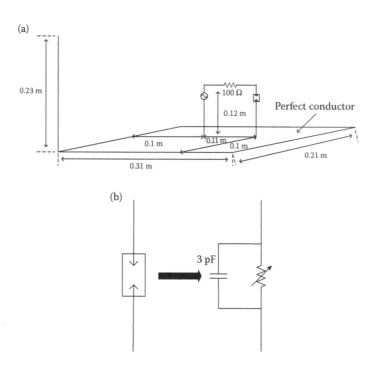

FIGURE 6.22
Simulation circuit for a PD characteristic: (a) simulation circuit and (b) PD model.

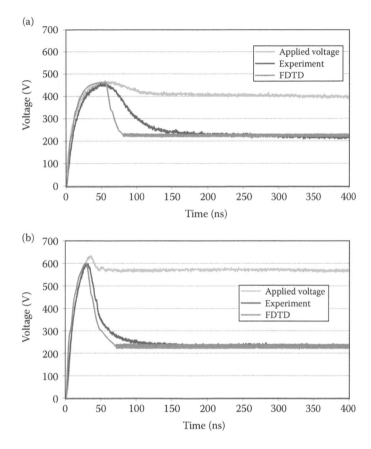

FIGURE 6.23
Experimental and simulation results of PD voltages: (a) an applied voltage of 400 V with a waveform of 50 ns and (b) an applied voltage of 600 V with a wave front of 50 ns.

Figure 6.23 shows an applied voltage and a comparison of the measured and simulation results. As shown in the figure, the initial rise corresponds well with the steady-state value. However, the waveform before the steady state shows a difference, which is caused by the approximation of the discharge characteristic.

6.3.4.3 Lightning Surges

6.3.4.3.1 Model Circuit

Figure 6.24 shows the experimental circuit. The voltage from the PG is applied to the circuit through a 3D 2-V cable. In the case of the current source, a voltage is applied to a resistor. The circuit is grounded on an aluminum plate. Experimental cases include individual grounding, common grounding, and a proposed method of individual grounding with a bypass installed on an appliance.

(a)

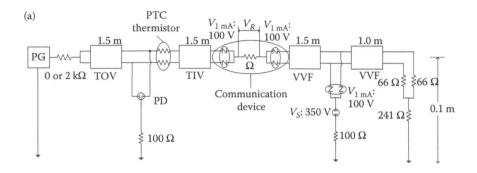

(b)

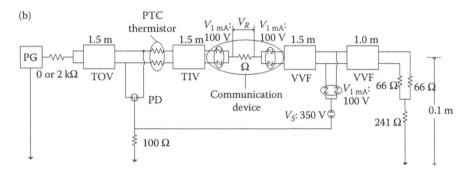

(c)

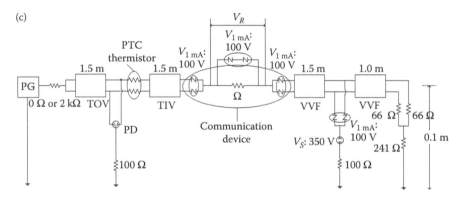

FIGURE 6.24
Experimental circuit for a transient: (a) individual grounding, (b) common grounding, and (c) the proposed grounding method.

6.3.4.3.2 Measured Results

Figure 6.25 shows the experimental results. In the case of the current source, the results of the common and individual methods are the same, so only the results of the individual method are shown. Figure 6.25a shows that the maximum and steady-state values of the common and proposed methods are lower than those of the individual method. Also, the maximum voltage

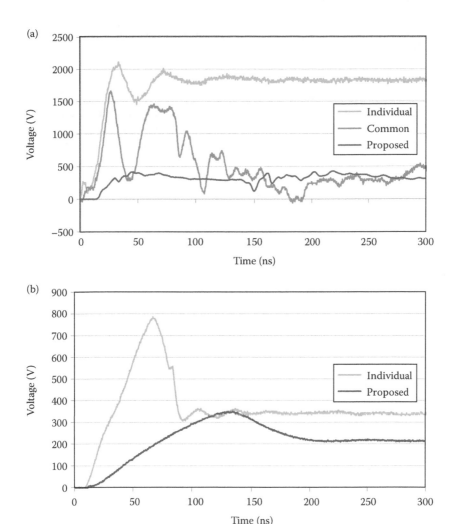

FIGURE 6.25
Experimental results of V_R in Figure 6.24: (a) voltage source case and (b) current source case.

of the proposed method is lower than that of the common method. In the common method, a larger current flows to the resistor, because the PD experiences a discharging lag. However, in the proposed method, the current flowing to the resistor rapidly decreases because the varistor works quickly. Therefore, the proposed method can reduce the maximum voltage more efficiently than the common method can.

Figure 6.25b indicates that the maximum and steady-state values of the proposed method are lower than those for individual grounding. Because the varistor has a capacitance of hundreds of picofarads, the initial rise slows down, which indicates that the proposed method can reduce an LS better than individual grounding can.

6.3.4.3.3 Simulation

Figure 6.26 shows the model circuit. The analytical space of the FDTD simulation is 5.85 m × 0.2 m × 0.21 m for the individual and common groundings, and 5.87 m × 0.2 m × 0.21 m for the proposed method. The cell size is 1 cm. The time step is 15 ps, and the absorbing boundary condition is the same as that in Section 6.3.4.2.2.

Figures 6.27 and 6.28 show the simulation results. According to the figures, the initial rise and steady-state values are in agreement. However, the waveform before the steady state exhibits a difference, because the characteristics could not be simulated well enough to develop the PD.

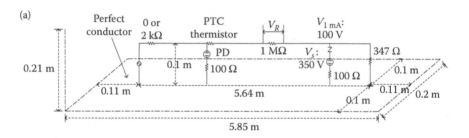

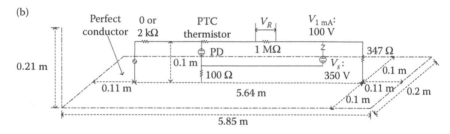

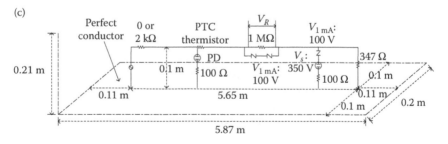

FIGURE 6.26
Model circuits for transient simulations: (a) individual grounding, (b) common grounding, and (c) the proposed grounding method.

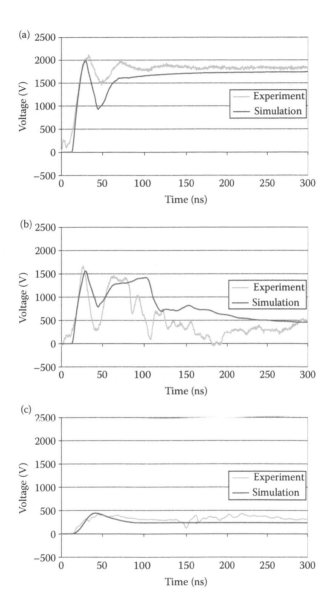

FIGURE 6.27
Effect of grounding on V_R for the voltage source case: (a) individual grounding, (b) common grounding, and (c) the proposed method.

6.3.4.4 Concluding Remarks

The characteristics of the PD are determined by the experimental results. The FDTD simulation results based on those characteristics corresponds well with the measured results. Therefore, the PD model can be used to simulate a communication system of a home appliance.

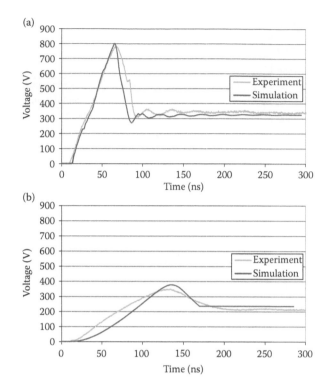

FIGURE 6.28
Effect of grounding on V_R for the current source case: (a) individual grounding and (b) the proposed method.

An experiment was conducted to investigate the characteristics of an LS as well. At the steady-state voltage, the common grounding and the proposed method can reduce the lightning voltage equally well. However, at maximum voltage, the proposed method can reduce the lightning overvoltage better than common grounding can. Therefore, the proposed method is well suited for protecting communication appliances. The FDTD simulation results agree well with the measured results. The FDTD simulation can thus be used to study LS characteristics from a communication line.

6.4 Analytical Method of Solving Induced Voltages and Currents

6.4.1 Introduction

Numerical simulation software is widely used by engineers, researchers, and university students and is a powerful tool for solving various problems. However, users of the software are frequently not well versed in it. Physically nonexistent input data may provide erroneous results, which can go unnoticed if the user does not understand practical problems that may arise. This lack of knowledge is partially due to the complexity of the software, which is highly advanced and does not require an understanding of the underlying mechanisms. Furthermore, the software technology is too intricate and complex to be easily explained by physical and engineering theories.

This section discusses the analytical formulation and the investigation of induced voltages and currents in conductors such as telephone lines and pipelines from power lines. There are many studies that discuss the induced voltages and currents: Carson and Sunde [35,36] are pioneers in this field, especially of induced voltages in telephone lines. Taflov and Dabkowski [37] developed an analytical method to predict the induced voltages in a buried pipeline based on the reflection coefficient method (details of this theory are described in References 38 and 39). This method was adopted in the CIGRE Guide [39]. There are many textbooks explaining the theory of electromagnetic coupling [40–45]. Additionally, many publications describe numerical simulations of voltages and currents through either a circuit-theory-based simulation tool, such as the EMTP, or a recent electromagnetic analysis method such as the finite-element method and the FDTD method [31,46–60]. Ametani [45] and Christoforidis et al. [58] have proposed an approach that combines numerical electromagnetic analysis with the circuit-theory-based simulation tool to calculate the impedance and admittance of a given circuit and the induced voltages and currents in the circuit. This hybrid approach is able to solve problems in which the impedance and admittance are either unknown or difficult to obtain and allows researchers to consider the physical dimensions of the phenomenon.

This section describes an analytical method of calculating the induced voltages and currents in a complex induced circuit, such as a cascaded pipeline with several power lines, based on a conventional four-terminal parameter (F-parameter) formulation. The F-parameter formulation itself is well known, so writing a theoretical formula of the F-parameter for a multiphase circuit [38,45] is relatively straightforward. Thus, calculation of the induced voltages and currents requires a computer, specifically software such as the EMTP. The method explained in this section replaces the multiphase F-parameter with a single-phase parameter by introducing an artificially induced current. The method is applied to a cascaded pipeline where the circuit parameters, the induced currents, and the boundary conditions are different in each section of the pipeline. The basic characteristics of the induced voltage and current distribution along the pipeline are explained based on the analytical results. The calculated results are compared with the EMTP simulations from Taflov and Dabkowski [37], the CIGRE Guide [39], and field results [60].

6.4.2 *F*-Parameter Formulation for Induced Voltages and Currents

6.4.2.1 *Formulation of F-Parameter*

Two basic approaches exist to handle electromagnetic induction from a power line (the inducing circuit) to a pipeline (the induced circuit) [35,36,40–45]. The first approach represents the inducing and induced circuits as a multiphase line system. The alternative method considers the induction from the inducing circuit as a voltage source [35,37] or as a current source [61] so that the system becomes a single-phase circuit source approach. As noted in Reference 46, a multiphase line approach becomes too complicated; one solution is to apply an *F*-parameter formulation to the current source [45,62]. The *F*-parameter can easily handle the cascaded line that appears frequently in real pipelines during electromagnetic induction.

In a distributed parameter line composed of a power line and a pipeline that are parallel to each other (as illustrated in Figure 6.29), the following solution for voltage V_x and current distance x from the sending end of the pipeline in Appendices 6A.2 and 6A.3 is obtained:

$$\left.\begin{array}{l} V_x = \cosh(\Gamma \cdot x) \cdot V_1 - V_0 \cdot \sinh(\Gamma \cdot x) \cdot (I_1 - I_0) = A_x \cdot V_1 - B_x \cdot (I_1 - I_0) \\ I_x - I_0 = -Y_0 \sinh(\Gamma \cdot x) \cdot V_1 + \cosh(\Gamma \cdot x) \cdot (I_1 - I_0) = -C_x \cdot V_1 + D_x \cdot (I_1 - I_0) \end{array}\right\} \quad (6.2)$$

where

$$I_0 = \frac{E}{F} = -z_m \frac{I_p}{z} \text{ is the artificially induced current} \quad (6.3)$$

$\Gamma = \sqrt{z \cdot y}$ is the propagation constant

$Y_0 = \sqrt{y/z} = 1/Z_0$ is the characteristic admittance

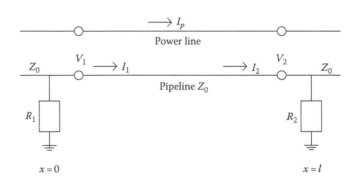

FIGURE 6.29
A pipeline parallel to a power line for $0 \leq x \leq l$.

z is the series impedance of the pipeline (Ω/m)

z_m is the mutual impedance between the power line and the pipeline (Ω/m)

y is the shunt admittance of the pipeline (S/m)

$E = -z_m I_p$ is the electromagnetically induced voltage (V/m)

I_p is the inducing current = power line current (A)

The following equation is obtained when written as a matrix:

$$\begin{pmatrix} V_x \\ I_x - I_0 \end{pmatrix} = \begin{bmatrix} A_x & -B_x \\ -C_x & D_x \end{bmatrix} \begin{pmatrix} V_1 \\ I_1 - I_0 \end{pmatrix} \tag{6.4}$$

where

$A_x = D_x = \cosh(\Gamma \cdot x)$

$B_x = Z_0 \sinh(\Gamma \cdot x)$

$C_x = Y_0 \sinh(\Gamma \cdot x)$

The following form is obtained when this equation is rewritten for $x = l$:

$$\begin{pmatrix} V_2 \\ I_2 - I_0 \end{pmatrix} = \begin{bmatrix} A & -B \\ -C & D \end{bmatrix} \begin{pmatrix} V_1 \\ I_1 - I_0 \end{pmatrix} \quad \text{or} \quad \begin{pmatrix} V_1 \\ I_1 - I_0 \end{pmatrix} = \begin{bmatrix} A & B \\ C & D \end{bmatrix} \begin{pmatrix} V_2 \\ I_2 - I_0 \end{pmatrix} \tag{6.5}$$

where

$$A = D = \cosh(\Gamma \cdot l), \quad B = Z_0 \sinh(\Gamma \cdot l), \quad C = Y_0 \sinh(\Gamma \cdot l) \tag{6.6}$$

This formulation is identical to the well-known *F*-parameter equation, except the current I is replaced by $I - I_0$ considering the induction from a power line.

6.4.2.2 Approximation of F-Parameters

The analytical evaluations of sinh and cosh functions in Equation 6.6 are difficult, and it is useful to adopt approximate *F*-parameters that can be easily evaluated by hand calculations. Similar to a conventional *F*-parameter approximation by assuming $|\Gamma \cdot l|$ to be much less than 1, the following result is given:

$$A = D \doteq 1, \quad B \doteq Z_0 \cdot \Gamma \cdot l = z \cdot l = z, \quad C \doteq 0 \quad \text{for } |\Gamma \cdot l| \ll 1 \tag{6.7}$$

6.4.2.3 Cascaded Connection of Pipelines

Assume a system consists of two cascaded sections of a pipeline as illustrated in Figure 6.30. In section 1, a pipeline is parallel to power line 1 with

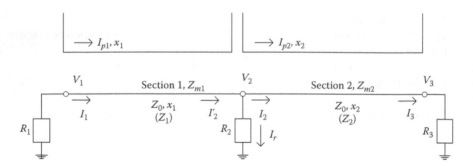

FIGURE 6.30
Cascaded connection of a pipeline.

current I_{p1} for the parallel length x_1. In section 2, the same pipeline is parallel to power line 2 with current I_{p2}. The mutual impedance is z_{m1} and z_{m2} in sections 1 and 2, respectively. Then, the following F-parameter equations are given:

$$\begin{pmatrix} V_1 \\ I_1 - I_{01} \end{pmatrix} = \begin{bmatrix} A_1 & B_1 \\ C_1 & D_1 \end{bmatrix} \begin{pmatrix} V_2 \\ I_2' - I_{01} \end{pmatrix}, \quad \begin{pmatrix} V_2 \\ I_2 - I_{02} \end{pmatrix} = \begin{bmatrix} A_2 & B_2 \\ C_2 & D_2 \end{bmatrix} \begin{pmatrix} V_3 \\ I_3 - I_{02} \end{pmatrix} \tag{6.8}$$

and

$$V_1 = -R_1 I_1, \quad V_2 = R_2(I_2' - I_2), \quad V_3 = R_3 I_3 \tag{6.9}$$

where
$I_{01} = E_1/z = z_{m1}, \, I_{P1}/z = Z_{m1} \, I_{P1}/Z_1, \, I_{02} = E_2/z = z_{m2}, \, I_{P2}/z = Z_{m2} \, I_{P2}/Z_2$
$Z_1 = z \cdot x_1, \, Z_{m1} = z_{m1} \cdot x_1, \, Z_2 = z \cdot x_2, \, Z_{m2} = z_{m2} \cdot x_2$
$A_i = \cosh(\Gamma \cdot x_i), \, B_i = Z_0 \cdot \sinh(\Gamma \cdot x_i), \, C_i = Y_0 \cdot \sinh(\Gamma \cdot x_i)$
$D_i = A_i; \, i = 1, 2$

Voltage V_i and current I_i in the pipeline are obtained by solving Equations 6.8 and 6.9.

6.4.3 Application Examples

6.4.3.1 Single Section Terminated by R_1 and R_2

Applying the same boundary conditions as Equation 6.9 to the circuit illustrated in Figure 6.31 produces the following relationship:

$$-R_1 I_1 = AR_2 I_2 + B(I_2 - I_0), \quad I_1 - I_0 = CR_2 I_2 + A(I_2 - I_0)$$

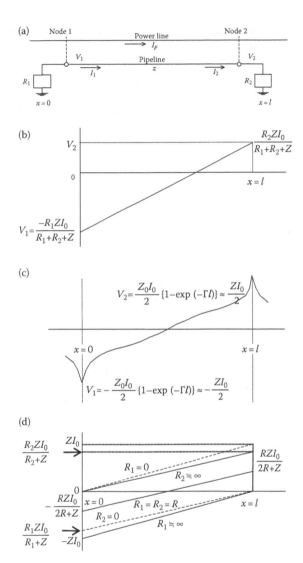

FIGURE 6.31
Voltage profile along a gas pipeline. (a) Circuit configuration, (b) $R_1 > R_2$, (c) $R_1 > R_2 = Z_0$, and (d) the effect of grounding resistance R_1 and R_2.

By solving these equations simultaneously, the following result is obtained:

$$
\left.
\begin{aligned}
&I_1 = \{B + R_2(A-1)\} I_0/K, \quad I_2 = \{B + R_1(A-1) I_0/K\} \\
&V_1 = -R_1 I_1, \quad V_2 = R_2 I_2 \\
&V_x = -(B_x + R_1 \cdot A_x) I_1 + B_x \cdot I_0, \quad I_x = (A_x + R_1 \cdot C_x) I_1 - (A_x - 1) I_0
\end{aligned}
\right\}
\tag{6.10}
$$

where $K = (R_1 + R_2) A + B + R_1 R_2 C$.

Substituting Equations 6.3 and 6.6 into these equations and rearranging results in the following:

$$
\left.
\begin{aligned}
V_x &= -Z_0 \sinh(\Gamma \cdot x) \cdot (I_1 - I_0) - R_1 \cosh(\Gamma \cdot x) \cdot I_1 \\
I_x &= \left\{ \cosh(\Gamma \cdot x) + \left(\frac{R_1}{Z_0} \right) \sinh(\Gamma \cdot x) \right\} I_1 - \{ \cosh(\Gamma \cdot x) - 1 \} I_0 \\
I_1 &= [Z_0 \sinh(\Gamma \cdot l) + R_1 \{ \cosh(\Gamma \cdot l) - 1 \}] \cdot \frac{I_0}{K} \\
I_2 &= [Z_0 \sinh(\Gamma \cdot l) + R_1 \{ \cosh(\Gamma \cdot l) - 1 \}] \cdot \frac{I_0}{K}
\end{aligned}
\right\}
\tag{6.11}
$$

where $K = (Z_0 + R_1 R_2 / Z_0) \sinh(\Gamma \cdot l) + (R_1 + R_2) \cosh(\Gamma \cdot l)$.

6.4.3.1.1 Approximation of V_x and I_x

It is hard to observe the characteristics of V_x and I_x based on Equation 6.11. By applying Equations 6.7 through 6.10, the following approximate solution is obtained:

$$
\left.
\begin{aligned}
V_x &= \{ (R_1 + R_2) \cdot z \cdot x - R_1 Z \} I_0 / (R_1 + R_2 + Z) \\
I_x &= I_1 = I_2 = Z I_0 / (R_1 + R_2 + Z)(18) \\
&\text{for } R_1 \neq 0, \quad R_2 \neq 0
\end{aligned}
\right\}
\tag{6.12}
$$

where $z \cdot l = Z$ (Ω), z (Ω/m), l (m).

The voltage along the pipeline in Figure 6.31a is drawn as in Figure 6.31b, where

$$
V_1 = \frac{-R_1 Z I_0}{(R_1 + R_2 + Z)}
$$

$$
V_2 = \frac{R_2 Z I_0}{(R_1 + R_2 + Z)}
\tag{6.13}
$$

$$
V_{\max} = |V_1| \quad \text{for } R_1 > R_2, \quad V_{\max} = V_2 \quad \text{for } R_2 > R_1
$$

It should be noted that the voltage profile along the pipeline appears to be linear due to the approximation given in Equation 6.7. If Equation 6.11 is used, the profile becomes nonlinear, as in Figure 6.31c, due to the nature of hyperbolic functions or exponential functions; however, it is continuous in the region of $0 \leq x \leq 1$. Thus, a position x exists where the voltage of the pipeline becomes zero (as shown in Figure 6.4 of Reference 37 or Figure 6.31b in this chapter). However, this is often forgotten by engineers. To understand the phase angle of the pipeline voltage (that is, the polarity at both ends), it

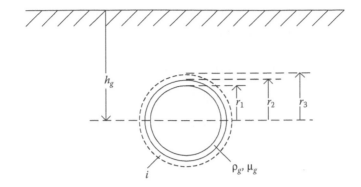

FIGURE 6.32
Configuration of a pipeline: $\rho_g = 1.5 \times 10^{-7}$ Ω-m, $\mu_g = 280$, $\varepsilon_i = 2.3$, $\rho_e = 50$ Ω-m, $r_1 = 19.13$ cm, $r_2 = 20.32$ cm, $r_3 = 20.64$ cm, and $h_g = -1.8$ m.

may be helpful to consider the physical nature of the induced voltage and how it relates to the theoretical analysis. However, this practice is conventionally neglected.

The applicable range of Equation 6.7 needs to be discussed. Figure 6.32 illustrates a typical gas pipeline used in Japan [59,60]. Its impedance and admittance can be easily evaluated by hand based on approximate formulas [45,59,63], which are given in Appendix 6A.4, and are provided here as

$$z = 0.613\angle 78.6° \,(\Omega/\text{km}), \quad y = 2.572\angle 90° \,(\text{ms}/\text{km})$$

$\Gamma \cdot l$ is evaluated for $l = 10$ km as

$$|\Gamma|\cdot l = \sqrt{z \cdot y} \cdot l = 0.03902 \ll 1 \tag{6.14}$$

It is clear that the approximation of $|\Gamma \cdot l| \ll 1$ is generally valid up to several kilometers. If a pipeline is long (around 10 km), then it is divided into sections so that each section satisfies this condition. In practice, the pipeline is divided into many sections corresponding to different power lines and groundings.

6.4.3.1.2 $R_1 = R_2 = Z_0$

$R_1 = R_2 = Z_0$ corresponds to the matching condition ($Z_0 =$ characteristic impedance) of a pipeline and is identical to the semi-infinite pipeline connected at nodes 1 and 2. The power line is parallel to the pipeline only for the section between the nodes. Although the CIGRE Guide [39] states that the pipeline extends several kilometers beyond the parallel route, these additional kilometers should be replaced by a semi-infinite pipeline or a length l of the parallel route that is much shorter than a few kilometers. With $R_1 = R_2 = Z_0$, Equation 6.11 is simplified as follows:

$$V_x = \left(\frac{Z_0 I_0}{2}\right)\left[\exp\{-\Gamma(l-x)-\exp(-\Gamma \cdot x)\}\right] \quad \text{for } 0 \le x \le 1$$

$$V_x = V_{max} \cdot \exp(-\gamma \cdot x) \quad \text{for } x\langle 0, x\rangle l$$

$$I_x = \left(\frac{I_0}{2}\right)[2-\exp\{-\Gamma(l-x)\}-\exp(-\Gamma \cdot x)]$$

$$V_{max} = -V_1 = V_2 = \left(\frac{Z_0 I_0}{2}\right) \cdot \{1-\exp(-\Gamma \cdot l)\}, \quad I_1 = I_2 = -\left(\frac{I_0}{2}\right)\{1-\exp(-\Gamma \cdot l)\}$$

The voltage profile along the pipeline is shown in Figure 6.31c. Equation 6.11 provides the same results as those in Reference 37 and the CIGRE Guide [39]. However, the guide does not deal with the effect of grounding resistances of a pipeline. These results are simplified by using the approximation in Equation 6.7 or $\exp(-\Gamma \cdot x) \fallingdotseq 1 - \Gamma \cdot x$ as

$$V_x = -\left(\frac{l}{2-x}\right)z I_0, \quad I_x = \frac{Z I_0}{2 Z_0}, \quad -V_1 = V_2 = \frac{Z I_0}{2}$$

$$V_{max} = \left|\frac{Z I_0}{2}\right|$$

It should be noted that the maximum induced voltage in this approximation is nearly the same as that of the accurate solution for $\exp(-\Gamma \cdot l) \ll 1$. The same assumption was also adopted in Reference 37 for a short pipeline and $\exp(-\Gamma \cdot l) \cong 0.1$ for a long line so as to make analysis possible.

6.4.3.1.3 Effect of Grounding Resistances of a Pipeline

The formulas in Table 6.13 for various grounding resistances are derived from the general solution of Equation 6.11. The voltages in Tables 6.13 and 6.14 are normalized by I_0.

TABLE 6.13

Voltages and Currents for Various Grounding Resistances: Exact Solution

R_1	R_2	$-V_x/I_0$	I_x/I_0
0	0	0	1
0	R	$-(Z_0 R/K_1) \cdot \sinh(\theta)$	$1 - (R/K_1)\cosh(\theta)$
R	0	$(Z_0 R/K_1) \cdot \sinh(\theta_0 - \theta)$	$1 - (R/K_1)\cosh(\theta_0 - \theta)$
R	R	$(2 Z_0 R K_4/K_3) \cdot \sinh(\theta_0/2 - \theta)$	$1 - (2 R K_4/K_3)\cosh(\theta_0/2 - \theta)$
∞	R	$[Z_0 \sinh(\theta_0-\theta) - R\{\cosh(\theta) - \cosh(\theta_0 - \theta)\}]Z_0/K_2$	$I_2 = (Z_0/K_2)\{\cosh(\theta_0) - 1\}, I_1 = 0$
R	∞	$[Z_0 \sinh(\theta) - R\{\cosh(\theta) - \cosh(\theta_0 - \theta)\}]Z_0/K_2$	$I_1 = (Z_0/K_2)\{\cosh(\theta_0) - 1\}, I_2 = 0$

$K_1 = Z_0 \sinh(\theta_0) + R \cosh(\theta_0)$, $K_2 = Z_0 \cosh(\theta_0) + R \sinh(\theta_0)$, $K_3 = Z_0 K_1 + R K_2$, $\theta_0 = \Gamma l$, $\theta = \Gamma x$
$K_4 = Z_0 \cosh(\theta_0/2) + R \sinh(\theta_0/2)$.

TABLE 6.14

Voltages and Currents for Various Grounding Resistances: Approximate Solution

R_1	R_2	$-V_x/I_0$	$-V_1/I_0$	$-V_2/I_0$	V_{max}/I_0
0	R	$-Rzx/(R+Z)$	0	$-RZ/(R+Z)$	$RZ/(R+Z)$
R	0	$Rz(l-x)/(R+Z)$	$RZ/(R+Z)$	0	$RZ/(R+Z)$
R	R	$2Rz(l/2-x)/(2R+Z)$	$RZ/(2R+Z)$	$-RZ/(2R+Z)$	$RZ/(2R+Z)$
∞	R	$Z(l-x)$	Z	0	Z
R	∞	$-z\,x$	0	$-Z$	Z
∞	∞	$Z(l/2-x)$	$Z/2$	$-Z/2$	$Z/2$

It is difficult to evaluate the voltage and current in a pipeline from these formulas without numerical calculations. By applying the approximation in Equation 6.7, the results listed in Table 6.14 and maximum voltage V_{max} are obtained. The analytical results from Table 6.14 are shown in Figure 6.31d.

It is clear from the table and Figure 6.31 that the severest voltage, ZI_0, appears when one end of a pipeline is open-circuited while the other end is grounded by a resistance R that is neither zero nor infinite. Considering $R = 0$ is practically impossible, both ends of the pipeline should be grounded so that the maximum voltage becomes less than half of the severest case, as provided by the ratio $R_2/(R_1 + R_2 + Z)$. If the pipeline length l is greater than 10 km, the ratio becomes less than one-third with $R_1 = R_2 \fallingdotseq 10\ \Omega$.

6.4.3.2 Two-Cascaded Sections of a Pipeline (Problem 6.1)

The induced voltage and current in the system in Figure 6.30 can be obtained by solving the equations in Equations 6.8 and 6.9 simultaneously. The solution is given in Appendix 6A.4. By using the approximation in Equation 6.7, Equation 6.8 is simplified, and the following equations are given in the system in Figure 6.31:

$$
\left.\begin{array}{l}
\begin{pmatrix} V_1 \\ I_1 - I_{01} \end{pmatrix} = \begin{bmatrix} 1 & 1 \\ 0 & 1 \end{bmatrix} \begin{pmatrix} V_2 \\ I_2' - I_{01} \end{pmatrix}, \quad
\begin{pmatrix} V_1 \\ I_2 - I_{02} \end{pmatrix} = \begin{bmatrix} 1 & Z_2 \\ 0 & 1 \end{bmatrix} \begin{pmatrix} V_3 \\ I_3 - I_{02} \end{pmatrix} \\[14pt]
\begin{pmatrix} V_2 \\ I_2' \end{pmatrix} = \begin{bmatrix} 1 & 0 \\ 1/R_2 & 1 \end{bmatrix} \begin{pmatrix} V_2 \\ I_2 \end{pmatrix}
\end{array}\right\} \tag{6.15}
$$

where $Z_1 = z \cdot x_1$, $Z_2 = z \cdot x_2$.

By solving Equations 6.15 and 6.9, the following results are easily obtained:

$$
\left.\begin{array}{ll}
I_1 = I_2' = \{(R_2 + R_3 + Z_2)Z_1 I_{01} + R_2 Z_2 I_{02}\}/K_1, & V_1 = -R_1 I_1 \\[4pt]
I_2 = I_3 = \{R_2 Z_1 I_{01} + (R_1 + R_2 + Z_1)Z_2 I_{02}\}K_1, & V_3 = R_3 I_3 \\[4pt]
I_r = I_2' - I_2 = \{(R_3 + Z_2)Z_1 I_{01} - (R_1 + Z_1)Z_2 I_{02}\}/K_1, & V_2 = R_2 I_r
\end{array}\right\} \tag{6.16}
$$

where $K_1 = R_1 R_2 + R_2 R_3 + R_3 R_1 + R_3 Z_1 + R_1 Z_2 + R_2 Z + Z_1 Z_2$, $Z = Z_1 + Z_2$.

One of the reasons for developing the F-parameter approach for cascaded connections of pipeline sections is to find effective grounding resistances and separation distances between them. Grounding resistance and separation distance play important roles when it comes to Japanese gas pipelines. Based on the approximate results, the characteristics of induced voltages and currents and the effect of grounding resistances can be discussed.

1. $I_{01} = I_{02} = I_0$: Equation 6.16 is further simplified as follows:

$$\left.\begin{array}{l} I_1 = I_2' = \{(R_3 + Z_2)Z_1 + R_2 Z\} I_0/K_1 \\ I_2 = I_3 = \{(R_1 + Z_1)Z_2 + R_2 Z\} I_0/K_1 \\ I_r = (R_3 Z_1 - R_1 Z_2) I_0/K_1 \end{array}\right\} \tag{6.17}$$

a. $R_2 = \infty$

$$I_1 = I_2 = I_3 = \frac{Z I_0}{(R_1 + R_3 + Z)}$$

$$V_1 = \frac{-R_1 Z I_0}{(R_1 + R_3 + Z)}, \quad V_3 = \frac{R_3 Z I_0}{(R_1 + R_3 + Z)}$$

$$V_2 = \frac{(R_3 Z_1 - R_1 Z_2) I_0}{(R_1 + R_3 + Z)}$$

This result is the same as Equation 6.13 for a single-section pipeline, considering that node 2 in Equation 6.13 now becomes node 3.

b. $R_3 = \infty$

$$I_1 = \frac{Z_1 I_0}{(R_1 + R_2 + Z_1)} = I_r, \quad I_2 = I_3 = 0$$

$$V_1 = \frac{-R_1 Z_1 I_0}{(R_1 + R_2 + Z_1)}$$

$$V_2 = \frac{R_2 Z_1 I_0}{(R_1 + R_2 + Z_1)}, \quad V_3 = Z I_0$$

These results show similar characteristics to those in Section 6.4.3.1.3 as far as V_1 and V_3 are concerned.

Equation 6.17 corresponds to a case of grounding an intermediate position of a pipeline, both ends of which are grounded through resistances R_1 and R_3. Thus, the effect of the grounding resistance R_2 on an induced voltage can be discussed in comparison to no grounding, that is,

$$R_2 = \infty: V_3 = -V_1 = \frac{RZI_0}{(2R+Z)} = V_{max}$$

The assumption of $R_1 = R_2 = R_3 = R$ leads to the following results:

$$R_2 \neq \infty: V_1 = -R\{R(Z_1 + Z) + Z_1 Z_2\}I_0/K_2$$

$$V_3 = -R\{R(Z_2 + Z) + Z_1 Z_2\} I_0/K_2 \quad K_2 = 3R^2 + 2RZ + Z_1 Z_2$$

Furthermore, assume that $x_1 = x_2 = l/2$, then

$$V_3 = -V_1 = \frac{RZI_0}{(2R+Z)}, \quad V_2 = 0 \quad \text{for } R_2 \neq \infty$$

This result is the same as that of a single-pipeline section in Section 6.4.3.1. This indicates that a grounding resistance (R_2 in Figure 6.30) at the middle of a pipeline is not effective enough to reduce the induced voltage if the circuit parameter and inducing current are constant; the voltage at the center of the pipeline $V(l/2)$ is zero when $R_1 = R_3 = R$. Even in the case of $R_1 \neq R_3$, the grounding resistance R_2 shows no reduction of the maximum voltage.

2. $I_{02} = 0$:

$$
\left.
\begin{aligned}
I_1 &= \frac{(R_2 + R_3 + Z_2)Z_1 I_0}{K_1}, \quad I_r = \frac{(R_3 + Z_2)Z_1 I_0}{K_1} \\
I_2 &= I_3 = \frac{R_2 Z_1 I_0}{K_1} \\
V_1 &= -R_1 I_1, \quad V_2 = R_2 I_r, \quad V_3 = R_3 I_3
\end{aligned}
\right\}
\tag{6.18}
$$

This case is a more general condition of Section 6.4.3.1 and the calculation examples discussed in the CIGRE Guide [39], in which the pipeline extends beyond the zone of influence but no grounding resistance is considered. Figure 6.33 shows the system diagram and analytical results of the induced voltages. Figure 6.33b is the result of Equation 6.18, where it should be noted that $V_2 > V_3$.

a. $R_2 = \infty: V_1 = \dfrac{-R_1 Z_1 I_0}{(R_1 + R_3 + Z)}$

$$V_2 = \frac{(R_3 + R_2)Z_1 I_0}{(R_1 + R_3 + Z)}$$

$$V_3 = \frac{R_3 Z_1 I_0}{(R_1 + R_3 + Z)}$$

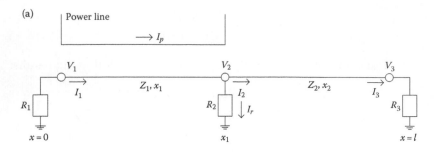

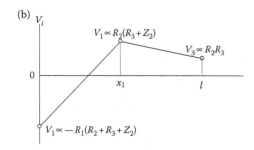

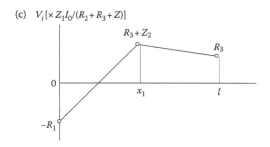

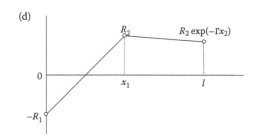

FIGURE 6.33
Case of $I_{02} = 0$. (a) System diagram, (b) $R_1 \neq R_2 \neq R_3 \neq 0$, (c) $R_2 = \infty$, and (d) $R_3 = \infty$.

b. $R_2 = \infty: V_1 = \dfrac{-R_1 Z_1 I_0}{(R_1 + R_2 + Z_1)}$

$$V_2 = V_3 = \dfrac{R_2 Z_1 I_0}{(R_1 + R_2 + Z_1)}$$

As explained in Section 6.4.3.1.1 and Figure 6.31c, all of the curves shown in Figure 6.33 are exponential or hyperbolic, not linear; therefore, V_3 should be expressed as

$$V_3 = V_2 \exp(-\Gamma x) \quad \text{for } x > x_1 \text{ in the case of } R_3 = \infty$$

The result of $R_2 = R_3 = \infty$ clearly equals that in Figure 6.31c if $R_1 = Z_0$, which corresponds to a semi-infinite pipeline to the left of node 1.

3. $I_{01} \neq I_{02}$: When I_{01} is much greater than I_{02}, results similar to those for $I_{02} = 0$ are obtained. Figure 6.34 shows analytical results in the case of (a) $I_{01} = 0.1$ $I_{02} = 0.5 I_0$ and (b) $I_{01} = 0.5 I_{02}$ under the condition that $x_1 = x_2 = 1/2$, that is, $Z_1 = Z_2 = Z/2$. Characteristics of the induced voltage that are similar to those explained for $I_{01} = I_{02} = I_0$ are observed when $I_{01} = 0.1 I_{02}$, that is, when I_{02} is far greater than I_{01}. On the contrary, the induced voltage characteristics for $I_{02} > I_{01} > 0.5 I_{01}$ are significantly different from those when I_{01} or I_{02} is zero, as shown in Figure 6.33.

    The proposed approach has proved effective in observing basic and qualitative characteristics of an induced voltage. It is easy to produce a computer code based on the proposed approach that can define the quantitative characteristics of the induced voltage.

4. *Effect of I_{01} relative to $I_{02} = I_0$:* Assume that $R_1 = R_2 = R_3 = R$, $x_1 = x_2$, and $I_{01} = m I_{02}$, where $m < 1$. Then, the following results of induced voltages V_1 to V_3 are derived from Equation 6.16:

$$\left. \begin{aligned} V_1 &= -R\{2R(1 + 2m) + mZ\} Z I_0 / (6R + Z)(2R + Z) \\ V_2 &= -R(1 - m) Z I_0 / (6R + Z) \\ V_3 &= R\{2R(2 + m) + Z\} Z I_0 / (6R + Z)(2R + Z) \end{aligned} \right\} \quad (6.19)$$

These analytical results are illustrated in Figure 6.34c. The result for $m = 0$ is the same as that in Equation 6.18, and that for $m = 1$ is identical to that in Figure 6.38. It is reasonable that the induced voltages $|V_1|$ and V_3 increase as the ratio m increases. On the contrary, the voltage V_2 reaches zero, corresponding to the characteristics explained in Section 6.4.3.1. The simplified formula in Equation 6.19 is useful for investigating the effect of inducing currents from many power lines that are observed quite often.

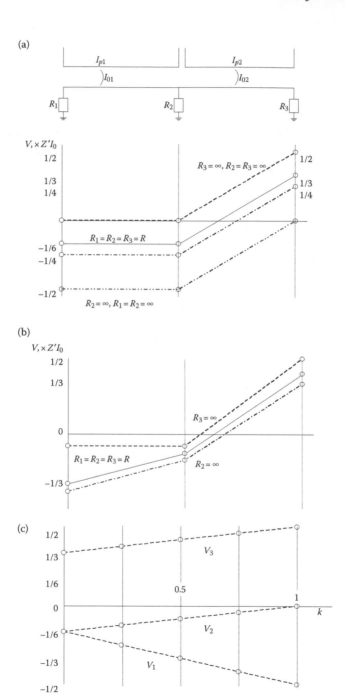

FIGURE 6.34
Case of $I_{01} \neq I_{02} = I_0$. (a) $I_{01} = 0.1I_{02}$, (b) $I_{01} = 0.5I_{02}$, and (c) the effect of I_{01}, relative to $I_{01}I_{02} = mI_{02}$.

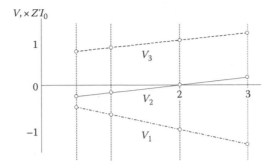

FIGURE 6.35
Effect of x_1 relative to x_2. $Z_1 = nZ_2 = nZ'$, $Z = (n + 1)Z'$.

5. *Effect of x_1 relative to x_2:* Assuming that $R_1 = R_2 = R_3 = R$, $I_{01} = 0.5I_{02} = 0.5I_0$, and $x_1 = nx_2$, that is, $Z_1 = nZ_2 = nZ'$, the following results are obtained from Equation 6.16:

$$
\left.
\begin{aligned}
V_1 &= -R\{(1+n)R + nZ'/2\}\,Z'I_0/K_2 \\
V_2 &= -R\{(1-n/2)R + nZ'/2\}\,Z'I_0/K_2 \\
V &= R\{(2+n/2)R + nZ'\}\,Z'I_0/K_2
\end{aligned}
\right\}
\tag{6.20}
$$

where $K_2 = 3R^2 + 2R(n + 1)Z' + n^2Z'$.

This analytical result is illustrated in Figure 6.35. It is clear from Equation 6.20 and Figure 6.35 that the induced voltages are linearly proportional to the length of x_1. Therefore, the induced voltages are proportional to the total mutual impedance $Z_{m1} = z_{m1} \cdot x_1$.

6.4.3.3 Three-Cascaded Sections of a Pipeline (Problem 6.1)

Let us now obtain currents $I_1, I_2, I_3, I_{r1}, I_{r2}$ and voltage V_1 in Figure 6.36 and draw voltage profiles along the line for (a) $I_{02} = 0$, $I_{01} = I_{03} = I_0$, $R_1 = R_2 = R_3 = R_4 = R$ and (b) $R_2 = R_3 = \infty$, $I_{01} > I_{02} > I_{03}$.

Figure 6.36 illustrates three-cascaded sections of a pipeline where each section is parallel to a different power line and has an induced current

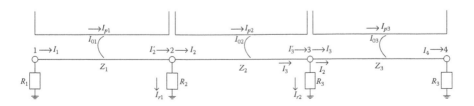

FIGURE 6.36
Three-cascaded sections of a pipeline. $Z_1 = z \cdot x_1 = Z'$, $Z_2 = z \cdot x_2$, and $Z_3 = z \cdot x_3$.

I_{0i} (i = 1–3). In this case, we use the approximation in Equation 6.7 to easily obtain the solution of the induced voltages and currents on each section. Based on the equations in Equation 6.15 and two more equations for nodes 3 and 4, the solutions of the currents and voltages can also be obtained.

6.4.4 Comparison with a Field Test Result

6.4.4.1 Comparison with EMTP Simulations

Table 6.15 shows a comparison of analytical results evaluated by an accurate formula from Equation 6.11 and by an approximate formula from Equation 6.13 to determine whether to include (power line) current $I_p = 1000\angle 0°$, separation distance y = 50–500 m, and pipeline lengths x = 1 and 10 km in Figure 6.31a. Grounding resistances R_1 and R_2 vary from zero to infinity, which corresponds to the highest induced voltage on the pipeline. The highest voltage is necessary to investigate the effect of pipe grounding. The cross section of the pipeline is $z = 0.613\angle 78.6°$ (Ω/km), and the mutual impedance to the power line is $Z_m = 0.0592\angle 49.9°$ for y = 500 m. Thus, the artificially induced current in this case is evaluated as $I_0 = 96.57\angle 28.7°$ (A/km) by Equation 6.4.

The analytical results show a maximum error of less than 5% in comparison with the EMTP simulation results in Table 6.15, and thus the accuracy of the analytical formula is satisfactory.

Table 6.16 shows a comparison of the approximate formula and the EMTP simulation results for the three-cascaded section of a pipeline in Figure 6.36 with a separation distance of y = 500 m. Case 10 in the table corresponds to Figure 6.37a, and case 32 corresponds to Figure 6.37b. The maximum error of the analytical results is 3.3% in case 30.

From this observation, the accuracy of the analytical method proposed in this chapter is satisfactory, and thus the method is useful in practice. It should be noted that the node voltages in Tables 6.15 and 6.16 are the maximum voltages on the pipeline, and thus the maximum voltage is accurately calculated even with the approximation.

6.4.4.2 Field Test Results

The field test results of induced voltages in an underground gas pipeline in Japan are given in Reference 60. Figure 6.38a illustrates the system configuration. The solid line represents the pipeline, and the dotted line represents an overhead transmission line. Figure 6.38b shows the configuration of a 500 kV vertical twin-circuit line. The system is simplified by five cascaded sections, as shown in Figure 6.38c. In a section where the separation distance y between the power line and the pipeline exceeds 300 m, the power line is neglected. Thus, the power line and its inducing current are considered only in sections 2 and 4. Both ends of the pipeline (nodes 1 and 6) are grounded by $R_1 = R_2$. The pipeline cross section is given in Figure 6.32. The 500 kV power

TABLE 6.15

Comparison with EMTP Simulation Results: Single Section ($I_p = 1000\angle 0°$ A)

				Accurate Formula				Approximate Formula				EMTP			
x (km)	y (m)	R_1 (Ω)	R_2 (Ω)	V_1	θ_1	V_2	θ_2	V_1	θ_1	V_2	θ_2	V_1	θ_1	V_2	θ_2
1	50	10	10	95.8	73.7	95.8	−106.3	95.8	73.7	95.8	−106.3	92.9	79.0	92.9	−101.0
1	100	10	10	75.8	69.8	75.8	−110.2	75.8	69.8	75.8	−110.2	73.5	75.2	73.5	−104.8
1	200	10	10	55.8	63.3	55.8	−116.7	55.9	63.3	55.9	−116.7	53.8	68.9	53.8	−111.1
1	500	10	10	30.6	45.4	30.6	−134.6	30.6	46.1	30.6	−133.9	29.4	53.8	29.4	−126.2
10	500	10	10	281.9	31.7	281.9	−148.3	279.1	32.0	279.1	−148.0	277.5	38.8	277.5	−141.2
1	500	10	∞	0.79	136.3	61.6	−132.2	0		61.5	−132.2	0.76	138.4	58.9	−125.2
1	500	∞	10	61.6	47.8	0.79	−43.7	61.5	47.8	0		58.9	54.8	0.76	−41.6
1	500	∞	∞	30.7	47.8	30.7	−132.2	30.8	47.8	30.8	−132.2	29.5	55.6	29.5	−124.4
1	500	0	10	0		60.7	−135.6	0		60.7	−135.6	0.0	52.8	58.5	−127.9

TABLE 6.16

Comparison with EMTP Simulation Results: Three Sections ($y = 500$ m)

Case No.	$I_{P1}/I_{P2}/I_{P3}$ (A)	$x_1/x_2/x_3$ (km)	$R_1/R_2/R_3/R_4$ (Ω)	Approximate Formula								EMTP Result							
				V_1	θ_1	V_2	θ_2	V_3	θ_3	V_4	θ_4	V_1	θ_1	V_2	θ_2	V_3	θ_3	V_4	θ_4
10	1000/0/1000	1/1/1	10	60.1	42.8	1.8	−60.3	1.8	119.7	60.1	−137.2	58.4	51.3	0.89	−44.6	0.89	135.4	58.4	−128.7
20	1000/100/1000	1/1/1	10	63.1	42.7	3.1	9.8	3.1	−170.2	63.1	−137.3	61.1	50.3	3.2	18.8	3.2	−161.2	61.1	−129.7
21	1000/100/1000	1/10/1	10	75.2	27.8	24.7	−14.4	24.7	165.6	75.2	−152.2	74.1	33.7	26.4	−7.7	26.4	172.3	74.1	−146.3
30	1000/500/1000	1/1/1	10	75.1	42.5	14.8	37.6	14.8	−142.4	75.1	−137.5	72.7	50.1	14.5	45.2	14.5	−134.8	72.7	−129.9
31	1000/500/1000	1/10/1	10	169.0	21.7	117.1	13.4	117.1	−166.6	169.0	−158.3	169.3	27.2	119.6	18.7	119.6	−161.3	169.3	−152.8
32	1000/500/1000	1/1/1	10/∞/∞10	75.2	42.8	14.9	39.3	14.9	−140.7	75.2	−137.2	72.8	50.4	14.5	46.9	14.5	−133.1	72.8	−129.6

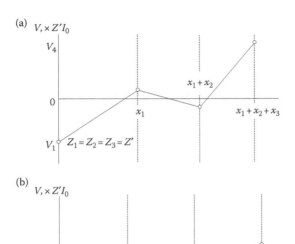

FIGURE 6.37
Solutions for the three-cascaded sections of a pipeline in Figure 6.36. (a) $I_{02} = 0$, $I_{01} = I_{03} = I_0$, $R_1 = R_2 = R_3 = R_4 = R$. (b) $R_2 = R_3 = \infty$, $I_{01} > I_{02} > I_{03}$.

line has a twin-circuit vertical configuration with two GWs (depicted in Figure 6.38b). Soil resistivity along the line ranges from 50 to 200 Ω-m.

As explained in Section 6.3, the node voltages are given in the following equation:

$$
\left.
\begin{aligned}
V_1 &= \frac{-R_1(Z_2 I_{01} + Z_4 I_{02})}{(R_1 + R_2 + Z)} \\
V_2 &= \frac{-(R_1 + Z_1)(Z_2 I_{01} + Z_4 I_{02})}{(R_1 + R_2 + Z)} \\
V_3 &= \frac{\{(R_2 + Z_3 + Z_4 + Z_5)Z_2 I_{01} - (R_1 + Z_1 + Z_2)Z_4 I_{02}\}}{(R_1 + R_2 + Z)} \\
V_4 &= \frac{\{(R_2 + Z_4 + Z_5)Z_2 I_{01} - (R_1 + Z_1 + Z_2 + Z_3)Z_4 I_{02}\}}{(R_1 + R_2 + Z)} \\
V_5 &= \frac{(R_2 + Z_5)(Z_2 I_{01} + Z_4 I_{02})}{(R_1 + R_2 + Z)} \\
V_6 &= \frac{R_2(Z_2 I_{01} + Z_4 I_{02})}{(R_1 + R_2 + Z)}
\end{aligned}
\right\}
\tag{6.21}
$$

where $Z = Z_1 + Z_2 + Z_3 + Z_4 + Z_5$.

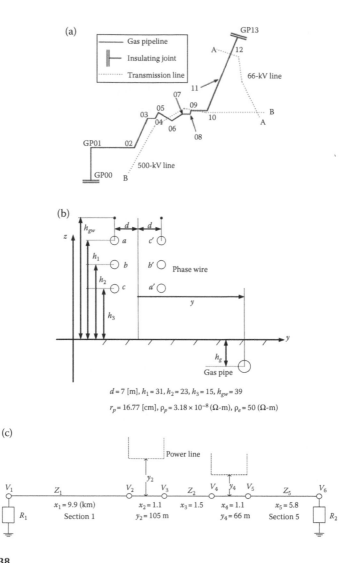

FIGURE 6.38
System of the configuration of a field measurement. (a) Field test circuit, (b) vertical twin-circuit 500-kV line, and (c) model circuit for an analytical calculation.

In Equation 6.21, as in Equation 6.3, the original induced currents I_{01} and I_{02} are given by the induced voltage E. The original induced voltage E is determined by the vector sum of the induced voltages due to the phase currents of the transmission line [60], or approximately by the zero-sequence current of a transmission line and the GW currents (as explained in the CIGRE Guide) [39]. The mutual impedance between a transmission line and a pipeline is calculated either numerically by the EMTP cable parameters [63] or analytically by the approximation formulas in Appendix 6A.4. The following is the EMTP result for section 4 in Figure 6.38c, with a soil resistivity of 50 Ω-m, where the separation between the center of the 500 kV power line and the pipeline is 66 m. The mutual impedance estimated from the curve in Figure 3.4 of the CIGRE Guide [39], agrees well with the EMTP result:

$Z_{m1} = 0.1428\angle 71.7°(\Omega/\text{km})$: phase 1 to pipeline

$Z_{m2} = 0.1446\angle 71.7°, Z_{m3} = 0.1459\angle 71.0°, Z_{m4} = 0.1530\angle 72.2°, Z_{m5} = 0.1557\angle 72.4°$

$Z_{m6} = 0.11578\angle 72.4°, Z_{m7} = 0.1407\angle 71.0°$: GW 1 to pipeline, $Z_{m8} = 0.1498\angle 72.1°$

The current in each phase of the 500-kV line is 1000 A. The GW currents are calculated in the same manner as those for a pipeline. The original induced voltage E in Equation 6.3 is analytically calculated using mutual impedance in the following manner:

$$E_i = z_{mi} \cdot I_{P1i}$$

For example, E_1 is calculated as $E_1 = z_{m1} \cdot I_{P1} = 0.1428\angle 71.7° \times 1000 = 142.8 \angle 71.7°$ (V/km).

The total induced voltage E is evaluated as the vector sum of the voltages.

$$E = \sum_{i=1}^{8} E_i = 2.029\angle 54.5°(V/\text{km})$$

Then, the original induced current I_{02} in section 4 of Figure 6.38b is given by

$$I_{a2} = \frac{E}{z} = 3.427\angle -23.7°(A)$$

where $z = 0.592\angle 78.2°$ (Ω/km) is the self-impedance of the pipeline with $\rho_e = 50$ (Ω-m).

Similarly, current I_{01} in section 2 is calculated as

$$I_{01} = \frac{E}{Z} = 0.778\angle -106.6°(A)$$

Substituting I_{01} and I_{02} into Equation 6.22 yields the following results:

1. $R_1 = Z_0 = 15.1\ \Omega,\ R_2 = \infty\ V_1 = V_2 = 0,$
   $V_3 = V_4 = 0.5047\angle 151.6°$
   $V_5 = V_6 = 2.340\angle -137.9°$
2. $R_1 = R_2 = Z_0$: $V_6 = -V_1 = 1.026\angle 23.\ 0°,$
   $V_2 = -1.17\angle 42.4°,\ V_3 = -1.159\angle 68.9°$
   $V_4 = 1.2101\angle -109.6°,\ V_5 = 1.092\angle -144.9°$

The measured result was 1.17–2.48 V in the system shown in Figure 6.38a [60]. The analytical results agree with the measured result, even with the approximation explained previously. A number of distribution lines (6.6 kV, 3.3 kV) exist near the pipeline, which may have contributed to the discrepancy.

6.4.5 Concluding Remarks

An analytical method to calculate induced voltages and currents in a complex pipeline system by applying an F-parameter is described in this section. The approach makes handling the system parameters easy, as inducing currents and boundary conditions differ in each section of the pipeline system. The profile of the induced voltages and currents along the pipeline can be evaluated by hand calculations, especially if a well-known approximation of hyperbolic functions of the F-parameter is used. Even with the approximation, the maximum induced voltage is accurately calculated. Analytical test results were compared with EMTP simulation and field test results, and it was concluded that even manual calculation agrees well with the test results.

Analytical results of the induced voltages clearly show that the maximum voltage appears at one end of a pipeline when the other end is grounded, and at both ends when they are grounded by the same resistance (as discussed in various reports and textbooks [36,37,39,40,44,45]). Also, a position exists where the pipeline-to-ground voltage becomes zero. Engineers in the field should keep this in mind. The effect of grounding resistance and induced currents on the induced voltages has been explained based on the analytical results.

The approach based solely on knowledge of electrical circuit theory, as taught in university undergraduate courses, is expected to be very useful for qualitative and predictive analyses of induced voltages and currents.

SOLUTION TO PROBLEM 6.1

$$
\left.
\begin{aligned}
I_1 &= \frac{\left\{(B \cdot C - R_3^2)Z_1 I_{01} + CR_2 Z_2 I_{02} + R_2 R_3 Z_3 I_{03}\right\}}{K_1} = I_2' \\
I_2 &= \frac{(CR_2 Z_1 I_{01} + ACZ_2 I_{02} + AR_3 Z_3 I_{03})}{K_1} + I_3' \\
I_3 &= \frac{\left\{R_2 R_3 Z_1 I_{01} + AR_3 Z_2 I_{02} + (AB - R_2^2)Z_3 I_{03}\right\}}{K_1} = I_4 \\
I_{r1} &= \frac{\left[\left\{C(R_3 + Z_2) - R_3^2\right\}Z_1 I_{01} - C(R_1 + Z_1)Z_2 I_{02} - R_3(R_1 + Z_1)Z_3 I_{03}\right]}{K_1} \\
I_{r2} &= \frac{\left[R_2(R_4 + Z_3)Z_1 I_{01} + A(R_4 + Z_3)Z_2 I_{02} - \left\{A(R_2 + Z_2) - R_2^3\right\}Z_3 I_{03}\right]}{K_1} \\
V_1 &= -R_1 I_1, \quad V_2 = R_2 I_{r1}, \quad V_3 = R_3 I_{r2}, \quad V_4 = R_4 I_4
\end{aligned}
\right\}
\tag{6.22}
$$

where

$$K_1 = A(BC - R_3^2) - CR_2^2 = C(AB - R_2^2) - AR_3^2$$
$$A = (R_1 + R_2 + Z_1), \quad B = (R_2 + R_3 + Z_2), \quad C = (R_3 + R_4 + Z_3)$$

Based on this equation, various investigations into induced voltages and currents can be conducted. It is clear that the assumptions $I_{01} = I_{02} = I_{03} = I_0$, $R_2 = R_3 = \infty$, and $Z_1 + Z_2 + Z_3 = Z$ yields the same formula as Equation 6.13, and the assumptions $I_{01} = I_{02}$, $R_2 = \infty$, and $Z_1 + Z_2$ replaced by Z_2 provide results identical to Equation 6.16.

Some examples are demonstrated in Figure 6.36b and c.

1. $I_{02} = 0$, $I_{01} = I_{03} = I_0$: In Figure 6.37a, $-V_1 = V_4 \fallingdotseq -Z'I_0$ for $R \gg Z'$. When $I_{02} = 0$, a similar result to Figure 6.36b is obtained, even in the case of $R_2 = R_3 = \infty$ due to the finite length of section 2 (i.e., x_2) and the fact that the polarity of the voltage at node 2 induced by current I_{01} is opposite to that at node 3 induced by current I_{03}. This clearly indicates the significance of the induced voltage polarity that has been neglected in most conventional studies [39]. When x_2 is very long or semi-infinite, the inducing currents I_{P1} (correspondingly I_{01}) and I_{P3} (I_{03}) do not affect each other, and thus a well-known characteristic explained in Section 6.4.3.1 for the single-section case is observed.

2. $R_1 = R_4 = \infty$: Figure 6.37b shows the analytical result in the case of $R_1 = R_4 = \infty$ with $I_{01} = I_0 > I_{02} > I_{03}$. The gradient of the voltage profile along the x-axis in section i is proportional to current I_{0i}.

Appendix 6A

6A.1 Test Voltage for Low-Voltage Control Circuits in Power Stations and Substations [16]

EMC Test Voltages (Unit: kV)

Circuit Category[a]	AC Withstand Voltage		Lightning Impulse Withstand Voltage			Between Contacts/Coil Terminals DC/AC	Oscillatory Wave		Immunity To Earth		Surge Immunity		Rectangular Impulse Wave[b]	
	To Earth	Between Terminals	To Earth	Between Terminals	VT/CT		To Earth Circuit	Between Terminals	I/O Signal Source Circuit	Circuit	To Earth Circuit	Between Terminals	To Earth Terminals	Between Terminals
1	2	2	7	4.5	4.5									
2-1	2	–	7	3		3								
2-2	2	–	5	3		3		Same as IEC 0225-22-1, 4 and 5				Only in Japan[b]			
2-3	1.5	–	5	3		3								
3	2	–	3	3		3								
4	2	2	4	4.5	3		2.5	–	1	1	2	1	1	1
5	2	–	4	3		3	2.5	2.5	1	2	2	1	1	1
6	2	–	4	–		–	–	–	0.5	1	–	–	–	–
7-1	2	–	–	–		–	–	–	0.5	1	–	–	–	–

(*Continued*)

			EMC Test Voltages (Unit: kV)												
	AC Withstand Voltage		Lightning Impulse Withstand Voltage				Oscillatory Wave		Immunity		Surge Immunity		Rectangular Impulse Wave[b]		
			Between Contacts/Coil Terminals						To Earth						
Circuit Category[a]	To Earth	Between Terminals	DC/AC Circuit	VT/CT	To Earth	Between Terminals	To Earth	Between Terminals	I/O Signal Circuit	Source Circuit	To Earth	Between Terminals	To Earth	Between Terminals	
7-2	1.5	–	–	–	–	–	–	–	0.5	1	–	–	–	–	
8	–	–	–	–	–	–	–	–	–	–	–	–	–	–	

Note: Breakdown accidents in control boards have occurred even when the control circuits were tested and confirmed according to category 4, 4 kV to earth. The surge voltage at the equipment side is higher than that at the board side. As a result, category 2-1, 7 kV to earth, is applied considering the dielectric strength capability of the equipment [16].

[a] Category: Main-circuit side 1 → control box side 8 (greater the number, lower the voltage)
  Category 1: Secondary circuit and tertiary circuit of an instrument transformer (VT or CT), which is equipped with the main circuit.
  Category 2: Control circuit of a CB and a DS.
  Category 2-1: Control circuit in case of the high dielectric strength required.
  In this case, 7 kV is applied as the test voltage stated in the table.
  Category 2-2: Control circuit with surge reduction countermeasures or no possibility of an excessive LS in case high dielectric strength is required.
  In this case, 5 kV is applied as the test voltage.
  For example, if control cables with the metallic sheath (CVVS) are used, both ends/terminals are to be grounded.
  Category 4: Secondary circuit and tertiary circuit of VT/CT in a direct control board, a protective relay board, a remote supervisory control board, other control devices, etc.
  In this case, 4 kV is applied as the test voltage, as stated in Table 6.16.
[b] See References 2,3 and 16.

6A.2 Traveling-Wave Solution

In a distributed-parameter line system composed of a power transmission line and a pipeline that are parallel to each other (as illustrated in Figure 6.1), the following differential equation is given, assuming that current I_p along the power line is constant [37,39,40]:

$$\frac{-dv_x}{dx} = z \cdot I_x - E, \qquad \frac{-dI_x}{dx} = y \cdot V_x \qquad (6.23)$$

where
z is the series impedance of the pipeline (Ω/m)
z_m is the mutual impedance between the power line and the pipeline (Ω/m)
y is the shunt admittance of the pipeline (S/m)
$E = -z_m I_p$ is the electromagnetically induced voltage (V/m)
I_p is the inducing current = power line current (A)
$V_x = V(x)$, $I_x = I(x)$ is the pipeline voltage and current at position x

Similar to an ordinary distribution line, the following traveling-wave solutions for voltage and current are easily obtained [45]:

$$\left.\begin{array}{l} V_x = k_1 \cdot \exp(\Gamma \cdot x) + k_2 \cdot \exp(-\Gamma \cdot x) \\ I_x - I_0 = -Y_0 \{k_1 \cdot \exp(\Gamma \cdot x) - k_2(-\Gamma \cdot x)\} \end{array}\right\} \qquad (6.24)$$

where k_1 and k_2 are the constants to be determined by boundary conditions

$$I_0 = \frac{E}{z} = \frac{-z_m I_p}{z} : \text{artificial induced content} \qquad (6.25)$$

where
$\Gamma = \sqrt{(z \cdot y)}$ is the propagation constant
$Y_0 = \sqrt{(y/z)} = 1/Z_0$ is the characteristic admittance

6A.3 Boundary Conditions and Solutions of Voltage and Current

In general, the boundary conditions in Figure 6.29 are given by

$$x = 0 : V(x = 0) = V_1, I(x = 0) = I_1 \qquad (6.26)$$

$$x = l : V(x = 1) = V_2, I(x = 1) = I_2 \qquad (6.27)$$

Substituting Equation 6.26 into Equation 6.24, the unknown constants k_1 and k_2 are given as functions of V_1 and I_1.

$$k_1 = \frac{\{V_1 - Z_0(I_1 - I_0)\}}{2}$$

$$k_2 = \frac{\{V_1 - Z_0(I_1 - I_0)\}}{2}$$

(6.28)

Similarly, k_1 and k_2 are defined as functions of V_2 and I_2 by applying Equation 6.27.

When no pipeline exists to the left of node 1 and the right of node 2, Equations 6.26 and 6.27 are rewritten as

$$V(x = 0) = -R_1 \cdot I_1, \quad V(x = 1) = R_2 \cdot I_2$$

(6.29)

This condition leads to the following results of k_1 and k_2:

$$k_1 = \frac{\{(1 + T_2)\exp(\Gamma \cdot l) - (1 + T_1)T_2\}Z_0 \cdot I_0}{2k_3}$$

$$k_2 = \frac{\{(1 + T_2)T_1 \exp(\Gamma \cdot l) - (1 + T_1)\exp(2\Gamma \cdot l)\} \times Z_0 \cdot I_0}{2k_3}$$

(6.30)

where
$k_1 = \exp(2\Gamma l) - T_1 \cdot T_2$
$T_1 = (R_1 - Z_0)/(R_1 + Z_0)$ is the reflection coefficient to the left of node 1
$T_2 = (R_2 - Z_0)/(R_2 + Z_0)$ is the reflection coefficient to the left of node 2

Equation 6.30 agrees with those provided in Reference 37 and the CIGRE Guide [39], although errors exist in the derivation of the formula in the appendix of the guide.

By substituting Equation 6.28 into Equation 6.24 and rewriting as a function of V_1 and I_1, the following equation is obtained:

$$V_x = \cosh(\Gamma \cdot x) \cdot V_1 - Z_0 \cdot \sinh(\Gamma \cdot x) \cdot (I_1 - I_0) = A_x \cdot V_1 - B_x \cdot (I_1 - I_0)$$

$$I_x - I_0 = -Y_0\sinh(\Gamma \cdot x) \cdot V_1 + \cosh(\Gamma \cdot x) \cdot (I_1 - I_0) = -C_x \cdot V_1 + D_x \cdot (I_1 - I_0)$$

(6.31)

6A.4 Approximate Formulas for Impedance and Admittance

1. Conductor internal impedance: See Equation 1.7
2. Earth-return impedance of an underground cable [45,64]

$$Z_{eij} = j\omega \left(\frac{\mu_0}{2\pi}\right) \left\{ \ln\left(\frac{S_{ij}}{d_{ij}}\right) + \frac{2(h_i + h_j)}{3h_e} - 0.077 \right\} (\Omega/m)$$

(6.32)

where

$$S_{ij} = \sqrt{(h_i + h_j + 2h_e)^2 + y_{ij}^2}, \quad d_{ij} = \sqrt{(h_i - h_j)^2 + y_{ij}^2}$$
$$d_{ij} = r_1 = r_2$$

h_i, h_j are the buried distances between cable i and j
y_{ij} is the separation distance between cable i and j
$h_e = \sqrt{\rho_e / j\omega\mu_0}$ is the complex penetration depth
ρ_e is the soil resistivity
μ_e is the free-space permeability for earth

3. Mutual impedance between overhead and underground cables [65]

$$Z_m = j\omega\left(\frac{\mu_0}{2\pi}\right) \cdot \ln\left(\frac{S_{ij}^*}{d_{ij}^*}\right) (\Omega/m) \tag{6.33}$$

where $S_{ij}^* = \sqrt{(h_i - h_j + 2h_e)^2 + y_{ij}^2}, \quad d_{ij}^* = \sqrt{(h_i - h_j)^2 + y_{ij}^2}$

4. Pipeline admittance

$$Y = G + j\omega C = \frac{j\omega 2\pi\varepsilon^*}{\ln(r_3/r_2)} (S/m) \tag{6.34}$$

where
$\varepsilon^* = \varepsilon_i + 1/j\omega\rho_i$ is the complex permittivity
ε_i is the coating permittivity
ρ_i is the coating resistivity

6A.5 Accurate Solutions for Two-Cascaded Sections

$$I_1 = \frac{(m_1 I_{01} + n_1 I_{02})}{K}, \quad I_2' = \frac{(m_2' I_{01} + n_2' I_{02})}{K},$$

$$I_2 = \frac{(m_2 I_{01} + n_2 I_{02})}{K}, \quad I_3 = \frac{(m_3 I_{01} + n_3 I_{02})}{K}$$

$$m_1 = \{R_2 R_3(A_1 C_2 + C_1 A_2) + (R_2 + R_3)A_1 A_2 + (R_2 C_1 + A_1)B_2\}$$
$$\times (B_1 + R_1 - R_1 A_1)/K + 1 - A_1$$

$$n_1 = \frac{R_2\{R_3 C_2 + A_2)B_4 + (R_3 A_2 + B_2)(1 - A_2)\}}{K}$$

$$m_2' = \frac{\{(R_2 + R_3)A_2 + B_2 + R_2 R_3 C_2\}(B_1 + R_1 - R_1 A_1)}{K}$$

$$n_2' = \frac{R_2(R_1 C_1 + A_1)\{(R_3 C_2 + A_2)B_2 + (R_3 A_2 + B_2)(1 - A_2)\}}{K}$$

$$m_2 = \frac{R_2(R_3C_2 + A_2)(B_1 + R_1 - R_1A_1)}{K}$$

$$n_2 = \frac{\{(R_1A_1 + B_1) + R_2(R_1C_1 + A_1)\}\{(R_3C_2 + A_2)B_2 + (R_3A_2 + B_2)(1 - A_2)\}}{K}$$

$$m_3 = \frac{R_2(B_1 + R_1 - R_1A_1)}{K}$$

$$n_3 = \frac{\left[(R_1A_1 + B_1)\{B_2 - R_2(1 - A_2)\} + R_2B_2(R_1C_1 + A_1)\right]}{K}$$

$$K = (R_1A_1 + B_1)\{(R_3A_2 + B_2) + R_2(R_3C_2 + A_2)\} + R_2(R_1C_1 + A_1)(R_3A_2 + B_2)$$

Considering the characteristic of hyperbolic functions, for example, $B_1C_2 = C_1B_2$ and $A_1B_2 + B_1A_2 = Z_0 \sinh \theta$, these equations can be rewritten in the form of hyperbolic functions as follows:

$$K = (R_1R_2 + R_2R_3 + R_3R_1)\cosh\theta_1 \cdot \cosh\theta_2 + R_1Z_0\cosh\theta_1 \cdot \sinh\theta_2$$
$$+ R_3Z_0\sinh\theta_1 \cdot \cosh\theta_2 + (Z_0^2 + R_1R_2 + R_2R_3)\sinh\theta_1 \cdot \sinh\theta_2$$
$$+ R_2\left(Z_0 + \frac{R_1R_3}{Z_0}\right)\sinh\theta$$

where $\theta_1 = \Gamma \cdot x_1$, $\theta_2 = \Gamma \cdot x_2$, $\theta = \Gamma(x_1 + x_2) = \Gamma \cdot l$.

These solutions become identical to those in Equation 6.16 when the approximation in Equation 6.7 is used.

References

1. Working Group (Chair: Agematsu, S.) Japanese Electrotechnical Research Association. 2002. Technologies of countermeasure against surges on protection relays and control systems. *ETRA Report* 57 (3) (in Japanese).
2. Matsumoto, T., Y. Kurosawa, M. Usui, K. Yamashita, and T. Tanaka. 2006. Experience of numerical protective relays operating in an environment with high-frequency switching surges in Japan. *IEEE Trans. Power Deliv.* 21(1):88–93.
3. Agematsu, S. et al. 2006. High-frequency switching surge in substation and its effects on operation of digital relays in Japan. *CIGRE 2006, General Meeting,* Paris, France, Paper C4-304.
4. Smith, B. and B. Standler. 1992. The effects of surge on electronic appliances. *IEEE Trans. Power Deliv.* 7(3):1275.
5. Imai, Y., N. Fujiwara, H. Yokoyama, T. Shimomura, K. Yamaoka, and S. Ishibe. 1993. Analysis of lightning overvoltages on low voltage power distribution lines due to direct lightning hits to overhead ground wire. *IEE Jpn.apan Trans.* PE 113-B:881–888.

6. Kawahito, M. 2001. Investigation of lightning overvoltages within a house by means of an artificial lightning experiment. *R&D News Kansai Electric Power* 32–33.

7. Nagai, Y. and H. Sato. 2005. Lightning surge propagation and lightning damage risk across electric power and communication system in residential house. *IEICE Japan. Research Meeting,*. Tokyo, Japan, EMC-05-18.

8. Hosokawa, T., S. Yokoyama, and T. Yokota. 2005. Study of damages on home electric appliances due to lightning. *IEE Jpn.apan Trans. PE.* 125-B (2):221–226.

9. Hosokawa, T., S. Yokoyama, and M. Fukuda. 2009. Trend of damages on home appliances due to lightning and future problems., *IEEJ Trans. PE* 129-B (8):1033–1038.

10. Ametani, A., H. Motoyama, K. Ohkawara, H. Yamakawa, and N. Sugaoka. 2009. Electromagnetic disturbances of control circuits in power stations and substations experienced in Japan. *IET GTD* 3(9):801–815.

11. Sonoda, T., Y. Takeuchi, S. Sekioka, N. Nagaoka, and A. Ametani. 2003. Induced surge characteristics from a counterpoise to an overhead loop circuit. *IEEJ Trans. PE* 123 (11):1340–1349.

12. Ametani, A. 2006. EMTP study on electro-magnetic interference in low-voltage control circuits of power systems. *EEUG 2006*, Dresden, Germany. Paper D-3 (EEUG-Proc.), pp. 24–27.

13. Ametani, A., T. Goto, S. Yoshizaki, and H. Motoyama. 2006. Switching surge characteristics in gas-insulated substation. *UPEC 2006*, Newcastle, U.K. Paper 12–19.

14. Ametani, A., T. Goto, N. Nagaoka, and H. Omura. 2007. Induced surge characteristics on a control cable in a gas-insulated substation due because of switching operation. *IEEJ Trans. PE* 127(12):1306–1312.

15. International Electrotechnical Commission. 2001. Standard for Electromagnetic Compatibility, IEC-61000-4.

16. Japanese Electrotechnical Commission. Test voltage for low-voltage control circuits in power stations and substations. JEC-0103-2004. *IEE Japan* (in Japanese).

17. IEEJ WG. 2002. The fact of lightning disturbances in a highly advanced ICT society and the subject to be investigated. *IEEJ Tech. Report* 902.

18. IEEE. 1980. Guide for Surge Voltages in Low-Voltage AC Power Circuits. ANSI/IEEE C62.41.

19. Murota, N. 1993. Characteristic of the lightning surge suppressors at low voltage responses. *IEICE. Japan Research Meeting*, Tokyo, Japan, EMCJ93-72.

20. Ideguchi, T. and M. Hatori. 1993. Measures for lightning protection of telecommunication equipment in the premises. *IEICE Jpn. Trans.* 13(1):16.

21. WG of Insulation Management. 1993. Study report on insulation management of low voltage circuits. *J. Elect. InstIn. Eng. Jpn.* 13 (11):1165.

22. Ametani, A., K. Matsuoka, H. Ohmura, and Y. Nagai. 2009. Surge voltages and currents into a customer due to nearby lightning. *Electr.ic Power Syst. Res.* 79:428–435.

23. Yokoyama, S. and H. Taniguchi. 1997. The third cause of lightning faults on distribution lines. *IEE Jpn. Trans. PE* 117-B (10):1332–1335.

24. Ametani, A., K. Hashimoto, N. Nagaoka, H. Omura, and Y. Nagai. 2005. Modeling of incoming lightning surges into a house in a low-voltage distribution system. *EEUG 2005*, Warsaw, Poland.

25. Nagai, Y. and N. Fukusono. 2004. Lightning surge propagation on an electric power facility connected with feeder lines from a pole transformer. KEPCO Research Committee of Insulation Condition Technologies, Osaka, Japan.

26. Nagai, Y. 2005. Lightning surge propagation into a model house from various places, in KEPCO Research Committee of Insulation Condition Technologies, Osaka, Japan.
27. Ametani, A., K. Kasai, J. Sawada, A. Mochizuki, and T. Yamada. 1994. Frequency-dependent impedance of vertical conductors and a multiconductor tower model. *IEE Proc. GTD* 141 (4):339–345.
28. Ametani, A., K. Shimizu, Y. Kasai, and N. Mori. 1994. A frequency characteristic of the impedance of a home appliance and its equivalent circuit. *IEE Japan. Annual Conference 1405*, Tokyo, Japan.
29. Soyama, D., Y. Ishibashi, N. Nagaoka, and A. Ametani. 2005. Modeling of a buried conductor for an electromagnetic transient simulation. *ICEE 2005*, Kunming, China. SM1-04.
30. Nayel, M. 2003. A study on transient characteristics of electric grounding systems. PhD thesis, Doshisha University, Kyoto, Japan.
31. Scott Meyer, W. 1982. *EMTP Rule Book*. Portland, OR: BPA.
32. Mozumi, T., T. Ikeuchi, N. Fukuda, A. Ametani, and S. Sekioka. 2002. Experimental formulas of surge impedance for grounding lead conductors in distribution lines. *IEEJ Trans. PE* 122-B (2):223–231.
33. Asakawa, S. et al. 2008. Experimental study of lightning surge aspect for the circuit mounted distribution and telecommunication and customer systems. *CRIEPI Research Report* H07011.
34. IEEJ WG (Convenor Ametani, A.). 2008. Numerical transient electromagnetic analysis method. IEEJ. ISBN 978-4-88686-263-1.
35. Carson, J. R. 1926. Wave propagation in overhead wires with ground return. *Bell Syst. Tech. J.* 5:539–554.
36. Sunde, E. D. 1951. *Earth Conduction Effect in Transmission System*. New York: Wiley.
37. Taflov, A. and J. Dabkowski. 1979. Prediction method for buried pipeline voltages due to 60 Hz AC inductive coupling, Part I: analysis'. *IEEE Trans. Power App. Syst.* 98(3):780–787.
38. Wedepohl, L. M. 1963. Application of matrix methods to the solution of traveling wave phenomenon in polyphase systems. *Proc. IEE* 110 (12):2200–2212.
39. CIGRE WG.36.02. 1995. *Guide on the Influence of High Voltage AC Power System on Metallic Pipeline*. Paris, France: CIGRE Publication 95.
40. Sakai, H. 1971. *Induction Interference and Shielding*. Tokyo, Japan: Nikkan Kogyo Pub.
41. Rickets, L. W., S. E. Bridges, and S. Mileta. 1976. *EMP Radiation and Protective Techniques*. New York: Wiley.
42. Degauque, P. and J. Homelin. 1993. *Electromagnetic Compatibility*. Oxford: Oxford University Press.
43. Paul, C. R. 1994. *Analysis of Multiconductor Transmission Lines*. New York: Wiley.
44. Koike, T. 1995. *Transmission and Distribution Engineering*. Tokyo, Japan: Yokohama Publication Company (in Japanese).
45. Ametani, A. 1990. *Distributed-Parameter Circuit Theory*. Tokyo, Japan: Corona Pub. Co.
46. Dommel, H. W. 1986. *EMTP Theory Book*. Portland, OR: BPA.
47. Tesche, F. M., M. V. Ianoz, and T. Karlsson, 1997. *EMC Analysis Methods and Computational Models*. New York: Wiley.
48. IEE Japan WG (Convenor Ametani, A.). 2002. Power system transients and EMTP analysis. *IEE Japan Technical Report 872*. Tokyo (in Japanese).

49. Harrington, R. F. 1968. *Field Computation by Moment Methods*. New York: Macmillan Company.
50. Uno, T. 1998. *Finite Difference Time Domain Method for Electromagnetic Field and Antenna*. Tokyo, Japan: Corona Publishing Company (in Japanese).
51. IEE Japan WG (Convenor Ametani, A.). 2006. Recent trends of power system transient analysis—A numerical electromagnetic analysis. *J. IEE Jpn.* 126(10):654–673 (in Japanese).
52. Frazier, M. J. 1984. Power line induced AC potential on natural gas pipelines for complex rights-off-way configurations. *EPRI. Report* EL-3106. AGA. Cat. L51418.
53. Dawalibi, F. P. and R. D. Southey. 1989. Analysis of electrical interference from power lines to gas pipelines. Part 1, Computation methods, *IEEE Trans. Power Deliv.* 1989, 4(3):1840–1846, Part 2, 1990. 5 (1):415–421.
54. Southey, R. D., F. P. Dawalibi, and W. Vukonichi. 1994. Recent advances in the mitigation of AC voltages occurring in pipelines located close to electric transmission lines. *IEEE Trans. Power Deliv.* 9 (2):1090–1097.
55. Dawalibi, F. P. and F. Dosono. 1993. Integrated analysis software for grounding EMF and EMI. *IEEE Comput. Appl. Power* 6 (2):19–24.
56. Haubrich, H. J., B. A. Flechner, and W. A. Machsynski. 1994. A universal model for the computation of the electromagnetic interference on earth return circuits. *IEEE Trans. Power Deliv.* 9 (3):1593–1599.
57. Safe Eng. Service & Tec. *HIFREQ User Manual*. 2002. Montreal, CA.
58. Christoforidis, G. C., D. P. Labridis, and P. S. Dokopoulos. 2005. Hybrid method for calculating the inductive interference caused by faulted power lines to nearby pipelines. *IEEE Trans. Power Deliv.* 20 (2):1465–1473.
59. Ametani, A., J. Kamba, and Y. Hosokawa. 2003. A simulation method of voltages and currents on a gas pipeline and its fault location. *IEE Jpn. Trans. PE* 123 (10):1194–1200 (in Japanese).
60. Isogai, H., A. Ametani, and Y. Hosakawa. 2006. An investigation of induced voltages to an underground gas pipeline from an overhead transmission line. *IEE Jpn. Trans. PE* 126 (1):43–50 (in Japanese).
61. Boker, H. and D. Oeding. 1996. Induced voltage in pipelines on right-of-way to high voltage lines. *Elektrizitatswirtschaft* 65:157–170.
62. Ametani, A. 2008. Four-terminal parameter formulation of solving induced voltages and currents on a pipeline system. *IET Sci. Meas. 2. Technol.* 2(2):76–87.
63. Ametani, A. 1994. *EMTP Cable Parameters Rule Book*. Portland, OR: BPA.
64. Wedepohl, L. M. and D. J. Wilcox. 1973. Transient analysis of underground power transmission systems. *Proc. IEE* 120123:253–260.
65. Ametani, A., T. Yoneda, Y. Baba, and N. Nagaoka. 2009. An investigation of earth-return impedance between overhead and underground conductors and its approximation. *IEEE Trans. EMC* 51(3):860–867.

7

Grounding

7.1 Introduction

The significance of grounding for electric power equipment and systems such as transformers, substations, buildings, home appliances, etc., is well-known, and a number of books and papers have been published on the subject [1–30]. Grounding is done for the following reasons:

1. To maintain the ground terminal voltage at zero which is, strictly speaking, soil potential.
2. It makes an abnormally large current, such as a lightning strike, dissipate into the ground.
3. To obtain a reference voltage to define (measure) voltage of equipment and systems.

Without grounding, human safety during maintenance work, for example, cannot be guaranteed, and the voltage of a circuit to be tested or to be simulated cannot be measured.

There are two basic grounding approaches:

1. Single-point grounding.
2. Multiple-point grounding.

When considering a distributed-parameter circuit such as an overhead or underground transmission line, it should be noted that single-point grounding and multiple-point grounding present entirely different phenomena.

The most significant difference between the two approaches is whether or not the current flows. Single-point grounding prevents a current from flowing in the conductor, while multiple-point grounding allows a current to circule between the grounding points on the conductor. The circulating current results in electrochemical corrosion of the conductor, a well-known occurrence in steel railroad lines and steel pipelines for oil and gas transportation [31]. Therefore, single-point grounding should be used in such systems. If the train rail and pipeline are very long, the pipeline, for example, is divided into a number of sections, each of which is isolated from adjacent sections. Thus, each isolated section has single-point grounding.

Multiple-point grounding functions differently. The metallic sheath (shield) of an underground cable is grounded at certain specific distances, usually about a few km. Most of the GWs (called earth wires or sky wires in different countries) of an overhead transmission line are grounded at every transmission tower. The purpose of the multiple-point grounding in the metallic sheath of the underground cable is to reduce any electromagnetic effects that might travel from the cable core to the external circuits, and also to reduce possible overvoltages that might affect on the sheath. A GW is grounded at every tower. Whenever lightning strikes the tower and/or the GW, the current of the lightning flows into the soil at the nearest grounding position. Thus, a large lightning current is prevented from flowing into a substation.

In this chapter, the practical grounding methods used in a gas pipeline, a transmission tower, GWs, an underground cable, etc., are explained in Section 7.2. In Section 7.3, modeling of the grounding is explained for a steady-state and transient analysis. First, the analytical and/or theoretical model of a grounding electrode is described. Then, the modeling methods used in EMTP simulations are described and the effect of the simulation model in an FDTD method is explained. Then, the measurement of a grounding impedance, measured results, theoretical investigations, the effect of the electrode shape, etc., on the grounding impedance are described in Section 7.4.

7.2 Grounding Methods

7.2.1 Gas Pipeline

Once an underground gas pipeline is installed, it is typically used for more than 30 years. In the case of a trunk line, its lifetime is estimated to be more than 50 years, because it is too costly to replace the pipeline. For example, a gas trunk line across Tokyo with a length of 40 km was constructed a few years ago: it is impossible to check the leakage current along the pipe and, furthermore, it is impossible to replace a part of the line, because it is installed underground at a depth of 50–100 m beneath the greater Tokyo megalopolis and it is completely encased in concrete. Therefore, it is imperative to try and prevent any current flows along the pipe. Figure 7.1 illustrates the basic configuration of a pipeline and its grounding. Every pipeline section with length x_0 is insulated from the adjacent sections and the pipeline itself is coated with an insulating material so as to isolate it from the soil as in Figure 7.2 [31,32]. The inner radius r_1 and the outer radius r_2 range from 15 to 50 cm.

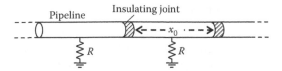

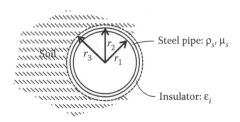

FIGURE 7.1
Gas pipeline and its grounding (single-point grounding): total length $n \cdot x_0$, where $n =$ the number of insulated pipeline sections.

FIGURE 7.2
Cross section of a pipeline.

The insulator coating thickness $(r_3 - r_2)$ is of 2–8 mm. The physical parameters ρ_s, μ_s, and ε_i are dependent on each pipeline and its manufacturer.

Over time, acidic water can cause a pinhole to develop within a coating insulator. When this occurs, a current can flow through the pinhole into the ground and return through the grounding resistance R. This circulating current results in electrochemical corrosion and, finally, a hole may be produced in the steel pipe [31].

Based on this scenario, it may seem wiser not to create any grounding. However, it is quite common for a power transmission line to run along a gas pipeline. This can generate an induced voltage along the pipeline. When the induced voltage exceeds a certain limit, it may cause a flashover that results in a gas explosion [31]. To prevent the induced voltage, a shielding conductor (steel) plate is occasionally installed in between the pipeline and the transmission line. However, this plate be limited to a very short length because of its installation cost and the potential for corrosion of the steel plate. Thus, for a long distance, it is common to adopt the method of grounding shown in Figure 7.1. A grounding rod is installed underground and embedded into a cubic hole filled with Mg powder. [31]. It should be noted that Mg is easily dissolved into soils so must therefore be carefully maintained (replaced) at regular intervals.

It should be noted that induced voltage described above cause a voltage difference between two sections of the pipeline as illustrated in Figure 7.1. If the voltage difference becomes higher than the flashover voltage across the insulating joint, a flashover occurs between the joint. This is another reason to install the grounding, that is, to prevent a flashover across an insulating joint between two sections of the pipeline.

7.2.2 Transmission Towers and GWs

When "grounding" is mentioned in the power-engineering field, the most common meaning is the grounding of a transmission tower when a GW is grounded to the tower. The impedance of the tower grounding is often called the "tower footing resistance (impedance)," and there are a number of publications discussing this [1–23]. Figure 7.3a illustrates the configuration of a typical tower structure and footing. Figure 7.3b is a simplified model circuit of the tower (structure) and the footing. For a steady state, the tower-footing impedance is represented simply by resistance, and a recommended value of resistance is specified in the standard of a transmission line. For example, in Japan, the resistance should be lower than 5 Ω for transmission voltage that is higher than 250 kV and 10 Ω for voltage lower than that [33,34]. For a transient analysis, the footing impedance is represented either by a resistor–capacitor circuit (RC circuit) or a resistor–inductor circuit (RL circuit), as in Figure 7.4 [1].

The tower footing is composed of steel structures and concrete. The length h_2 of the footing is about 5 m of a tower with a height of 30 m. In general,

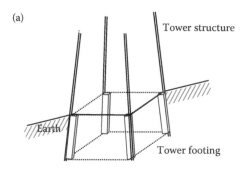

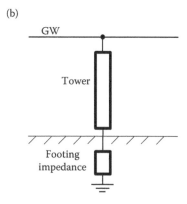

FIGURE 7.3
Tower footing. (a) Tower structure and footing and (b) simplified model circuit of (a).

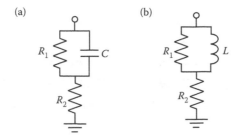

FIGURE 7.4
Model circuit of tower-footing impedance. (a) RC circuit and (b) RL circuit.

the ratio of the tower height h_1 and h_2 is about 6:1. This structure is heavily dependent on the mechanical strength of the tower and must be designed to take into account the natural environment, for example, the highest possible wind velocity.

7.2.3 Underground Cable

Figure 7.5 illustrates a cable system with grounding. Figure 7.5a is called "single-bonded cable," which means that the metallic sheaths of three-phase cables are grounded only at a single point. Figure 7.5b shows a "solidly-bonded (normal-bond) cable," in which both ends of the cable are grounded. Figure 7.6 shows the cross-bonded cables, in which the steady-state currents induced to the metallic sheaths are cancelled out. There are two methods of cross-bonding: (a) sheath cross-bonding, and (b) core cross-bonding.

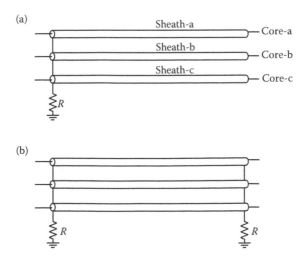

FIGURE 7.5
Grounding of a cable. (a) Single-bonded cable and (b) solidly bonded cable.

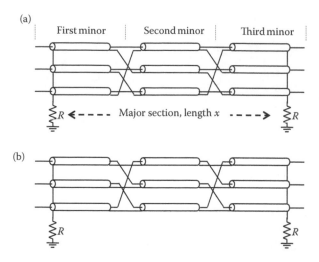

FIGURE 7.6
Cross-bonded cable. (a) Sheath cross-bonding and (b) core cross-bonding.

The choice between sheath cross-bonding or core cross-bonding varies by country and/or utility company.

In a cross-bonded cable, both ends of three-phase sheaths are short-circuited and are always grounded at every major section, because the cross-bonding becomes a transition point of traveling waves during a transient, and a high overvoltage appears on the sheath. For the grounding of the sheaths, long grounding rods and/or a small grounding mesh are adopted. However, it is not easy to get a low-enough impedance, especially in a mountainous area. The sending and/or receiving ends are grounded to a substation grounding mesh.

7.2.4 Buildings

Figure 7.7 illustrates grounding options for buildings. Figure 7.7a shows the grounding of an ordinary building in which the steel poles of the building's structure become part of the grounding. If necessary, a grounding mesh is installed and a down conductor attached to a lightning rod on the top of the building is connected to the mesh. Figure 7.7b shows the grounding of an earthquake-proof building. The earthquake-proof equipment is electrically isolated from the building's foundation, and therefore a copper (Cu) conductor connects the steel poles of the building's structures and its foundation structures.

7.2.5 Distribution Lines and Customer's House

Figure 7.8 illustrates a distribution system composed of (a) the distribution pole and (b) a customer's house. A distribution pole is made up of a steel pipe, a steel frame with concrete cement, or wood with a down conductor. The ratio

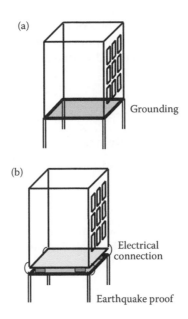

FIGURE 7.7
Grounding of a building. (a) Building and (b) earthquake-proof building.

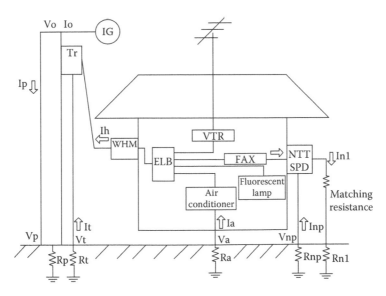

FIGURE 7.8
A distribution system: distribution pole and customer's house.

of the pole height h_1 and the basement length h_2 is about 1:3. A customer house involves the groundings of home appliances and a telephone line, that are made up of either copper or a steel rod with a length of about 1 m.

7.3 Modeling for Steady-State and Transient Analysis

7.3.1 Analytical/Theoretical Model

7.3.1.1 Steady-State

In a steady-state analysis, a grounding impedance is almost always modeled by the grounding resistance R. There are a number of resistance formulas proposed in various References 1 through 5. Among these, Sunde's formulas are well-known and widely used [3,21]:

1. *Single vertical grounding electrode:* The grounding resistance and capacitance of a single grounding electrode (rod) with length x and radius r within soil with resistivity ρ and permittivity ε is given in the following formula (Formula 7.1):

$$R = \frac{A_1 \rho}{2\pi x}\ (\Omega), \quad C = \frac{2\pi\varepsilon x}{A_1}\ (\text{F}) \tag{7.1}$$

   where

$$A_1 = \ln\left[\frac{1+\sqrt{1+a^2}}{a}\right] + a - \sqrt{1+a^2}, \quad a = \frac{r}{2x} \tag{7.2}$$

$$\varepsilon = \varepsilon_0\varepsilon_r, \varepsilon_r:\text{ earth relative permittivity}$$

   When x is far greater than r, Equation 7.2 is simplified as

$$A_1 = \ln\left(\frac{2}{a}\right) - 1 = \ln\left(\frac{4x}{r}\right) - 1 \quad \text{for } a \ll 1 \tag{7.3}$$

2. *Single horizontal electrode:* When a grounding electrode with length x and radius r is buried horizontally in soil at depth h, Sunde's formula for resistance R and capacitance C is modified as

$$R = \frac{A_2 \rho}{\pi x}\ (\Omega), \quad C = \frac{\pi\varepsilon x}{A_2}\ (\text{F}) \tag{7.4}$$

where

$$A_2 = \ln\left[\frac{(1+\sqrt{1+b^2})}{b}\right] + b - \sqrt{1+b^2}, \quad b = \frac{\sqrt{2rh}}{x} \tag{7.5}$$

When b is far smaller than 1,

$$A_2 = \ln\left(\frac{2x}{\sqrt{2rh}}\right) - 1 \quad \text{for } b \ll 1 \tag{7.6}$$

3. *Multiple vertical electrodes:* When the number of vertical electrodes described in Equation 7.1 is n, the total resistance of the electrodes is given by Sunde as follows:

$$R = \left(\frac{\rho}{2\pi xn}\right)\left[\ln\left(\frac{4x}{r}\right) - 1 + \left(\frac{2xn}{\pi D}\right)\ln\left(\frac{2n}{\pi}\right)\right] \tag{7.7}$$

where D: diameter of a circumferential circle of n vertical rods.

4. *Plate electrode with cross section S:* In a theoretical analysis of a grounding mesh, a plate electrode representation is adopted. Assuming that the cross section area of the mesh is S and the buried depth below ground level is h, the following equivalent resistance is derived [21,28]:

$$R = \left(\frac{\rho}{4\pi}\right)\left(1 - \frac{4h}{nr}\right) \quad \text{for } 4h \ll \pi r \tag{7.8}$$

where

$$r = \sqrt{\frac{S}{\pi}}$$

Neglecting the buried depth h, the above formula agrees with Sunde's approximation.

$$R = \frac{\rho}{4r} \tag{7.9}$$

7.3.1.1.1 Transient

When the transient phenomenon on a grounding electrode is solved analytically, a RC parallel circuit (as in Figure 7.4a) or an RL circuit (as in Figure 7.4b) is often adopted. The value of the resistances in Figure 7.4 is given in Equations 7.1 through 7.4 with the following relations:

$$R = R_1 + R_2$$

If a horizontal grounding electrode shows a capacitive nature, the value of C in Figure 7.4a is necessary and is given by Sunde as in Equation 7.4. If a horizontal grounding electrode shows an inductive nature, inductance L in Figure 7.4b derived by Sunde is modified [28]:

$$L = \frac{\mu_0 A x}{\pi} \tag{7.10}$$

When the electrode is at ground level, that is, $h = 0$, the following formula is obtained:

$$L = \left(\frac{\mu_0 x}{2\pi}\right)\left[\ln\left(\frac{2x}{r}\right) - 1\right] \tag{7.11}$$

The above assumes a lumped-circuit equivalence of a grounding electrode. When a frequency component involved in a transient is high and the wave length is shorter than the electrode length, it is impossible to assume the lumped equivalence. Instead, we have to adopt a distributed-parameter line to represent the grounding electrode [26,28].

7.3.2 Modeling for EMTP Simulation

Because any EMTP-type simulation tool is based on circuit theory, the formulas and values explained in the previous section are adopted. If the distributed-parameter nature of a horizontal or vertical electrode is to be taken into account, a model circuit such as that in Figure 7.9 can be used [22,24].

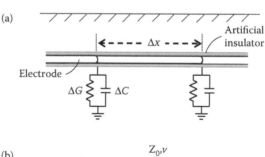

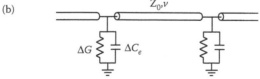

FIGURE 7.9
Distribution line model of a grounding electrode. (a) Representation of a horizontal electrode with length $x = n \cdot \Delta x$ and (b) model circuit.

The electrode is represented by an underground cable composed of a core conductor (electrode) and an artificial insulator. This makes it possible to evaluate the series impedance and the capacitance of the electrode by a subroutine called "Cable Constants" of the EMTP [35–37]. Then, the equivalent capacitance in Figure 7.9b is defined as

$$C_e = C - C_c \tag{7.12}$$

where
  $C = \Delta C \cdot x$ given in Equation 7.4
  $C_c \fallingdotseq \Delta C_c \cdot x$ calculated by "Cable Constants"
  $C_e \fallingdotseq \Delta C_e \cdot x$ equivalent capacitance

G is the grounding electrode conductance that is given by

$$G = \frac{1}{R} = \frac{\pi x}{A_2 \rho} \tag{7.13}$$

7.3.3 Numerical Electromagnetic Analysis

A transient response on a grounding electrode is easily calculated in the time domain by an FDTD method as explained in Chapter 5 of this book. However, it should be noted that the accuracy of an FDTD simulation is highly dependent on the sizes of a working space and cells used in the simulation, as is well-known. Additionally, the simulation's accuracy depends on the absorbing boundary conditions. Figure 7.10 shows the effect of stabilization coefficient d and the floating point operation of FDTD simulation results of current waveforms, when a lumped voltage source with amplitude $E = 2$ kV and a rise time of 1 µs is applied to the center of an infinite horizontal conductor with height $h = 50$ m and radius $r = 115$ mm above a perfectly

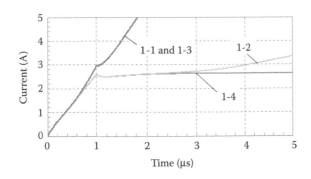

FIGURE 7.10
Effect of the stabilization coefficient d on current waveforms with the lumped voltage source for Cases 1-1 to 1-4. Case 1-1: $d = 7.5 \times 10^{-3}$, single precision Case 1-2: $d = 0$, single precision Case 1-3: $d = 7.5 \times 10^{-3}$, double precision Case 1-4: $d = 0$, double precision.

conducting earth. Because the surge impedance of the conductor is evaluated as $Z_s = 60 \ln(2\,h/r) = 406\ \Omega$, the current amplitude should be $I = E/2Z_s = 2.46$ A. The results in Figure 7.10 are far from the theoretical estimation with the exception of case 1-4, $d = 0$ with double precision. In fact, an FDTD simulation depends heavily on the absorbing boundary conditions, the accuracy of which is a function of the stabilization coefficient, floating point operation, time step, etc. [38], when the working space and cell size are fixed. Therefore, preliminary simulations to investigate the effect of the working space and cell sizes are inherent in an FDTD simulation.

A frequency response, that is, a steady-state analysis, of a grounding electrode can be performed by an FEM, NEC, MoM, etc. [38]. However, a user should be reminded that the accuracy is highly dependent on the simulation conditions and the model circuit, as explained above for the FDTD simulation.

7.4 Measurement of Transient Responses on Various Grounding Electrodes

In general, it is not easy to measure transient responses in a test field, because there is no power source (AC voltage source) and no voltage reference (no ground terminal). Therefore, one has to prepare all the apparatus required for the measurement. For example, one needs:

1. An AC voltage source to supply power to the measuring equipment.
2. At least ten grounding rods for grounding the AC source, an oscilloscope, etc., and for a voltage reference.
3. IG, oscilloscopes, voltage and current probes, and lead wire(s) (more than 100 m depending on the space of a test site).
4. Insulating rods for suspending the lead wires for source application, voltage, and current measurement.
5. Related handicraft and manufacturing tools.

It is particularly difficult to perform a transient measurement of grounding electrodes, because a grounding electrode is the measuring object but the voltage reference itself is another grounding electrode. Thus, it can occur that one measures the voltage of the voltage reference, but not the target electrode. Also, the length of the grounding electrode is short, often less than a few meters in the case of a vertical electrode, thus the observed time period is very small. Assume that the electrode length is 1 m; that gives the traveling time of about 10 ns with the wave propagation velocity at a soil of 100 m/µs. Then, the dominant transient frequency is higher than 10 MHz.

If a lead wire from the target electrode to the oscilloscope is 1 m, then the lead wire inductance becomes about 1 μH; that results in the impedance of $\omega L = 2\pi f \cdot L = 2\pi \times 25 \times 10^6 \times 1 \times 10^{-6} = 157\ \Omega$. This impedance is much higher than the grounding electrode impedance, and thus the measured voltage might be the voltage drop due to the lead wire's inductance. Figure 7.11 shows photos of transient measurements on grounding electrodes in a field.

In the following section, typical examples of transient response measurements on grounding electrodes are explained.

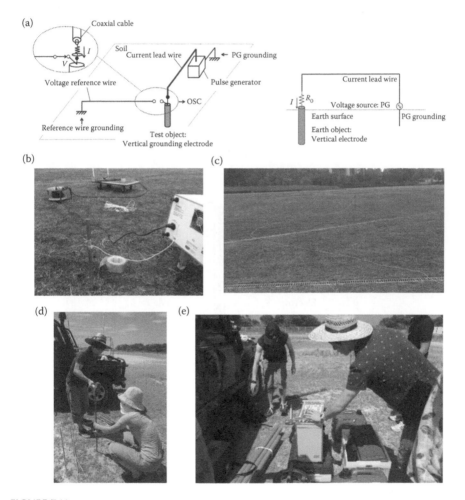

FIGURE 7.11
Test equipment and arrangements for transient measurement on a grounding electrode. (a) Basic circuit for measurement: test electrode, voltage source, current lead wire, voltage reference wire, oscilloscope, voltage probe, high-voltage probe, current probe, soil resistivity meter, wire reel, color corn, and grounding rods, (b) PG (voltage source) grounding, (c) voltage reference wire runs from the right bottom corner of the photo (c) to the top, the other side of the field in the photo, (d) installing grounding rods, and (e) power source and PG.

7.4.1 Transient Response Measurements on Multiple Vertical Electrodes and FDTD Simulations

7.4.1.1 Experimental Conditions

Figure 7.12 illustrates an experimental setup for measuring transient voltages at the top of an electrode [30]. The radius of the electrode is 10 mm, and the length is 1 and 1.5 m. An output voltage from a PG is applied to the electrode through a resistance of 5 kΩ so that the source can be regarded as a current source. Figure 7.13 shows an injected current waveform. A current lead wire and a voltage reference wire are kept perpendicular to each other and to the electrode so as to avoid mutual coupling between those as much as possible. A vinyl-covered conductor is used for the wires. The PG is set at the distance of 30 m from the electrode.

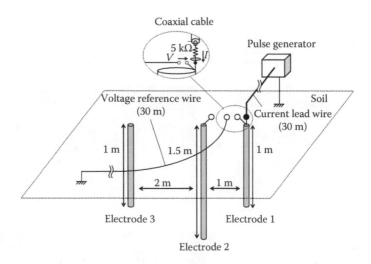

FIGURE 7.12
Experimental setup.

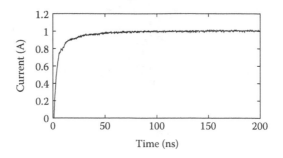

FIGURE 7.13
Injected current.

TABLE 7.1

Experimental and Simulation Conditions and Results.

(a) Voltage V_1

	x_1 (m)	x_2 (m)	y (m)	Measured Voltage V_1 (V) Peak/Time (ns)	$t = 200$ (ns)	Simulated Voltage V_1 (V) Peak/Time (ns)	$t = 200$	Sunde R_s (Ω)
Case								
Case 1-1	1	1.5	1	31.1/5.4	8.9	28.0/7.2	8.7	9.1
Case 1-2	1.5	1	1	30.1/6.2	7.4	28.5/7.1	6.2	6.6
Case 2-1	1	1.5	2	29.5/5.5	9.2	28.7/7.1	8.7	9.1
Case 2-2	1.5	1	2	28.6/5.6	7.7	29.2/7.8	6.1	6.6

(b) Voltage V_2

	Measured Voltage V_2 (V) Negative Peak/time (ns)	Positive Peak/time (ns)	$t = 200$ (ns)	Simulated Results V_2 (V) Negative Peak/time (ns)	Positive Peak/time (ns)	$t = 200$ (ns)	Sunde R_m (Ω)
Case							
Case 1-1	−8.1/16.1	6.0/30.5	1.1	−8.4/11.0	5.8/25.1	0.9	1.4
Case 1-2	−7.4/16.2	5.0/30.8	0.9	−7.9/10.8	5.8/25.2	0.9	1.4
Case 2-1	−10.2/16.6	3.9/32.4	0.5	−9.3/11.0	5.4/30.8	0.4	0.8
Case 2-2	−15.2/14.5	5.7/29.8	0.4	−10.4/11.1	6.7/30.9	0.4	0.8

Electrode radius $r = 1$ cm, $\rho = 11.5$ Ωm, applied current: rise time $T_f = 10$ ns, and amplitude $I_0 = 1$ A.

Table 7.1 summarizes the experimental conditions, the parameters of the electrodes, the separation distance, and the peak voltage. The peak voltage corresponds to the impedance for the applied current is 1 A with the rise time $T_f \fallingdotseq 10$ ns, that is, nearly a step function. In the measurement, I is the injected current, V_1 is the potential difference between the voltage of the inducting electrode "1" and the voltage reference wire, and V_2 is the potential difference between the voltage of the inducted electrode "2" and the reference wire.

The soil resistivity of the test site was measured by Wenner's four-electrode method. It was found that the test site was bilayered. The resistivity of the upper layer with a depth of 1.9 m was measured at 11.5 Ωm, and that of the second layer, from a depth of 1.9 m to infinity, was 345 Ωm. The steady-state resistance of a vertical grounding electrode with a length of 1 m was measured to be 9.4 Ω by a Yokogawa resistivity meter, and that with the length of 1.5 m was 7.1 Ω.

7.4.1.2 Measuring Instruments

1. PG: Noiseken INS-4040, input impedance 50 Ω.
2. Oscilloscope: Tektronix DPO-4104, frequency band DC to 1 GHz.
3. Voltage probe: Tektronix P6139A, frequency band DC to 500 MHz, input capacitance 8 pF.

4. Current probe: Tektronix CT-1, frequency band 25 kHz to 1 GHz, sensitivity 5 mV/mA.

5. Resistivity meter: Yokogawa M&C, IM3244.

7.4.1.3 Measured Results

Figure 7.14 illustrates the measured results together with the FDTD simulation results. Transient voltage V_1 of the inducing electrode (electrode 1) shows a steep voltage rise at the beginning due to the electrode inductance, and converges to the steady-state voltage determined by the product of the

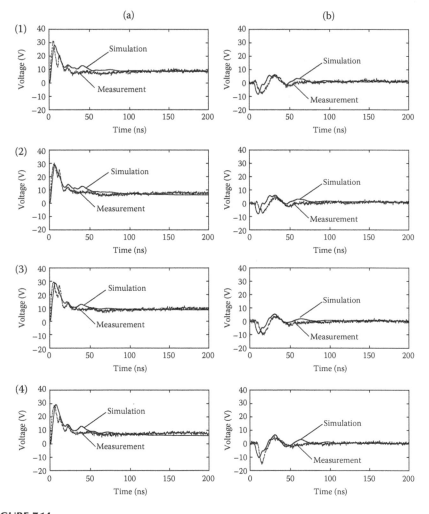

FIGURE 7.14
Measured and FDTD simulation results for the cases in Table 7.1. (a) V_1 and (b) V_2. (1) Case 1-1 (x_1: 1 m, x_2: 1.5 m, y: 1 m); (2) Case 1-2 (x_1: 1.5 m, x_2: 1 m, y: 1 m); (3) Case 2-1 (x_1: 1 m, x_2: 1.5 m, y: 2 m); (4) Case 2-2 (x_1: 1.5 m, x_2: 1 m, y: 2 m).

steady-state grounding resistance and the current. The steady-state resistance is given in Equation 7.1.

The steady-state resistance R_s evaluated by Equation 7.1 is given in Table 7.1a. The resistance in the table is 9.1 Ω for the 1-m electrode, and 6.6 Ω for the 1.5-m electrode. A difference of R_s from the measured result is 15% at most. This difference seems to be caused by (1) the fact that Sunde neglects the earth permittivity in the resistance formula, and (2) the nonhomogeneous real earth and the measuring system. In Reference 28, a number of measured results were compared with Sunde's formulas and it was concluded that Sunde's formulas involve 20%–100% difference from the measured results.

Induced voltage V_2 to electrode 2 becomes negative at the wavefront and converges to a steady-state after some oscillations. The negative voltage is greater with a separation distance of 2 m than with that of 1 m, as observed in Figure 7.14b. The negative voltage is caused by a potential rise of a voltage reference wire and a lead wire to a probe. The details of this phenomenon are investigated in Appendix 7A.1. As the separation distance between two electrodes becomes larger, the induced voltage becomes smaller. Because the voltages of the reference wire and the lead wire are not dependent on the separation distance, the negative voltage is more pronounced in the case of greater separation.

The mutual resistance R_m is defined by Sunde in the following form [3]:

$$R_m = \frac{\rho}{4\pi x_1 x_2} \left\{ f(x_1 + x_2) - \frac{f(x_1 - x_2) + f(x_2 - x_1)}{2} - \sqrt{(x_1 + x_2)^2 + y^2} \right.$$
$$\left. - \sqrt{(x_1 - x_2)^2 + y^2} \right\}$$

(7.14)

$$f(x) = x \ln \frac{\sqrt{x^2 + y^2} + x}{y}$$

where x_1: the length of the inducing electrode (electrode 1), x_2: the length of the induced electrode (electrode 2), and y: separation distance.

The mutual resistance R_m evaluated by Equation 7.14 is given in Table 7.1b. A difference of 30%–50% is observed between the measured results and R_m in Equation 7.14.

Figure 7.15a shows the measured results of induced voltage V_2 for Cases 1-1 and 1-2, and Figure 7.15b for Cases 2-1 and 2-2. The results agree with each other except for the negative voltage at around $t = 20$ ns, and thus, the symmetry between the two electrodes is clear.

7.4.1.4 Experiments and FDTD Simulations

7.4.1.4.1 Simulation Conditions

Figure 7.16 shows a working space for an FDTD simulation. The dimensions of the space are $x = 5.0$ m, $y = 6.0$ m, $z = 4.1$ m, and the cell size is 0.02 m.

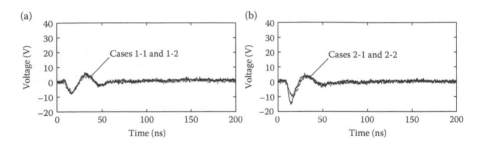

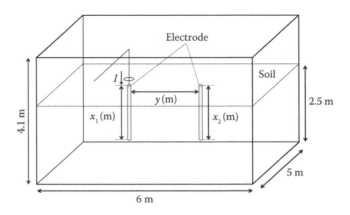

FIGURE 7.16
FDTD simulation setup.

Liao's second-order absorbing boundary is set at the boundaries of the working space. The soil is modeled by two layers, the upper layer with resistivity of 11.5 Ωm to a depth of 1.9 m, and the lower layer with resistivity of 345 Ωm from a depth of 1.9 m. The relative permittivity is taken to be 10. The grounding electrodes are assumed to be a perfect conductor, and are arranged in the same manner as the experimental setup. The current lead wire and the measuring system are represented as a conductor with a radius of 1 mm, covered by vinyl with a thickness of 0.5 mm by a thin-wire model [38]. A current waveform in Figure 7.13 is applied to the top of the electrode as illustrated in Figure 7.16. The time step Δt of an FDTD simulation is taken to be 0.027 ns, and the maximum observation time is 1000 ns. VSTL (Visual Surge Test Lab) [39] is used for the FDTD simulation.

7.4.1.4.2 Comparison with Measured Results

Figure 7.14 shows measured voltage waveforms for (a) Case 1-1, (b) Case 1-2, (c) Case 2-1, and (d) Case 2-2, respectively together with simulation results. Figure 7.17 shows the simulation results for Case 1-1 up to $T_{max} = 1000$ ns. It

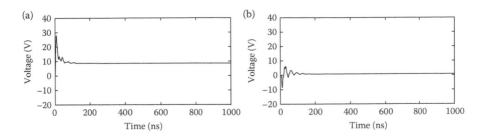

FIGURE 7.17
FDTD simulation results for Case 1-1 ($T_{max} = 1$ μs). (a) V_1 and (b) V_2.

is observed that the voltage is in a steady state at $t = 200$ ns. Thus, the results in Figure 7.14 are shown only for the time of 200 ns. Table 7.1 summarizes the transient and steady-state voltages, and self-resistance R_s and mutual resistance R_m of the electrode calculated by Sunde's formulas given in Equations 7.1 and 7.4.

We can observe in Figure 7.14 and Table 7.1 that the simulation results agree well with the measured results, with the exception of the negative voltage in V_2. (This detail is explained in Appendix 7A.1.) Therefore, FDTD simulation can be used to investigate the mutual coupling between the vertical grounding electrodes in the following section.

7.4.1.5 A Study of Mutual Coupling by FDTD Simulations

7.4.1.5.1 Simulation Conditions

Figure 7.18 illustrates a working space for an FDTD simulation. The dimensions of the working space are $x = 40$ m, $y = 30$ m, and $z = 20.2$ m with the cell size $\Delta s = 0.1$ m. The soil resistivity is taken to be $\rho = 100$ Ωm and the relative permittivity $\varepsilon_r = 10$. An electrode voltage is calculated by integrating the electric field from the absorbing boundary without considering a measuring system. The remaining conditions are the same as those given in Section 7.4.1.4. Simulations are carried out by varying the electrode length x_1 and x_2 and separation distance y from 1 to 10 m. A current source with the waveform of Figure 7.19 is applied through 5 kΩ from the top of the lead wire.

7.4.1.5.2 Relation between Inducing Electrode Length and Induced Voltage

Figure 7.20 shows the FDTD simulation results when the separation distance y from 1 to 10 m with the length $x_2 = 1$ m of electrode 2 (the induced electrode) is varied. Figure 7.21 shows the simulation result with $x_2 = 1$ m for $x_1 = 1$–10 m. Table 7.2 summarizes the results.

It is clear from Figure 7.20b that the induced voltage V_2 becomes smaller as separation distance y becomes greater. This characteristic is the same as that of an overhead line. On the other hand, voltages V_1 and V_2 become smaller

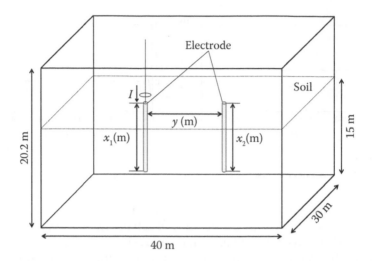

FIGURE 7.18
A model circuit for FDTD simulation.

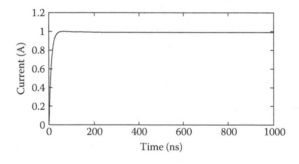

FIGURE 7.19
Applied current waveform.

as the electrode length x_1 increases. It is reasonable that V_1 decreases as x_1 increases, because the electrode impedance decreases as x_1 increases as was demonstrated in Equation 7.1. This characteristic is the basis of grounding. However, the initial rise of the voltage is kept somewhat independent of the electrode length, because of the electrode inductance.

It is observed in Figure 7.21b that the induced voltage V_2 at $t = 1\,\mu s$ is decreased by 60% in the case of $x_1 = 10$ m from that in the case of $x_1 = 1$ m. However, the peak value is relatively unchanged, as we saw with V_1.

For y greater than 5 m, no significant influence of the electrode length is observed.

In summary, the induced voltage is noticeably influenced by the separation distance, and tends to be decreased inversely proportional to the electrode length unless the separation is not large.

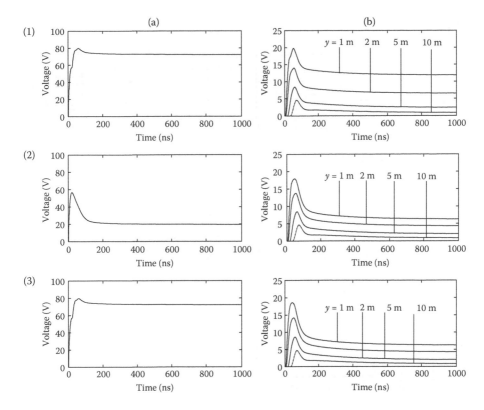

FIGURE 7.20
Effect of separation distance y. (a) V_1 at electrode 1 and (b) V_2 at electrode 2. (1) $x_1 = x_2 = 1$ m, (2) $x_1 = 5$ m, $x_2 = 1$ m, and (3) $x_1 = 1$ m, $x_2 = 5$ m.

7.4.1.5.3 Difference from an Overhead Conductor

Figure 7.20 shows the simulation results for varying the separation distance y. It has been made clear that the steady-state induced voltage is inversely proportional to the electrode length, as seen in Figures 7.20 and 7.21. However, the induced voltage in an overhead line is proportional to the line length.

$$V_m = -z_m \cdot x \cdot I_0 \qquad (7.15)$$

The electrostatic (capacitive) induced voltage is independent of the line length:

$$\overline{V}_s = \frac{C_{12}}{C_2 + C_{12}} \overline{V}_1 \qquad (7.16)$$

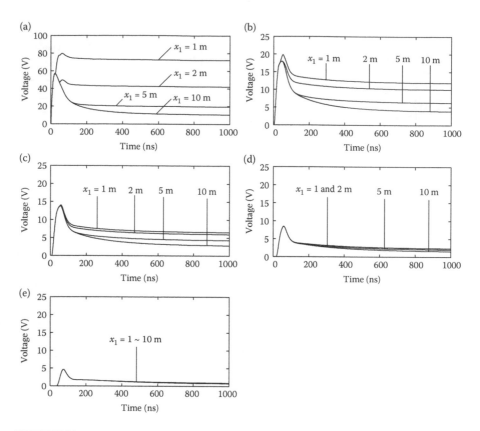

FIGURE 7.21
Effect of electrode length x_1 ($x_2 = 1$ m). (a) V_1, (b) V_2 with $y = 1$ m, (c) V_2 with $y = 2$ m, (d) V_2 with $y = 5$ m, and (e) V_2 with $y = 10$ m.

where z_m: mutual impedance (Ω/m), x: line length, I_0: inducing current, V_1: inducing voltage, C_{12}: mutual capacitance (F/m), and C_2: capacitance of the induced line to the soil (F/m).

Let us investigate the mutual resistance of grounding electrodes in Equation 7.14. By assuming $x_1 = x_2 = x$,

$$R_m = \left(\frac{\rho}{2\pi x}\right)\left[\ln\left\{\frac{\left(\sqrt{4x^2 + y^2} + 2x\right)}{y}\right\} - \left\{\frac{\left(\sqrt{4x^2 + y^2} - y\right)}{2x}\right\}\right] \quad (7.17)$$

It should be clear in the above equation that the resistance is inversely proportional to the electrode length. It is the same for self-resistance. This characteristic is exactly opposite to that of an overhead line, and is the basic principle of the grounding electrode.

TABLE 7.2

Parameters Used for FDTD Simulations and Simulation Results

Case	ε_r	x_1 (m)	x_2 (m)	y (m)	Simulation Results of V_2 (V) Peak/Time (ns)	$t = 1$ (μ s)	Sunde R_m (Ω)
Case 111	10	1	1	1	19.8/52.84	11.87	13.1
Case 112	10	1	1	2	14.04/55.41	6.53	7.44
Case 113	10	1	1	5	8.44/59.45	2.40	3.14
Case 114	10	1	1	10	4.64/70.77	0.90	1.58
Case 211	10	2	1	1	17.97/45.97	10.01	11.2
Case 212	10	2	1	2	13.81/52.44	5.97	6.9
Case 213	10	2	1	5	8.47/59.45	2.34	3.09
Case 214	10	2	1	10	4.65/70.91	0.90	1.58
Case 311	10	5	1	1	17.92/45.97	6.28	7.34
Case 312	10	5	1	2	13.78/52.71	4.32	5.23
Case 313	10	5	1	5	8.45/59.45	2.05	2.8
Case 314	10	5	1	10	4.64/70.91	0.85	1.53
Case 411	10	10	1	1	17.92/45.97	3.88	4.77
Case 412	10	10	1	2	13.78/52.71	2.86	3.68
Case 413	10	10	1	5	8.45/59.45	1.58	2.3
Case 414	10	10	1	10	4.64/70.91	0.75	1.4
Case 131	10	1	5	1	18.58/43.41	6.27	7.34
Case 132	10	1	5	2	14.19/52.17	4.31	5.23
Case 133	10	1	5	5	8.61/59.45	2.04	2.8
Case 134	10	1	5	10	4.71/70.91	0.85	1.53

Electrode radius $r = 1$ cm, applied current: rise time $T_f \fallingdotseq 10$ ns, amplitude $I_0 = 1$ A, and $\rho = 100$ Ωm.

When the separation y becomes very large, Equation 7.17 becomes

$$R_m \cong \frac{\rho}{2\pi y} \tag{7.18}$$

The above result indicates that the induced voltage and the mutual resistance are inversely proportional to y. Thus, the electrode length has no influence.

The reason for this significant difference between an insulated conductor (overhead line) system and a grounding system is that currents flowing into the grounding electrode penetrate soils and, after a certain distance, no current flows through the electrode. This phenomenon has been experimentally confirmed in Reference 23. Correspondingly, it is shown in Reference 26 that a grounding electrode can be regarded as an open-circuited conductor. It should be noted that there are inductive, capacitive, and conductive coupling between the grounding electrodes, while only the former two exist in the isolated conductor system. This fact is a cause of the difference observed between the grounding electrode and the isolated conductor system and, because of this, the grounding electrode shows a traits of a nonuniform conductor [26].

7.4.1.5.4 Reciprocity between the Electrodes

Figure 7.22 shows the simulation results of induced voltage V_2 for two different electrodes lengths, that is, $x_1 = 5$ m and $x_2 = 1$ m (Cases 311–314) with $y = 1$–10 m, together with Cases 131–134 ($x_1 = 1$ m, $x_2 = 5$ m). It is clear that the voltages of the inducing and induced electrodes are the same for the different electrode lengths, that is, there exists reciprocity between the electrodes, just as with an insulated conductor system.

7.4.1.5.5 Effect of the Separation Distance

Figure 7.23 shows the simulation results of induced voltage V_2 for varying the separation distance y. It can be seen that the induced voltage is inversely proportional to the separation distance, which is the same as an isolated conductor system.

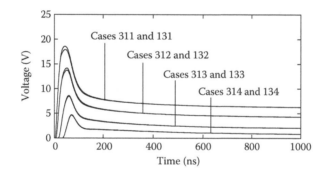

FIGURE 7.22
Comparison between $x_1 = 1$ m, $x_2 = 5$ m and $x_1 = 5$ m, $x_2 = 1$ m.

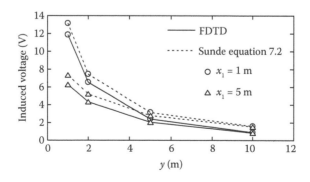

FIGURE 7.23
Characteristic of the induced voltage at 1 μs versus the separation distance ($x_2 = 1$ m).

7.4.1.5.6 Effect of Relative Soil Permittivity

Figure 7.24 shows induced voltage V_2 for $\varepsilon_r = 30$ and $x_1 = x_2 = 1$ m. When the separation distance y is greater than 2 m no influence of permittivity is observed, as in Figure 7.20.1b. Figure 7.25 shows the propagation velocity for the soil relative permittivity $\varepsilon_r = 1$, 10, and 30. The propagation velocity v is evaluated by

$$v = \frac{y}{\tau} \tag{7.19}$$

where τ: time of voltage V_2 appearance at the electrode top.

It is observed that the velocity becomes independent of soil permittivity for the separation distance greater than 6 m in Figure 7.25.

In the FDTD simulations, the intrinsic propagation constant is defined as $\gamma = \sqrt{j\omega\mu(\sigma + j\omega\varepsilon)}$ by (σ: soil conductivity, $\varepsilon = \varepsilon_r \varepsilon_0$: soil permittivity, μ: soil permeability, and ω: angular frequency). Thus, the propagation velocity is

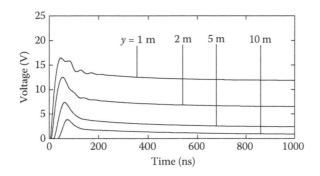

FIGURE 7.24
Transient voltage V_2 with $\varepsilon_r = 30$ ($x_1 = x_2 = 1$ m).

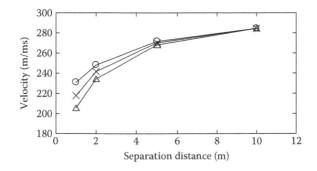

FIGURE 7.25
Time delay characteristic with $\varepsilon_r = 1$, 10, and 30, $\bigcirc$: $\varepsilon_r = 1$, $\times$: $\varepsilon_r = 10$, and $\triangle$: $\varepsilon_r = 30$.

dependent on σ, ε, and μ. When the separation distance becomes large, currents in the soil concentrate at ground level, and thus, the propagation velocity converges to the light velocity in free space.

7.4.2 Theoretical Analysis of Transient Response

A model circuit of a grounding electrode composed of a RC parallel circuit and a distributed-parameter line is illustrated in Figure 7.26a [28]. The impedance ωL being far greater than the resistance R of the distributed line in general, the model circuit of Figure 7.26a is simplified as Figure 7.26b where L_0 is the series inductance of the distributed line and L' is the inductance of a lead wire when measuring an electrode voltage $v'(t)$. In practice, a measured voltage $v(t)$ by a voltage probe differs from the electrode voltage $v'(t)$. This is true for any voltage of power equipment in a substation and/or a power facility. The main concern of this section is $v(t)$ rather than $v'(t)$.

7.4.2.1 Analytical Formula of Electrode Voltage

1. Impulse current application

   An electrode voltage is derived when the following impulse current is applied:

   $$i(t) = I_0\{\exp(-\alpha t) - \exp(-\beta t)\} \tag{7.20}$$

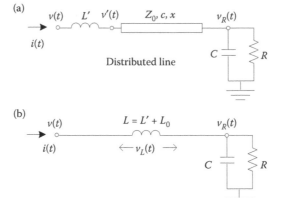

L': lead wire inductance, L_0: grounding electrode series inductance
x: length, Z_0: surge impedance, c: propagation velocity
$v(t)$: voltage at measuring point, $v'(t)$: electrode sending-end voltage

FIGURE 7.26
A simple equivalent circuit of a grounding electrode. (a) An original model and (b) a simplified model.

The frequency response of the above is given by using Laplace operator s.

$$I(s) = I_0 \left[\frac{1}{(s+\alpha)} - \frac{1}{(s+\beta)} \right] \tag{7.21}$$

The impedance seen from the sending end in Figure 7.26b is

$$Z(s) = sL + \frac{1}{(s+\gamma)C} \tag{7.22}$$

where $\gamma = 1/\tau$, $\tau = RC$: time constant.

A product of Equations 7.21 and 7.22 gives $V(s)$, that is transformed into a time domain. Thus,

$$v(t) = v_L(t) + v_R(t) = v_L(t) + v_{R1}(t) + v_{R2}(t) \tag{7.23}$$

$$v_L(t) = LI_0 \left[\beta \exp(-\beta t) - \alpha \exp(-\alpha t) \right] \tag{7.24}$$

$$\left. \begin{aligned} v_{R1}(t) &= RI_0 \left\{ \frac{\exp(-\alpha t)}{A} - \frac{\exp(-\beta t)}{B} \right\} \\ v_{R2}(t) &= -RI_0 \left(\frac{1}{A} - \frac{1}{B} \right) \exp(-\gamma t) \\ \text{where } A &= 1 - \alpha\tau, \quad B = 1 - \beta\tau \end{aligned} \right\} \tag{7.25}$$

It is clear in the above equations that measured voltage $v(t)$ consists of three exponential functions, $\exp(-\alpha t)$, $\exp(-\beta t)$, and $\exp(-\gamma t)$. α and β are dependent on the applied current, and $\gamma = 1/\tau$, is a function of R and C that represents the soil.

2. The case of $\alpha = 0$

For simplicity, when $\alpha = 0$ in Equation 7.20 is assumed, that is, $i(t) = I_0\{1 - \exp(-\beta t)\}$, the following result is obtained:

$$v_L(t) = LI_0\{\delta(t) + \beta\exp(-\beta t)\} \tag{7.26}$$

$$\left. \begin{aligned} v_{R1}(t) &= RI_0 \left\{ \frac{1 - \exp(-\beta t)}{B} \right\} \\ v_{R2}(t) &= RI_0 \left(\frac{\beta\tau}{B} \right) \exp(-\gamma t) \end{aligned} \right\} \tag{7.27}$$

where $\delta(t) = \infty$: $t = 0$, $\delta(t) = 0$: $t \neq 0$ Dirac's delta function.

The second term of Equation 7.26 is produced by inductance L of the circuit and corresponds to Equation 7.24 when an impulse current is applied. The maximum voltage of $v_L(t)$ appears theoretically at $t = 0$, as is clear from Equations 7.24 through 7.26. In the circuit of Figure 7.26a with a distributed-parameter line, a voltage given by $L'I_0\delta(t)$ is produced by the inductance L' of a lead wire. In reality, the maximum voltage appears after $t = 0$ because of the surge impedance Z_0' of the lead wire and the resistance of the circuit.

$$V_{Lmax} = v_L(T_{V1}) : T_{V1} > 0 \tag{7.28}$$

Equation 7.27 corresponds to a charged voltage of a RC circuit, as is well-known. Its maximum voltage is given by

$$V_{Rmax} = V_R(\infty) = RI_0 \quad \text{for } t = \infty$$

7.4.2.2 Transient Voltage Waveform

7.4.2.2.1 Maximum Current I_p and Time of Its Appearance T_i

By substituting zero into the time derivative of Equation 7.20, the maximum current I_p and the time of its appearance T_i are given as

$$T_i = \frac{\ln(\beta/\alpha)}{(\beta - \alpha)} \tag{7.29}$$

$$I_P = i(T_i) = I_0\{\exp(-\alpha T_i) - \exp(-\beta T_i)\}$$

In general, the above T_i is greater than the wavefront duration T_f of an impulse waveform expressed by double exponential functions.

$$T_f < T_i \tag{7.30}$$

For example, in the case of $T_f = 1\,\mu s$ and $T_t = 70\,\mu s$, that is, a standard impulse waveform, T_i in Equation 7.29 is given as $T_i = 1.8\,\mu s$.

7.4.2.2.2 Maximum Value of $v_L(t)$ and Time of Its Appearance T_V

From the time derivative of Equation 7.20, the following result is obtained:

$$t = T_{V2} = \frac{2\ln(\beta/\alpha)}{(\beta - \alpha)} = 2T_i \tag{7.31}$$

$$V_{LP} = LI_0\{\beta\exp(-\beta T_{V2})-\alpha\exp(-\alpha T_{V2})\} < 0$$

It is interesting to note that

$$V_2(T_{V2}) = V_{LP} \quad \text{at } t = T_{V2} = 2T_i : \text{negative peak} \tag{7.32}$$

$$V_L(T_i) = 0 \quad \text{at } t = T_i$$

The above relation is reasonable for $V_L(t) = Ldi/dt$. The following result is clear from Equation 7.24 at $t = 0$:

$$V_L(0) = LI_0(\beta - \alpha) = V_{Lmax} \quad \text{at } t = 0 \tag{7.33}$$

It is readily determined from Equations 7.31 and 7.33 that

$$V_L(0) > |V_{LP}|$$

The above relation means that $V_L(t)$ takes its maximum value at $t = 0$. This is just theoretical. In reality, $V_L(t)$ becomes the maximum at $t = T_{V1} > 0$, as in Equation 7.28.

7.4.2.2.3 Maximum Value of $v_R(t)$ and the Time of Its Appearance

As is clear from Equation 7.25, v_R is given as a sum of v_{R1} due to resistance R and v_{R2} with the time constant τ determined by capacitance C. Because $\alpha\tau$ and $\beta\tau$ are far smaller than 1, A and B become nearly equal to 1. Thus,

$$v_R(t) \approx v_{R1}(t) = Ri(t) \tag{7.34}$$

The above equation means that the transient response of the RC circuit in Figure 7.26 is dominated by R, and the maximum voltage V_{Rmax} and its time of appearance are given by

$$V_{Rmax} \approx V_{R1}(T_i) = R \cdot I_0 = RI_0\{\exp(-\alpha T_i)-\exp(\beta T_i)\} \quad \text{at } t = T_i \tag{7.35}$$

7.4.2.2.4 Maximum Value of $v(t)$ and the Time of Appearance

Considering $V_R(0)$ and Equation 7.32, the maximum value of $v(t)$ and the time of its appearance are given either by Equations 7.33 or 7.35.

$$V_{max} = \max[V_{Lmax}, V_{Rmax}] \tag{7.36}$$

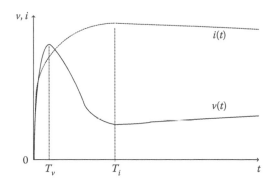

FIGURE 7.27
Qualitative wave-shapes of transient current and voltages.

7.4.2.2.5 Transient Voltage Waveforms

Considering the above, the voltage waveform is qualitatively drawn from Equations 7.23 and 7.26 as in Figure 7.27.

The dotted line in Figure 7.27 is the current given in Equation 7.20.

7.4.2.3 Analytical Investigation

The first term of Equation 7.26 becomes infinite when a pulse waveform is applied to the circuit in Figure 7.26b. However, this is just mathematical. In reality, $v'(t)$ in Figure 7.26a at around $t = 0$ is determined by

$$v'(t) = Z_0 i(t) \tag{7.37}$$

It is not easy to obtain the transient voltage in Figure 7.26a because of the distributed-parameter line. Assuming that a pulse-like current ($\alpha = 0$ in Equation 7.20) is applied, the transient voltages for $0 \leq t \leq 2\,T$ ($T = x/c$: traveling time along an electrode with the length of x) are obtained analytically by adopting a traveling-wave theory

$$0 \leq t < T \quad v(t) = v'(t) + v'_L(t),\ v_R(t) = 0$$

$$v'_L(t) = L' I_0 \{\delta(t) + \beta \exp(-\beta t)\}$$

$$v'(t) = Z_0 i(t) = Z_0 \{1 - \exp(-\beta t)\} \tag{7.38}$$

$$0 \leq t \leq 2T \quad v_R(t) = R I_0 \left\{ \frac{1 - \exp(-\beta t')}{B + (\beta \tau / B)\exp(-\gamma t')} \right\}$$

where $t' = t - T.$

The above solution involves $\delta(t)$ similarly to Equation 7.26 in the case of Figure 7.26b. That is, $v(0)$ at the measured position becomes very large due to $L'I_0\{\delta(0) + \beta\}$, although $v'(0)$ at the sending end of the electrode is $Z_0 i(0)$. Again, this is merely mathematical, because there exists no ideal current source. In practice, a current source is represented by a voltage source and a large resistance. Also, a lightning channel involves an impedance. Thus, the internal impedance of a current source representing a lightning current can never be infinite. Thus, the term $L'I_0\delta(0)$ never appears in reality, but $v'(0)$ is roughly given by $\beta L' I_0$. This is also true in Figure 7.26b.

In summary, Figure 7.26b gives the same solution as that in Figure 7.26a if the time delay due to traveling-wave propagation along a distributed line is neglected. Thus, Figure 7.26b can be used as an approximation of Figure 7.26a.

In Figure 7.26a, if a lead wire is represented by its surge impedance Z_0' rather than the inductance, $v(t)$ at around $t = 0$ is given by

$$v(t) = Z_0' i(t) \tag{7.39}$$

The above surge impedance Z_0' is not easily obtained because of its finite length. Instead, Z_0' is approximately given from the formulation explained above by

$$Z_0' \approx (\beta - \alpha)L' \tag{7.40}$$

From the above investigation, the characteristics of transient voltage $v(t)$ of a grounding electrode are theoretically summarized as follows:

1. The maximum voltage appears due to an inductive component of a circuit including a lead wire inductance as

$$V_{Lmax} \approx (\beta - \alpha)LI_0 \quad \text{at } t = T_1 \approx T_{v1}$$

2. The following voltage appears due to R and C components of the circuit:

$$V_{Rmax} \approx RI_0\{\exp(-\alpha T_i) - \exp(-\beta T_i)\} \quad \text{at } t \approx T_i$$

3. The maximum electrode voltage is determined either by V_{Lmax} when the inductance is relatively large or by V_{Rmax} when R is relatively large. Because of $T_{v1} < T_i$ in general, the electrode shows an inductive characteristic when V_{Lmax} gives the maximum electrode voltage, and a resistive/capacitive characteristic when V_{Rmax} becomes the maximum.

7.4.2.4 Circuit Parameters

7.4.2.4.1 Electrode

To evaluate the above-explained characteristics, the well-known Sunde's steady-state formulas described in Section 7.3.1.1 are adopted for R and C of the electrode.

1. Vertical electrode

$$R = \frac{K_1 \rho}{2\pi x}, \quad C = \frac{2\pi \varepsilon x}{K_1} \tag{7.41}$$

$$(7.41) = (7.1)$$

$$K_1 = \ln\left(\frac{4x}{r}\right) - 1, \quad \text{for } x \gg r$$

2. Horizontal electrode

$$R = \frac{K_2 \rho}{\pi x}, \quad C = \frac{\pi \varepsilon x}{K_2} \tag{7.42}$$

$$(7.42) = (7.4)$$

$$K_2 = \ln\left(\frac{2x}{\sqrt{rh}}\right) - 1 \quad \text{for } x \gg r$$

where
   x: electrode length
   r: radius
   h: buried depth

Time constant τ of the RC circuit is thus obtained.

$$\tau = RC = \rho_e \varepsilon_e = \rho_e \varepsilon_i \varepsilon_0 = \left(\frac{\rho_e \varepsilon_i}{36\pi}\right) \times 10^{-9} \approx 8.8 \times 10^{-12} \varepsilon_i \rho_e \tag{7.43}$$

It is clear from the above equation that $\tau = RC$ is independent of the shape and configuration of an electrode, but is determined only by the soil resistivity ρ and permittivity ε. This is the assumption and/or the condition when Sunde derived resistance R and capacitance C of the electrode. The same is applied to a propagation velocity c along the electrode:

$$c = \frac{1}{\sqrt{\varepsilon \mu_0}} = \frac{c_0}{\sqrt{\varepsilon_r}}, \quad c_0 = \frac{1}{\sqrt{\varepsilon \mu_0}} : \text{light velocity in free space} \tag{7.44}$$

The above formula is quite often used to determine the soil permittivity:

$$\varepsilon_r = \left(\frac{c_o}{c}\right)^2 \quad c: \text{measured value of the velocity} \tag{7.45}$$

Equation 7.44 corresponds to the propagation constant in the case of neglecting series resistance R and shunt conductance G of the electrode, and series impedance Z of the electrode in this case becomes

$$Z = j\omega L_0, \quad L_0 = \frac{\mu_0 K_2 x}{\pi} \tag{7.46}$$

Sunde gave the following inductance:

$$L_0 = \left(\frac{\mu_0 x}{2\pi}\right)\left\{\ln\left(\frac{2x}{r}\right) - 1\right\} \tag{7.47}$$

$$(7.47) = (7.11)$$

The above equation corresponds to Equation 7.46 under the assumption that the buried depth $h = r/2$ where r is the electrode radius, and the inductance becomes one half of that buried completely underground.

In general, the soil resistivity ranges from 10 to 5000 Ωm and the relative permittivity ranges from 1 to 40. Then, the time constant ranges are in the following region:

$$10^{-11} \le \tau \le 10^{-7} \tag{7.48}$$

7.4.2.4.2 Current Waveform

The coefficients α and β of the current waveform in Equation 7.20 are determined by its wavefront duration T_f and tail T_t in the following form:

$$\alpha = \frac{a}{T_t}, \quad \beta = \frac{b}{T_f}$$

where a and b are given as a function of $k = T_t/T_f$ in a book of *High-Voltage Engineering*, and range in the following region:

$$0.5 < a < 1.0, \quad 1.8 < b \le 3.2$$

Assuming that $0.01\ \mu s \le T_f \le 5\ \mu s$, $1\ \mu s \le T_t \le 100\ \mu s$, then α and β are given by

$$\frac{0.5}{T_t} \le \alpha \le \frac{2}{T_t}, \quad \frac{1.8}{T_f} \le \beta \le \frac{3.2}{T_f} \tag{7.49}$$

7.4.2.4.3 Coefficients of Voltage Formula

From Equations 7.48 and 7.49, the following relation is obtained:

$$\frac{0.5\times10^{-11}}{T_f} \le \alpha\tau \le \frac{1.0\times10^{-11}}{T_t}, \quad \frac{1.8\times10^{-11}}{T_f} \le \beta\tau \le \frac{3.2\times10^{-11}}{T_t}$$

Thus, $0.01\ \mu s \le T_f \le 5\ \mu s$: $\beta\tau \ll 1$

$$1\ \mu s \le T_t \le 100\ \mu s : \alpha\tau \ll 1, \quad \text{and} \quad \alpha\tau < \beta\tau \tag{7.50}$$

It is confirmed that $A \approx B \approx 1$ in Section 7.4.2.2 and thus, the theoretical investigation in Section 7.4.2.3 is appropriate.

7.4.2.5 Wave Propagation Characteristic

Based on Equations 7.42 and 7.46, the wave propagation characteristic on a grounding electrode is discussed.

7.4.2.5.1 Surge Impedance Z_s

At the time nearly equal to zero, or in a very high-frequency (VHF) region, the following surge impedance is given as an approximation of characteristic impedance $Z_0(\omega)$ [40]:

$$Z_s = \lim_{\omega\to\infty} Z_0(\omega) = \sqrt{\frac{L_0}{C}} = 120\frac{K_2}{\sqrt{\varepsilon_r}} \tag{7.51}$$

7.4.2.5.2 Characteristic Impedance Z_0

The characteristic impedance Z_0 is defined as a function of frequency by

$$Z_0 = Z_0(\omega) = \sqrt{\frac{Z}{Y}} = \sqrt{\frac{R_0 + j\omega L_0}{G + j\omega C}} \tag{7.52}$$

The series resistance R_0 being far smaller than ωL_0, the following equation is obtained from Equations 7.42 and 7.46:

$$Z_0(\omega) = Z_s\sqrt{\frac{j\omega}{\sigma/\varepsilon + j\omega}} \tag{7.53}$$

where $\sigma = 1/\rho$.

The above equation is approximated by

$$f \gg f_e: Z_0(\omega) = Z_s \tag{7.54}$$

$$f \ll f_e: Z_0(\omega) = (1 + j)Z_{0e}$$

where

$$Z_{0e}(\omega) = 20 \, K_2 \sqrt{\rho f} \times 10^{-9}$$

$$f_e = \frac{\rho}{2\pi\varepsilon} = \frac{18 \times 10^9}{\rho\varepsilon_r} : \text{critical frequency}$$

7.4.2.5.3 Propagation Constant Γ

Neglecting the series resistance R_0,

$$\Gamma(\omega) = \left(\frac{x}{c}\right)\sqrt{j\omega\left(j\omega + \frac{\sigma}{\varepsilon}\right)} \tag{7.55}$$

The above equation is approximated as

$$f \gg f_e: \Gamma(\omega) = \frac{j\omega x}{c}, \quad c = \frac{c_0}{\sqrt{\varepsilon_r}}$$

$$f \ll f_e: \Gamma(\omega) = (\alpha + j\beta)x, \quad \alpha = \beta = 2\pi\sqrt{\sigma f} \times 10^{-7} \tag{7.56}$$

where $c = \omega/\beta = \sqrt{\rho f \times 10^7}$.

It is observed in Equations 7.54 and 7.56 that the surge impedance and propagation velocity are functions of the soil permittivity in a frequency region higher than the critical frequency f_e. However, in a frequency region lower than f_e, those are dependent only on $\sqrt{\rho f}$. This is reasonable because $\omega\varepsilon_e \ll \sigma_e = 1/\rho_e$ for $f \ll f_e$. Thus, when discussing a transient characteristic for $f \ll f_e$, the effect of soil permittivity on electrode capacitance C can be neglected approximately. This frequency region is given as

$$4.5 \, \text{MHz} < f < 180 \, \text{MHz} \quad \text{for } \rho_e = 100 \, \Omega\text{m}, \quad \varepsilon_r = 1 \sim 40$$

$$0.22 \, \text{MHz} < f < 9 \, \text{MHz} \quad \text{for } \rho_e = 2000 \, \Omega\text{m}, \quad \varepsilon_r = 1 \sim 40$$

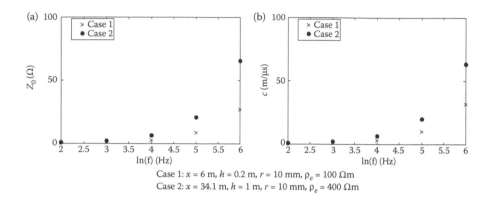

FIGURE 7.28
Approximate frequency responses of characteristic impedance and propagation velocity.
(a) Surge impedance Z_{0e} and (b) velocity c.

Considering the above equation, Figure 7.28 is drawn from Equations 7.54 and 7.55. The results in the figure agree with measured results in References 26 and 28, and thus, the analytical investigation in this section is confirmed to be adequate. The characteristic impedance is dependent on the coefficient K that is a function of electrode length x, radius r, and buried depth h. However, K ranges from 6 to 7 for $10 \leq x \leq 100$ m, $2 \leq r \leq 10$ mm, and $0.5 \leq d \leq 2$ m. Therefore, it is clear that K is significantly dependent on ρ and f. The propagation constant is not dependent on K.

7.4.2.6 Concluding Remarks

This section derives an analytical equation for the transient voltage and current at the sending end of a grounding electrode by adopting an approximate circuit of the electrode. Based on the equation, transient voltage and current responses of the electrode are explained analytically.

It is made clear that the so-called inductive characteristic of a grounding impedance is caused, in many cases, by the inductance of a current lead wire used in the measurement. The wave propagation characteristic on the electrode is determined by the soil permittivity in a VHF region, but is determined by $\sqrt{\rho f}$ in a lower-frequency region, where ρ is the soil resistivity and f is the frequency. Although the characteristic impedance of an electrode is proportional to $\ln(x/\sqrt{rh})$ where r is the electrode radius, h the buried depth, and x the length, the effect of $\sqrt{\rho f}$ is more pronounced.

7.4.3 Investigation of Various Measured Results

7.4.3.1 Test Circuit

Figure 7.29 illustrates a typical circuit to measure a transient response on a grounding electrode generalized from those in References 26 and 40 through 45.

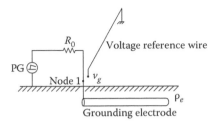

FIGURE 7.29
Experimental circuit.

In all the collected cases, including a scaled-down experiment and a real field test, the electrode length x ranges from 2 to 100 m. The electrode radius r is from 2 to 10 mm with a hard copper rod in most cases. The buried depth h ranges from 0.1 to 1 m. As the source, a current IG was used in the case of a high-amplitude current, and a PG was often used in the case of a low-amplitude current. In most experiments, a resistance of some kΩ is connected to the IG or PG to represent a lightning current. As a current lead wire, a strand copper wire covered with vinyl insulation layer (IV) was most common, but a coaxial cable was occasionally used. In all the measured results, the soil resistivity was clearly specified, but only a few cases specified the soil permittivity.

7.4.3.2 Measured Results

Table 7.3 summarizes the experimental conditions and results in the collected data [41–45]. The parameters in the table are as follows:

1. Current

    I_p: maximum value, T_i: the time of I_p appearance

    T_f/T_t: wavefront and wave tail duration following the 30/90% definition

    T_s: maximum observation time specified in the reference

    $i(t)$: current amplitude at time t

    In some cases, the wave tail duration T_t is not given, that is, $T_t > T_s$. In such a case, T_t is determined by extrapolation.

2. Voltage

    T_v: the time of the first peak appearance of a voltage waveform, which is not necessarily the time corresponding to the maximum voltage

    $v(t)$: voltage at time t

In Table 7.3, all the voltages and currents except I_p are normalized by I_p, and thus correspond to the case in which a unit of 1 A is applied. Based on

TABLE 7.3

Test Conditions and Results.

(a) Horizontal Electrode

	Electrode				Applied Current							Measured Voltage			Reference No.
Case No.	r (mm)	d (m)	x (m)	ρ_e (Ωm)	I_p (A)	T_t/T_t (μs)	T_i (μs)	$i(T_v)$ (A)	$i(T_s)$ (A)	T_s (μs)	T_v (μs)	$v(T_v)$ (V)	$v(T_i)$ (V)	$v(T_s)$ (V)	
H1-1	5.64	0.5	10	100–200/2.5 m 1500 –1800/18 m 200–500/40 m	2	0.01/1	0.01	1	1	12	1.6 (0.01)	60 (52)	52	64	1
H1-2					2	1/79	1	1	0.9	12	3	62	48	60	
H1-3					2	3/79	3	1	0.9	12	5	62	54	60	
H1-4					350	0.9/79	0.9	1	0.8	35	3	60	25	44	
H2	12	0.6	15	70 ($\varepsilon_r = 15$)	36	0.55/70	0.55	0.685	1	0.7	0.167	17.6	11.1	8.9	2
H3-1	4.37	1	34.6	200	1	0.04/18	0.04	1	0.95	4	0.04	10	10	3.8	3
H3-2			8.65		1	0.04/18	0.04	0.95	0.9	4	0.03	63	60	40	
H4-1	4.37	1	5	100–300	2.4	0.2/20	0.2	0.9	0.83	4	0.03	29.2	25	31.2	3
H4-2			10			0.2/20	0.2	0.9	0.83	4	0.06	28.8	26.7	26.1	
H4-3			15			0.2/20	0.2	0.9	0.83	4	0.06	28.5	26.7	20.1	
H4-4			30			0.2/20	0.2	0.83	0.83	4	0.1	28.1	25.8	14.3	
H4-5	10	0.2	6	100 ($\varepsilon_r = 9$)	1.55	0.055/1	0.055	0.967 (0.855)	1	1	0.044 (0.011)	47 (41)	45.5	37	4
H6-1	2.52	0.3	88.5	200	5.14	3.5/20	5.8	0.47	0.67	50	1.67	8.21	35	1.33	5
H6-2			18.5		5	3.5/20	5	1	0.65	50	5	12.5	12.5	13.8	

(Continued)

TABLE 7.3 (*Continued*)

Test Conditions and Results.

(a) Horizontal Electrode

Case No.	Electrode				Applied Current						Measured Voltage				Reference No.
	r (mm)	d (m)	x (m)	ρ_e (Ωm)	I_P (A)	T_f/T_t (μs)	T_i (μs)	$i(T_v)$ (A)	$i(T_s)$ (A)	T_s (μs)	T_v (μs)	$v(T_v)$ (V)	$v(T_i)$ (V)	$v(T_s)$ (V)	
H7-1	1	0.01	6	100	1	0.03/1	0.03	0.867	1	0.05	0.009	108	71	50	6
H7-2	1	0.2	6	1	0.03/1	0.03	0.9	1	0.05	0.012	108	71	50	
H7-3	10	0.01	6	1	0.03/1	0.03	0.9	1	0.05	0.012	91.3	65	46	
H7-4	10	0.2	6	1	0.03/1	0.03	0.7	1	0.05	0.004	78.3	50	32.6	

(b) Vertical Electrode

Case No.	Electrode			Applied Current						Measured Voltage				Reference No.
	r (mm)	x (m)	ρ_e (Ωm)	I_P (A)	T_f/T_t (μs)	T_i (μs)	$i(T_v)$ (A)	$i(T_s)$ (A)	T_s (μs)	T_v (μs)	$v(T_v)$ (V)	$v(T_i)$ (V)	$v(T_s)$ (V)	
V1-1	12.5	30.85	100–200/2.5 m	2	0.01/1	0.01	1	1	5	0.01	48	48	8	1
V2-2			1500	2	1/79	1	0.85	1	5	0.6	19	13	8	
V2-3			–1800/18 m	2	3/79	3	0.75	1	5	1.4	12	8	8	
V2-4			200–500/40 m	350	0.9/79	0.9	0.4	0.9	14	0.4	22	16	6.5	
V2-1	125	2.2	Same as Case 1	2	0.01/1	0.01	1	1	12	2.3	70	70	72	1
V2-2				2	1/79	1	0.95	0.9	12	5	70	52	64	
V2-3				2	3/79	3	1	0.9	12	6	70	58	64	
V2-4				350	0.9/79	0.9	1	0.85	20	2.0	70	60	63	
V3	8	6	50 ($\varepsilon_r = 15$)	33.5	0.75/70	0.75	0.742	0.791	3	0.38	12.3	8.96	5.57	2
V4	7	1	100 (9)	0.73	0.038/1	0.038	0.75	1	0.55	0.0125	53.4	30.1	42.5	4

the theory explained in Section 7.4.2, estimated various parameters are given in Table 7.4. In the table, the parameters are defined as follows:

$$R(T_i) = \frac{v(T_i)}{i(t_i)}, \quad R(T_s) = \frac{v(T_s)}{i(T_s)} \tag{7.57}$$

$$L = \frac{V_{Lmax}}{(\beta - \alpha)I_0} \quad \text{for } T_v < T_i \tag{7.58}$$

$$= \frac{v(t)}{\left.\dfrac{di(t)}{dt}\right|_{i=T_i}} \quad \text{for } T_v \geq T_i$$

Here, the following approximation is adopted:

$$\left.\frac{di(t)}{dt}\right|_{t=Ti} \approx \frac{i(T_i) - i(0)}{T_i} = \frac{i(T_i)}{T_i}$$

In this case, L is given as

$$L = T_i \frac{v(T_i)}{i(T_i)} \tag{7.59}$$

As is clear from Equation 7.58, $(\beta - \alpha)$ corresponds to $1/T_i$ or $1/T_v$ in Equation 7.59, and also corresponds to $\delta(t)$ when a pulse current is applied. If $k = \beta/\alpha > 70$, β becomes much greater than α, and Equation 7.58 is approximated by

$$L = \frac{V_{Lmax}}{\beta I_0} T_v < T_i \tag{7.60}$$

7.4.3.3 Discussion

Based on the analytical study in the previous sections, the characteristic of a grounding electrode are estimated as follows:

1. Inductive: $V_{max} = V_{Lmax} > V_{Rmax}$, and $T_v < T_i$ (or $R_v > R_i$, R_s and $T_v < T_i$)
2. Resistive: $V_{max} = V_{Rmax} > V_{Lmax}$, V_s ($R_i > R_v$, R_s)
3. Capacitive: $V_{max} = V_s > V_{Lmax}$, V_{Rmax} ($R_s > R_v$, R_i)

where

$$V_s = v(T_s), \; R_v = R(T_v), \; R_i = R(T_i), \; R_s = R(T_s)$$

TABLE 7.4

Estimated Parameters and Characteristics.

(a) Horizontal Electrode

Case No.	Electrode x (m)	Waveform T_f/T_t (μs)	α (×10⁴)	β (×10⁶)	T_i (μs)	T_v (μs)	Transient Resistance $R(T_i)$ (Ω)	$R(T_v)$ (Ω)	$R(T_s)$ (Ω)	Inductance $L(T_i)$ (μH)	$L(T_v)$ (μH)	Characteristic	Sunde's Resistance (Ω)
H1-1	10	0.01/1	70	310	0.01	1.6 (0.01)	52	60 (52)	64	0.52	96 (0.52)	Capacitive/resistive	58/400 Ωm
H1-2		1/79	0.899	3.07	1	3	48	62	66.7	48	144	Capacitive/resistive	58/400 Ωm
H1-3		3/79	0.949	2.85	3	5	54	62	66.7	162	310	Capacitive/resistive	58/400 Ωm
H1-4		0.9/79	0.886	3.17	0.9	3	25	60	55	49.5	180	Inductive/capacitive	58/400 Ωm
H2	15	0.55/70	0.1	5.18	0.55	0.167	11.1	25.7	8.9	6.11	4.29	Inductive	6.7
H3-1	34.1	0.04/18	3.8	80	0.04	0.04	10	10	3.8	0.4	0.4	Resistive	4.0
H3-2	8.65	0.04/18	3.8	80	0.04	0.03	60	66.3	40	2.4	1.99	Inductive	31.1
H4-1	5	0.2/20	3.5	15.5	0.2	0.03	25	32.4	31.2	5.0	0.97	Inductive/capacitive	53.5/200 Ωm
H4-2	10	0.2/20	3.5	15.5	0.2	0.06	26.7	32.0	26.1	5.34	1.92	Inductive	31.2
H4-3	15	0.2/20	3.5	15.5	0.2	0.06	26.7	31.7	20.1	5.34	1.90	Inductive	22.8
H4-4	30	0.2/20	3.5	15.5	0.2	0.1	25.8	33.9	14.3	6.78	3.39	Inductive	11.6
H5	6	0.055/1	78	49	0.055	0.044 (0.011)	45.5	48.6 (48.0)	37	2.50	2.14 (0.53)	Inductive	22.5
H6-1	88.5	3.5/20	5.25	0.557	5.8	1.67	35	11.1	1.99	203	18.54	Inductive	5.3
H6-2	18.5	3.5/20	5.25	0.557	5	5	12.5	12.5	21.2	62.5	62.5	Capacitive	20.2
H7-1	6	0.03/1	74	96	0.03	0.009	71	124.6	50	2.13	1.12	Inductive	36.5
H7-2	6	0.03/1	74	96	0.03	0.012	71	120.0	50	2.13	1.44	Inductive	28.6
H7-3	6	0.03/1	74	96	0.03	0.012	65	101.4	46	1.95	1.22	Inductive	30.5
H7-4	6	0.03/1	74	96	0.03	0.004	50	111.6	32.6	1.5	0.45	Inductive	22.5

(Continued)

TABLE 7.4 (*Continued*)

Estimated Parameters and Characteristics.

(b) Vertical Electrode

Case No.	Electrode x (m)	Waveform T_f/T_t (μs)	α (×10⁴)	β (×10⁶)	T_i (μs)	T_v (μs)	Transient Resistance $R(T_i)$ (Ω)	$R(T_v)$ (Ω)	$R(T_s)$ (Ω)	Inductance $L(T_i)$ (μH)	$L(T_v)$ (μH)	Characteristic	Sunde's Resistance (Ω)
V1-1	30.58	0.01/1	70	310	0.01	0.01	48	48	8	0.48	0.48	Inductive	33.8/400 Ωm
V1-2		1/79	0.899	3.07	1	0.6	13	22.3	8	13	13.4	Inductive	33.8/400 Ωm
V1-3		3/79	0.949	2.85	3	1.4	8	16.0	8	24	22.4	Inductive	33.8/400 Ωm
V1-4		0.9/79	0.886	3.17	0.9	0.4	16	55	7.2	14.4	22	Inductive	33.8/400 Ωm
V2-1	2.2	0.01/1	3.170	310	0.01	2.3	70	70	72	0.7	161	Capacitive	70.6/150 Ωm
V2-2		1/79	0.899	3.07	1	5	52	73.7	71	52	368	Inductive/capacitive	70.6/150 Ωm
V2-3		3/79	0.949	2.85	3	6	58	70	71	174	420	Capacitive	70.6/150 Ωm
V2-4		0.9/79	0.886	3.17	0.9	2.0	60	70	74	54	140	Capacitive	70.6/150 Ωm
V3	6	0.75/70	0.1	4.13	0.75	0.38	8.96	16.6	7.04	6.72	6.31	Inductive	18.6
V4	1	0.038/1	2.8575	75	0.038	0.0125	30.1	71.2	42.5	1.143	0.89	Inductive	170.2

Thus, the maximum voltage is produced predominantly by the inductive component of an electrode circuit when T_v is smaller than T_i, and the voltage converges to that determined by the electrode resistance. On the contrary, when the RC component is dominant, the maximum voltage appears at T_v which is greater than T_i. The case of $T_v \approx T_i$ corresponds to $v = Ri$, and thus indicates the resistive characteristic. T_v being greater than T_i means that the time constant $\tau = RC$ is greater and, thus, the influence of the capacitance is dominant, that is, the electrode impedance is capacitive.

The above theoretical investigation agrees with most of the measured results in Tables 7.3 and 7.4 and, thus, a model circuit in Figure 7.26b is useful to analyze the impedance characteristics of a grounding electrode.

The characteristics of some cases in Table 7.4 are not clear, and are explained as follows:

1. Only cases H4-1 and V2-2 show $T_v < T_i$ but $R_v > R_s > R_i$, and a capacitive characteristic in a longer time period. All other cases show an inductive characteristic when $T_v < T_i$.

2. Cases H1 and H6-2 show $T_i \leq T_v$ and a capacitive characteristic when $R_s > R_v$, R_i, while H1-1 to H1-3 look resistive for $R_v \approx R_s$. Case H1-4 shows both inductive and capacitive characteristics for $R_v > R_s > R_i$.

   In the case of an impulse-like current waveform, $v(t)$ starts to decrease corresponding to a current decrease for $t > T_f$. As a result, R_s becomes smaller than R_i and R_v. Thus, it becomes hard to observe a capacitive characteristic. In other words, the capacitive characteristic is clearly observed only in the case of a step-function waveform of a current. The above observation agrees with the theoretical analysis in Section 7.4.2.5.

3. When T_f is small, β becomes large and $v_L = L di/dt \approx BLI_0$. Thus, it looks inductive except for Case H-1. However, it is not clear if inductance L is due to an electrode or a lead wire when measuring the voltage. If the latter, then the electrode impedance may not be inductive. In this case, it is necessary to confirm if the measured result is correct, and to carry out a measurement with a lead wire of which the inductance is far smaller than the electrode inductance. For example, Case H-7 in Table 7.4 shows an inductive characteristic independently of the electrode radius and the buried depth. T_f of Case H-7 is 30 ns, and the time to the crest is about 10 ns. The derivative of the current at this time period is $di/dt = 10$ ns $= 0.5$ A/10 ns $= 0.5 \times 10^7$. Assuming that the length of a lead wire to the electrode is 0.5–1 m, which yields $L' = 0.5$ μH, the voltage produced by this inductance becomes $V_L' = L' \cdot di/dt = 0.5 \times 10^{-6} \times 0.5 \times 10^7 = 25$ V. When this voltage is deducted from the peak voltage V_{Lmax} in Table 7.4, the inductive characteristic does not become clearer. If the length is 1 m and $L' = 1$ μH, $V_L' = 50$ V. Then $V_{Lmax} \approx V_{Rmax}$, and thus

it becomes resistive. The above observation has indicated that the electrode impedance in Case H-7 may not be inductive.

7.4.3.4 Effect of Lead Wire

To confirm the theoretical analysis in Section 7.4.2 and the observation in Section 7.4.3, an NEA, which is the most accurate method for analyzing a transient [38], is carried out by adopting an FDTD method [20,39].

Figure 7.30 illustrates an experimental circuit with electrode length $x = 8$ m, radius $r = 1$ mm, and buried depth $h = 0.22$ m. The soil resistivity was measured to be 140 Ωm. An impulse current with the amplitude $I_0 = 1$ A and the wavefront duration $T_f = 20$ ns is applied. To make an FDTD simulation easy, the same conductor as the electrode is used as a lead wire with length x_1. Figure 7.31 shows a comparison of the simulation results for $x_1 = 0.22$ m with a measured result. It should be clear that the FDTD simulation gives a satisfactory accuracy.

Figure 7.32 shows FDTD simulation results for $x_1 = 0.22$, 0.5, and 1.0 m. It is clear in the figure that the maximum voltage becomes greater and thus

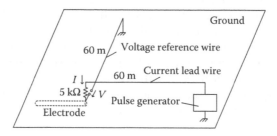

FIGURE 7.30
A detail of an experimental setup.

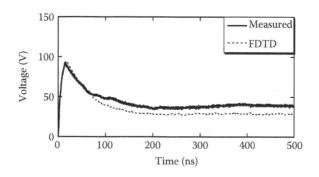

FIGURE 7.31
Comparison of an FDTD simulation with measured results: $r = 1$ mm, electrode length $x = 8$ m, buried depth $h = 0.2$ m, $\rho = 140$ Ωm, and lead wire length $x_1 = 0.22$ m.

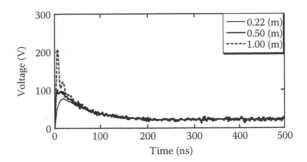

FIGURE 7.32
Influence of the lead wire length x_1 from 0.22 to 1.0 m.

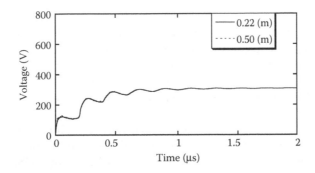

FIGURE 7.33
Influence of the lead wire length for $\rho = 2000\ \Omega m$.

the inductive characteristic becomes more noticeable as the lead wire length increases.

Figure 7.33 shows a simulation result for $\rho = 2000\ \Omega m$. As is well-known, electrode impedance tends to be capacitive in the case of a higher soil resistivity, and thus the lead wire does not affect the characteristic.

Figure 7.34 shows a comparison of simulation results for the case of $T_f = 20$ and 300 ns. It is clear that the inductive characteristic for $T_f = 20$ ns is no longer observed for $T_f = 300$ ns. Instead, it becomes capacitive. It should be noted that the voltage at $t = 500$ ns is the same, 24 V, for both $T_f = 20$ ns and 300 ns. The observation clearly indicates that electrode impedance shows an inductive characteristic due to the lead wire's inductance when wavefront duration T_f of an applied current is small.

7.4.4 Reduction of Grounding Impedance: Effect of Electrode Shape

7.4.4.1 Introduction

There are a number of publications that discuss the reduction of grounding impedances. In recent years it has become common to install digital circuits

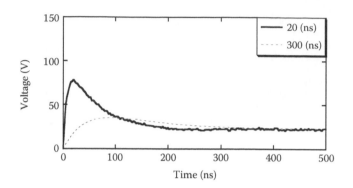

FIGURE 7.34
Influence of the wavefront duration

not only in power system control equipment but also in buildings and home appliances. Thus, it becomes more important to reduce the grounding impedances to guarantee reliable operation of the digital circuits for both steady and transient states as observed in References 47 through 50. From the above, KEPCO carried out a series of field tests to measure the grounding impedances.

A measured result involves noise due to mutual coupling between a current lead wire and a voltage reference wire, and also the unknown parameters such as soil resistivity and permittivity. An NEA method [38] can simulate a transient only by the geometrical and physical parameters of a given system, and thus, the NEA method is very effective to analyze the transient responses of a grounding electrode buried in soil of which the resistivity and permittivity are unknown.

This section shows field test results of transient and steady-state responses on grounding electrodes with various shapes, that is, a vertical rod, a rectangular plate, a circular plate, etc. [29]. The field tests were carried out in different seasons over a period of 3 years to obtain reliable results. Then, NEA simulations were carried out using VSTL [39] that is based on an FDTD method. A comparison with the measured results is first made to confirm the accuracy of the VSTL. Then, the VSTL is applied to investigate the reduction of the grounding impedances with various electrodes.

7.4.4.2 Field Measurements

7.4.4.2.1 Experimental Setup

Field measurements were carried out three times in a test yard located at the foot of a mountain in different years and seasons. The test field is placed in Hyogo Prefecture, Japan and owned by KEPCO. The soil resistivity of the test yard varies from 80 to 200 Ωm. Figure 7.35 illustrates an experimental setup. A PG (PG1, 2.6 kV/rise time 50 ns) and a noise simulator (PG2, 1.9 kV/rise time less than 1 ns) were used as a voltage source, and were terminated

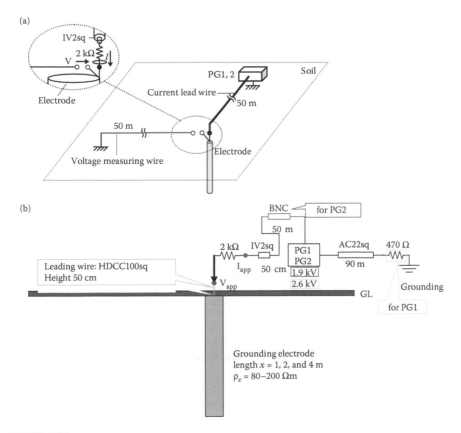

FIGURE 7.35
Circuit configuration of experiments. (a) Overview and (b) circuit configuration.

by a matching resistance of 470 Ω to a grounding electrode for the PG. The PG was connected to a test electrode through a resistance of 2 kΩ to represent a lightning current. An applied current was measured by a surge CT, and electrode voltage was measured by a voltage probe. The specifications of the PGs, the CTs, and the probe are given in Table 7.5. A current lead wire (IV2sq 50 m + AC22sq 90 m), a voltage reference wire (IV2sq 150 m), and the test electrode were set to be perpendicular to each other. The shapes of the test electrodes are shown in Figure 7.36.

7.4.4.2.2 Measured Results

Table 7.6 shows the shapes of tested electrodes and the corresponding measured results of peak voltages and converged voltages at the end of the measurement (PG1: $t = 4.5\ \mu s$, PG2: $t = 450$ ns) normalized by the peak current. Figures 7.37 and 7.38 show measured current and voltage waveforms in the cases of PG1 ($T_f \approx 0.5\ \mu s$) and PG2 ($T_f \approx 10$ ns), respectively. It is observed in Figure 7.37 that the grounding impedance is mostly resistive. In Figure 7.38,

TABLE 7.5

Specifications of Voltage Sources and Measuring Instruments

Measuring Equipment	Specifications
Voltage source	(Pulse Generator: PG1)
	Manufacturer: Cosmotec, maximum output voltage: 5 kV, waveform: 0.05/100 μs, power capacity: 50 VA
	(Noise Simulator: PG2)
	Manufacturer: NoiseKen, model: INS-4040, maximum output voltage: 4 kV, pulse rise time: <1 ns, maximum pulse width: 1 μs
Oscilloscope	Manufacturer: Tektronix, model: TPS2024, isolated channels: 4 ch, sample rate: 2 GS/s, bandwidth: DC ~200 MHz
Voltage probe	Manufacturer: Yokogawa Electric, model: 701944, attenuation ratio 100:1, bandwidth: DC ~400 MHz, maximum input voltage: 1000 Vrms. 6000-V peak
CT (1)	Manufacturer: Pearson, model: 2100, output: 1 V/1 A, rise time: <20 ns, maximum peak current: 500 A, option: BNC Cable (50 Ω), feed-through terminal (50 Ω)
CT (2)	Manufacturer: Pearson, model: 2877, output: 1 V/1 A, rise time: <2 ns, maximum peak current: 100 A, option: BNC Cable (50 Ω), feed-through terminal (50 Ω)

it shows a capacitive characteristic. In fact, the steady-state resistance R_g measured separately is greater than the converged value at $t = 450$ ns as observed in Table 7.6b.

It is clear that the peak and converged (called "steady-state" hereafter) values of the grounding resistance are significantly reduced when a rectangular plate is used (Cases 2 and 3 in Figure 7.36). For example, the peak and steady-state resistances of Cases 2-1 and 3-1 ($x = 1$ m) are reduced to 1/2 to 1/3 of those of the rod case, Case 1-1. The steady-state resistance of Case 6 is half of that of Case 4-3 (pipe electrode) as observed in Table 7.6b and Figure 7.38. However, the circular plate used in Case 7 shows no reduction of the grounding resistance. It is reasonable that the grounding resistance becomes smaller as the electrode length becomes longer. For Cases 1 and 4 in Table 7.6, the steady-state resistance evaluated by Sunde's formula [3] is given as a reference. For the evaluation of the resistance for the L-type rod in Case 1, it is represented by an equivalent circular cylinder with the radius of r_e that gives a surface area equivalent to that of the L-type rod. This equivalence has been confirmed to be reasonable from a comparison of the measured and simulation results. It looks an interesting subject to develop a theoretical formula of the grounding impedance for the electrodes in Cases 2-7 in Figure 7.36.

Any measured result of a grounding impedance involves unknown physical parameters, such as soil resistivities along the earth's surface, depth at which the impedance is buried, irregular waveforms of an applied current, and possible noise during a measurement. As a result, it is not possible to discuss uniformity of the measured results. A numerical simulation, however, is a very effective way to determine uncertainties.

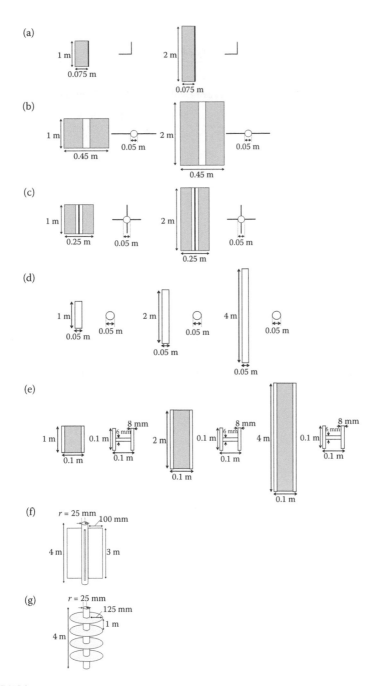

FIGURE 7.36
Electrode shapes. (a) L-type rod: Case 1, (b) rectangular plate (width 0.45 m): Case 2, (c) rectangular plate (width 0.25 m): Case 3, (d) vertical pipe (radius 0.025 m): Case 4, (e) H-type rod: Case 5, (f) four plates (fin-type): Case 6, and (g) four circular plates: Case 7.

TABLE 7.6

Transient Peak Voltages and the Converged Values.

(a) PG1—Converged Value at $t = 4.5\ \mu s$

Case	Electrode Shape	Length x (m)	Measured Voltage (V/A)		Simulation (V/A)		Sunde R (Ω)
			Peak	Converged	Peak	Converged	
1-1	L-type rod	1	106	99	98	83	82.4
1-2		2	44	42	40	35	49.3
2-1	Rectangular plate width (w) 0.45 m	1	33	31	48	40	
2-2		2	23	21	28	23	
3-1	Rectangular plate width (w) 0.25 m	1	42	40	40	35	
3-2		2	27	26	24	20	

(b) PG2—Converged Value at $t = 450$ ns

Case	Electrode Shape	Length x (m)	Measured Result				Simulation (V/A)			Sunde R (Ω)
			Steady-State R_g	First Peak		Converged Value (V/A)	First Peak		Converged Value (V/A)	
				V (V)/I (A)	t (ns)		V (V)/I (A)	t (s)		
4-1	Vertical pipe $r = 25$ mm	1	118	37/0.45	17	107.5	39.7/0.76	20	76.8	97.6
4-2		2	42	28/0.45	17	34.1	39.1/0.74	20	42.2	57.0
4-3		4	32	34/0.45	17	24.1	39.1/0.74	20	22.2	32.6
5-1	H-type rod Figure 7.36d	1	105	37/0.35	11	93.8	33.7/0.75	20	63.2	
5-2		2	40	43/0.45	14	34.1	33.6/0.75	20	35.3	
5-3		4	34	33/0.45	14	26.8	33.0/0.74	20	18.7	
6	4 plates/ Figure 7.36f	4	36	31.5/0.35	11	30	35.2/0.74	20	19.3	
7	4 circular/ Figure 7.36g	4	19	25/0.3	7	16.3	35.6/0.74	20	17.1	

7.4.4.3 FDTD Simulation

7.4.4.3.1 Simulation Model

Figure 7.39 illustrates an example of a model circuit for an FDTD simulation using VSTL [39]. For $x = 1$ m, the analytical space is taken to be $x_1 = 1.82$ m. $x_2 = z = y = 1.42$ m with a cell size of $\Delta s = 0.01$ m. The boundaries of the space are represented by Liao's second-order absorbing boundary [38]. The electrode is assumed to be perfectly conducting. The voltage of the electrode

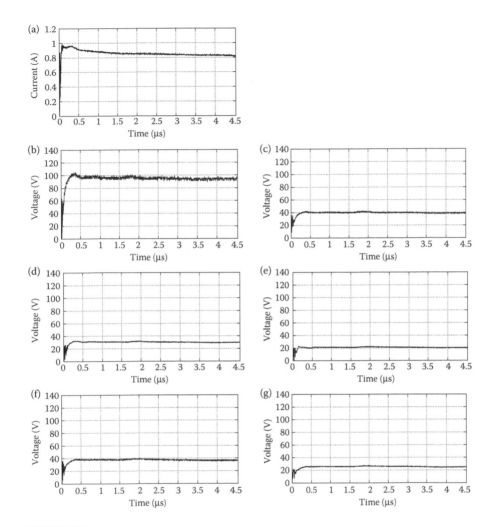

FIGURE 7.37
Measured results of electrode transient voltages with a PG: Cases 1-3. (a) Injected current, (b) Case 1-1: L-type rod, $x = 1$ m, (c) Case 1-2: L-type rod, $x = 2$ m, (d) Case 2-1: Plate (w 0.45 m), $x = 1$ m, (e) Case 2-2: Plate (w 0.45 m), $x = 2$ m, (f) Case 3-1: Plate (w 0.25 m), $x = 1$ m, and (g) Case 3-2: Plate (w 0.25 m), $x = 2$ m.

at the sending end is calculated by integrating the electric field from the absorbing boundary.

7.4.4.3.2 *Comparison with Measured Results*

Figures 7.40 and 7.41 show FDTD simulation results corresponding to the measured results in Figures 7.37 and 7.38. The simulation results of peak and steady-state voltages are given in Table 7.6. In the simulation, the soil resistivity is set at 150 Ωm and the relative permittivity is 10. It can be observed in

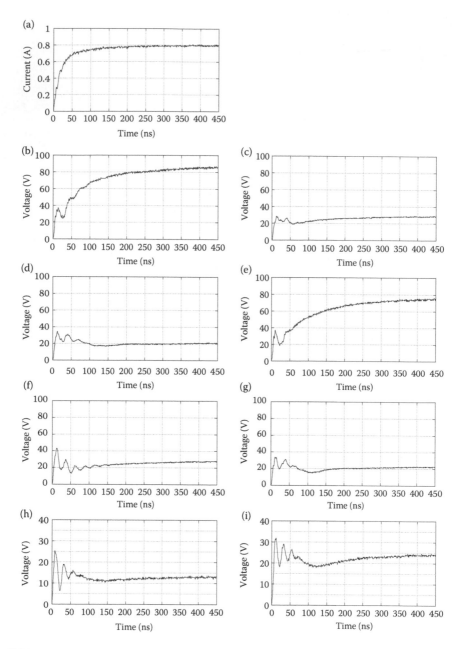

FIGURE 7.38
Measured results of electrode transient voltages with a noise simulator: Cases 4-7. (a) Injected current, (b) Case 4-1: Pipe, $x = 1$ m, (c) Case 4-2: Pipe, $x = 2$ m, (d) Case 4-3: Pipe, $x = 4$ m, (e) Case 5-1: H-type rod, $x = 1$ m, (f) Case 5-2: H-type rod, $x = 2$ m, (g) Case 5-3: H-type rod, $x = 4$ m, (h) Case 6: four plates (fin-type), and (i) Case 7: 4 circular plates.

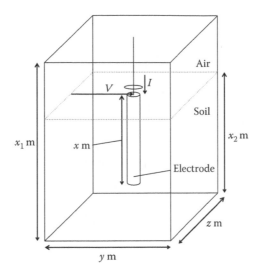

FIGURE 7.39
A model circuit for an FDTD simulation.

the table and figures that the simulation results agree reasonably well with the measured results. Thus, it is possible to investigate the effect of electrode shapes on the reduction of grounding impedances.

7.4.4.3.3 Effect of Electrode Shapes on Grounding Impedance Reduction

Figure 7.42 illustrates the shapes in FDTD simulations. The electrode length is set to $x = 1$ m, the soil resistivity is $\rho_e = 42$ Ωm, and the relative permittivity is $\varepsilon_r = 10$ as a reference case. Figure 7.43 shows an injected current waveshape with the amplitude 1 A and rise time $T_f = 10$ ns. Table 7.7 gives the simulation conditions, the peak, and the steady-state values of calculated transient voltages.

1. *Vertical rectangular plate:* Figure 7.44 shows the simulation results for the electrodes of Case 0 (vertical rod) in Figure 7.42a and the rectangular plates in Figure 7.42b–e. It is clear from Table 7.7 and Figure 7.44a that the peak and steady-state voltages in the plate cases are significantly reduced in comparison with those in the rod case. When the width of the plate is $w = 0.6$ m (Case A1), both the peak and the steady-state voltages are reduced to 1/3 of those of the rod case (Case 0). Even with $w = 0.3$ m (Case A0), those are reduced to 1/2 of the rod case.

   Case A2 further reduces the voltages by about 10% in comparison with those in Case A1. Case A3 with four plates reduces the voltage to 1/4 of those in Case 0. The reduction effect is observed to saturate when the number of plates and the surface area are increased.

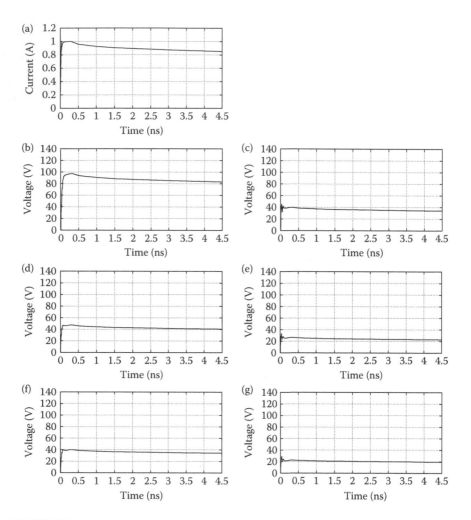

FIGURE 7.40

FDTD simulation results corresponding to Figure 7.37. (a) Injected current, (b) Case 1-1: L-type rod, $x = 1$ m, (c) Case 1-2: L-type rod, $x = 2$ m, (d) Case 2-1: Plate (w 0.45 m), $x = 1$ m, (e) Case 2-2: Plate (w 0.45 m), $x = 2$ m, (f) Case 3-1: Plate (w 0.25 m), $x = 1$ m, and (g) Case 3-2: Plate (w 0.25 m), $x = 2$ m.

2. *Horizontal circular plate:* Figure 7.45a shows a comparison of transient impedance between the horizontal circular plate electrodes in Figure 7.42g–i. The peak and steady-state values are given in Table 7.7. It is observed that the steady-state resistance is reduced to nearly the same value as that in the vertical rectangular plate case in Figure 7.44. However, the transient peak value demonstrates little reduction. In fact, those in Cases B11 and B21 match those of the rod electrode. The reason for this is that the impedance seen from the

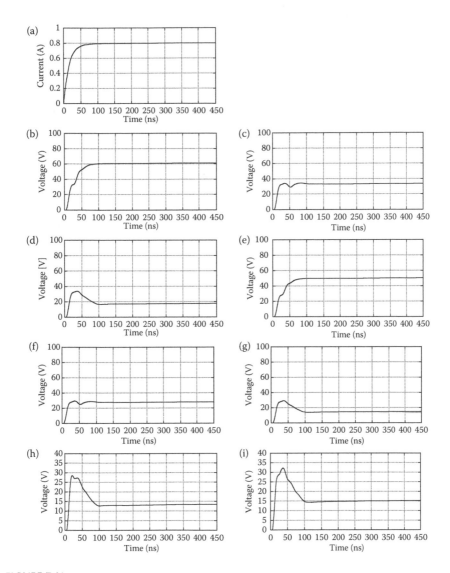

FIGURE 7.41

FDTD simulation results corresponding to Figure 7.38. (a) Injected current, (b) Case 4-1: Pipe, $x = 1$ m, (c) Case 4-2: Pipe, $x = 2$ m, (d) Case 4-3: Pipe, $x = 4$ m, (e) Case 5-1: H-type rod, $x = 1$ m, (f) Case 5-2: H-type rod, $x = 2$ m, (g) Case 5-3: H-type rod, $x = 4$ m, (h) Case 6: four plates (fin-type), and (i) Case 7: four circular plates.

current injected terminal at $t = 0$ in Cases B11 and B21 is the same as that of the rod electrode. The reduction effect of the horizontal circular plate appears after the current reaches the plate, that is, after $t = h/c$, where h is the depth of the plate from ground level and c is the traveling velocity. The same characteristic is observed in Cases B12

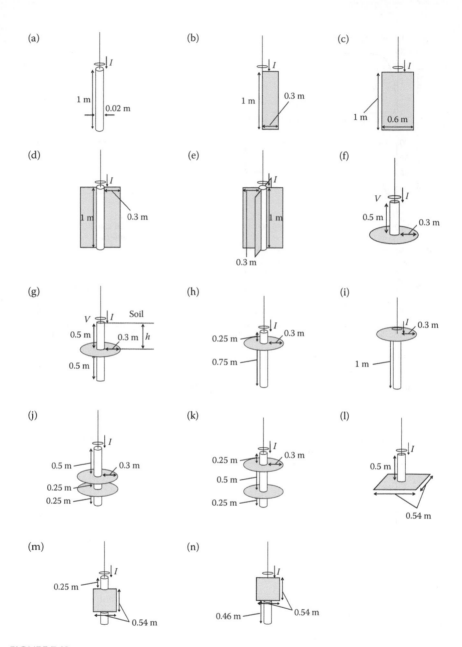

FIGURE 7.42
Electrode shapes investigated. (a) Case 0: Vertical rod only, (b) Case A0: Vertical rectangular plate, (c) Case A1: Rectangular plate with $w = 0.6$ m, (d) Case A2: Vertical rod + 2 rectangular plates, (e) Case A3: Vertical rod + 4 rectangular plates, (f) Case B0: Horizontal circular plate, (g) Case B11: Vertical rod + a circular plate at the center, (h) Case B12: Vertical rod + a circular plate at an upper position, (i) Case B13: Vertical rod + ground level circular plate rod, (j) Case B21: Vertical rod + 2 circular plates at a lower position, (k) Case B22: Vertical rod + 2 circular plates at upper and lower positions, (l) Case C0, (m) Case C12, and (n) Case C13.

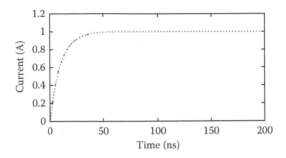

FIGURE 7.43
Injected current with $T_f = 20$ ns.

TABLE 7.7

Transient Peak Voltages (Peak) and the Steady-State Voltages (Converged) for Each Electrode Shape

Case	Electrode Shape	Simulation (V/A)	
		Peak	Converged
0	Vertical rod only	48.1	26.7
A0	Vertical rectangular plate with $w = 0.3$ m	24.3	12.9
A1	Vertical rectangular plate with $w = 0.6$ m	17.3	9.00
A2	Vertical rod with 2 rectangular plates	17.0	8.80
A3	Vertical rod with 4 rectangular plates	11.1	6.71
B0	Horizontal circular plate	49.8	13.0
B11	Vertical rod with a circular plate at the center	49.8	11.8
B12	Vertical rod with a circular plate at an upper	57.3	12.5
B13	Vertical rod with a circular plate at-ground level	19.8	16.5
B21	Vertical rod with 2 circular plates at the lower	57.3	8.18
B22	Vertical rod with 2 circular plates at an upper and lower	49.8	11.8
C0	Horizontal square plate	19.8	12.4
C12	Vertical rod with a square plate at an upper	57.0	12.4
C13	Vertical rod with a square plate at-ground level	15.2	13.7

and B11 in Figure 7.45a and in Cases B21 and B22 in Figure 7.45b. For example, in Case B12, the voltage starts to decrease at $t = 6.8$ ns due to negative reflection from the plate, and the peak voltage therefore stays at about 28 V, while in Case B11, the peak voltage has already reached 35 V, when the negative reflection comes back at $t = 11.8$ ns.

It is obvious from the above observation that the horizontal plate should be installed as near to ground level as possible, that is, the buried depth h should be as small as possible. This is clearly observed in Case B13 in Figure 7.45a, that shows a resistive characteristic with

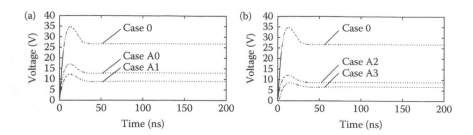

FIGURE 7.44
Reduction of a grounding resistance by a vertical rectangular plate: Case A. (a) Vertical plate and (b) vertical rod and plate.

a far smaller peak voltage in comparison with those of Cases B11 and B12. However, it should be noted that the steady-state resistance is nearly double of those of Cases B11 and B12. The reason for this is that the surface area facing the soil in Case B13 (only the lower face) is half of that in Cases B11 and B12 (both the upper and lower faces).

3. *Multiple horizontal plates:* Figure 7.45b shows the transient voltage of an electrode composed of two horizontal circular plates. The voltage waveform is similar to that of an electrode with one horizontal plate as shown in Figure 7.45a. Thus, it is concluded that an electrode with multiple horizontal plates has no practical value.

4. *Rectangular plate versus a circular plate:* Currents flowing into the ground from an electrode tend to be proportional to the surface area facing the soil [21,28]. An electrode with a rectangular plate with a surface area the same as that of a circular plate, is estimated to show a similar transient characteristic to that circular plate. Figure 7.46 is a comparison of the rectangular and the circular plates; it is obvious that they are almost identical. The observation has made it clear that the ground impedance tends to be proportional to the surface area of an electrode facing soils independent of its shape, which is either circular or rectangular.

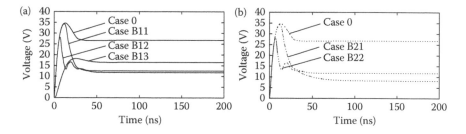

FIGURE 7.45
Reduction of a grounding resistance by a horizontal circular plate: Case B. (a) A single circular plate and (b) two circular plates.

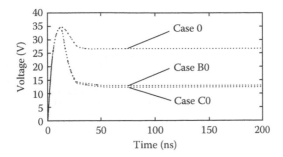

FIGURE 7.46
Comparison of a horizontal rectangular plate with a circular plate.

The above discussion is based on a horizontal plate. It was explained in Section 7.4.4.2 that the grounding impedance of a vertical rectangular plate is far smaller than that of a horizontal circular plate. Figure 7.47 shows a comparison of the impedance of horizontal circular plates, Case B12 in Figure 7.42h and Case B13 in Figure 7.42i, and those of equivalent (same surface area) vertical plates, Case C12 in Figure 7.42m and Case 13 in Figure 7.42n. The vertical plate shows a smaller impedance than that of a horizontal plate if the surface areas are the same, that is, the equivalent surface area of a vertical plate is smaller than that of a horizontal plate that gives the same impedance.

Thus, it is recommended to make a plate electrode vertical rather than horizontal when an additional plate is installed to reduce the grounding impedance.

7.4.4.3.4 *Effect of Soil Resistivity and Rise Time of a Current*
In the previous investigations, the soil resistivity was 42 Ωm. Figure 7.48 shows transient voltage waveforms for a rod electrode (Case 0) and vertical

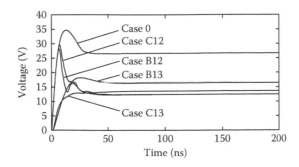

FIGURE 7.47
Comparison of vertical and horizontal plates.

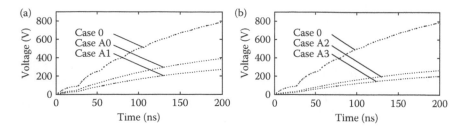

FIGURE 7.48
A transient voltage for $\rho_e = 2000$ Ωm. (a) Vertical plate and (b) a vertical rod with a vertical plate.

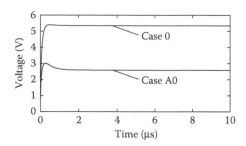

FIGURE 7.49
Transient voltage when $T_f = 1$ μs.

plate electrodes (Cases A0–A3) in the case of soil resistivity of 2000 Ωm. A capacitive characteristic of the impedance is observed in all the cases in Figure 7.48. The impedance of the plate electrode (Cases A0–A3) is less than 1/3 of that of the rod electrode (Case 0). This means that the vertical plate is very effective in reducing the grounding impedance even in the case of high soil resistivity. The soil permittivity has almost no influence on the reduction of the impedance. Figure 7.49 shows a comparison of transient voltages in the case of current rise time $T_f = 1$ μs. It is observed that the grounding impedance is reduced by nearly half by a vertical plate in comparison with the reduction achieved by a rod electrode.

7.4.4.4 Summary

This section has investigated grounding impedance reduction using various shapes of grounding electrodes in comparison with a conventional vertical rod based on field measurements and FDTD simulations. We have found that a vertical rectangular conducting plate is effective in reducing the transient and steady-state impedance to less than half of the reduction achieved by a vertical rod. A circular conducting plate parallel to the soil surface is also effective in reducing the steady-state impedance, but the transient impedance is nearly the same as that of a vertical rod. The rate of the reduction is somehow proportional to the surface area of the electrode, but tends to saturate as the area increases. The transient impedance is somewhat independent of the soil resistivity and permittivity, but tends to be inversely proportional to the wavefront duration of an applied current.

7.4.5 Transient Induced Voltage to Control Cable from Grounding Mesh

7.4.5.1 Introduction

Electromagnetic interference becomes more and more significant in control circuits of generator stations and substations as the amount of digitally controlled equipment increases. A survey of failures and malfunctions of low voltage control equipment in generator stations and substations shows that nearly 70% of the failures are caused by lightning surges [46–48]. The lightning current flows into a grounding mesh in a station and the current causes an induced transient voltage on a control cable. Because the metallic sheath of the control cable is grounded to the mesh as recommended by a standard and a guideline for electromagnetic interference [48–50], the cable sheath voltage becomes the same as that of the grounding mesh. If both the ends of the metallic sheath are grounded following IEC/CIGRE recommendations [48,49], a current is circulating in a closed loop composed of the cable sheath and the grounding mesh. The circulating current induces a voltage to the cable core, and the core voltage is given as a vector sum of the induced voltages from the lightning current along the grounding mesh and the circulating current. The lightning current traveling along conductors of the mesh, flows into the soil, and becomes a function of the length of the conductor [51]. Thus, the induced voltage to the cable is a function of the conductor's length.

The above fact seems to be poorly understood, and a detailed analysis of the components, that is, (1) the induced voltage due to mutual coupling from the lightning current along the grounding mesh, (2) the counterpoise voltage transferred to the cable sheath because of the sheath grounding, and (3) the circulating current due to sheath both-end grounding, has not been done. This is because all the measured results have involved the components named above [51–54].

This section investigates induced voltages to a control cable based on EMTP simulations. It is easy to evaluate the above-mentioned components separately using the EMTP [35–37]. That is, a simulation neglecting mutual coupling between a control cable and a grounding mesh gives voltages on the cable by grounding the cable sheath to the mesh. An induced voltage from the lightning current along the grounding mesh is also investigated, taking into account the current flowing into the soil. The effect of cable sheath grounding and a grounding lead is also discussed.

7.4.5.2 Model Circuit

Figure 7.50 illustrates a model circuit for investigating transient voltages and currents on a control cable. A counterpoise representing a part of a grounding mesh is buried at the depth of $h_g = 0.3$ m from the ground level. The counterpoise is a copper cylinder with an outer radius of $r_g = 2.5$ cm, and the soil resistivity is 100 Ωm. As a control cable, a 3D2 V cable, of which the cross section and the physical parameters are given in Figure 7.51, is suspended at a height of $h_c = 0.1$ m above ground level. A step function current with the amplitude of 1 A is applied to the sending end of the counterpoise as in Figure 7.50. Transient voltages and currents on the control cable and the counterpoise are calculated by EMTP.

A counterpoise is represented by a model circuit illustrated in Figure 7.52a that is composed of a distributed line with surge impedance Z_0, propagation velocity v_0, and shunt admittance Y_g as explained in Section 7.3.2 [22,24]. Variables "m" and "n" in Figure 7.52a are set as 1 and 5, respectively.

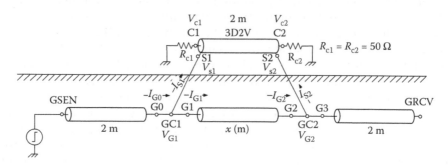

FIGURE 7.50
A model circuit.

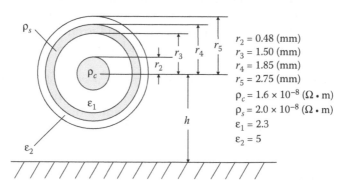

FIGURE 7.51
A control cable (3D2 V).

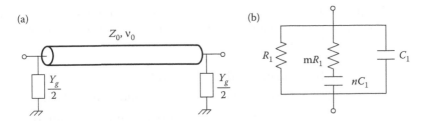

FIGURE 7.52
A model circuit of a counterpoise. (a) Counterpoise model and (b) Y_g.

Figure 7.53 shows a comparison of the experimental and EMTP simulation results using the model circuit of Figure 7.52 for transient voltages and currents along the counterpoise with the total length of 6 m in Figure 7.50. The figure shows that the accuracy of the counterpoise model in Figure 7.52 is satisfactory in comparison with the measured results.

In a simulation of an induced voltage to an overhead control cable from a counterpoise, the control cable and the counterpoise are represented as a distributed-parameter line in the EMTP [35–37]. The parameters of the line models are evaluated by the EMTP Cable Parameters (CP) [37]. First, the model system is evaluated as an overhead line system by the CP with a negative sign of the depth of the counterpoise. Initially, the input data, the CP gives the self-impedance/admittance of the overhead cable and the mutual impedance to the counterpoise. Then, the self-impedance/admittance of the counterpoise is calculated as an underground cable. Finally, the self-impedance/admittance of the counterpoise in the first calculation is replaced by those in the second.

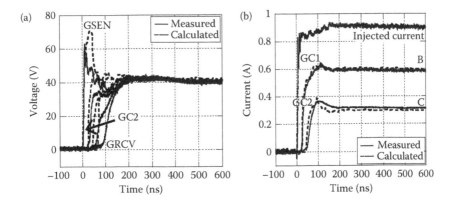

FIGURE 7.53
Comparison of measured and simulation results of counterpoise voltages and currents. (a) Voltage and (b) current.

Table 7.8 gives the simulation conditions, such as the grounding of the metallic sheath of a control cable, and results of the maximum voltages and currents. In the table, case *Xi-Lj* represents

X = A: $x = 2$ (m), B: $x = 10$ (m), and C: $x = 20$ (m)

D: $x = 2$ (m) and two more counterpoises connected to node GC1

$i = 0$: no mutual coupling between the control cable and the counterpoise

$i = 1$: mutual coupling between the control cable and the counterpoise

$L = 0$: no inductance considered for a grounding lead

$L = 1$:5 µH for a lead wire

j for grounding of the metallic sheath to a counterpoise

$j = 1$: no grounding, $j = 2$: sending end grounded

$j = 3$: receiving end, $j = 4$: both ends

Figures 7.54 through 7.61 show the simulation results of transient voltages and currents. The peak values are summarized in Table 7.8.

7.4.5.3 No Mutual Coupling (Case X0-Lj)

7.4.5.3.1 Case A0-0j: x = 2 m, L = 0

Figure 7.54 is the case of neglecting the mutual coupling between a control cable and a counterpoise with the parallel length $x = 2$ m. Thus, the voltages and currents on the control cable in the figure are caused only by connecting (grounding) the metallic sheath of the cable to the counterpoise.

No voltage and no current appear on the control cable in Figure 7.54a, for there is neither connection nor mutual coupling between the cable and the counterpoise. When one end of the sheath is grounded to the counterpoise, a voltage and a current appear on the grounded node. In Case A0-02 (sending end sheath to the counterpoise), the sheath voltage at the sending end S1 becomes nearly the same as the counterpoise voltage of about 20 V at the grounded node G1, and the core voltage at the sending end C1 reaches 4 V as observed in Figure 7.54b.2; a small voltage of about 1 V is observed at the core receiving end. The core voltages are produced by electrostatic and magnetic coupling between the cable core and the sheath during an initial transient period up to about 300 ns as a small current, about 0.1 A, on the cable sheath at the sending end observed in Figure 7.54b-(1) during the time period. A similar trend is observed for the receiving end sheath grounded to the counterpoise (Case A0-03), that is, the sheath voltage becomes nearly the same as the counterpoise voltage at the receiving end. Small voltages appear on both ends of the core and a very small current is observed on the sheath in Figure 7.54c. It should be noted that the core voltages and the sheath current are much smaller than those in Case A0-02.

TABLE 7.8

Simulation Conditions and Results.

(a) No Mutual Coupling (Case X0-Lj)

			Connection		Maximum Voltage (V)						Maximum Current (A)		
Case	x (m)	L (μH)	S1	S2	C1	C2	S1	S2	G1	G2	G0	S1	S2
A0-01	2	–	–	–	0.00	0.00	0.00	0.00	19.40	16.68	0.729	–	–
A0-02		0	on	–	4.46	1.38	19.26	20.70	19.26	16.64	0.728	0.116	–
A0-03			–	on	0.633	1.30	16.76	16.64	19.26	16.64	0.735	–	0.039
A0-04			on	on	4.93	–3.61	17.00	16.48	17.00	16.48	0.743	0.263	–0.262
A0-12		5	on	–	2.74	1.51	27.25	29.07	18.5	16.8	0.723	0.087	
A0-13			–	on	0.822	1.32	19.10	18.32	19.34	16.80	0.741		0.044
A0-14			on	on	2.701	–1.28	19.08	17.27	18.05	16.71	0.735	0.104	–0.105
B0-01	10	–	–	–	0.00	0.00	0.00	0.00	19.40	6.17	0.826	–	–
B0-02		0	on	–	8.13	2.28	17.32	19.38	17.32	6.21	0.828	0.211	–
B0-03			–	on	0.340	0.723	6.348	6.210	19.40	6.21	0.827	–	0.021
B0-04			on	on	8.13	–6.00	15.74	6.70	15.74	6.70	0.840	0.229	–0.230
C0-01	20		–	–	0.00	0.00	0.00	0.00	19.40	1.42	0.839		
C0-02		0	on	–	8.15	2.13	15.78	17.85	15.78	1.32	0.838	0.207	
C0-03			–	on	0.104	0.289	1.176	1.319	19.40	1.32	0.839		0.008
C0-04			on	on	8.15	–6.07	15.78	3.321	15.78	3.32	0.871	0.209	–0.204
C0-12		5	on	–	6.87	1.82	17.47	19.73	17.16	1.29	0.841	0.174	
C0-13			–	on	0.100	0.321	1.079	1.280	19.39	1.296	0.838		0.009
C0-14			on	on	6.87	–4.84	14.46	7.02	17.16	2.83	0.867	0.179	–0.185
D0-04	2	0	on	on	2.73	–1.99	9.98	9.91	9.98	9.91	0.934	0.154	–0.151

(b) With Mutual Coupling (Case X1-Lj)

A1-01	2	–	–	–	2.34	–2.53	2.82	–2.70	19.31	16.54	0.711	–	–
A1-02		0	on	-	4.52	–1.96	19.29	16.61	19.29	16.49	0.709	0.111	–
A1-03			–	on	3.10	–2.10	16.81	16.49	18.93	16.49	0.716	–	0.042
A1-04			on	on	5.47	–4.36	18.04	16.47	18.04	16.47	0.711	0.243	–0.239
A1-12		5	on	–	4.46	–1.66	26.15	24.29	18.63	16.51	0.707	0.077	–
A1-13			–	on	3.18	–1.46	21.93	19.03	19.10	16.51	0.722	–	0.052
A1-14			on	on	4.42	–2.67	20.97	17.30	18.66	16.55	0.712	0.075	–0.075
B1-01	10	–	–	–	2.53	–2.57	2.96	–2.89	19.38	6.019	0.821	–	–
B1-02		0	on	–	8.41	–2.52	17.69	16.46	17.69	6.025	0.822	0.199	
B1-03			–	on	2.88	–2.28	8.280	6.025	19.38	6.025	0.822		0.030
B1-04			on	on	8.41	–6.56	16.29	6.371	16.29	6.371	0.830	0.236	–0.233
C1-01	20	–	–	–	2.23	–2.25	2.60	–2.551	19.37	1.495	0.837	–	–
C1-02		0	on	–	8.45	–2.32	16.35	15.85	16.35	1.459	0.839	0.198	–
C1-03			–	on	2.48	–2.21	5.071	1.459	19.37	1.459	0.837	–	0.028
C1-04			on	on	8.45	–6.63	16.35	2.783	16.35	2.783	0.853	0.200	–0.200
C1-12		5	on	–	7.50	–2.60	17.85	16.24	17.49	1.509	0.843	0.163	–
C1-13			–	on	2.59	–2.17	5.511	1.725	19.37	1.509	0.837	–	0.023
C1-14			on	on	7.47	–5.60	15.53	6.595	17.49	2.148	0.850	0.163	–0.162
D1-04	2	0	on	on	2.94	–2.33	10.00	9.914	10.00	9.914	0.924	0.148	–0.147

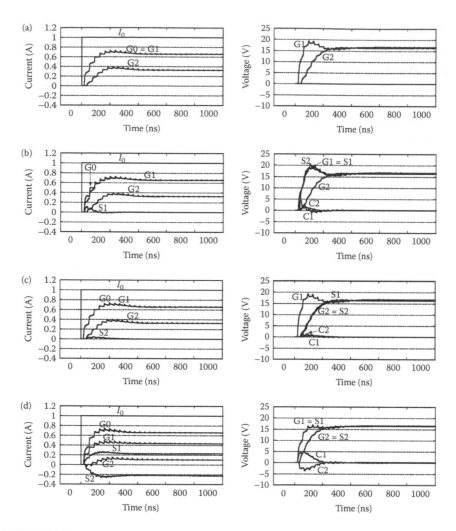

FIGURE 7.54
Simulation results neglecting mutual coupling between the control cable and the counterpoise ($x = 2$ m). (a) Sheath not grounded to the counterpoise (Case A0-01), (b) sending end sheath grounded to the counterpoise (Case A0-02), (c) receiving end sheath grounded to the counterpoise (Case A0-03), and (d) sheath both the ends grounded to the counterpoise (Case A0-04): (1) current and (2) voltage.

When both ends of the cable sheath are grounded to the counterpoise a large current, nearly 0.25 A, which is about 30% of the current I_{G0} flowing into node GC1 of the counterpoise, flows into the sheath as in Figure 7.54d-(1), and results in nearly 5 V at the core sending end and −3.5 V at the receiving end as observed in Figure 7.54d-(2). The core voltages are generated by electromagnetic coupling between the sheath and the core. It should be

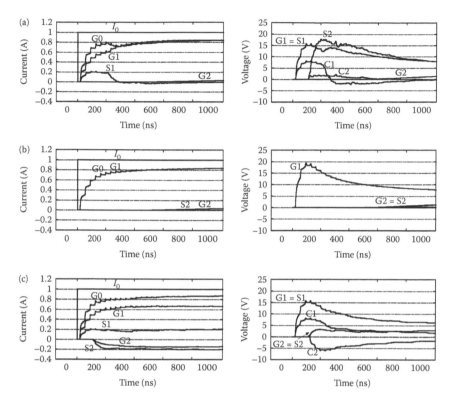

FIGURE 7.55
Effect of parallel length x with no mutual coupling ($x = 20$ m). (a) Sending end sheath grounded to the counterpoise (Case C0-02), (b) receiving end sheath grounded to the counterpoise (Case C0-03), and (c) Both sheath ends grounded to the counterpoise (Case C0-04): (1) current and (2) voltage.

noted that the counterpoise voltages and currents are relatively unaffected by the sheath grounding.

In summary, grounding the metallic sheath of a control cable to a counterpoise results in a current flowing into the sheath and generating core voltages due to electromagnetic and static coupling between the cable core and the sheath. The grounding of both ends of the sheath to the counterpoise, in particular, produces a sheath current of about 30% of the counterpoise current and the voltage proportional to the current appears on the core during the initial time period of a transient. Grounding the sending end sheath to the counterpoise also produces a rather high voltage to the core at the sending end.

7.4.5.3.2 Effect of Length x

Figure 7.55 shows the case of no mutual coupling with the parallel length $x = 20$ m between a control cable and a counterpoise. It is clear that currents

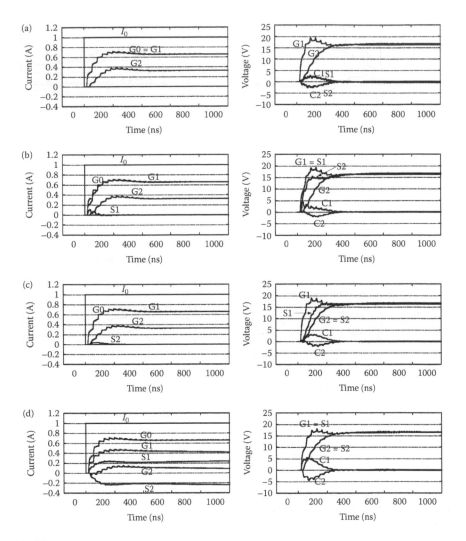

FIGURE 7.56

Simulation results considering mutual coupling between the control cable and the counterpoise ($x = 2$ m). (a) Sheath not grounded to the counterpoise (Case A1-01), (b) sending end sheath grounded to the counterpoise (Case A1-02), (c) receiving end sheath grounded to the counterpoise (Case A1-03), and (d) sheath both the ends grounded to the counterpoise (Case A1-04): (1) current and (2) voltage.

at the sending end of the counterpoise are greater, and the voltages smaller, than those in Figure 7.54 for $x = 2$ m, because the longer counterpoise length decreases its impedance seen from the sending end. Also, the currents along the counterpoise show a smooth increasing characteristic, while the voltage at the sending end shows a peak at around $t = 250$ ns and then converges to a certain value. The characteristics explained above for the counterpoise agree with those for the measured results shown in References 51 through 54.

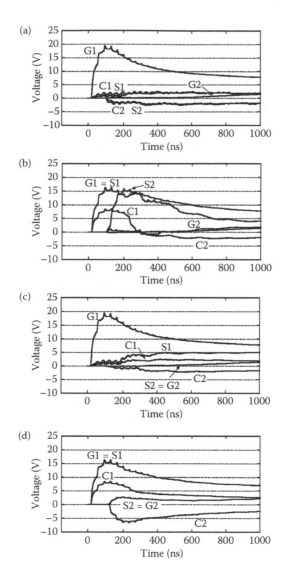

FIGURE 7.57
Effect of parallel length ($x = 20$ m). (a) Sheath not grounded to the counterpoise (Case C1-01), (b) sending end sheath grounded to the counterpoise (Case C1-02), (c) receiving end sheath grounded to the counterpoise (Case C1-03), and (d) sheath both the ends grounded to the counterpoise (Case C1-04).

The sheath currents in Figure 7.55 for $x = 20$ m show a remarkable difference from those in Figure 7.54 for $x = 2$ m. The current in Figure 7.55a-(1) for the sending end sheath grounded to the counterpoise is nearly twice that in Figure 7.54b-(1), and is sustained up to 160 ns, corresponding to twice the propagation time of a traveling wave on the sheath from the sending end to

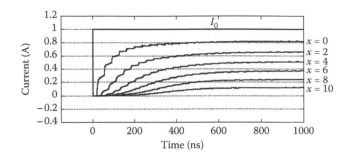

FIGURE 7.58
Currents along the counterpoise at the distance x from the sending end.

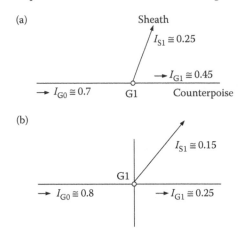

FIGURE 7.59
Crossing counterpoises at node G1. (a) A single counterpoise and (b) counterpoises connected at node G1.

the receiving end. Corresponding to this fact, the sending end core voltage reaches 8 V, which is about two times that in Figure 7.54b. A similar trend is observed in Figure 7.55c for $x = 20$ m in comparison with Figure 7.54d for $x = 2$ m. It is quite clear in Figure 7.55 c-(1) that there exists a circulating current between the control cable sheath and the counterpoise, that is, current S1 at the sheath sending end and current S2 at the receiving end. The sheath current produces the core voltages of about 8 V at the sending end and −6 V at the receiving end, much greater than those in Figure 7.54d-(2). The cable voltages and current in Figure 7.55b, where the receiving end sheath is connected to the counterpoise, are much lower than those in Figure 7.54c, because the counterpoise current at the receiving end becomes much smaller due to current penetration into the soil. It should be noted that the voltage difference between the cable core and the counterpoise is large at the sending end as in Table 7.8a, but becomes much smaller at the receiving end in comparison with the case of $x = 2$ m.

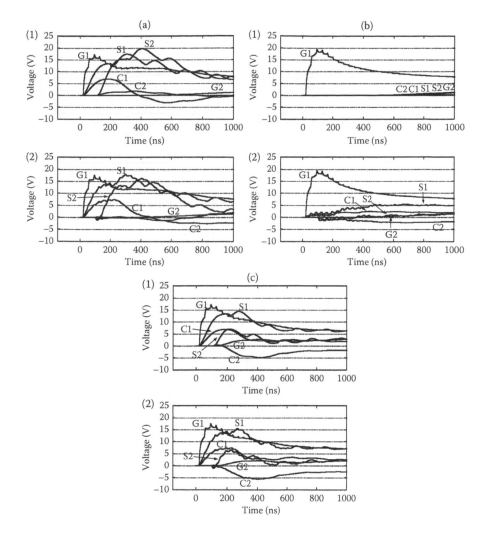

FIGURE 7.60
Effect of lead wire inductance on cable voltages. (a) sending end grounding, (1) No mutual coupling (Case C0-12), (2) mutual coupling (Case C1-12), (b) receiving end grounding, (1) no mutual coupling (Case C0-13), (2) mutual coupling (Case C1-13), (c) both-end grounding, (1) no mutual coupling (Case C0-14), (2) mutual coupling (Case C1-14).

7.4.5.4 With Mutual Coupling (Case X1-Lj)

7.4.5.4.1 Case A1-0j: $x = 2\ m$, $L = 0$

Figure 7.56 shows the simulation results with mutual coupling between a control cable and a counterpoise corresponding to Figure 7.54 with no mutual coupling. It is clear in Figure 7.56a that only an induced voltage of about 2.5 V, due to a counterpoise current, appears on the core and the sheath, because

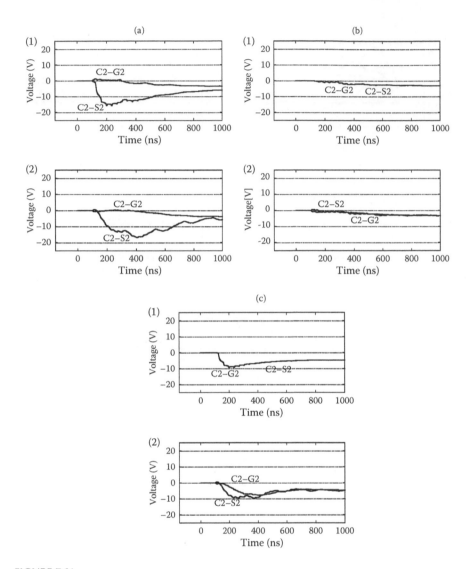

FIGURE 7.61
Core voltages to the sheath and counterpoise. (a) Sending end grounding, (1) No lead wire inductance (Case C1-02) and (2) lead wire inductance considered (Case C1-12). (b) Receiving end grounding, (1) No lead wire inductance (Case C1-03) and (2) Lead wire inductance considered (Case C1-13). (c) Grounding at both ends, (1) No lead wire inductance (Case C1-04) and (2) Lead wire inductance considered (Case C1-14).

the sheath is not grounded to the counterpoise. The polarity at the receiving end is opposite to the sending end as in Reference 40. It can be observed in Figure 7.56b that the current on the sheath is nearly the same as that in Figure 7.54b and the core voltages at both the ends are given as the sum of those in Figure 7.54b and the induced voltages in Figure 7.56a. The same observation is made for Figure 7.56c in comparison with Figure 7.54c. It should be noted that Figure 7.56d is nearly the same as Figure 7.54d. This means that the core voltage in Figure 7.56d is generated by a circulating current between the sheath and the counterpoise, although the induced voltage, 2.5 V. The current on the counterpoise is decreased by the grounding at both ends. If the sheath is grounded to another counterpoise on which no lightning current flows, the sheath grounding at both ends results in the smallest voltage on the cable core as recommended by standards and guides [47–50]. The recommendation is correct only in the case of no lightning current on a counterpoise to which the sheath is grounded. No significant difference is observed for the counterpoise voltages and currents between Figures 7.54 and 7.56.

In summary, the sending end voltage is the largest in the case of the sending end and both the ends that are grounded. The receiving end voltage becomes the largest in the case in which both ends are grounded because of a circulating current in a closed loop composed of the cable metallic sheath and the counterpoise. Thus, receiving end grounding is preferable to sending-end or grounding at both ends. The observations qualitatively agree with the measured results in References 46 and 51 through 53, especially in Reference 53 for the same 3D2 V cable.

7.4.5.4.2 Effect of Length x

Figure 7.57 shows the simulation results of voltages for $x = 20$ m. Current waveforms are not shown because those are nearly the same as Figure 7.55-(1). It is clear in Figure 7.57a for the sheath not grounded to the counterpoise that the induced voltages to the core and sheath from the counterpoise are far greater than those in Figure 7.56a for $x = 2$ m. This is reasonable because an induced voltage is proportional to the parallel length of the inducing circuit, as is well-known [40]. The length for Figure 7.57 being 20 m, a 10-times-greater induced voltage than that in Figure 7.56 is expected if it is in a steady-state, but the induced voltage at node C1 is about 2 V in Figure 7.57a, which is smaller than that in Figure 7.56. The reason for the smaller-than-expected induced voltage based on the theory of steady-state induction is that the inducing current on the counterpoise decreases as the distance from the sending end increases, as shown in Figure 7.58. It is clear in Figure 7.58 that the current along the counterpoise is decreasing exponentially as the distance increases. Furthermore, the rise time of the current, that is, di/dt, decreases rapidly. The transient induced voltage is given as

$$v_m(t) = M \frac{di}{dt} \tag{7.61}$$

where M: mutual inductance.

It should be clear from the above equation that the transient induced voltage decreases rapidly in proportion to the rise time decrease of the inducing current.

The core voltages in Figure 7.57b–d are given as the sum of the induced voltage in Figure 7.57a and the voltage in Figure 7.55 due to the sheath grounding to the counterpoise. Thus, the sending end core voltage reaches nearly 8 V in Figure 7.57b and d, and the receiving end core voltage is about −7 V in Figure 7.57d, and −6 V in Figure 7.57c.

7.4.5.4.3 *Effect of Crossing Counterpoises (Part of a Mesh)*

In practice, the metallic sheath of a control cable may be grounded to a node of a grounding mesh where mesh branches, that is, counterpoises, are connected to each other as illustrated in Figure 7.59. The impedance at node G1 seen from the left in Figure 7.59b becomes about 1/3 of that in Figure 7.59a on which all the previous investigations are based. The impedance variation in Figure 7.59b results in current distribution different from that in Figure 7.59a, and thus, a current flowing into the cable sheath connected to node G1 differs. An example is presented in the figure. It is observed in Figure 7.59 that the current flowing into the sheath is about 30% of the original current I_{G0} on the counterpoise when the sheath is grounded to an intermediate node of a single counterpoise as in 7.59a, while it is less than 20% when the sheath is grounded to the mesh node as in 7.59b. This difference results in dissimilar transient voltages and currents on a control cable as shown in Table 7.8.

A comparison of Case D1-04 with Case A1-04 and D0-04 with A0-04 in Table 7.8 makes it clear that the cable voltages are reduced to less than half by grounding the sheath to the node of a grounding mesh where counterpoises are connected to each other. The result is reasonable because a current flowing into the sheath (I_{S1}) and a current flowing through the counterpoise to the right of node G1 (I_{G1}) become smaller. I_{G1} is the inducing current to produce the cable voltage when the sheath is not grounded to the counterpoise.

The above result indicates that an analysis of the induced voltage to the control cable can be carried out assuming a single counterpoise as the severest case. Also, it should be mentioned that counterpoises that are a part of the grounding mesh parallel to a control cable induce voltages on the cable. The voltage induced by the counterpoise nearest to the cable is the largest, and those due to the other counterpoises are much smaller because of the distance to the cable, and are somehow cancelled out because of the symmetrical configuration of the counterpoises to the cable. Therefore, it is expected that only the induced voltage from the nearest counterpoise is enough to be considered in the transient analysis of the induced voltage.

7.4.5.4.4 *Effect of Lead Wire Inductance*

It is well-known that a transient response in a circuit is significantly influenced by the impedance of a lead wire used for grounding, connecting circuits, and measurements as explained in Section 7.4.2. In Reference 47, it is said that the grounding of the metallic sheath of a control cable may not be effective at all during a high-frequency transient because of the grounding lead inductance.

Figure 7.60 shows the effect of the grounding lead inductance assuming 1 μH/m on voltages and currents of a control cable. It is observed in Table 7.8 that the inductance of 5 μH, corresponding to the 5 m grounding lead, reduces a current flowing into the sheath and the core voltages to nearly half of that seen in the case without mutual coupling. However, Figure 7.60 with mutual coupling shows that the cable voltages do not differ much from those in Figure 7.57. The reason for this is readily explained by the fact that the high impedance of the grounding lead due to the inductance decreases the current flowing into the sheath and thus the current on the counterpoise increases, which induces higher voltages on the cable core. In a manner similar to that discussed in Section 7.4.5.3, the largest core voltage at the receiving end is observed when both ends of the sheath are grounded. Thus, it is concluded that the receiving end grounding is preferred even when lead wire inductance is considered.

The above observations agree with those explained in Reference 47 that sheath grounding does not become effective for a transient due to the inductance of a lead wire.

7.4.5.4.5 *Core Voltages to Sheath and Counterpoise*

The previous sections studied voltages to the zero-potential surface to investigate the effect of sheath grounding. In practice, a voltage difference from the core of a control cable to the sheath or to a counterpoise is used as the core voltage. Figure 7.61 shows the voltage difference. It is clear that the voltage difference from the core to the sheath is nearly the same as that to the counterpoise independently from the lead wire inductance. However, the voltage difference to the sheath is entirely dependent on the sheath grounding, and is greater than that to the counterpoise. The voltage difference is greatest in the case of the sheath sending end grounding, and the receiving end grounding shows the smallest difference. The results again suggest that the receiving end grounding is better than grounding at both ends.

7.4.5.5 **Conclusion**

The Section 7.4.5 has investigated the effect of sheath grounding on a control cable based on EMTP simulations, when a lightning current flows into counterpoises representing part of a grounding mesh. Voltages and currents observed on the control cable are given as a superposition of a well-known induced voltage due to the lightning current on the counterpoise, of a voltage transferred from the counterpoise to which the cable metallic sheath is grounded, and of a lightning current circulating in a closed loop composed of the metallic sheath and the counterpoise when both ends of the cable are grounded. The induced voltage from the counterpoise is not necessarily dominant even in a long cable because the current along the counterpoise decreases rapidly as the distance from the sending end increases. However, the sheath grounding decreases the node voltage due to the induced voltage, but at the same time increases voltage transferred from the counterpoise. The inductance of a grounding lead reduces a current flowing into the cable sheath during a high frequency transient, and thus more current flows through the counterpoise. This results in a higher induced voltage to the control cable from the counterpoise.

Appendix 7A

7A.1 Negative Voltage at the Front of an Induced Voltage

1. *Investigation of voltage measuring wire arrangement:* Figure 7A.1 shows various voltage measuring wire arrangements.

   a. *Connection A:* The measuring wire is connected only with the ground of a voltage probe for an induced electrode as in Figure 7A.1a.

   b. *Connection B:* The measuring wire is branched into two wires and each branched wire is connected with the ground of voltage probes for the inducing and induced voltages as in Figure 7A.1b.

   Figure 7A.2 shows the measured results of inducing and induced voltages for the above two arrangements of the measuring wire connection. It is observed that the negative induced voltage is greater in Connection A than in Connection B. The reason for the difference is clear, that is, Connection B cancels out noise induced by the voltage measuring wire.

2. *Investigation of the negative voltage:* Measured results in Section 7.4.1.3 were obtained by a simultaneous measurement of inducing and induced voltages. Figure 7A.3 shows a measured result of the induced voltage when only the induced voltage is measured. No negative voltage is observed in Figure 7A.3.

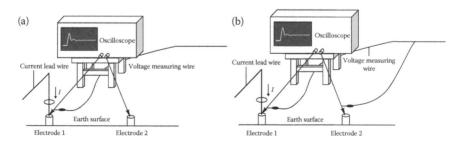

FIGURE 7A.1
Position of a voltage measuring wire. (a) Connection A and (b) Connection B.

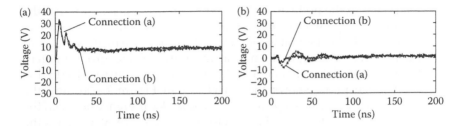

FIGURE 7A.2
Effect of a voltage measuring wire. (a) V_1 and (b) V_2.

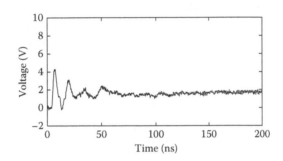

FIGURE 7A.3
Experimental result of V_2.

Thus, it should be clear now that the negative voltage is caused by noise induced by the voltage measuring wire.

3. *Investigation by FDTD simulations:* Figure 7A.4 shows an FDTD simulation model of a measuring system. Figure 7A.5 shows the simulation results of voltages at nodes A to E in Figure 7A.4. The voltages are calculated by integrating the electric field from the absorbing

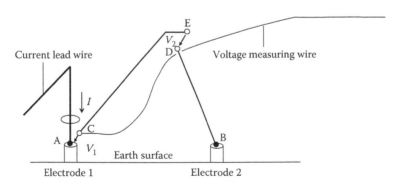

FIGURE 7A.4
The detail of a measurement system.

boundary to the node. It is observed in Figure 7A.5a and b that the induced voltage waveform is similar to the inducing voltage waveform. Similarly, the voltage is induced to the measuring wire as in Figure 7A.5c. Owing to the measuring wire voltage, ground potential rise, and electromagnetic induction from the current lead wire, the voltage probe shows the waveform in Figure 7A.5d, and the ground of the oscilloscope in Figure 7A.5e. A difference between the voltages V_P and V_E gives the waveform in Figure 7A.5f, which is the same as that in Figure 7.14b.

Figure 7A.6 shows the effect of the current lead wire height, which is 10 and 60 cm. The lower height gives a lower negative voltage at the front part of the induced voltage. Thus, it should be clear that the negative part of the induced voltage is caused by coupling of the current lead wire, the voltage measuring wire, and the probe wire.

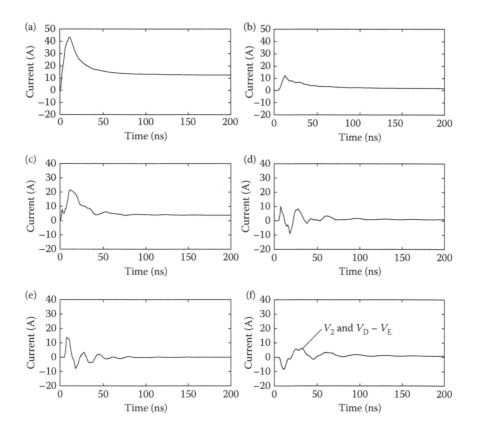

FIGURE 7A.5
Simulation result at various nodes in Figure 7A.4. (a) Point A (top of electrode 1), (b) Point B (top of electrode 2), (c) Point C (end of the measuring wire), (d) Point D (end of voltage probe 1), (e) Point E (the ground of an oscilloscope) and (f) Comparison between V_2 and $V_D - V_E$.

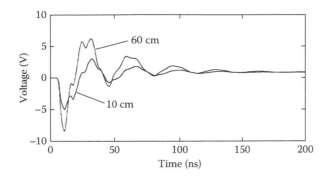

FIGURE 7A.6
Effect of a current lead wire.

References

1. Bewley, L. V. 1951. *Traveling Waves on Transmission Systems*. New York: Dover.
2. Tagg, G. F. 1964. *Earth Resistances*. London: G. Newnes.
3. Sunde, E. D. 1968. *Earth Conduction Effects in Transmission Systems*. New York: Dover.
4. Meliopoulos, A. P. Sakis. 1988. *Power System Grounding and Transients*. New York: Marcel Dekker.
5. Takahashi, T. 1990. *Introduction to Designing Grounding System*. Tokyo: Ohm Pub. Co. (in Japanese).
6. Belashi, P. L. 1941. Impulse and 60-cycle characteristics of a driven ground. *AIEE Trans.* 62:334.
7. Berger, K. 1946. Das verhalten von erdungen unter hohen stosstoemen. *Bull. Assoc. Suisse Eleck.* 197.
8. Schwarz, S. J. 1954. Analytical expressions for the resistance of grounding systems. *IEEE Transactions on Power Apparatus and Systems*. 73(2): 1011–1016.
9. Ryabkova, E. Y. and V. M. Mishikin. 1967. Impulse characteristics of earthings for transmission line towers. *Elektichestvo*. 8:67.
10. Korsuntcev, A. V. 1958. Application of the theory of similitude to the calculation of concentrated earth electrodes. *Elektichestvo*. 5:31–35 (in Russian).
11. Liew, A. C. and M. Darveniza. 1974. Dynamic model of impulse characteristics of concentrated earths. *Proc. IEE* 121:124.
12. Verma, R. and D. Mukhedkar. 1980. Impulse impedance of ground wire. *IEEE Trans. Power App. Syst.* 99:2003–2007.
13. Gupta, B. R. and B. Thapar. 1980. Impulse impedance of grounding grids. *IEEE Trans. Power App. Syst.* 99:2357–2362.
14. Rogers, E. J. 1981. Impedance characteristics of large tower footings to a 100 μs wide square wave of current. *IEEE Trans. Power App. Syst.* 100:66–71.
15. Chisholm, W. A. and W. Janischewskyj. 1989. Lightning surge response of ground electrodes. *IEEE Trans. Power Del.* 4:1329–1337.
16. Menter, F. E. and L. Grcev. 1994. EMTP-based model for grounding system analysis. *IEEE Trans. Power Del.* 9:1838–1849.

17. Grcev, L. and F. Dawalibi. 1990. An electromagnetic model for transients in grounding systems. *IEEE Trans. Power Del.* 5:1773–1781.
18. Hembach, M. and L. D. Grcev. 1997. Grounding system analysis in transients programs applying electromagnetic field approach. *IEEE Trans. Power Del.* 12:186–193.
19. Tanabe, K. 2000. A method for computational analysis of transient resistance for grounding systems based on the FD-TD method. *IEEJ Trans. PE* 120:1119–1126 (in Japanese).
20. Tanabe, K. 2001. Novel method for analyzing dynamic behavior of grounding systems based on the finite-difference time-domain method. *IEEE Power Eng. Rev.* 21:55–57.
21. IEEJ WG (Convener: Ametani, A.). 2002. Power system transients and EMTP analysis. *IEEJ Technical Report* 872 (in Japanese).
22. Ametani, A., D. Soyama, Y. Ishibashi, N. Nagaoka, and S. Okabe. 2004. Modeling of a buried conductor for an electromagnetic transient simulation. *IEEJ Trans. EIS* 123:1–6.
23. Motoyama, H. 2005. Experimental investigations of transients resistance characteristics of various grounding electrodes. *CRIPEI Research Report* H01010 (in Japanese).
24. Ametani, A., N. Taki, D. Miyazaki, N. Nagaoka, and S. Okabe. 2007. Lightning surges on a control cable incoming through a grounding lead. *IEEJ Trans. PE* 127:267–275.
25. Hirai, T., T. Miyazaki, S. Okabe, and K. Aiba. 2007. A study on an analysis model of grounding electrode for power distribution lines considering transient behavior. *IEEJ Trans. PE* 127:833–839 (in Japanese).
26. Mishra, A. K., A. Ametani, N. Nagaoka, and S. Okabe. 2007. Nonuniform characteristics of a horizontal ground electrode. *IEEE Trans. Power Del.* 22:2327–2334.
27. Ahmeda, M., N. Ullah, N. Harid, H. Griffiths, and A. Haddad. 2009. Current and voltage distribution in a horizontal earth electrode under impulse conditions. *Proceedings of UPEC 2009*, Glasgow, UK. Paper PS8–6.
28. Ametani, A., H. Morii, and T. Kubo. 2011. Derivation of theoretical formulas and an investigation on measured results of grounding electrode transient responses. *IEEJ Trans. PE* 131: 205–214 (in Japanese).
29. Ametani, A., Y. Fujita, H. Morii, and T. Kubo. 2012. A study on a method for reducing a grounding impedance. *IEEJ Trans. PE* 132: 284–291 (in Japanese).
30. Fujita, Y., A. Ametani, K. Nakamura, and K. Yamabuki. 2012. A study on mutual coupling between vertical grounding electrodes. *IEEJ Trans. PE* 132: 969–977.
31. *Handbook of Corrosions and the Countermeasures*. 2000. Tokyo: Maruzen Pub. Co.
32. Ametani, A., J. Kanba, and T. Hosokawa. 2003. A simulation method of voltages and currents on a gas pipeline and its fault location. *IEEJ Trans. PE* 123:1194–1200.
33. Japanese Electrotechnical Committee. 1994. Standard of High Voltage Testing. *JEC-102-1994*. Tokyo: IEE Japan (in Japanese).
34. Technical Recommendation for Electrical Facilities. 2012. *Ministerial Ordinance No.68*, Japanese Ministry of Economy, Trade and Industry (in Japanese).
35. Scott-Mayer, W. 1982. *EMTP Rule Book*. Portland: Bonneville Power Administration.
36. Dommel, H. W. 1984. *EMTP Theory Book*. Portland: Bonneville Power Administration.
37. Ametani, A. 1994. *Cable Parameters Rule Book*. Portland: Bonneville Power Administration.

38. CIGRE WG C4-501 (Convener: Ametani, A.). 2013. Guideline for Numerical Electromagnetic Analysis Method and Its Application to Surge Phenomena. *CIGRE-TB* 543.
39. CRIEPI: "Visual Surge Test Lab. (VSTL)," http://criepi.or.jp 2007.
40. Ametani, A. 1990. *Distributed-Parameter Circuit Theory*. Tokyo: Corona Pub. Co.
41. Rochereau, H. 1998. Response of earth electrodes when fast fronted currents are flowing out. *EDF Buletin de la Direction des Etudes et Recherches* B(2): 13–22.
42. Ametani, A., M. Nayel, S. Sekioka, and T. Sonoda. 2002. Basic investigation of wave propagation characteristics on an underground bare conductor. *Proceedings of ICEE 2002*, Jeju, Korea: 2141–2146.
43. Ametani, A., T. Chikara, Y. Baba, N. Nagaoka, and S. Okabe. 2008. A characteristics of a grounding electrode on the earth surface. *Proceedings of IWHV 2008*, Kyoto, Japan. Paper HV-08-76.
44. Sekioka, S. 2009. An investigation of grounding for a wind farm. *IEEJ Research Meeting*, Nagasaki, Japan. HV-09–50 (in Japanese).
45. Yutthagowith, P., A. Ametani, N. Nagaoka, and Y. Baba. 2010. Influence of a measuring system to a transient voltage on a vertical conductor. *IEEJ Trans. EEE* 5:221–228.
46. Ametani, A., H. Motoyama, K. Ohkawara, H. Yamakawa, and N. Suga. 2009. Electromagnetic disturbances of control circuits in power stations and substations experienced in Japan. *IET-GTD* 3:801–815.
47. Working Group (Chair: Agematsu, S.), Japanese Electrotechnical Research Association. 2002. Technologies of countermeasures against surges on protection relays and control systems. *ETRA Report* 57(3) (in Japanese).
48. CIGRE WG C4–208. 2013. EMC within power plants and substations. *CIGRE-TB* 543.
49. IEC. 2001. Electromagnetic Compatibility Part 6–5 Generic Std. Immunity for Power Station and Substation Environments. *IEC TC 61000-6-5*.
50. Japanese Electotechnical Committee. 2005. *Standard of Test Voltages for Low-Voltage Control Circuits in Power Stations and Substations. JEC-0103-2005*, Tokyo: IEE Japan (in Japanese).
51. Ametani, A., M. Nayel, S. Sekioka, and T. Sonoda. 2002. Basic investigation of wave propagation characteristics of an underground bare conductor. *ICEE 2002*, Jeju, Korea, Proceedings: 2141–2146.
52. Ametani, A., T. Okumura, N. Nagaoka, and N. Mori. 2003. Experimental investigation of a transient induced voltage to an overhead control cable from a grounding circuit. *CIRED 2003*, Barcelona, Spain. Paper No. 2.4.
53. Sonoda, T., Y. Takeuchi, S. Sekioka, N. Nagaoka, and A. Ametani. 2003. Induced surge characteristics from a counterpoise to an overhead circuit. *IEEJ Trans. PE* 123:1340–1349.
54. Motoyama, H. 2005. Lightning surge characteristics of low voltage and control circuits (I). *CRIEPI Report* H04011 (in Japanese).

8

Problems and Application Limits of Numerical Simulations

Because the electromagnetic transients program (EMTP) is based on circuit theory assuming transverse electromagnetic (TEM) mode propagation, it cannot give an accurate solution for a high-frequency transient that involves non-TEM mode propagation. Additionally, the EMTP cannot deal with a circuit of unknown parameters. A numerical electromagnetic analysis (NEA) method, on the other hand, can deal with a transient associated with both TEM and non-TEM mode propagations. Furthermore, it requires no circuit parameters. However, this method results in numerical instability if the analytical space, the boundary conditions, the cell size, etc., are not appropriate. Also, it requires massive computer resources, and the existing codes are not generalized enough to deal with various types of transients, especially in a large network.

8.1 Problems with Existing Impedance Formulas Used in Circuit Theory-Based Approaches

8.1.1 Earth-Return Impedance

8.1.1.1 Carson's Impedance

The reason Carson's impedance is very popular and widely used is simply due to its asymptotic expression [1]. During the early days of computing, the calculation of Pollaczek's infinite integral [2] was very difficult because of limited computation capability. Therefore, Carson's asymptotic formula was the only possible way to evaluate earth-return impedance [3]. However, the asymptotic expression inherently necessitates formulas for a small variable—for example, low frequency—and for a large variable, and this results in a discontinuity of the calculated impedance as a function of frequency. Also, the accuracy in the boundary region is not high enough. The same holds true for Schelkunoff's formula for the internal impedance of a conductor [4].

In this day and times, advancements in computing capabilities make it possible to calculate an infinite integral and various methods of evaluating

Pollaczek's impedance have been proposed. A typical example is the work by Noda [5]. This author, however, doubts even Pollaczek's formula.

8.1.1.2 Basic Assumptions of Impedance

Pollaczek's and Carson's formulas were derived under the assumption that

$$\text{length } x \gg \text{height } h \gg \text{radius } r. \tag{8.1}$$

It should be noted that most formulas for capacitance and inductance of conductors given in textbooks are based on this condition. It can be easily confirmed that any capacitance formula gives an erroneously large value when the radius equals the height. Correspondingly, the inductance of an infinite conductor becomes larger than that of a real finite conductor [6,7].

Furthermore, the formulas neglect displacement currents

$$\frac{1}{\rho_e} \gg \omega \varepsilon_e \quad \text{or} \quad f \ll \frac{1}{2\pi \omega \varepsilon_e \rho_e} \tag{8.2}$$

where
ρ_e is the earth resistivity
ε_e is the permittivity
$\omega = 2\pi f$

For example, the applicable range of a frequency in the case of $\rho_e = 1000 \ \Omega$ m and $\varepsilon_e = \varepsilon_0$ is given by

$$f \ll 18\,\text{MHz} \quad \text{or} \quad t \gg 50\,\text{ns}$$

Even in the case of $\rho_e = 100 \ \Omega$ m, a transient of a 10 ns time region cannot be simulated by Pollaczek's and Carson's impedances [8–11]. It should be noted that most frequency-dependent line models are not applicable in these models because they are based on Pollaczek's and Carson's impedances.

Under conditions in which Equations 8.1 and 8.2 are not satisfied, only Kikuchi's and Wedepohl's impedance formulas are applicable [8–10]. These require more advanced numerical integration than that applied to Pollaczek's formula.

8.1.1.3 Nonparallel Conductors

Pollaczek's and Carson's impedances are for a horizontal conductor. In reality, there are a number of nonhorizontal conductors, such as vertical and inclined ones. Although many papers have been published on the impedance of vertical conductors such as transmission towers, it is still not clear if the proposed formulas are correct. The empirical formula in Reference 12 is almost identical to an analytical formula [13], which agrees quite well with the measured results. However, the analytical formula requires further investigation to confirm if the derivation is correct.

Impedance formulas for inclined and nonparallel conductors have been proposed in References 6, 7, and 14. Since the formulas have been derived from the idea of a complex penetration depth [15] using Neumann's inductance formula, they require further theoretical analyses.

8.1.1.4 Stratified Earth

Earth is stratified, as is well-known, and its resistivity varies significantly at the top layer depending on the weather and climate. The earth-return impedance of an overhead conductor above the stratified earth was derived in Reference 16, and the stratified-earth effect was investigated in Reference 17. The stratified-earth effect may be far more significant than the accurate evaluation of the homogenous earth-return impedance of Pollaczek and Carson, and this requires further investigation.

8.1.1.5 Earth Resistivity and Permittivity

Earth resistivity, as mentioned earlier, is weather/climate dependent. The resistivity after the rains is lower than that measured during dry days. Also, it may be frequency dependent. The frequency dependence of earth permittivity may be far more significant than that of earth resistivity. Furthermore, water (H_2O), which is a dominant factor for earth permittivity, is extremely temperature dependent [18]. As a result, the error due to the uncertainty of earth resistivity and permittivity might be far greater than that due to the incompleteness of the earth-return impedance derived by Carson and Pollaczek. This should be remembered as a physical reality that is important in engineering practice.

8.1.2 Internal Impedance

8.1.2.1 Schelkunoff's Impedance

Schelkunoff's impedance was derived under the condition that a conductor must be in a free space corresponding to Equation 8.1. Therefore, the impedance is not applicable to finite-length conductors in proximity. This suggests that the internal impedance of such conductors is yet to be derived.

8.1.2.2 Arbitrary Cross-Section Conductor

Schelkunoff's impedance assumes that a conductor is circular or cylindrical. In reality, many conductors exist withcross sections are not circular or cylindrical. The internal impedance of a conductor with an arbitrary cross section was derived in Reference 19, which has been implemented in the EMTP cable parameters program [20].

Reference 21 shows an approximation of a conductor with a T or hollow rectangular shape by a cylindrical-shaped conductor. Although the internal impedance of a conductor with an arbitrary cross section can be accurately evaluated by a finite-element method of numerical calculation, this requires a great deal of time and computer memory. Either an analytical formula or an efficient numerical method needs to be developed.

8.1.2.3 Semiconducting Layer of Cables

It is well-known that a semiconducting layer exists on the surface of a cable conductor, which occasionally produces a significant effect on a cable transient. The impedance of the semiconducting layer was derived in Reference 22 and may be implemented into a cable-impedance calculation. It should be noted that the admittance of the semiconducting layer is far more important than its impedance, from a transient analysis viewpoint.

8.1.2.4 Proximity Effect

The significance of the proximity effect on conductor impedance is well-known. There are a number of papers that derive a theoretical formula of impedance and admittance [23–29] and discuss impedance variation due to proximity based on numerical simulations [30–33]. The proximity effect may be very important in a steady-state power system's performance from a power loss viewpoint; some quantitative results at a frequency of 50 or 60 Hz have been published [34–37].

It has been pointed out that the proximity effect is also significant in a transient state because a surge waveform is noticeably distorted by the increase in conductor resistance due to the proximity effect. Unfortunately, almost no data exist investigating the proximity effect on a transient [33].

A formula is available that considers the proximity or the eccentricity of a conductor enclosed within a conducting-pipe enclosure [24]; that is, a PT cable [20,38]. However, there is no formula that considers the proximity between two conductors above the earth.

8.1.3 Earth-Return Admittance

Earth-return impedance has been well discussed, and its effect on the wave-propagation characteristic and the transient waveform is well-known, as is clear from a number of publications. Earth-return admittance [8,9,39,40], however, is neglected in most studies on wave propagation and surge characteristics, and its significant effect is not well understood [8,9,40–43].

It has been pointed out in References 8,9,40, and 43 that attenuation starts to decrease at a critical frequency that is inversely proportional to the earth's resistivity and the conductor's height. This phenomenon is caused by negative conductance and corresponds to the transition between the TEM mode propagation, called "earth-return wave," and the transverse magnetic (TM) mode propagation, called "surface wave," as discussed by Kikuchi in 1957 [9]. When earth-return admittance is neglected, attenuation increases monotonously as the frequency increases. The wave-propagation velocity and the characteristic impedance become greater when earth-return admittance is considered. The study of earth-return admittance may be another challenging and prospective field for transient analysis, including the transition among TEM, TM, and TE modes of propagation [44].

8.2 Existing Problems in Circuit Theory-Based Numerical Analysis

8.2.1 Reliability of a Simulation Tool

Quite often, a problem appears unexpectedly for a user but not for the developers of a simulation tool; it is hard for developers to predict such problems at the development stage. These problems are caused quite often by the misuse of the tool by the user. Therefore, reliability and severity tests of simulation tools are very important. For example, it took nearly 10 years to carry out reliability and severity tests on tens of thousands of cases with EMTP cable constants. It should be noted that the reliability of a tool (that is, the probability of a problem occurring) is proportional to the number of elements (that is, the number of subroutines and options) although each individual element has very high reliability. Input data often cause numerical instability when the data physically do not exist; this problem is related to the assumption of formulas adopted in the simulation tool as explained in Section 8.1. To avoid such a problem, a "KILL CODE" is prepared in the EMTP. The kill code judges whether the input data are beyond the limits of assumption. It may be noteworthy that nearly half of the EMTP codes are kill codes. This may be considered by developers in another simulation tool.

8.2.2 Assumptions and Limits of a Simulation Tool

It should be noted that most of the existing or well-known formulas of conductor impedances and admittances are derived based on the assumption of an infinitely long conductor. The frequency of discovery of new electrical phenomena is increasing year after year, corresponding to the advancement in measuring equipment. For example, the sampling frequency of an oscilloscope, which is 1 GHz today, was approximately 10 MHz 10 years ago. The length is inversely proportional to the frequency, and therefore it becomes necessary to deal with a transient on a 1-m conductor whose natural resonant frequency is on the order of 100 MHz. Schelkunoff's, Pollaczek's, and Carson's impedances adopted in any circuit theory-based simulation tool, such as the EMTP, may not be applied [3]. The limits of assumption should be clearly explained in the rule book of a simulation tool, and the kill codes corresponding to the impedance and admittance limits should be prepared in the tool. Problems often appear when the user adopts a commercial software, unless a developer or a user group gives a guide for its usage. Even in the case of a well-known simulation tool such as the EMTP, problems occur frequently. The best solution to avoid such problems is for electrical engineers to realize that such phenomena need to be simulated in physical terms to be clearly understood—that is engineering. We are not computer engineers, nor information technology (IT) engineers.

8.2.3 Input Data

As was mentioned in above, a simulation tool user should be careful about input data. Quite often, input data beyond the limits of assumption of the tool are used, and users then complain that the tool gives erroneous output—this was the author's experience as a developer of the original EMTP beginning in 1976. At the same time, both the user and the developer should recognize that there are a number of uncertain physical parameters, typically dealing with earth resistivity, that vary along a transmission line and also along the depth of the earth [16,17]. The stratified-earth effect on a transient may be far more influential than the accuracy of numerical calculations of Pollaczek's and Carson's earth-return impedances, assuming a homogenous earth. It is interesting to note that, since 1978, the stratified-earth option of the EMTP cable constants has never been used. In addition, data on stray capacitances and residual inductances of a power apparatus are, in general, not available from a manufacturer. The same is the case with regard to the nonlinear characteristic of the apparatus and the resistivity and permittivity of a cable insulator and a semiconducting layer [18].

8.3 NEA for Power System Transients

The numerical electromagnetic analysis (NEA) method [45–50] is becoming one of the most promising approaches to solve transient phenomena that are very hard to solve using existing circuit theory-based simulation tools such as the EMTP. The existing circuit theory-based approaches cannot solve a three-dimensional (3-D) transient or a transient involving a sphere-wave propagation and a scattered field, such as a transient across an archon, a wave front transient at a transmission tower due to lightning, or the voltage and current at the corner or across the spacer of a gas-insulated bus due to a switching surge. Also, the circuit theory-based approach makes it difficult to solve a transient in a complex medium, such as the transient on a grounding electrode and that on a semiconducting layer of a cable Furthermore, the circuit theory approach cannot be applied if the circuit parameters are not known. The NEA method can solve such problems, because it calculates Maxwell's equation directly.

A working group of the IEE Japan was founded in April of 2004, and it carried out an investigation on the NEA and its applications. The results derived by the working group were published as a book by IEE Japan [49]. CIGRE working group, WG C4. 501, was established [50] in 2009, and a CIGRE technical brochure (TB) has been published [38].

The NEA method is useful in dealing with power system transients, such as in the following topics:

- Surge characteristics of overhead transmission-line towers.
- Surge characteristics of vertical grounding electrodes and horizontally placed square-shaped grounding electrodes.
- Surge characteristics of air-insulated substations.
- Lightning-induced surges on overhead distribution lines.
- Surge characteristics of a wind-turbine tower struck by lightning and its interior transient magnetic field.
- Very fast transients in gas-insulated switchgears.
- Three-dimensional (3-D) electromagnetic field analysis.

The details of the NEA are explained in Chapter 5.

In summary, NEA methods can provide greater accuracy when compared to simulation results obtained using circuit theory-based approaches.

However, as massive computation resources are, in general, required, NEA methods can be considered useful tools to set reference cases and study specific problems. Also, a perfect conductor assumption in a finite-difference time-domain (FDTD) method, for example, results in the difficulty in analyzing TEM, TM, and TE transition of wave propagation along a lossy conductor above a lossy earth [8,9,43,44].

References

1. Carson, J. R. 1926. Wave propagation in overhead wires with ground return. *Bell Syst. Tech. J.* 5:539–554.

2. Pollaczek, F. 1926. Uber das Feld einer unendlich langen wechselstromdurch-flossenen Einfachleitung. *ENT* 9(3):339–359.

3. Dommel, H. W. 1986. *EMTP Theory Book*. Portland, OR: BPA.

4. Schelkunoff, S. A. 1934. The electromagnetic theory of coaxial transmission line and cylindrical shields. *Bell Syst. Tech. J.* 13:532–579.

5. Noda, T. 2006. Development of accurate algorithms for calculating ground-return and conductor-internal impedances CRIEPI Report H05003 by Central Research Institute of Electric Power Industries in Japan (CRIPEI Tokyo) Report H05003.

6. Ametani, A. and A. Ishihara. 1993. Investigation of impedance and line parameters of a finite-length multiconductor system. *Trans. IEE Jpn.* 113-B(8):905–913.

7. Ametani, A. and T. Kawamura. 2005. A method of a lightning surge analysis recommended in Japan using EMTP. *IEEE Trans. Power Deliv.* 20(2):867–875.

8. Kikuchi, H. 1955. Wave propagation on the ground return circuit in high frequency regions. *J. IEE Jpn.* 75(805):1176–1187.

9. Kikuchi, H. 1957. Electro-magnetic field on infinite wire at high frequencies above plane-earth. *J. IEE Jpn.* 77:721–733.

10. Wedepohl, L. M. and A. E. Efthymiais. 1978. Wave propagation in transmission line over lossy ground—A new complete field solution. *IEEE Proc.* 125(6):505–510.

11. Ametani, A., T. Yoneda, Y. Baba, and N. Nagaoka. 2009. An investigation of earth-return impedance between overhead and underground conductors and its approximation. *IEEE Trans. EMC* 51(3):860–867.

12. Hara, T., O. Yamamoto, M. Hayashi, and C. Uenosono. 1989. Empirical formulas of surge impedance for single and multiple vertical conductors. *Trans. IEE Jpn.* 110-B:129–136.

13. Ametani, A., Y. Kasai, J. Sawada, A. Mochizuki, and T. Yamada. 1994. Frequency-dependent impedance of vertical conductors and a multiconductor tower model. *IEE Proc. Generat. Transm. Distrib.* 141(4):339–345.

14. Ametani, A. 2002. Wave propagation on a nonuniform line and its impedance and admittance. *Sci. Eng. Rev. Doshisha Univ.* 43(3):135–147.

15. Deri, A. et al. 1981. The complex ground return plane: A simplified model for homogeneous and multi-layer earth return. *IEEE Trans. Power App. Syst.* 100(8):3686.

16. Nakagawa, M., A. Ametani, and K. Iwamoto. 1973. Further studies on wave propagation in overhead lines with earth return—Impedance of stratified earth. *Proc. IEE* 120(2):1521–1528.

17. Ametani, A. 1974. Stratified effects on wave propagation—Frequency-dependent parameters. *IEEE Trans. Power App. Syst.* 93(5):1233–1239.

18. Ametani, A. 2000. Problems and countermeasures of cable transient simulations. *EMTP J.* 5:3–11.

19. Ametani, A. and I. Fuse. 1992. Approximate method for calculating the impedances of multi conductors with cross-section of arbitrary shapes. *Elect. Eng. Jpn.* 111(2):117–123.

20. Ametani, A. 1994. *Cable Parameters Rule Book*. Portland, OR: BPA.

21. Ametani, A., N. Nagaoka, R. Koide, and T. Nakanishi. 1999. Wave propagation characteristics of iron conductors in an intelligent building. *Trans. IEE Jpn.* B-120(1):271–277.

22. Ametani, A., Y. Miyamoto, and N. Nagaoka. 2004. Semiconducting layer impedance and its effect on cable wave-propagation and transient characteristics. *IEEE Trans. Power Deliv.* 19(4):523–531.
23. Tegopoulos, J. A. and E. E. Kriezis. 1971. Eddy current distribution in cylindrical shells of infinite length due to axial currents, Part II—Shells of infinite thickness. *IEEE Trans. Power App. Syst.* 90:1287–1294.
24. Brown, G. W. and R. G. Rocamora. 1976. Surge propagation in three-phase pipe-type cables, Part I—Unsaturated pipe. *IEEE Trans. Power App. Syst.* 95:88–95.
25. Dugan, R. C. et al. 1977. Surge propagation in three-phase pipe-type cables, Part II—Duplication of field test including the effects of neutral wires and pipe saturation. *IEEE Trans. Power App. Syst.* 96:826–833.
26. Schinzinger, R. and A. Ametani. 1978. Surge propagation characteristics of pipe enclosed underground cables. *IEEE Trans. Power App. Syst.* 97:1680–1687.
27. Dokopoulos, P. and D. Tampakis. 1984. Analysis of field and losses in three-phase gas cable with thick walls: Part I. Field analysis. *IEEE Trans. Power App. Syst.* 103(9):2728–2734.
28. Dokopoulos, P. and D. Tampakis. 1985. Part II calculation of losses and results. *IEEE Trans. Power App. Syst.* 104(1):9–15.
29. Poltz, J., E. Kuffel, S. Grzybowski, and M. R. Raghuveer. 1982. Eddy-current losses in pipe-type cable systems. *IEEE Trans. Power App. Syst.* 101(4):825–832.
30. Fortin, S., Y. Yang, J. Ma, and F. P. Dawalibi. 2005. Effects of eddy current on the impedance of pipe-type cables with arbitrary pipe thickness. *ICEE 2005*, Gliwice, Poland. Paper TD2-09.
31. Chien, C. H. and R. W. G. Bucknall. 2009. Harmonic calculation of proximity effect on impedance characteristics in subsea power transmission cables. *IEEE Trans. Power Deliv.* 24(2):2150–2158.
32. Gustavsen, B., A. Bruaset, J. J. Bremnes, and A. Hassel. 2009. A finite element approach for calculating electrical parameters of umbilical cables. *IEEE Trans. Power Deliv.* 24(4):2375–2384.
33. Ametani, A. et al. 2013. Wave propagation characteristics on a pipe-type cable in particular reference to the proximity effect. *IEE Journal of High Voltage Engineering Conference*, Kyoto, Japan. Paper HV-13-005.
34. Ishikawa, T., K. Kawasaki, and I. Okamoto. 1976. Eddy current losses in cable sheath (1). *Dainichi Nihon Cable J.* 61:34–42.
35. Ishikawa, T., K. Kawasaki, and O. Okamoto. 1977. Eddy current losses in cable sheath (2). *Dainichi Nihon Cable J.* 62:21–64.
36. Kawasaki, K., M. Inami, and T. Ishikawa. 1981. Theoretical consideration on eddy current losses on non-magnetic and magnetic pipes for power transmission systems. *IEEE Trans. Power App. Syst.* 100(2):474–484.
37. Mekjian, A. and M. Sosnowski. 1983. Calculation of altering current losses in steel pipe containing power cables. *IEEE Trans. Power App. Syst.* 102(2):382–388.
38. Ametani, A. 1980. A general formulation of impedance and admittance of cables. *IEEE Trans. Power App. Syst.* 99(3):902–910.
39. Wise, W. H. 1948. Potential coefficients for ground return circuits. *Bell Syst. Tech. J.* 27:365–371.
40. Nakagawa, M. 1981. Further studies on wave propagation along overhead transmission lines: Effects of admittance correction. *IEEE Trans. Power App. Syst.* 100(7):3626–3633.

41. Rachidi, F., C. A. Nucci, and M. Ianoz. 1999. Transient analysis of multi-conductor lines above a lossy ground. *IEEE Trans. Power Deliv.* 14(1):294–302.
42. Hashmi, G. M., M. Lehtonen, and A. Ametani. 2010. Modeling and experimental verification of covered conductors for PD detection in overhead distribution networks. *IEE J. Trans. PE* 130(7):670–678.
43. Ametani, A., M. Ohe, Y. Miyamoto, and K. Tanabe. 2012. The effect of the earth-return admittance on wave propagation along an overhead conductor in a high-frequency region. *EEUG Proceedings*, Zwickau, Germany, 1:6–22.
44. Sommerfeld, A. 1964. *Partial Differential Equation in Physics*. New York: Academic Press.
45. Yee, K. S. 1966. Numerical solution of initial boundary value problems involving Maxwell's equations in isotropic media. *IEEE Trans. Antenn. Propag.* 14(3):302–307.
46. Uno, T. 1998. *FDTD Method for Electromagnetic Fields and Antennas*. Tokyo, Japan: Corona Pub. Co.
47. Taflove, A. and S. C. Hagness. 2000. *Computational Electromagnetics. The Finite-Difference Time-Domain Method*. Norwood, MA: Artech House.
48. CRIEPI by Central Research Institute of Electric Power Industry, Tokyo, Japan, 2007. Visual Test Lab. (VSTL). http://cripei.denken.or.jp/jp/electric/substance/09.pdf.
49. IEE Japan WG. Working Group of Numerical Transient Electromagnetic Analysis (Convenor: A. Ametani) 2008. *Numerical Transient Electromagnetic Analysis Methods*. IEE Japan.
50. Ametani, A., T. Hoshino, M. Ishii, T. Noda, S. Okabe, and K. Tanabe. 2008. Numerical electromagnetic analysis method and its application to surge phenomena. *CIGRE 2008 General Meeting*, Paris, France. Paper C4-108.

Index